D0008236

Methods in Enzymology

Volume 100

RECOMBINANT DNA

Part B

METHODS IN ENZYMOLOGY

EDITORS-IN-CHIEF

Sidney P. Colowick   Nathan O. Kaplan

Methods in Enzymology

Volume 100

Recombinant DNA

Part B

EDITED BY

Ray Wu
SECTION OF BIOCHEMISTRY
MOLECULAR AND CELL BIOLOGY
CORNELL UNIVERSITY
ITHACA, NEW YORK

Lawrence Grossman
DEPARTMENT OF BIOCHEMISTRY
THE JOHNS HOPKINS UNIVERSITY
SCHOOL OF HYGIENE AND PUBLIC HEALTH
BALTIMORE, MARYLAND

Kivie Moldave
DEPARTMENT OF BIOLOGICAL CHEMISTRY
COLLEGE OF MEDICINE
UNIVERSITY OF CALIFORNIA
IRVINE, CALIFORNIA

1983

ACADEMIC PRESS

A Subsidiary of Harcourt Brace Jovanovich, Publishers

New York London
Paris San Diego San Francisco São Paulo Sydney Tokyo Toronto

ACADEMIC PRESS, INC.
111 Fifth Avenue, New York, New York 10003

United Kingdom Edition published by
ACADEMIC PRESS, INC. (LONDON) LTD.
24/28 Oval Road, London NW1 7DX

Library of Congress Cataloging in Publication Data

Main entry under title:

Recombinant DNA.

Pt. edited by Ray Wu, Lawrence Grossman, Kivie Moldave.
Includes bibliographical references and indexes.
1. Recombinant DNA. I. Wu, Ray. II. Grossman, Lawrence, Date III. Moldave, Kivie, Date IV. Series: Methods in enzymology ; v. 68, etc.
[DNLM: 1. DNA, Recombinant. W1 ME9615K v. 68, etc. / QU 58 R312 1979]
QP601.M49 vol. 68, etc. 574.1'925 s 79-26584
[QH442] [574.87'3282]
ISBN 0-12-182000-9 (v. 100)

PRINTED IN THE UNITED STATES OF AMERICA

83 84 85 86 9 8 7 6 5 4 3 2 1

Table of Contents

Section III. Proteins with Specialized Functions Acting at Specific Loci

Section IV. New Methods for DNA Isolation, Hybridization, and Cloning

Section V. Analytical Methods for Gene Products

Section VI. Mutagenesis: *In Vitro* and *in Vivo*

Contributors to Volume 100

Article numbers are in parentheses following the names of contributors.
Affiliations listed are current.

A. BECKER (12), *Department of Medical Genetics, University of Toronto, Toronto, Ontario M5S 1A8, Canada*

MICHAEL D. BEEN (8), *Department of Microbiology and Immunology, School of Medicine, University of Washington, Seattle, Washington 98195*

GERALD A. BELTZ (19), *Department of Cellular and Developmental Biology, The Biological Laboratories, Harvard University, Cambridge, Massachusetts 02138*

H. C. BIRNBOIM (17), *Radiation Biology Branch, Atomic Energy of Canada Limited, Chalk River, Ontario KOJ 1JO, Canada*

ROBERT BLAKESLEY (1, 26), *Bethesda Research Laboratories, Inc., Gaithersburg, Maryland 20877*

DAVID BOTSTEIN (31), *Department of Biology, Massachusetts Institute of Technology, Cambridge, Massachusetts 02139*

CATHERINE A. BRENNAN (2), *Department of Biochemistry, School of Basic Medical Sciences and School of Chemical Sciences, University of Illinois, Urbana, Illinois 61801*

BONITA J. BREWER (8), *Department of Genetics, University of Washington, Seattle, Washington 98195*

DAVID R. BROWN (16), *Department of Developmental Biology and Cancer, Albert Einstein College of Medicine, Bronx, New York 10461*

HANS BÜNEMANN (27), *Institut für Genetik, Universität Düsseldorf, D-4000 Düsseldorf, Federal Republic of Germany*

MALCOLM J. CASADABAN (21), *Department of Biophysics and Theoretical Biology, University of Chicago, Chicago, Illinois 60637*

JAMES J. CHAMPOUX (8), *Department of Microbiology and Immunology, School of Medicine, University of Washington, Seattle, Washington 98195*

PETER T. CHERBAS (19), *Department of Cellular and Developmental Biology, The Biological Laboratories, Harvard University, Cambridge, Massachusetts 02138*

JOANY CHOU (21), *Department of Biophysics and Theoretical Biology, University of Chicago, Chicago, Illinois 60637*

R. JOHN COLLIER (25), *Department of Microbiology and The Molecular Biology Institute, University of California, Los Angeles, California 90024*

NICHOLAS R. COZZARELLI (11), *Department of Molecular Biology, University of California, Berkeley, California 94720*

ALBERT E. DAHLBERG (23), *Division of Biology and Medicine, Brown University, Providence, Rhode Island 02912*

GUO-REN DENG (5), *Section of Biochemistry, Molecular and Cell Biology, Cornell University, Ithaca, New York 14853*

ALAN DIAMOND (30), *Sidney Farber Cancer Institute and Harvard Medical School, Boston, Massachusetts 02115*

JOHN E. DONELSON (6), *Department of Biochemistry, University of Iowa, Iowa City, Iowa 52242*

K. DORAN (26), *Bethesda Research Laboratories, Inc., Gaithersburg, Maryland 20877*

BERNARD DUDOCK (30), *Department of Biochemistry, State University of New York, Stony Brook, New York 11794*

THOMAS H. EICKBUSH (19), *Department of Biology, University of Rochester, Rochester, New York 14627*

STUART G. FISCHER (29), *Department of Biological Sciences, Center for Biological Macromolecules, State University of New York, Albany, New York 12222*

ERICH FREI (22), *Department of Cell Biology, Biocenter of the University, CH-4056 Basel, Switzerland*

ROY FUCHS (1), *Corporate Research and Development, Monsanto Company, St. Louis, Missouri 63166*

JAMES I. GARRELS (28), *Cold Spring Harbor Laboratory, Cold Spring Harbor, New York 11724*

M. GOLD (12), *Department of Medical Genetics, University of Toronto, Toronto, Ontario M5S 1A8, Canada*

PETER GOWLAND (22), *Department of Cell Biology, Biocenter of the University, CH-4056 Basel, Switzerland*

LAWRENCE GREENFIELD (25), *Cetus Corporation, Berkeley, California 94710*

MANUEL GREZ (20), *Department of Microbiology, University of Southern California School of Medicine, Los Angeles, California 90033*

RICHARD I. GUMPORT (2), *Department of Biochemistry, School of Basic Medical Sciences and School of Chemical Sciences, University of Illinois, Urbana, Illinois 61801*

LI-HE GUO (4), *Section of Biochemistry, Molecular and Cell Biology, Cornell University, Ithaca, New York 14853*

DOUGLAS HANAHAN (24), *Department of Biochemistry and Molecular Biology, Harvard University, Cambridge, Massachusetts 02138, and Cold Spring Harbor Laboratory, Cold Spring Harbor, New York 11724*

JAMES L. HARTLEY (6), *Bethesda Research Laboratories Inc., Gaithersburg, Maryland 20877*

HANSJÖRG HAUSER (20), *Gesellschaft für Biotechnologische Forschung, Mascheroder Weg 1, D-3300 Braunschweig, Federal Republic of Germany*

C. J. HOUGH (26), *Bethesda Research Laboratories, Inc., Gaithersburg, Maryland 20877*

TAO-SHIH HSIEH (10), *Department of Biochemistry, Duke University Medical Center, Durham, North Carolina 27710*

JERARD HURWITZ (16), *Department of Developmental Biology and Cancer, Albert Einstein College of Medicine, Bronx, New York 10461*

KENNETH A. JACOBS (19), *Department of Cellular and Developmental Biology, The Biological Laboratories, Harvard University, Cambridge, Massachusetts 02138*

CORNELIS VICTOR JONGENEEL (9), *Department of Biochemistry/Biophysics, University of California, San Francisco, San Francisco, California 94143*

FOTIS C. KAFATOS (19), *Department of Cellular and Developmental Biology, The Biological Laboratories, Harvard University, Cambridge, Massachusetts 02138*

DONALD A. KAPLAN (25), *Cetus Corporation, Berkeley, California 94710*

KENNETH N. KREUZER (9), *Department of Biochemistry/Biophysics, University of California, San Francisco, San Francisco, California 94143*

JUDY H. KRUEGER (33), *Department of Biology, Massachusetts Institute of Technology, Cambridge, Massachusetts 02139*

HARTMUT LAND (20), *Center of Cancer Research, Massachusetts Institute of Technology, Cambridge, Massachusetts 02139*

ABRAHAM LEVY (22), *Friedrich-Meischer-Institut, Ciba-Geigy, CH-4058 Basel, Switzerland*

WERNER LINDENMAIER (20), *Gesellschaft für Biotechnologische Forschung, Mascheroder Weg 1, D-3300 Braunschweig, Federal Republic of Germany*

LEROY F. LIU (7), *Department of Physiological Chemistry, Johns Hopkins University Medical School, Baltimore, Maryland 21205*

ALICE E. MANTHEY (2), *Department of Biochemistry, School of Basic Medical Sciences and School of Chemical Sciences,*

University of Illinois, Urbana, Illinois 61801

SUSAN R. MARTIN (8), *Genetic Systems Corp., 3005 First Avenue, Seattle, Washington 98121*

ALFONSO MARTINEZ-ARIAS (21), *Department of Biophysics and Theoretical Biology, University of Chicago, Chicago, Illinois 60637*

BETTY L. MCCONAUGHY (8), *Department of Genetics, University of Washington, Seattle, Washington, 98195*

WILLIAM K. MCCOUBREY, JR. (8), *Department of Microbiology and Immunology, School of Medicine, University of Washington, Seattle, Washington 98195*

MATTHEW MESELSON (24), *Department of Biochemistry and Molecular Biology, Harvard University, Cambridge, Massachusetts 02138*

HOWARD A. NASH (15), *Laboratory of Neurochemistry, National Institute of Mental Health, Bethesda, Maryland 20205*

MARKUS NOLL (22), *Department of Cell Biology, Biocenter of the University, CH-4056 Basel, Switzerland*

LYNN OSBER (14), *Departments of Human Genetics, Yale University School of Medicine, New Haven, Connecticut 06510*

RICHARD OTTER (11), *Department of Molecular Biology, University of California, Berkeley, California 94720*

W. PARRIS (12), *Department of Medical Genetics, University of Toronto, Toronto, Ontario M5S 1A8, Canada*

CHARLES M. RADDING (14), *Departments of Human Genetics and of Molecular Biophysics and Biochemistry, Yale University School of Medicine, New Haven, Connecticut 06510*

RANDALL R. REED (13), *Department of Genetics, Harvard Medical School, Boston, Massachusetts 02115*

DANNY REINBERG (16), *Department of Developmental Biology and Cancer, Albert Einstein College of Medicine, Bronx, New York 10461*

PAUL J. ROMANIUK (3), *Department of Biochemistry, University of Illinois, Urbana, Illinois 61801*

THOMAS SCHMIDT-GLENEWINKEL (16), *Department of Developmental Biology and Cancer, Albert Einstein College of Medicine, Bronx, New York 10461*

GÜNTHER SCHÜTZ (20), *Institut für Zell- und Tumorbiologie, Deutsches Krebsforschungszentrum, Im Neuenheimer Feld 280, D-6900 Heidelberg, Federal Republic of Germany*

STUART K. SHAPIRA (21), *Committee on Genetics, University of Chicago, Chicago, Illinois 60637*

TAKEHIKO SHIBATA (14), *Department of Microbiology, The Institute of Physical and Chemical Research, Saitama 351, Japan*

DAVID SHORTLE (31), *Department of Microbiology, State University of New York, Stony Brook, New York 11794*

MICHAEL SMITH (32), *Department of Biochemistry, Faculty of Medicine, University of British Columbia, Vancouver, British Columbia V6T 1W5, Canada*

EDMUND J. STELLWAG (23), *Department of Microbiology, University of Minnesota, Minneapolis, Minnesota 55455*

PATRICIA S. THOMAS (18), *Genetic Systems Corporation, 3005 First Avenue, Seattle, Washington 98121*

J. A. THOMPSON (26), *Bethesda Research Laboratories, Inc., Gaithersburg, Maryland 20877*

OLKE C. UHLENBECK (3), *Department of Biochemistry, University of Illinois, Urbana, Illinois 61801*

GRAHAM C. WALKER (33), *Department of Biology, Massachusetts Institute of Technology, Cambridge, Massachusetts 02139*

ROBERT D. WELLS (26), *Department of Biochemistry, Schools of Medicine and Dentistry, University of Alabama, Birming-*

ham, University Station, Birmingham, Alabama 35294

PETER WESTHOFF (27), *Botanik IV, Universität Düsseldorf, D-4000 Düsseldorf, Federal Republic of Germany*

RAY WU (4, 5), *Section of Biochemistry, Molecular and Cell Biology, Cornell University, Ithaca, New York 14853*

LISA S. YOUNG (8), *Institute of Molecular Biology, University of Oregon, Eugene, Oregon 97403*

STEPHEN L. ZIPURSKY (16), *Division of Biology, California Institute of Technology, Pasadena, California 90025*

MARK J. ZOLLER (32), *Department of Biochemistry, Faculty of Medicine, University of British Columbia, Vancouver, British Columbia V6T 1W5, Canada*

Preface

Exciting new developments in recombinant DNA research allow the isolation and amplification of specific genes or DNA segments from almost any living organism. These new developments have revolutionized our approaches to solving complex biological problems and have opened up new possibilities for producing new and better products in the areas of health, agriculture, and industry.

Volumes 100 and 101 supplement Volumes 65 and 68 of *Methods in Enzymology*. During the last three years, many new or improved methods on recombinant DNA or nucleic acids have appeared, and they are included in these two volumes. Volume 100 covers the use of enzymes in recombinant DNA research, enzymes affecting the gross morphology of DNA, proteins with specialized functions acting at specific loci, new methods for DNA isolation, hybridization, and cloning, analytical methods for gene products, and mutagenesis: *in vitro* and *in vivo*. Volume 101 includes sections on new vectors for cloning genes, cloning of genes into yeast cells, and systems for monitoring cloned gene expression.

Ray Wu
Lawrence Grossman
Kivie Moldave

METHODS IN ENZYMOLOGY

EDITED BY

Sidney P. Colowick and Nathan O. Kaplan

VANDERBILT UNIVERSITY
SCHOOL OF MEDICINE
NASHVILLE, TENNESSEE

DEPARTMENT OF CHEMISTRY
UNIVERSITY OF CALIFORNIA
AT SAN DIEGO
LA JOLLA, CALIFORNIA

I. Preparation and Assay of Enzymes
II. Preparation and Assay of Enzymes
III. Preparation and Assay of Substrates
IV. Special Techniques for the Enzymologist
V. Preparation and Assay of Enzymes
VI. Preparation and Assay of Enzymes (*Continued*)
Preparation and Assay of Substrates
Special Techniques
VII. Cumulative Subject Index

METHODS IN ENZYMOLOGY

EDITORS-IN-CHIEF

Sidney P. Colowick Nathan O. Kaplan

VOLUME VIII. Complex Carbohydrates
Edited by ELIZABETH F. NEUFELD AND VICTOR GINSBURG

VOLUME IX. Carbohydrate Metabolism
Edited by WILLIS A. WOOD

VOLUME X. Oxidation and Phosphorylation
Edited by RONALD W. ESTABROOK AND MAYNARD E. PULLMAN

VOLUME XI. Enzyme Structure
Edited by C. H. W. HIRS

VOLUME XII. Nucleic Acids (Parts A and B)
Edited by LAWRENCE GROSSMAN AND KIVIE MOLDAVE

VOLUME XIII. Citric Acid Cycle
Edited by J. M. LOWENSTEIN

VOLUME XIV. Lipids
Edited by J. M. LOWENSTEIN

VOLUME XV. Steroids and Terpenoids
Edited by RAYMOND B. CLAYTON

VOLUME XVI. Fast Reactions
Edited by KENNETH KUSTIN

VOLUME XVII. Metabolism of Amino Acids and Amines (Parts A and B)
Edited by HERBERT TABOR AND CELIA WHITE TABOR

VOLUME XVIII. Vitamins and Coenzymes (Parts A, B, and C)
Edited by DONALD B. MCCORMICK AND LEMUEL D. WRIGHT

VOLUME XIX. Proteolytic Enzymes
Edited by GERTRUDE E. PERLMANN AND LASZLO LORAND

VOLUME 75. Cumulative Subject Index Volumes XXXI, XXXII, and XXXLIV–LX
Edited by EDWARD A. DENNIS AND MARTHA G. DENNIS

VOLUME 76. Hemoglobins
Edited by ERALDO ANTONINI, LUIGI ROSSI-BERNARDI, AND EMILIA CHIANCONE

VOLUME 77. Detoxication and Drug Metabolism
Edited by WILLIAM B. JAKOBY

VOLUME 78. Interferons (Part A)
Edited by SIDNEY PESTKA

VOLUME 79. Interferons (Part B)
Edited by SIDNEY PESTKA

VOLUME 80. Proteolytic Enzymes (Part C)
Edited by LASZLO LORAND

VOLUME 81. Biomembranes (Part H: Visual Pigments and Purple Membranes, I)
Edited by LESTER PACKER

VOLUME 82. Structural and Contractile Proteins (Part A: Extracellular Matrix)
Edited by LEON W. CUNNINGHAM AND DIXIE W. FREDERIKSEN

VOLUME 83. Complex Carbohydrates (Part D)
Edited by VICTOR GINSBURG

VOLUME 84. Immunochemical Techniques (Part D: Selected Immunoassays)
Edited by JOHN J. LANGONE AND HELEN VAN VUNAKIS

VOLUME 85. Structural and Contractile Proteins (Part B: The Contractile Apparatus and the Cytoskeleton)
Edited by DIXIE W. FREDERIKSEN AND LEON W. CUNNINGHAM

VOLUME 86. Prostaglandins and Arachidonate Metabolites
Edited by WILLIAM E. M. LANDS AND WILLIAM L. SMITH

VOLUME 87. Enzyme Kinetics and Mechanism (Part C: Intermediates, Stereochemistry, and Rate Studies)
Edited by DANIEL L. PURICH

VOLUME 88. Biomembranes (Part I: Visual Pigments and Purple Membranes, II)
Edited by LESTER PACKER

VOLUME 89. Carbohydrate Metabolism (Part D)
Edited by WILLIS A. WOOD

VOLUME 90. Carbohydrate Metabolism (Part E)
Edited by Willis A. Wood

VOLUME 91. Enzyme Structure (Part I)
Edited by C. H. W. HIRS AND SERGE N. TIMASHEFF

VOLUME 92. Immunochemical Techniques (Part E: Monoclonal Antibodies and General Immunoassay Methods)
Edited by JOHN J. LANGONE AND HELEN VAN VUNAKIS

VOLUME 93. Immunochemical Techniques (Part F: Conventional Antibodies, Fc Receptors, and Cytotoxicity)
Edited by JOHN J. LANGONE AND HELEN VAN VUNAKIS

VOLUME 94. Polyamines (in preparation)
Edited by HERBERT TABOR AND CELIA WHITE TABOR

VOLUME 95. Cumulative Subject Index Volumes 61–74 and 76–80 (in preparation)
Edited by EDWARD A. DENNIS AND MARTHA G. DENNIS

VOLUME 96. Biomembranes (Part J: Membrane Biogenesis: Assembly and Targeting (General Methods; Eukaryotes)) (in preparation)
Edited by SIDNEY FLEISCHER AND BECCA FLEISCHER

VOLUME 97. Biomembranes (Part K: Membrane Biogenesis: Assembly and Targeting (Prokaryotes, Mitochondria, and Chloroplasts)) (in preparation)
Edited by SIDNEY FLEISCHER AND BECCA FLEISCHER

VOLUME 98. Biomembranes (Part L: Membrane Biogenesis (Processing and Recycling)) (in preparation)
Edited by SIDNEY FLEISCHER AND BECCA FLEISCHER

VOLUME 99. Hormone Action (Part F: Protein Kinases) (in preparation)
Edited by JACKIE D. CORBIN AND JOEL G. HARDMAN

VOLUME 100. Recombinant DNA (Part B)
Edited by RAY WU, LAWRENCE GROSSMAN, AND KIVIE MOLDAVE

VOLUME 101. Recombinant DNA (Part C) (in preparation)
Edited by RAY WU, LAWRENCE GROSSMAN, AND KIVIE MOLDAVE

VOLUME 102. Hormone Action (Part G: Calmodulin and Calcium-Binding Proteins) (in preparation)
Edited by ANTHONY R. MEANS AND BERT W. O'MALLEY

Gerhard Schmidt
1901–1981

Gerhard Schmidt (1901–1981)

This hundredth volume of *Methods in Enzymology* is dedicated to the memory of a dear friend and colleague whose pioneering work on the nucleic acids was important to the development of the techniques described in this and related volumes. Gerhard Schmidt was among the first to recognize the power of a combined chemical and enzymatic approach to the analysis of the structure of the nucleic acids. The importance of his work was belatedly recognized by his election to the National Academy of Sciences in 1976. In his classic work in 1928, while in Frankfurt in Embden's laboratory, he demonstrated the deamination of "muscle adenylic acid" by a highly specific enzyme which fails to deaminate "yeast adenylic acid." He speculated (correctly) that the two adenylic acids differed in the position of the phosphate group. He is probably best known for his development in 1945, while at the Boston Dispensary, of the method for determining the RNA, DNA, and phosphoproteins in tissues by phosphorus analysis (the Schmidt–Thannhauser method). He made many other contributions in the nucleic acid field, beginning with his studies with P. A. Levene at the Rockefeller Institute in 1938–1939 on the enzymatic depolymerization of RNA and DNA, and extending into the 1970s when he published some of the first definitive work on the nature of DNA–histone complexes.

Schmidt's research was by no means limited to the nucleic acids. He was almost equally involved in studies on the structure and measurement of the complex lipids. He also made important observations on the accumulation of inorganic polyphosphates in living cells. During the period between his forced flight from Germany in 1933 when the Nazis came to power and his employment by Thannhauser at the Boston Dispensary in 1940, he had a variety of research fellowships in Italy, Sweden, Canada, and the United States, including one in 1939–1940 in the laboratory of Carl and Gerty Cori in St. Louis, where he worked on the enzymatic breakdown of glycogen by the muscle and liver phosphorylases.

It was during this St. Louis period that one of us (SPC), then a graduate student in the Cori laboratory, came to know Gerhard intimately. In the mid-1940s, the other one of us (NOK), then a postdoctoral fellow with Fritz Lipmann at the Massachusetts General Hospital, also developed close scientific and personal ties with Gerhard. In the early 1950s, when we had joined the McCollum–Pratt Institute, Gerhard was invited to participate in the Symposia on Phosphorus Metabolism where he presented a monumental review on the polyphosphates and metaphosphates, and was also a central figure in the discussions on the nucleic acids. In the late 1950s and the 1960s, when NOK returned to Boston to be on the Brandeis faculty, the close ties with Gerhard were renewed. In the early 1960s, shortly after SPC joined the Vanderbilt faculty, Gerhard was invited there as a visiting professor and gave a series of memorable lectures on the nucleic acids which also formed the basis for his typically thorough chapter on that subject which appeared in *Annual Reviews of Biochemistry* for 1964.

During all the years from 1940 on, Gerhard did his research at the Boston Dispensary where Thannhauser had established a clinical chemistry laboratory. Throughout that time, Gerhard also held a joint appointment in biochemistry at the Tufts University School of Medicine where he participated in the teaching of medical students and the training of graduate students. He enjoyed a good relationship with the successive Chairmen of that department, three of whom, Alton Meister, Morris Friedkin, and Henry Mautner, were especially helpful. Dr. Mautner was instrumental in establishing the Gerhard Schmidt Memorial Lectureship which was initiated in December, 1981.

Gerhard was one of the most universally beloved figures in biochemistry. Perhaps this was because he lacked the "operator" gene. He would never have been comfortable as Chairman of a department or as President of a genetic engineering company. He liked to laugh, especially at himself. He identified with Laurel and Hardy, and once injured his jaw while rocking with laughter at one of their movies. He had a delightful collection of anecdotes, which, like his lectures, were carefully constructed and overly lengthy, but always well received by the Schmidt-story afficionados. He was enthusiastic about many things in addition to science, but he attacked with special gusto the playing of good chamber music or the eating of a good Liederkranz.

We present this dedication to his wife, Edith, and his sons, Michael and Milton, all of whom he loved very much, perhaps even more than his science, his music, and his Liederkranz.

SIDNEY P. COLOWICK
NATHAN O. KAPLAN

Section I

Use of Enzymes in Recombinant DNA Research

[1] Guide to the Use of Type II Restriction Endonucleases

By ROY FUCHS and ROBERT BLAKESLEY

Type II restriction endonucleases are DNases that recognize specific oligonucleotide sequences, make double-strand cleavages, and generate unique, equal molar fragments of a DNA molecule. By the nature of their controllable, predictable, infrequent, and site-specific cleavage of DNA, restriction endonucleases proved to be extremely useful as tools in dissecting, analyzing, and reconfiguring genetic information at the molecular level. Over 350 different restriction endonucleases have been isolated from a wide variety of prokaryotic sources, representing at least 85 different recognition sequences.[1,2] A number of excellent reviews detail the variety of restriction enzymes and their sources,[2,3] their purification and determination of their sequence specificity,[4,5] and their physical properties, kinetics, and reaction mechanism.[6] Here we provide a summary, based on the literature and our experience in this laboratory, emphasizing the practical aspects for using restriction endonucleases as tools. This review focuses on the reaction, its components and the conditions that affect enzymic activity and sequence fidelity, methods for terminating the reaction, some reaction variations, and a troubleshooting guide to help identify and solve restriction endonuclease-related problems.

The Reaction

Despite the diversity of the source and specificity for the over 350 type II restriction endonucleases identified to date,[1,2] their reaction conditions are remarkably similar. Compared to other classes of enzymes these conditions are also very simple. The restriction endonuclease reaction (Table I) is typically composed of the substrate DNA incubated at 37° in a solution buffered near pH 7.5, containing Mg^{2+}, frequently Na^{+}, and the selected restriction enzyme. Specific reaction details as found in the liter-

[1] R. Blakesley, *in* "Gene Amplification and Analysis," Vol. 1: "Restriction Endonucleases" (J. G. Chirikjian, ed.), p. 1. Elsevier/North-Holland, Amsterdam, 1981.

[2] R. J. Roberts, *Nucleic Acids Res.* **10,** r117 (1982).

[3] J. G. Chirikjian, "Gene Amplification and Analysis," Vol. 1: "Restriction Endonucleases." Elsevier/North-Holland, Amsterdam, 1981.

[4] R. J. Roberts, *CRC Crit. Rev. Biochem.* **4,** 123 (1976).

[5] This series, Vol. 65, several articles.

[6] R. D. Wells, R. D. Klein, and C. K. Singleton, *in* "The Enzymes" (P. D. Boyer, ed.), 3rd ed., Vol. 14, Part A, p. 157. Academic Press, New York, 1981.

ISBN 0-12-182000-9

TABLE I
GENERALIZED REACTION CONDITIONS FOR RESTRICTION ENDONUCLEASES

Conditions	Reaction type	
	Analytical	Preparative
Volume	20–100 μl	0.5–5 ml
DNA	0.1–10 μg	10–500 μg
Enzyme	1–5 units/μg DNA	1–5 units/μg DNA
Tris-HCl (pH 7.5)	20–50 m*M*	50 m*M*
$MgCl_2$	5–10 m*M*	10 m*M*
2-Mercaptoethanol	5–10 m*M*	5–10 m*M*
Bovine serum albumin	50–500 μg/ml	200–500 μg/ml
Glycerol	<5% (v/v)	<5% (v/v)
NaCl	As required	As required
Time	1 hr	1–5 hr
Temperature	37°	37°

ature for the more frequently used enzymes are listed in Table II. Note that in most cases these data do not represent optimal reaction conditions.

By convention, a unit of restriction endonuclease activity is usually defined as that amount of enzyme required to digest completely 1 μg of DNA (usually of bacteriophage lambda) in 1 hr.[4] This definition was chosen for convenience, since the useful, readily measurable end result of a restriction endonuclease reaction is completely cleaved DNA. However, a unit defined in this manner measures enzyme activity by an end point rather than by the classical initial rate term. Thus, traditional kinetic arguments based upon substrate saturating (initial rate) conditions cannot be applied to restriction endonucleases defined in this (enzyme saturating) manner.

One reason why there are few proper kinetic data on restriction endonucleases lies in the difficulty in measuring restriction enzyme activities during the linear portion of the reaction when using the standard enzyme assay.[7] The strong emphasis placed on their use as research tools in molecular biology rather than on investigation of their biochemical properties also contributed to the deficiency. Hence we lack good experimental data on conditions for optimal activity. For most newly isolated restriction endonucleases, assay buffers were selected for convenience during enzyme isolation rather than for optimal reactivity. These conditions have persisted as dogma. Thus, the implied precision and unique-

[7] P. A. Sharp, B. Sugden, and J. Sambrook, *Biochemistry* **12,** 3055 (1973).

ness of these values, e.g., pH 7.2 vs pH 7.4, is frequently without experimental basis. In fact, where investigated, restriction endonucleases usually show relatively broad activity profiles for the various reaction parameters.[8-10]

The fact that restriction endonucleases are active under a variety of conditions indicates that, similar to other nucleases, they are rather hardy enzymes. From an enzymologist's viewpoint, these enzymes can be mishandled and still demonstrate activity. But to achieve reproducible, efficient, and specific DNA cleavages, certain factors concerning restriction enzyme reactions should be considered. From our experience the most important factors for proper restriction endonuclease use are (*a*) the purity and physical characteristics of the substrate DNA; (*b*) the reagents used in the reaction; (*c*) the assay volume and associated errors; and (*d*) the time and temperature of incubation.

In the following sections each of these reaction parameters is discussed in detail. General conclusions are drawn in order to provide the researcher a framework in which properly to use restriction endonucleases. However, one must always be cognizant of the fact that each restriction endonuclease represents a unique enzymic protein. Any kinetic or biochemical generalization applied to the over 350 restriction enzymes will find exceptions.

DNA

The single most critical component of a restriction endonuclease reaction is the DNA substrate. DNA products generated in the reaction are directly affected by the degree of purity of the DNA substrate. Improperly prepared DNA samples will be cleaved poorly, if at all, producing partially digested DNA. In addition to DNA purity, other DNA-associated parameters that affect the products of the restriction endonuclease reaction include: DNA concentration, the specific sequence at and adjacent to the recognition site (including nucleotide modifications), and the secondary/tertiary DNA structure. Physical data pertaining to the DNA to be cleaved, if known, can guide one in choosing appropriate reaction conditions or prereaction treatments. Conversely, the response of a DNA of unknown physical properties to a standard restriction endonuclease digest can suggest certain characteristics of the DNA, e.g., the extent of methylation (see below).

[8] R. W. Blakesley, J. B. Dodgson, I. F. Nes, and R. D. Wells, *J. Biol. Chem.* **252,** 7300 (1977).

[9] P. J. Greene, M. S. Poonian, A. L. Nussbaum, L. Tobias, D. E. Garfin, H. W. Boyer, and H. M. Goodman, *J. Mol. Biol.* **99,** 237 (1975).

[10] B. Hinsch and M.-R. Kula, *Nucleic Acids Res.* **8,** 623 (1980).

TABLE II
REACTION CONDITIONS FOR CERTAIN RESTRICTION ENDONUCLEASES SELECTED FROM THE LITERATURE

Enzyme	Temperature (°C)	pH	Tris-HCl (m*M*)	$MgCl_2$ (m*M*)	NaCl (m*M*)	2-Mercapto-ethanol (m*M*)	Notes	Reference*
*Alu*I	37	7.9	6	6	—	6	—	1
*Asu*I	37	7.5	20	10	100	—	—	2
*Ava*I	37	7.5	20	10	100	20	—	3
*Ava*II	37	7.5	20	10	100	20	—	3
*Bal*I	37	7.9	6	6	—	6	—	4
*BamH*I	37	7.3	10	13[a]	50–100[a]	—	—	5
*BamH*I·1[b]	37	8.5	20	10	—	2	*c*	6
*Bcl*I	50[a]	7.4	12	12	—	0.5 m*M* DTT[b]	—	7
*Bgl*I	30[a]	9.5[a]	20 m*M* GOH[b]	20[a]	150[a]	7	—	8
*Bgl*II	30[a]	9.5[a]	20 m*M* GOH[b]	10[a]	—	—	*d, e*	8
*Bsp*I	37	8[a]	25	20[a]	50[a]	—	*e, f*	9
*Bst*I	37–50[a]	7–9.5[a]	100	0.5–2[a]	—	—	*g*	10
*Bst*I*[b]	37	9	100	>10	—	—	*h*	10
*Bst*15031	65[a]	7.8[a]	10[a]	0.2[a]	—	6.6	—	11
*Bsu*I	37	7.4[a]	10	10[a]	150[a]	1 m*M* DTT[b]	—	12
*Bsu*I*[b]	37	8.5	25	10	—	5	*i*	12
*Cla*I	37	7.4	6	6	50	6	—	13
*Dde*I	37	7.5	100	5	100	—	*j*	14
*Dpn*I	37	7.5	50	5	50	—	*k, l*	15
*Dpn*II	37	7.5	50	5	50	—	*k, l*	15
*Eca*I	37	8	10	10	—	—	*l*	16
*Eco*RI	37	7.1–7.5[a]	100	5[a]	50[a]	—	—	17
*Eco*RI*[b]	37	8.5	25	2	—	—	*e*	18
*Eco*RII	37	7.4	25	5	—	—	—	19
*Fnu*DII	37	7.9	6	6	50–150[a]	6	—	20
*Hae*II	37	7.9	6	6	—	6	—	21

*Hae*III	70[a]	7.5[a]	50[a]	5[a]	—	0.5 m*M* DTT	—	22
*Hga*I	37	7.6	10	5[a]	—	7	*m*	23
*Hgi*AI	30–45[a]	7.5–8.5[a]	10	10	100–150[a]	10	—	24
*Hha*I	37	7.9	6	6	—	6	—	25
*Hinc*II	37	7.9	10	6.6	60	6	—	26
*Hind*III	37	8.5[a]	10	10[a]	60[a]	7	*e, n*	27
*Hin*fI	37	7.5	6.6	10	50	6.6	*j*	28
*Hpa*I	45[a]	7.7–8.1[a]	10	5[a]	—	10	*k, o*	29
*Hpa*II	37	7.5	10	10	—	10	*k, p*	29
*Hph*I	37	7.4	10	10	6 m*M* KCl	10	—	30
*Kpn*I	37	7.9	6	6	—	6	—	31
*Mbo*I	37	7.9	6	6	—	6	—	32
*Mbo*II	37	7.9	6	6	—	6	—	32
*Mla*I	40	7.4	6.7	6.7	60 m*M* KCl	6.7	—	33
*Msp*I	37	8.0	20	10	5	10	*q*	34
*Nci*I	37	7.5	6	6	6	6	*k*	35
*Ngo*II	55[a]	8.5[a]	100	1[a]	20	—	*k*	36
*Pst*I	37	7.4	6.6	6.6	50	—	—	37
*Pvu*I	37	7.9	6	6	—	6	—	38
*Pvu*II	37	7.9	6	6	—	6	—	38
*Rsa*I	34[a]	7.9[a]	10	6	—	0.5 m*M* DTT[b]	*o*	39
*Rsh*I	30–37[a]	7.9	10	6	—	0.5 m*M* DTT[b]	*l*	40
*Sal*I	37	7.9	6	6	—	6	—	41
*Sau*3AI	30	7.5	6	15	60	6	—	42
*Sau*96I	30	7.4	6	15	60	6	—	43
*Sma*I	37	7.5	10	10	50	—	—	44
*Sph*I	37	7.3–7.8[a]	6	6[a]	50[a]	6	*l*	45
*Sst*I	37	7.5	10	10	100	10	—	46
*Stu*I	37	7.9	10	10	100	—	—	47
*Taq*I	37	7.4	10	10	—	10	*r*	48
*Tha*I	60[a]	7.4	10	10	—	—	—	49
*Tth*I	60[a]	7.5–8.5[a]	20	5	50	10	*e, s*	50
*Tth*111, I	65[a]	7.4	8	8[a]	50[a]	8	*t*	51

(*continued*)

TABLE II (*continued*)

Enzyme	Temperature (°C)	pH	Tris-HCl (m*M*)	$MgCl_2$ (m*M*)	NaCl (m*M*)	2-Mercapto-ethanol (m*M*)	Notes	Reference*
*Tth*111, II	65[a]	7.4	6	6	120–150[a]	6	*r*	52
*Xba*I	37	7.9	6	6	—	6	—	53
*Xho*I	37	7.9	6	6	—	6	—	54
*Xma*I	37	7.9	6	6	—	6	—	44
*Xor*II	37	7.4	6	12–24[a]	—	6	*e*	55

[a] Optimal condition.
[b] Abbreviations: DTT, dithiothreitol; GOH, glycine-NaOH; *Bam*HI.1, *Bst*I*, *Bsu*I*, *Eco*RI*, the secondary, "star" activities of *Bam*HI, *Bst*I, *Bsu*I, and *Eco*RI, respectively.
[c] In addition, 36% (v/v) glycerol and >20 × excess of enzyme are needed.
[d] Activity is stimulated twofold with 200 m*M* NaCl.
[e] Mn^{2+} can substitute for Mg^{2+}.
[f] Zn^{2+} inhibits activity.
[g] Activity is inhibited 50% by 50 m*M* NaCl.
[h] In addition, >5% glycerol is needed.
[i] In addition, 25% glycerol and 20–40 × excess of enzyme are needed.
[j] In addition, 500 μg of bovine serum albumin are needed per milliliter.
[k] In addition, 100 μg of bovine serum albumin are needed per milliliter.
[l] Activity is inhibited by ≥100 m*M* NaCl.
[m] Active to 500 m*M* KCl.
[n] Activity is inhibited by >250 m*M* NaCl or pH <7.
[o] Active to 200 m*M* NaCl.
[p] Activity is inhibited by >60 m*M* NaCl.
[q] In addition, 50 μg of bovine serum albumin are needed per milliliter.
[r] Active at 70°.
[s] Active to 300 m*M* NaCl.
[t] Activity is inhibited by >200 m*M* NaCl.
* Key to references:
1. R. J. Roberts, P. A. Myers, A. Morrison, and K. Murray, *J. Mol. Biol.* **102,** 157 (1976).

2. S. G. Hughes, T. Bruce, and K. Murray, *Biochem. J.* **185,** 59 (1980).
3. S. G. Hughes, and K. Murray, *Biochem. J.* **185,** 65 (1980).
4. R. E. Gelinas, P. A. Myers, G. A. Weiss, R. J. Roberts, and K. Murray, *J. Mol. Biol.* **114,** 433 (1977).
5. B. Hinsch, and M. Kula, *Nucleic Acids Res.* **8,** 623 (1980).
6. George, R. W. Blakesley, and J. G. Chirikjian, *J. Biol. Chem.* **255,** 6521 (1980).
7. A. H. A. Bingham, T. Atkinson, D. Sciaky, and R. J. Roberts, *Nucleic Acids Res.* **5,** 3457 (1978).
8. T. A. Bickle, V. Pirrotta, and R. Imber, this series, Vol. 65, p. 132.
9. P. Venetianer, this series, Vol. 65, p. 109.
10. C. M. Clarke, and B. S. Hartley, *Biochem. J.* **177,** 49 (1979).
11. J. F. Catterall, and N. E. Welker, this series, Vol. 65, p. 167.
12. S. Bron, and W. Horz, this series, Vol. 65, p. 112.
13. H. Mayer, R. Grosschedl, H. Schutte, and G. Hobom, *Nucleic Acids Res.* **9,** 4833 (1981).
14. R. A. Makula, and R. B. Meagher, *Nucleic Acids Res.* **8,** 3125 (1980).
15. S. A. Lacks, this series, Vol. 65, p. 138.
16. G. Hobom, E. Schwarz, M. Melzer, and H. Mayer, *Nucleic Acids Res.* **9,** 4823 (1981).
17. R. A. Rubin, and P. Modrich, this series, Vol. 65, p. 96.
18. B. Polisky, P. Greene, D. E. Garfin, B. J. McCarthy, H. M. Goodman, and H. W. Boyer, *Proc. Natl. Acad. Sci. U.S.A.* **72,** 3310 (1975).
19. S. G. Hattman, and S. Hattman, *J. Mol. Biol.* **98,** 645 (1975).
20. A. C. P. Lui, B. C. McBride, G. F. Vovis, and M. Smith, *Nucleic Acids Res.* **6,** 1 (1979).
21. R. J. Roberts, J. B. Breitmeyer, N. F. Tabachnik, and P. A. Myers, *J. Mol. Biol.* **91,** 121 (1975).
22. R. W. Blakesley, J. B. Dodgson, I. F. Nes, and R. D. Wells, *J. Biol. Chem.* **252,** 7300 (1977).
23. M. Takanami, *Methods Mol. Biol.* **7,** 113 (1974).
24. N. L. Brown, M. McClelland, and P. R. Whitehead, *Gene* **9,** 49 (1980).
25. R. J. Roberts, P. A. Myers, A. Morrison, and K. Murray, *J. Mol. Biol.* **103,** 199 (1976).
26. A. Landy, E. Ruedisueli, L. Robinson, C. Foeller, and W. Ross, *Biochemistry* **13,** 2134 (1974).
27. H. O. Smith and G. M. Marley, this series, Vol. 65, p. 104.
28. K. N. Subramanian, S. M. Weissman, B. S. Zain, and R. J. Roberts, *J. Mol. Biol.* **110,** 297 (1977).
29. J. L. Hines, T. R. Chauncey, and K. L. Agarwal, this series, Vol. 65, p. 153.
30. D. G. Kleid, this series, Vol. 65, p. 163.
31. J. Tomassini, R. Roychoudhury, R. Wu, and R. J. Roberts, *Nucleic Acids Res.* **5,** 4055 (1978).
32. R. E. Gelinas, P. A. Myers, and R. J. Roberts, *J. Mol. Biol.* **114,** 169 (1977).
33. M. Duyvesteyn, and A. de Waard, *FEBS Lett.* **111,** 423 (1980).
34. O. J. Yoo, and K. L. Agarwal, *J. Biol. Chem.* **255,** 10559 (1980).

TABLE II (*continued*)

35. R. Watson, M. Zuker, S. M. Martin, and L. P. Visentin, *FEBS Lett.* **118,** 47 (1980).
36. D. J. Clanton, W. S. Riggsby, and R. V. Miller, *J. Bacteriol.* **137,** 1299 (1979).
37. D. I. Smith, F. R. Blattner, and J. Davies, *Nucleic Acids Res.* **3,** 343 (1976).
38. T. R. Gingeras, L. Greenough, I. Schildkraut, and R. J. Roberts, *Nucleic Acids Res.* **9,** 4525 (1981).
39. S. P. Lynn, L. K. Cohen, S. Kaplan, and J. F. Gardner, *J. Bacteriol.* **142,** 380 (1980).
40. S. P. Lynn, L. K. Cohen, J. F. Gardner, and S. Kaplan, *J. Bacteriol.* **138,** 505 (1979).
41. J. R. Arrand, P. A. Myers, and R. J. Roberts, *J. Mol. Biol.* **118,** 127 (1978).
42. J. S. Sussenbach, C. H. Monfoort, R. Schiphof, and E. E. Stobberingh, *Nucleic Acids Res.* **3,** 3193 (1976).
43. J. S. Sussenbach, P. H. Steenbergh, J. A. Rost, W. J. van Leeuwen, and J. D. A. van Embden, *Nucleic Acids Res.* **5,** 1153 (1978).
44. S. A. Endow and R. J. Roberts, *J. Mol. Biol.* **112,** 521 (1977).
45. L. Y. Fuchs, L. Covarrubias, L. Escalante, S. Sanchez, and F. Bolivar, *Gene* **10,** 39 (1980).
46. A. Rambach, this series, Vol. 65, p. 170.
47. H. Shimotsu, H. Takahashi, and H. Saito, *Gene* **11,** 219 (1980).
48. S. Sato, C. A. Hutchison, III, and J. I. Harris, *Proc. Natl. Acad. Sci. U.S.A.* **74,** 542 (1977).
49. D. J. McConnell, D. Searcy, and G. Sutcliffe, *Nucleic Acids Res.* **5,** 1979 (1978).
50. A. Venegas, R. Vicuna, A. Alonso, F. Valdes, and A. Yudelevich, *FEBS Lett.* **109,** 156 (1980).
51. T. Shinomiya and S. Sato, *Nucleic Acids Res.* **8,** 43 (1980).
52. Shinomiya, M. Kobayashi, and S. Sato, *Nucleic Acids Res.* **8,** 3275 (1980).
53. B. S. Zain, and R. J. Roberts, *J. Mol. Biol.* **115,** 249 (1977).
54. T. R. Gingeras, P. A. Myers, J. A. Olsen, F.A. Hanberg, and R. J. Roberts, *J. Mol. Biol.* **118,** 113 (1978).
55. R. Y. H. Wang, J. G. Shedlarski, M. B. Farber, D. Kuebbing, and M. Ehrlich, *Biochim. Biophys. Acta* **606,** 371 (1980).

Depending upon the subsequent use of the cleaved DNA, the demands on the purity of the DNA may vary. Generally, RNA and/or DNA contamination does not significantly interfere with the apparent restriction reaction rate as measured by digest completion. This is in spite of the fact that nonspecific binding to nucleic acids reduces the effective concentration of a restriction endonuclease. Contaminating nucleic acids more often interfere by obscuring the detection or selection of reaction products. For example, positive clones screened by rapid lysis methods[11] may be difficult to identify if the insert DNA excised by restriction endonuclease cleavage migrates in the same region as the intense broad tRNA band upon agarose gel electrophoresis. In such cases, treatment with DNase-free RNase or purification with a quick minicolumn using RPC-5 ANALOG[12] is recommended. On the other hand, sequencing protocols, e.g., the M13mp7 dideoxy method,[13] require highly purified DNA as restriction cleavage products. Protein contaminations are tolerated in a restriction reaction as long as the products eventually are protein-free. It should be noted, however, that the presence of other nucleases will reduce the integrity of the product, whereas proteins tightly bound to the DNA may lessen or block the cleavage reaction. DNAs are customarily deproteinized by phenol extraction prior to restriction endonuclease treatment.

Compounds involved in DNA isolation should be rigorously removed by dialysis or by ethanol precipitation and drying prior to addition of the DNA sample to the restriction endonuclease reaction. For example, Hg^{2+}, phenol, chloroform, ethanol, ethylene(diaminetetraacetic) acid (EDTA), sodium dodecyl sulfate (SDS), and NaCl at high levels interfere with restriction reactions, and some can alter the recognition specificity of restriction endonucleases. Drugs frequently used in DNA studies, e.g., actinomycin and distamycin A,[14] also influence restriction endonuclease activity.

In a typical reaction, the restriction endonuclease is in considerable molar excess of the substrate DNA. Therefore, consideration of DNA concentration usually is not required. In fact, it was necessary to dilute *Hae*III[8] or *Bam*HI[15] approximately 1000-fold from typical unit assay conditions in order to observe a substrate cleavage rate proportional to the

[11] R. W. Davis, M. Thomas, J. Cameron, T. P. St. John, S. Scherer, and R. A. Padgett, this series, Vol. 65, p. 404.

[12] J. A. Thompson, R. W. Blakesley, K. Doran, C. J. Hough, and R. D. Wells, this volume [26].

[13] J. Messing, R. Crea, and P. H. Seeburg, *Nucleic Acids Res.* **9,** 309 (1981).

[14] V. V. Nosikov, E. A. Braga, A. V. Karlishev, A. L. Zhuze, and O. L. Polyanovsky, *Nucleic Acids Res.* **3,** 2293 (1976).

[15] J. George, unpublished results, 1981.

amount of enzyme added to the reaction. Further, caution must be exercised when attempting to extrapolate the amount of enzyme required for a complete digest based upon the number of recognition sites in a particular DNA. Preliminary observations using the enzyme-saturated, end point-dependent unit assay indicates that apparently no general correlation exists between recognition site density and restriction enzyme units required.[16]

By exception, the concentration of the substrate DNA did influence the apparent reaction rate for *Hin*dIII under enzyme-saturating conditions. A typical reaction for unit determination contains 1 μg of lambda DNA in a 50-μl reaction volume (20 μg/ml). One unit, but not 0.5 unit, of *Hin*dIII completely cleaves 1 μg of lambda DNA. One unit of *Hin*dIII also completely cleaves 4 μg (80 μg/ml) of lambda DNA under these conditions.[16] This peculiar response in *Hin*dIII activity cannot be attributed to enzyme : DNA concentration ratios, but is assumed to reflect the absolute DNA concentration dependence of *Hin*dIII. In contrast to the increased *Hin*dIII activity in the presence of increased DNA, 10 units of *Hpa*I, *Kpn*I, or *Sau*3AI proved to be insufficient to cleave completely 4 μg (80 μg/ml) of lambda DNA in a 15-hr reaction.[16] This phenomenon may be attributed to the viscosity produced by high concentrations of high molecular weight DNA (e.g., lambda DNA), which can inhibit enzyme diffusion and, therefore, inhibit some enzyme activities. These apparently anomalous results point out that one cannot directly compare units determined by titrating enzyme with those obtained by titrating (changing the concentration of) DNA. Further, DNA concentrations near or below the K_m of a restriction enzyme (1–10 n*M*[6]) could also inhibit apparent enzyme cleavage. However, for lambda DNA the K_m is approximately 1000-fold less than the concentration used in the standard reaction for unit determination. From these observations it is recommended that the DNA concentration be at or near that used in the unit assay reaction for the particular restriction endonuclease.

Restriction endonucleases probably show their greatest sensitivity to the DNA sequence. Obviously, the sequence of the recognition site is essentially invariant, as this distinguishes type II restriction endonucleases from other nucleases. The stringent sequence requirement frequently can be relaxed by alterations of the reaction environment, generating the "star" activity (see below) observed for a number of enzymes, *Eco*RI being the most notable. Sequences adjacent to the recognition site also influence the rate of cleavage. A nearly 10-fold difference in reaction rate was observed between two of the *Eco*RI sites in lambda DNA.[17] A

[16] This laboratory, unpublished results, 1981.

[17] M. Thomas and R. W. Davis, *J. Mol. Biol.* **91,** 315 (1975).

TABLE III
EFFECT OF BASE ANALOG SUBSTITUTIONS IN DNA ON RESTRICTION ENDONUCLEASE ACTIVITY

Enzyme	Recognition sequence	Relative activity of base analogs[a–c]				
		HMC	GHMC	U	HMU	BrdU
*Bam*HI[d]	GGATCC		−	++	+	+
*Eco*RI[d–f]	GAATTC	++	−	++	+	+
*Hae*II[d]	PuGCGCPy		−		+	+
*Hha*I[d]	GCGC		−		++	
*Hin*dII[d,e]	GTPyPuAC	−	−	+	+	
*Hin*dIII[d–f]	AAGCTT	−	−	+	+	+
*Hpa*I[d,g]	GTTAAC		−	+	+	+
*Hpa*II[d]	CCGG		−		++	+
*Mbo*I[g]	GATC					Enhanced 5-fold

[a] *Activity symbols:* ++, full activity; +, diminished activity; −, no activity; blank, not tested.
[b] In these studies, HMC or GHMC were in place of cytosine, while U, HMU, or BrdU replaced thymidine in the tested DNAs.
[c] Abbreviations used: HMC, 5-hydroxymethylcytosine; GHMC, glucosylated 5-hydroxymethylcytosine; U, uridine; HMU, 5-hydroxymethyluridine; BrdU, 5-bromodeoxyuridine; Py, pyrimidine; Pu, purine.
[d] K. L. Berkner and W. R. Folk, *J. Biol. Chem.* **254,** 2551 (1979).
[e] D. A. Kaplan and D. P. Nierlich, *J. Biol. Chem.* **250,** 2395 (1975).
[f] M. A. Marchionni and D. J. Roufa, *J. Biol. Chem.* **253,** 9075 (1978).
[g] J. Petruska and D. Horn, *Biochem. Biophys. Res. Commun.* **96,** 1317 (1980).

similar effect was reported for *Pst*I.[18] In addition, thymine substituted by 5-bromodeoxyuridine prevented cleavage of some *Sma*I sites in the DNA tested, even though the 5-bromodeoxyuridine was not part of the canonical recognition sequence (CCCGGG).[19]

Nucleotide changes within the recognition sequence more directly affect the restriction endonuclease reaction (Tables III and IV). For *Eco*RI, cleavage was unaffected by 5-hydroxymethylcytosine substitution for cytosine[20] or by the absence or the presence of the 2-amino group of guanine.[21] Glycosylation of 5-hydroxymethylcytosine, however, made the DNA resistant to cleavage by *Eco*RI as well as by *Hpa*I, *Hin*dII, *Hin*dIII, *Bam*HI, *Hae*II, *Hpa*II and *Hha*I.[22] Substitution of thymine with

[18] K. Armstrong and W. R. Bauer, *Nucleic Acids Res.* **10,** 993 (1982).
[19] M. A. Marchionni and D. J. Roufa, *J. Biol. Chem.* **253,** 9075 (1978).
[20] P. Modrich and R. A. Rubin, *J. Biol. Chem.* **252,** 7273 (1977).
[21] D. A. Kaplan and D. P. Nierlich, *J. Biol. Chem.* **250,** 2395 (1975).
[22] K. L. Berkner and W. R. Folk, *J. Biol. Chem.* **254,** 2551 (1979).

5-hydroxymethyluridine diminished activities of enzymes with AT-containing sites, whereas a differential effect was observed for uridine and 5-bromodeoxyuridine substitutions.[22] Methylation of nucleotides within restriction endonuclease recognition sequences, occurring almost exclusively as 5-methylcytosine or N^6-methyladenine, prevented most

TABLE IV
METHYLATED DNAS AS SUBSTRATES FOR RESTRICTION ENDONUCLEASES[a]

	Sequences containing 5-methylcytosine or N^6-methyladenine[b]		
Enzyme	Cleaved	Not cleaved	References
*Aos*II	—	GPu^{m}CGPyC	*d, e*
*Ava*I	—	CPy^{m}CGPuG	*f*
*Ava*II	—	GG(A)C^{m}C (T)	*g*
*Bst*NI	C^{m}C(A)GG (T)	—	*h*
*Eco*RII	—	C^{m}C(A)GG (T)	*h, i*
*Hae*II	—	PuG^{m}CGCPy	*e, f*
*Hae*III	GGC^{m}C	GG^{m}CC	*j, k*
*Hap*II	—	C^{m}CGG	*e, l*
*Hha*I	—	G^{m}CGC	*f, j*
*Hpa*II	^{m}CCGG	C^{m}CGG	*j, l*
*Msp*I	C^{m}CGG	^{m}CCGG	*l, m*
*Pst*I	—	^{m}CTGCAG	*n*
*Pvu*II	—	^{m}CAGCTG	*n*
*Sal*I	—	GT^{m}CGAC	*d, e*
*Sma*I	—	CC^{m}CGGG	*e, u*
*Taq*I	T^{m}CGA	—	*o*
*Xho*I	—	CT^{m}CGAG	*d, e*
*Xma*I	CC^{m}CGGG	—	*u*
*Bam*HI	GG^{m}ATCC	—	*g, p*
*Bgl*II	AG^{m}ATCT	—	*p*
*Dpn*I	G^{m}ATC[c]	—	*o, q*
*Dpn*II	—	G^{m}ATC	*q*
*Eco*RI	—	GA^{m}ATTC	*r*
*Fnu*EI	G^{m}ATC	—	*s*
*Hin*dII	—	GTPyPu^{m}AC	*k*
*Hin*dIII	—	^{m}AAGCTT	*k*
*Hpa*I	—	GTTA^{m}AC	*t*
*Mbo*I	—	G^{m}ATC	*p, s*
*Mbo*II	—	GAAG^{m}A	*g*
*Sau*3AI	G^{m}ATC	—	*o, p*
*Taq*I	—	TCG^{m}A	*d, o*

enzymes from cleaving. In Table IV are listed the responses of a variety of restriction enzymes to DNA methylation. Several enzymes were found to vary in their response to hemimethylated DNAs, where only one of the two strands is methylated (Table IV).[23]

Modification of all or the vast majority of certain base types within the DNA of certain bacteriophages has, as expected, more drastic effects on the ability and rate of restriction endonuclease cleavage than modifications that occur solely within the recognition sequences described above.

[23] Y. Gruenbaum, H. Cedar, and A. Razin, *Nucleic Acids Res.* **9,** 2509 (1981).

[a] The enzymes *Bst*NI, *Hinc*II, *Hinf*I, *Hpa*I, and *Taq*I have been reported to cleave hemimethylated DNA (i.e., only one DNA strand contains ^{m}C). In addition *Msp*I, *Sau*3A, and *Hae*III nick the unmethylated strand of the hemimethylated DNA [R. E. Streeck, *Gene* **12,** 267 (1980); and Y. Gruenbaum, H. Cedar, and A. Razin, *Nucleic Acids Res.* **9,** 2509 (1981)].

[b] Abbreviations used: —, not determined; Pu, purine; Py, pyrimidine; ^{m}C, 5-methylcytosine; ^{m}A, N^6-methyladenine.

[c] Methylation is required for cleavage.

[d] L. H. T. van der Ploeg and R. A. Flavell, *Cell* **19,** 947 (1980).

[e] M. Ehrlich and R. Y. H. Wang, *Science* **212,** 1350 (1981).

[f] A. P. Bird and E. M. Southern, *J. Mol. Biol.* **118,** 27 (1978).

[g] K. Bachman, *Gene* **11,** 169 (1980).

[h] S. Hattman, C. Gribbin, and C. A. Hutchison, III, *J. Virol.* **32,** 845 (1979).

[i] M. S. May and S. Hattman, *J. Bacteriol.* **122,** 129 (1975).

[j] M. B. Mann and H. O. Smith, *Nucleic Acids Res.* **4,** 4211 (1977).

[k] P. H. Roy and H. O. Smith, *J. Mol. Biol.* **81,** 427 (1973).

[l] C. Waalwijk and R. A. Flavell, *Nucleic Acids Res.* **5,** 3231 (1978).

[m] T. W. Sneider, *Nucleic Acids Res.* **8,** 3829 (1980).

[n] A. P. Dobritsa and S. V. Dobritsa, *Gene* **10,** 105 (1980).

[o] R. E. Streeck, *Gene* **12,** 267 (1980).

[p] B. Dreiseikelman, R. Eichenlaub, and W. Wackernagel, *Biochim. Biophys. Acta* **562,** 418 (1979).

[q] S. Lacks and B. Greenberg, *J. Biol. Chem.* **250,** 4060 (1975).

[r] A. Dugaiczyk, J. Hedgepeth, H. W. Boyer, and H. M. Goodman, *Biochemistry* **13,** 503 (1974).

[s] A. C. P. Lui, B. C. McBride, G. F. Vovis, and M. Smith, *Nucleic Acids Res.* **6,** 1 (1979).

[t] L.-H. Huang, C. M. Farnet, K. C. Ehrlich, and M. Ehlich, *Nucleic Acids Res.* **10,** 1579 (1982).

[u] H. Youssoufian and C. Mulder, *J. Mol. Biol.* **150,** 133 (1981).

When 30 type II restriction endonucleases were separately incubated with *Xanthomonas oryzae* phage XP12 DNA, all cytosine residues of which are modified to 5-methylcytosine, only *Taq*I cleaved efficiently. When bacteriophage T4 DNA, which contains only 5-hydroxymethylcytosine, but not cytosine, was tested, again only *Taq*I cleaved, although inefficiently. The complete substitution of thymine residues with 5-hydroxymethyluracil in the genome of *Bacillus subtilis* phages SP01 and PBS1 either had no effect or for, some of the restriction enzymes, only reduced cleavage efficiency. The substitution of thymine by phosphogluconated or glucosylated 5-(4′,5′-dihydroxy)pentyluracil in *B. subtilis* phage SP15 DNA precluded cleaving by most of the restriction endonucleases tested.[24] *Dde*I, *Taq*I, *Tha*I, and *Bst*NI did cleave this DNA very poorly. Complete nucleotide substitutions cause drastic alterations not only in the recognition sequences for these restriction enzymes, but also in the secondary and tertiary DNA structures.

The proximity of the recognition site to the terminus of a DNA can also influence cleavage. *Hpa*II and *Mno*I required at least one base preceding the 5′ end of the recognition sequence for cleavage.[25] The minimal duplex hexanucleotide recognition sequences for *Eco*RI (GAATTC), *Bam*HI (GGATCC), and *Hin*dIII (AAGCTT) were resistant to cleavage. However, *Eco*RI will cleave if the sequence is extended by one base to GAATTCA.[26] On the other hand, when *Hha*I (GCGC) cleaved poly(dG-dC), about 85% of the product was the limit tetranucleotide.[27]

Secondary and tertiary structure of the recognition/cleavage site also affects the restriction endonuclease reaction rate. Restriction enzymes typically require the substrate cleavage site to be in a duplex form for cleavage as shown for *Hae*III,[8] *Eco*RI,[9] and *Msp*I.[28] *Hin*dIII apparently requires at least two uninterrupted turns of the double helix for cleavage.[26] Certain restriction endonucleases (*Bsp*RI, *Hae*III, *Hha*I, *Hin*fI, *Mbo*I, *Mbo*II, *Msp*I, and *Sfa*I) will cleave "single-stranded" viral DNAs of bacteriophages ϕX174, M13, or f1 whose cleavage sites are in the duplex form. Even though *Hpa*II was reported to cleave a single strand,[29] there is no conclusive evidence that a bona fide single-stranded restriction site is cleaved. The fact that certain enzymes do not cleave the "single-stranded" viral DNAs indicates that properties in addition to the DNA

[24] L.-H. Huang, C. M. Farnet, K. C. Ehrlich, and M. Ehrlich, *Nucleic Acids Res.* **10,** 1579 (1982).
[25] B. R. Baumstark, R. J. Roberts, and U. L. RajBhandary, *J. Biol. Chem.* **254,** 8943 (1979).
[26] Y. A. Berlin, N. M. Zvonok, and S. A. Chuvpilo, *Bioorg. Khim.* **6,** 1522 (1980).
[27] R. J. Roberts, P. A. Myers, A. Morrison, and K. Murray, *J. Mol. Biol.* **103,** 199 (1976).
[28] O. J. Yoo, and K. L. Agarwal, *J. Biol. Chem.* **255,** 10559 (1980).
[29] K. Horiuchi, and N. D. Zinder, *Proc. Natl. Acad. Sci. U.S.A.* **72,** 2555 (1975).

recognition sequence are required for restriction endonucleolytic cleavage (for review, see Wells and Neuendorf[30]).

Cleavage of RNA · DNA hybrid molecules were described for several restriction endonucleases.[31] The fate of the RNA was not followed, but presumably RNA was degraded to small oligonucleotides. This would not be surprising since restriction endonucleases are frequently not assayed for, or purified from, ribonucleases. It is difficult unequivocally to conclude that true RNA · DNA hybrids were cleaved, since the remaining DNA strand could potentially self-hybridize, as in the "single-stranded" viral DNAs, to provide the appropriate duplex substrate. This must await further experimentation.

Another DNA structural variant frequently encountered in restriction endonuclease reactions is superhelicity. Generally, larger amounts of restriction enzyme are required to cleave supercoiled plasmid or viral DNAs completely than for linear DNA. A comparison of the relative cleavage efficiencies for several supercoiled and linear DNAs are presented in Table V. If a supercoiled DNA (e.g., pBR322 plasmid DNA) is first linearized with a restriction endonuclease or relaxed with topoisomerase,[32] frequently less enzyme is needed for complete cleavage (Table VI).

Reagents

The components of a restriction endonuclease buffer system should be of the highest quality available. Contaminants, e.g., heavy metals in buffer components, should be looked for and avoided. Reagents should be free of enzyme activities, especially nucleases. Filter or heat-sterilize all reagent stocks, then store frozen and replace frequently in order to maintain quality and integrity. For convenience several of the reagents can be mixed together as a 10-fold concentrated stock solution. When added to the final reaction mixture, an appropriate single dilution into sterile water is made. These precautions will help to ensure the desired quality in the DNA product of the reaction.

A number of buffers are available to maintain the assay pH between 7 and 8. Tris(hydroxymethyl)aminomethane (Tris), the most widely used and least noxious, has a large temperature coefficient that should be considered when preparing and using this buffer. The pH of Tris buffers also

[30] R. D. Wells, and S. K. Neuendorf, *in* "Gene Amplification and Analysis," Vol. I: "Restriction Endonucleases" (J. G. Chirikjian, ed.), p. 101. Elsevier/North-Holland, Amsterdam, 1981.

[31] P. L. Molloy, and R. H. Symons, *Nucleic Acids Res.* **8,** 2939 (1980).

[32] J. LeBon, C. Kado, L. Rosenthal, and J. G. Chirikjian, *Proc. Natl. Acad. Sci. U.S.A.* **74,** 542 (1977).

TABLE V
RELATIVE ACTIVITIES OF CERTAIN RESTRICTION ENDONUCLEASES ON SEVERAL DNA SUBSTRATES[a]

	Enzyme units required for complete cleavage of specified DNA[b,c]				
Enzyme[d]	Lambda	Ad-2	pBR322	ϕX174RF	SV40
*Bam*HI	1	2	3	—	4
*Eco*RI	1	1	2.5	—	3
*Hha*I	1	10	4	1	2
*Hind*III	1	3	2.5	—	10
*Hin*fI	1	1	1	1	1
*Hpa*II	1	1	2	1	10
*Pst*I	1	2	1.5	1	1
*Pvu*II	1	4	4	—	4
*Sau*3AI	1	2	2.5	—	1.5
*Taq*I	1.5	1	10	1	0.5
*Xor*II	1	1	>10	—	—

[a] H. Belle Isle, unpublished results, 1981.

[b] Activity was measured by incubation of 1 μg of the specified DNA with various amounts of the respective restriction endonucleases under appropriate standard reaction conditions. These values represent the minimum number of units of enzyme required for complete digestion of the specified DNA as monitored by agarose gel electrophoresis [P. A. Sharp, B. Sugden, and J. Sambrook, *Biochemistry* **12,** 3055 (1973)]. Enzyme activity units are defined as the minimum amount of enzyme required to digest completely 1 μg of lambda (or ϕX174 RF for *Taq*I, or Ad-2 for *Xor*II) DNA under standard reaction conditions.

[c] Abbreviations used: lambda, bacteriophage lambda CI857 Sam7; Ad-2, Adenovirus type 2; pBR322, supercoiled plasmid pBR322; ϕX174 RF, supercoiled bacteriophage ϕX174 replicative form; SV40, supercoiled simian virus 40; and —, recognition sequence for this enzyme not present in this DNA.

[d] All enzymes and DNAs were from Bethesda Research Laboratories, Inc.

varies with concentration and should therefore be reset upon dilution. Glycine is useful as a restriction endonuclease buffer for reactions at pH >9. Phosphate is an excellent buffer for assays between pH 6.0 and 7.5 and has a minimal temperature coefficient. But phosphate buffers should be used only if no subsequent enzyme reactions are to be performed that

TABLE VI
EFFECT OF DNA SUPERHELICITY ON RESTRICTION ENZYME ACTIVITY[a]

Enzyme[c]	Enzyme units required for complete cleavage[b]	
	Supercoiled pBR322 DNA	Linear pBR322 DNA[d]
*Bam*HI	2	1
*Eco*RI	2.5	1
*Hin*dIII	2.5	2.5
*Sal*I	7.5	3

[a] H. Belle Isle, unpublished results, 1981.

[b] Activity was measured by incubation of 1 μg of pBR322 DNA with various amounts of the respective restriction endonucleases under appropriate standard reaction conditions. These values represent the minimum number of units of enzyme required for complete digestion of the DNA as monitored by agarose gel electrophoresis [P. A. Sharp, B. Sugden, and J. Sambrook, *Biochemistry* **12,** 3055 (1973)]. Enzyme activity units are defined as the minimum amount of enzyme required to digest completely 1 μg of lambda DNA under standard reaction conditions.

[c] All enzymes and DNAs were from Bethesda Research Laboratories, Inc.

[d] Linear form III pBR322 DNA was prepared by incubation of supercoiled form I DNA with *Pst*I, followed by phenol extraction and ethanol precipitation.

are inhibited by the phosphate ion, e.g., DNA end-labeling[33] or ligation.[34] Typical methods of phenol extraction or ethanol precipitation will not significantly reduce the phosphate ion content in a DNA sample. Dialysis or multiple ethanol precipitations with 2.5 *M* ammonium acetate are, on the other hand, effective. Citrate and other biological buffers that chelate Mg^{2+} cannot be used.

The selected buffer concentration must be sufficient to maintain the proper pH of the final reaction mixture. Buffer concentrations greater than 10 m*M* are recommended to provide the appropriate buffering capacity under conditions where the pH of most distilled water supplies are low. In addition, the reaction pH should not be altered when a relatively large volume of an assay component, e.g., the DNA substrate, is added. In general, the reaction rate is not significantly affected by the concentra-

[33] G. Chaconas and J. H. van de Sande, this series, Vol. 65, p. 75.

[34] A. W. Hu, manuscript in preparation (1982).

tion of Tris buffer above 10 m*M* e.g., *Hae*III demonstrated <20% variance in reactivity between 15 and 120 m*M*.[8]

Many restriction enzymes have significant activity over a rather broad pH range. *Hae*III has an activity optimum at pH 7.5, but retains at least 50% of its activity when assayed at 1.5 units above or below pH 7.5.[8] Some other enzymes studied, *Bst*I,[35] *Hae*II,[8] *Hgi*AI,[36] *Hha*I,[8] *Ngo*II,[37] *Sph*I,[38] and *Tth*I[39] showed similar profiles. Selected enzymes such as *Eco*RI are sensitive to altered pH. Not only does *Eco*RI activity significantly decrease,[40] but an altered activity (see Secondary Activity below) appears when the pH is increased from 7.2 to 8.5.[41] Thus, the pH should be maintained at the recommended value by a buffer with adequate capacity.

Type II restriction endonucleases require Mg^{2+} as the only cofactor. Complete chelation of Mg^{2+} by EDTA can thus effectively stop the reaction. Restriction enzyme activities are relatively insensitive to the Mg^{2+} concentration; similar rates are observed from 5 to 30 m*M*.[8,42] Similar to other nucleic acid enzymes, some restriction endonucleases accept Mn^{2+} as a substitute for Mg^{2+}, although with varying results. *Eco*RI and *Hin*dIII change their recognition specificity with such replacement.[43,44] *Hae*III is approximately 50% as active with $MnCl_2$ as with $MgCl_2$,[8] while *Xor*II[42] and *Tth*I[39] are equally active with Mg^{2+} or Mn^{2+}.

Whereas *Eco*RI functions, although inefficiently, with other divalent cations (Mn^{2+}, Co^{2+}, Zn^{2+}), Mg^{2+} cannot be replaced by other divalent cations (Cu^{2+}, Ba^{2+}, Cr^{2+}, Co^{2+}, Zn^{2+}, and Ni^{2+}) in the *Hae*III reaction.[8] *Bam*HI showed secondary "star" activity when Zn^{2+} or Co^{2+} replaced Mg^{2+} at pH 6, but no activity at pH 8.5.[15] *Bsp*I is quite active with Mn^{2+}, but completely inhibited with Zn^{2+}.[45] It is unclear at this point whether metal ions such as Zn^{2+} contribute to restriction endonuclease structural

[35] C. M. Clarke and B. S. Hartley, *Biochem. J.* **177,** 49 (1979).

[36] N. L. Brown, M. McClelland, and P. R. Whitehead, *Gene* **9,** 49 (1980).

[37] D. J. Clanton, W. S. Riggsby, and R. V. Miller, *J. Bacteriol.* **137,** 1299 (1979).

[38] L. Y. Fuchs, L. Covarrubias, L. Escalante, S. Sanchez, and F. Bolivar, *Gene* **10,** 39 (1980).

[39] A. Venegas, R. Vicuna, A. Alonso, F. Valdes, and A. Yuldelevich, *FEBS Lett.* **109,** 156 (1980).

[40] R. A. Rubin and P. Modrich, this series, Vol. 65, p. 96.

[41] B. Polisky, P. Greene, D. E. Garfin, B. J. McCarthy, H. M. Goodman, and H. W. Boyer, *Proc. Natl. Acad. Sci. U.S.A.* **72,** 3310 (1975).

[42] R. Y.-H. Wang, J. G. Shedlarski, M. B. Farber, D. Kuebbing, and M. Ehrlich, *Biochim. Biophys. Acta* **606,** 371 (1980).

[43] T. I. Tikchonenko, E. V. Karamov, B. A. Zavizion, and B. S. Naroditsky, *Gene* **4,** 195 (1978).

[44] M. Hsu and P. Berg, *Biochemistry* **17,** 131 (1978).

[45] P. Venetianer, this series, Vol. 65, p. 109.

stability as demonstrated with other nucleic acid enzymes, such as *Escherichia coli* DNA polymerase I.[46] Unless metal chelators such as EDTA, EGTA, or *o*-phenanthroline are present in the reaction, one need not be concerned about adding to the reaction metal ions other than Mg^{2+} for the activity or fidelity of restriction endonucleases.

Restriction endonucleases show a wide diversity in their responses to ionic strength (Table II). Most enzymes do not absolutely require specific monovalent cations, but rather are stimulated by the corresponding ionic strength. *Sma*I, however, does have an absolute requirement for K^+.[16] Many enzymes are stimulated by 50–100 m*M* NaCl or KCl (e.g., *Sph*I,[38] *Mlu*I[47]), whereas others are drastically inhibited at concentrations >20 m*M* (e.g., *Fok*I,[47] *Hin*dII,[48] and *Fnu*DI[49]). Other cations, e.g., NH_4^+, can in some cases provide the stimulating ionic strength.[50] Loss of restriction enzyme activity (Table VII) and recognition specificity[41,51] can result from inappropriate monovalent cation concentrations. Recommended concentrations (see Table II or VII) should therefore be closely followed. Special caution also should be used in selecting the appropriate buffers for multiple enzyme digestions (see Other Reaction Considerations).

Sulfhydryl reagents such as 2-mercaptoethanol and dithiothreitol are routinely used in restriction enzyme reactions. Historically, 2-mercaptoethanol was added to restriction enzyme preparations and reactions as a general precaution based on the labilities of other nucleic acid enzymes. Nath demonstrated that not all restriction endonucleases require such reagents.[52] *Bgl*II, *Eco*RI, *Hin*dIII, *Hpa*I, *Sal*I, and *Sst*II activities are insensitive, and *Ava*I, *Bam*HI, *Pvu*I, and *Sma*I activities are inhibited by the sulfhydryl reactive compounds *p*-mercuribenzoate and *N*-ethylmaleimide. In other studies, *Hpa*I and *Hpa*II,[53] and *Eco*RI[43] were unaffected, whereas the "star" activity of *Eco*RI (*Eco*RI*)[43] was sulfhydryl sensitive. Where not required, the sulfhydryl reagents should be omitted from the reaction to prevent stabilization of possible contaminating activities. When used, only freshly prepared stocks of 2-mercaptoethanol and dithiothreitol at final reaction concentrations of no greater than 10 and 1.0 m*M*, respectively, should be employed.

Bovine serum albumin (BSA) or gelatin is frequently used in restric-

[46] A. Kornberg, "DNA Replication." Freeman, San Francisco, California, 1980.

[47] H. Sugisaki and S. Kanazawa, *Gene* **16,** 73 (1981).

[48] H. O. Smith and G. M. Marley, this series, Vol. 65, p. 104.

[49] A. C. P. Lui, B. C. McBride, G. F. Vovis, and M. Smith, *Nucleic Acids Res.* **6,** 1 (1979).

[50] D. I. Smith, F. R. Blattner, and J. Davies, *Nucleic Acids Res.* **3,** 343 (1976).

[51] R. A. Makula and R. B. Meagher, *Nucleic Acids Res.* **8,** 3125 (1980).

[52] K. Nath, *Arch. Biochem. Biophys.* **212,** 611 (1981).

[53] J. L. Hines, T. R. Chauncey, and K. L. Agarwal, this series, Vol. 65, p. 153.

TABLE VII

RESTRICTION ENDONUCLEASE ACTIVITY IN CORE BUFFER[a]

	Relative enzyme activity (percent) in core buffer with[b,c]		
Enzyme[d]	0 mM NaCl	50 mM NaCl	100 mM NaCl
*Acc*I	200	100	50
*Alu*I	140	100	40
*Ava*I	75	125	40
*Ava*II	150	125	50
*Bal*I	27	14	5
*Bam*HI	67	117	33
*Bcl*I	120	120	80
BglI	33	67	89
*Bgl*II	50	88	100
*Bst*EII	40	160	120
*Cfo*I	50	20	5
*Cla*I	86	43	3
*Dde*I	50	150	200
*Dpn*I	133	133	117
*Eco*RI[e]	10	10	10
*Eco*RII	50	75	100
*Hae*II	200	100	25
*Hae*III	114	100	50
*Hha*I	33	56	67
*Hinc*II	25	75	100
*Hin*dIII	160	200	120
*Hin*fI	100	100	100
*Hpa*I	25	12	5
*Hpa*II	71	43	7
*Kpn*I	67	33	6
*Mbo*I	75	100	50
*Mbo*II	50	30	10
*Mnl*I	85	95	75
*Msp*I	100	33	17
*Nci*I	67	22	6
*Pst*I	100	125	100
*Pvu*II	33	67	50
*Sal*I	10	25	150
*Sau*3AI	40	20	10
*Sau*96	53	80	53
*Sma*I[f]	0	0	0
*Sph*I	10	20	40
*Sst*I	57	71	29

TABLE VII (*continued*)

	Relative enzyme activity (percent) in core buffer with[b,c]		
Enzyme[d]	0 m*M* NaCl	50 m*M* NaCl	100 m*M* NaCl
*Taq*I	12	25	50
*Tha*I	83	67	33
*Xba*I	100	70	50
*Xho*I	117	150	150
*Xma*III	50	67	33
*Xor*II	50	25	5

[a] A. MarSchel, unpublished results, 1981.

[b] Core buffer is 50 m*M* Tris-HCl (pH 8.0), 10 m*M* $MgCl_2$, 1 m*M* dithiothreitol, 100 μg of bovine serum albumin per milliliter, and an appropriate amount of NaCl.

[c] Enzyme activity was measured by incubation of an appropriate DNA with various amounts of the respective endonuclease in the standard reaction buffer, in core buffer, or in core buffer supplemented with NaCl. The standard buffer was either that listed in Table II or, in some cases, the listed buffer as modified by this laboratory to give greater activity. One enzyme unit is defined as the minimum amount of enzyme required to digest completely 1 μg of lambda (or adenovirus type 2 for *Bcl*I, *Eco*RII, *Sal*I, *Sau*96I, *Sma*I, *Sst*I, *Xba*I, *Xho*I, *Xma*III, and *Xor*II; φX174 RF for *Taq*I; SV40 form I for *Mbo*I, and *Mbo*II; or pBR322 for *Dpn*I and *Mnl*I) DNA as monitored by agarose gel electrophoresis [P. A. Sharp, B. Sugden, and J. Sambrook, *Biochemistry* **12,** 3055 (1973)]. The unit concentration of each enzyme determined in the core buffer, or in core buffer with NaCl is listed as a percentage of the unit concentration determined in the standard buffer (designated 100% activity).

[d] All enzymes and DNAs were from Bethesda Research Laboratories, Inc.

[e] *Eco*RI has a narrow pH optimum range for enzyme activity. When the pH was lowered to 7.2, the following relative enzymic activities were obtained: 44%, 89%, and 67% in core buffer supplemented with 0, 50, and 100 m*M* NaCl, respectively.

[f] *Sma*I has an absolute requirement for K^+, which is absent from the core buffer. When the core buffer contains 15 m*M* KCl, the following relative enzymic activities were obtained: 50%, 25%, and 7% in core buffer supplemented with 0, 50, and 100 m*M* NaCl, respectively.

tion endonuclease preparations to stabilize enzyme activity in long-term incubation or storage. *Hpa*I and*Hpa*II are quite unstable when the protein concentration is <20 μg/ml.[53] Addition of exogenous proteins protects the restriction endonucleases from proteases, nonspecific adsorption, and harmful environmental factors such as heat, surface tension, and chemicals, that cause denaturation. Only sterile solutions of nuclease-free BSA or heavy metal-free gelatin should be added to restriction enzyme reactions. In general, little harm results from addition of these proteins to the reaction. Occasionally, excess BSA binding to DNA causes band smearing during gel electrophoresis. This is eliminated by the addition of SDS to the sample followed by heating to 65° for 5 min prior to sample loading.

The importance of water quality should not be overlooked. Glass-distilled water free of ions and organic compounds should be used for all buffers and reaction components. Deionized water is satisfactory provided the content of organic material is not significant.

Glycerol added to restriction endonuclease stocks stabilizes the enzymes and prevents freezing at low temperature (−20°) during long-term storage. A number of restriction enzymes show reduced recognition specificity in the presence of glycerol (see below). In general, restriction enzyme reactions should contain <5% (v/v) glycerol (final concentration).

Core Buffer System

Many laboratories stock a large panel of individual buffer systems appropriate for the many restriction endonucleases in use (see Table II). Identification of one or a few primary buffer systems that would take advantage of the similarities of the restriction enzymes, while reflecting as closely as possible the optima for each enzyme, would provide a valuable convenience for restriction endonuclease use. For example, reaction conditions for the enzymes reported by Roberts and co-workers (e.g., *Alu*I, *Bal*I, and *Xho*I; see reviews by Roberts[2,4]) were based on a single buffer system, the "6/6/6" [6 m*M* Tris-HCl (pH 7.9), 6 m*M* $MgCl_2$, and 6 m*M* 2-mercaptoethanol]. Although this system suffices for those enzymes, it can be improved upon by consideration of more recent data on restriction endonuclease reactions.

In application of several facts described in the preceding section, we devised a basic assay system, the "core buffer" [50 m*M* Tris-HCl (pH 8.0), 10 m*M* $MgCl_2$, 1 m*M* dithiothreitol, and 100 μg of BSA per milliliter] to which is added 0, 50, or 100 m*M* NaCl depending upon an individual enzyme's greatest activity. In Table VII are compared the relative activities for a number of commonly used restriction endonucleases assayed both in this core buffer system and in the standard buffer. More than 60% of the enzymes tested were at least 80% as active in the core buffer as in the standard buffer. In fact, 34% of the enzymes exhibited higher activity in the core buffer, demonstrating that many of the standard buffers are suboptimal. As expected for any class of enzymes this large, some enzymes are not amenable to the core buffer reaction conditions. *Bal*I, *Hpa*I, *Sau*3A, and *Sph*I lost more than 50% of their activities under the core buffer conditions. These enzymes should continue to be used as described in Table II. Although the present core buffer system fails to identify the optima that are obtained from initial rate studies, it permits a rational consolidation and a practical solution to the variety of buffer systems currently in use.

Volume

Although restriction endonucleases exhibit activity over wide concentration ranges, the reaction volume should be carefully selected. Analytical reactions (<50 μl) are especially susceptible to significant concentration errors. Pipetting of small volumes of reaction components can introduce significant error, especially when using repeating pipettes outside their tolerance limits. Viscous solutions, e.g., the restriction endonuclease stocks, are especially difficult to dispense accurately in volumes of less than 5 μl. Significant variation in the extent of reaction can be observed with inadvertent delivery of insufficient enzyme. Positive displacement or calibrated glass micropipettes are recommended for measuring critical volumes. Alternatively, samples should be diluted so that ≥5 μl can be pipetted.

Component concentrations in small volume reactions (<50 μl) can also be altered significantly during incubation. This is especially apparent in long-term (>1 hr) or high-temperature (>37°) incubations, which evaporate a considerable percentage of the water. Reactions in capped microfuge tubes can trap the water, but the collected moisture should be centrifuged into the reaction volume occassionally. Overlayering the reaction volume with mineral oil for high-temperature incubations offers another solution; however, one must be careful during retrieval of reaction products.

Large-volume reactions (>0.5 ml) can on occasion fail to give complete DNA digestion. Scaled-up reactions should take into account final DNA and enzyme concentrations. Viscous DNA solutions inhibit enzyme diffusion and can significantly reduce apparent enzyme activity. For troublesome digestions, sometimes 20 0.5-ml reactions are more successful than a single 10-ml reaction.

Incubation Time and Temperature

The restriction endonuclease unit presents a practical, though unusual, enzyme activity definition based on complete digestion of the substrate. Frequently used is the equation

$$a\ (\mu\text{g of DNA cleaved}) = b\ (\text{units of enzyme}) \times c\ (\text{hours of incubation})$$

This equation is sometimes useful as a guide, but extrapolation of incubation time or amount of enzyme from this definition can lead to erroneous results. For example, one unit of restriction enzyme may or may not represent sufficient enzyme molecules to cleave 2 μg of DNA completely in 2 hr of incubation. The extrapolation assumes that the enzyme activity is stable and linear over the entire incubation period. From our experience not all restriction enzymes remain completely active at their reaction temperature for 1 hr or longer. An exception is *Bal*I, which remains active for at least 16 hr.[54] Although long (overnight) incubations can occasionally be successful in saving on the amount of restriction enzyme used, it is not recommended. From experience, contaminating nonspecific exonucleases and endonucleases usually survive better than the specific restriction endonucleases in long incubations. Thus, even slight nonspecific nuclease contamination can, given enough time, destroy the precision and uniqueness of fragments generated by restriction enzyme cleavage. The most reliable results are obtained by maintaining reaction conditions as defined for unit activity.

Most restriction endonuclease activities are determined at 37°. Restriction enzymes isolated from thermophilic bacteria are more stable and more active at temperatures higher than 37°.[35,39,55] Selected restriction endonucleases were studied to ascertain the effect of assay temperature on their activity. *Eco*RI is inactive above 42°,[9] while *Bam*HI loses significant activity at 55°.[10] Curiously, *Hae*III (from a nonthermophilic bacterium) is fully active at 70°, whereas its companion enzyme *Hae*II is inactivated above 42°.[8] High temperature reactivity of restriction endonucleases can be used advantageously, e.g., as probes of DNA secondary structure[8] or for suppression of contaminating enzymic activities.[55] Below 37° most restriction endonucleases remain active, although at reduced rates. Thus, DNA cleavage will occur once all necessary reaction components are present, even though the reaction vessel remains on the bench top or in an ice bath. For example, *Eco*RI was demonstrated to cleave a

[54] R. E. Gelinas, P. A. Myers, G. A. Weiss, R. J. Roberts and K. Murray, *J. Mol. Biol.* **114,** 433 (1977).

[55] S. Sato, C. A. Hutchison III, and J. I. Harris, *Proc. Natl, Acad. Sci. U.S.A.* **74,** 542 (1977).

duplex octanucleotide in the temperature range from 5° to 30°.[9] Hence, the order of addition of components to the reaction mixture should place the enzyme last, at which point the reaction is deemed to have started.

Stopping Reactions

Restriction endonuclease reactions can be stopped by one of several different methods. The method chosen depends upon the subsequent use of the DNA products. For reactions performed solely for the purpose of analyzing the DNA fragments by gel electrophoresis, chelation of Mg^{2+} by EDTA is an effective method to terminate cleavage. If desired, the reaction can be reestablished readily by addition of excess Mg^{2+}. Dissociation and/or denaturation of the restriction endonuclease by adding 0.1% SDS also stops the reaction. For ease and efficiency we add to the reaction one-tenth volume of a solution containing 50% (v/v) glycerol, 100 m*M* Na_2 EDTA (pH 8), 1% (w/v) SDS, and 0.1% (w/v) bromophenol blue. Incubation of this mixture at 65° for 5 min just prior to gel application ensures distinct, reproducible DNA fragment patterns by dissociating bound proteins (e.g., BSA) and reducing DNA·DNA associations, such as the "sticky ends" of lambda DNA.

When the products of the reaction are to be used subsequently for kinasing, ligation, or sequencing, the reaction can be terminated, in some cases, by heat inactivation of the enzyme, or more reliably by phenol extraction of the DNA fragments. Some enzymes such as *Eco*RI[9] or *Hae*II[8] are irreversibly inactivated by exposure to 65° for 5 min, whereas *Tth*I[39] and *Hin*dIII[48] remain active after this treatment. Therefore, we suggest extraction of the DNA from the reaction mixture with an equal volume of phenol freshly saturated with 0.1 *M* Tris-HCl (pH 8). An ether extraction to remove the residual phenol is followed by two consecutive precipitations of the DNA with one-half volume of 7.5 *M* ammonium acetate and two volumes of ethanol for 30 min at −70°. Suspension of the DNA in appropriate buffer provides restriction fragments free of restriction reaction components, phenol, and the enzyme.

Detection of Reaction Products

Total DNA Mass

Upon completion of a restriction endonuclease reaction, the DNA fragments are typically separated by agarose or polyacrylamide gel electrophoresis.[7,56] Usually, the resolved fragments are detected by direct

[56] E. Southern, this series, Vol. 68, p. 152.

staining. Fluorescence of ethidium bromide bound to DNA is the most frequently used method to observe the DNA fragments. In agarose gels a sensitivity of about 20 ng per band is expected. Native, single-stranded DNA and RNA will also fluoresce, but with relatively less intensity. Methylene blue, acridine orange, and Stains-All[57] also can be used. Since Stains-All employs 50% formamide as a solvent, this stain is very useful for detecting DNA fragments in gels run under denaturing conditions.[58] Ethidium bromide, on the other hand, stains very poorly, if at all, under these conditions. Uniform radioactive labeling of DNA also provides a means to detect the total mass of each DNA band by autoradiography.[56]

The fragments of a DNA generated by a restriction enzyme reaction are equimolar with respect to one another. Thus, detection of DNA by mass provides a direct correlation between stain intensity and fragment length. Conversely, the relative molarities of restriction fragments of known lengths can be determined from their relative intensities. If used quantitatively a standard curve must be employed, as the linear relationship between intensity and mass is valid only over a narrow range.[59] Note also that especially small DNA fragments (<75 base pairs) may be difficult to detect by this method.

Total DNA Ends

The intensity of radioactively end-labeled DNA restriction fragments following gel electrophoretic separation and autoradiography[56] is molarity dependent. In contrast to the mass-dependent measurement, this method visualizes each DNA fragment equally, regardless of size. Short oligonucleotides are easily detectable by this technique. End-labeling methods are also several orders of magnitude more sensitive than direct staining. The 5′-phosphate end generated by almost all restriction endonucleases[4] (*Nci*I was found to generate 5′-hydroxyl and 3′-phosphate ends[34]) is radioactively labeled (^{32}P) by treatment of the fragments with alkaline phosphatase followed by incubation with polynucleotide kinase and [γ-^{32}P]ATP.[33] Alternatively, the 3′ end is labeled by one of several enzymic procedures.[5]

Detection of Specific DNA Sequences

A specific DNA sequence can be detected among a complex mixture of DNA sequences by using a radiolabeled DNA or RNA probe comple-

[57] A. E. Dahlberg, C. W. Dingman, and A. C. Peacock, *J. Mol. Biol.* **41,** 139 (1969).
[58] T. Maniatis and A. Efstratiadis, this series, Vol. 65, p. 299.
[59] A. Prunell, this series, Vol. 65, p. 353.

mentary to the desired DNA sequence. Southern or blot hybridization[56] is highly sensitive and specific, capable of detecting a single specific DNA sequence in the midst of a tremendous excess of nonspecific DNA sequences. Restriction endonuclease fragments radiolabeled by nick translation or radiolabeled synthetic polynucleotides can serve as effective hybridization probes.

Other Reaction Considerations

In addition to the conditions described above, other reaction parameters pertinent to the use of restriction endonucleases and the generated products include (*a*) the extent of methylation of the DNA substrate and the selection of the appropriate restriction endonucleases to cleave methylated DNA; (*b*) those conditions that elicit the expression of secondary (star) activities of specific restriction endonucleases; (*c*) the parameters required to generate partial digestion of DNAs; (*d*) the ability to perform multiple digestions; and (*e*) the level of contaminating endonuclease and exonuclease activities.

Methylation

In bacterial systems, methylation usually occurs at either an adenine residue (N-6 position) or a cytosine residue (5 position) within the recognition sequence(s) for the specific endogenous restriction endonuclease(s).[60] Methylation of eukaryotic DNA is almost exclusively restricted to the 5 position of cytosine and primarily (>90%) to the cytosine residues present in the dinucleotide CpG.[61] Findings in eukaryotic systems have suggested the involvement of methylation in numerous functions which include: transcriptional regulation, differentiation, influence of chromosomal structure, DNA repair and recombination, and designation of sites for mutation (reviewed by Ehrlich and Wang[60]). The sensitivity of restriction endonuclease cleavage to methylation (Table IV) can be used advantageously to deduce the patterns and the extent of methylation in DNA. For example, the differential reactivity of the isoschizomers *Msp*I and *Hpa*II to ^{m}CG was used to identify gross tissue specific differences in methylation patterns and, more important, to identify the methylation status of cleavage sites within a specific gene or genetic region.[62]

[60] M. Ehrlich and R. Y.-H. Wang, *Science* **212,** 1350 (1981).

[61] A. Razin and A. D. Riggs, *Science* **210,** 604 (1980).

[62] C. Waalwijk and R. A. Flavell, *Nucleic Acids Res.* **5,** 4631 (1978).

Secondary (Star) Activity of Restriction Endonucleases

Secondary (star) activity of a restriction endonuclease refers to the relaxation of the strict canonical recognition sequence specificity resulting in the production of additional cleavages within a DNA. One prominent example is *Eco*RI, where the usual hexanucleotide sequence (GAATTC) is reduced to a tetranucleotide sequence (AATT) for *Eco*RI* (*Eco*RI "star").[41] Several parameters responsible either individually or in combination for generating star activities include (*a*) glycerol concentration; (*b*) ionic strength; (*c*) pH; (*d*) the presence of organic solvents; (*e*) divalent cations; and (*f*) high enzyme-to-DNA ratios. Restriction enzymes that have been shown to express secondary activities under these conditions are listed in Table VIII (and in Table II). Cleavage in the presence of a high glycerol concentration in the reaction mixture represents the most commonly recognized factor associated with secondary activities.

At restriction enzyme-to-DNA ratios of 50 units/μg, glycerol concentrations as low as 7.5% (v/v) can cause the generation of star activities.[63,64] At lower enzyme-to-DNA ratios (10 units/μg), glycerol concentrations of 20% (v/v) or greater are required before restriction enzyme star activities are observed.[63,64] Relatively low levels of organic solvents such as dimethyl sulfoxide (DMSO), ethanol, ethylene glycol, and dioxane, can also produce similar losses in cleavage specificity.[43,63,64] DMSO at concentrations of 1–2% (v/v) in the final reaction mixture can cause star activities.[43,63]

For restriction enzymes that require high salt concentrations (100 m*M* or greater) in the reaction mixture, a reduction in the salt concentration can result in the generation of secondary activities.[16] *Bam*HI, for example, at enzyme-to-DNA ratios of 100 units/μg cleaves pBR322 at one site in reactions containing 100 m*M* NaCl, at two sites in reactions containing 50 m*M* NaCl, and at eight sites in reactions in the absence of NaCl.[15,63] Additional factors such as the substitution of Mn^{2+} for Mg^{2+} as the divalent cation has also been reported to stimulate star activities of both *Eco*RI[43] and *Hin*dIII.[44] Increasing the assay pH from pH 7.5 to 8.5 also increases *Eco*RI* activity.[41]

Although the generation of secondary activities can provide restriction enzymes with new sequence specificities (no isoschizomers are known for *Eco*RI*) that may prove to be useful in some instances, these activities rarely result in complete or equal cleavage of all possible secondary recognition sites. Therefore, to eliminate or minimize the expression of restriction enzyme secondary activities, all restriction enzyme assays

[63] J. George, R. W. Blakesley, and J. G. Chirikjian, *J. Biol. Chem.* **255,** 6521 (1980).

[64] E. Malyguine, P. Vannier, and P. Yot, *Gene* **8,** 163 (1980).

TABLE VIII
REACTION CONDITIONS THAT INDUCE SECONDARY "STAR" ACTIVITY IN CERTAIN RESTRICTION ENDONUCLEASES

Enzyme	Alterations of standard reaction conditions[a]	References
*Ava*I	A, B, D	*b, c*
*Bam*HI	A, B, C, D, E, H	*b, c, d, e*
*Bst*I	B, D	*b, f*
*Bsu*I	B, D, F	*g*
*Eco*RI	A, B, D, E, F	*b, h, i, j*
*Hae*III	B, D	*b*
*Hha*I	B, D, G	*b, e*
*Hind*III	E	*k*
*Hpa*I	A, B, D	*b, c*
*Pst*I	A, B, D, G	*c, e*
*Pvu*II	B, D	*l*
*Sal*I	A, B, D, G	*b, c, e*
*Sst*I	B, D, G	*b, e*
*Sst*II	B, D	*b*
*Xba*I	B, D, G	*b, e*

[a] Abbreviations used: A, ethylene glycol (45%); B, glycerol (12–20%); C, ethanol (12%); D, high enzyme : DNA ratio (>25 units/μg); E, Mn^{2+} substituted for Mg^{2+}; F, pH 8.5; G, dimethyl sulfoxide (8%); and H, absence of NaCl.

[b] J. George and J. G. Chirikjian, *Proc. Natl. Acad. Sci. U.S.A.* **79,** 2432 (1982).

[c] K. Nath and B. A. Azzolina, *in* "Gene Amplification and Analysis," Vol. 1: "Restriction Endonucleases" (J. G. Chirikjian, ed.), p. 113. Elsevier/North-Holland, Amsterdam, 1981.

[d] J. George, R. W. Blakesley, and J. G. Chirikjian, *J. Biol. Chem.* **255,** 6521 (1980).

[e] E. Malyguine, P. Vannier, and P. Yot, *Gene* **8,** 163 (1980).

[f] C. M. Clarke and B. S. Hartley, *Biochem. J.* **177,** 49 (1979).

[g] K. Heininger, W. Horz, and H. G. Zachau, *Gene* **1,** 291 (1977).

[h] B. Polisky, P. Greene, D. E. Garfin, B. J. McCarthy, H. M. Goodman, and H. W. Boyer, *Proc. Natl. Acad. Sci. U.S.A.* **72,** 3310 (1975).

[i] C. J. Woodbury, Jr., O. Hagenbuchle, and P. H. von Hippel, *J. Biol. Chem.* **255,** 11534 (1980).

[j] T. I. Tikchonenko, E. V. Karamov, B. A. Zavizion, and B. S. Naroditsky, *Gene* **4,** 195 (1978).

[k] M. Hsu, and P. Berg, *Biochemistry* **17,** 131 (1978).

[l] H. Belle Isle, unpublished results, 1981.

should be performed under the recommended standard assay conditions especially in regard to pH, ionic strength, and divalent cation concentration. The amount of glycerol introduced into the assay should be kept below 5% (v/v), and prolonged incubation with high enzyme-to-DNA ratios should be avoided. In addition, the introduction of additional components via the DNA substrate, especially DNA previously exposed to organic solvents, can be minimized by dialyzing the DNA prior to restriction enzyme cleavage.

Partial Digestion of DNA

Partial digestion refers to incomplete cleavage of the DNA, observed as fragments of higher molecular weight than the final cleavage products. These usually disappear by increasing incubation time or the amount of enzyme added. When DNA fragments generated by restriction endonuclease cleavage (e.g., *Sau*3A) are used in "shotgun" cloning experiments, partial digestion of the DNA substrate is frequently desirable. Under this condition internal recognition sequences for the selected restriction endonuclease remain intact at a frequency nearly dependent upon the amount of enzyme added and the incubation condition used. Partial digestions also could be obtained by substitution of other divalent cations (e.g., Mn^{2+} or Zn^{2+} [65]) for Mg^{2+} (see Table II) to slow the reaction or by addition of DNA binding ligands, such as actinomycin[14,66] and 6,4′-diamidino-2-phenylindole.[67] Each of these methods, however, is nonrandom, showing a hierarchy of cleavage rates for the various sites within the DNA. A more effective technique to generate random partial digests is partially to methylate the DNA prior to restriction endonuclease cleavage.[68]

Multiple Digestions

Mapping analysis or isolation of particular DNA fragments frequently requires the digestion of DNA by more than one restriction endonuclease. When sufficient quantities of DNA are available, the safest procedure for multiple digestion involves independent restriction enzyme digestions separated by phenol extraction and ethanol precipitation. However, when DNA substrate quantities are limited and where the selected restriction endonucleases have similar assay requirements (e.g., pH, [Mg^{2+}], [NaCl], buffer), two consecutive or simultaneous digestions can proceed with no buffer alterations. This consideration was important in establishing the

[65] T. A. Bickle, V. Pirrotta, and R. Imber, this series, Vol. 65, p. 132.

[66] M. Goppelt, J. Langowski, A. Pingoud, W. Haupt, C. Urbanke, H. Mayer, and G. Maass, *Nucleic Acids Res.* **9,** 6115 (1981).

[67] J. Kania and T. G. Fanning, *Eur. J. Biochem.* **67,** 367 (1976).

[68] E. Ferrari, D. J. Henner, and J. A. Hoch, *J. Bacteriol.* **146,** 430 (1981).

core buffer system (see the section The Reaction). But even where identical reaction conditions are recommended for two enzymes, digestions should be performed consecutively, rather than simultaneously, to ensure that each enzyme cleaves completely. When double digestions require restriction enzymes with different recommended assay conditions, each reaction should be performed under its optimal conditions. For example, to perform a *Kpn*I, *Hin*fI double digestion where both enzymes have identical assay requirements except for NaCl concentration (Tables II, and VII), one should first cleave to completion with *Kpn*I in *Kpn*I assay buffer, then increase the NaCl concentration and cleave with *Hin*fI. For enzymes with significantly different pH, buffer, salt, or Mg^{2+} requirements, the assay buffer can be changed effectively and the DNA quantitatively recovered by a 2- to 3-hr dialysis in a microdialyzer prior to digestion with the second restriction endonuclease. Use of the recommended reaction conditions for each restriction enzyme ensures production of the appropriate restriction enzyme fragments.

Contaminating Activities

Because restriction endonucleases are used essentially as reagents in DNA cleaving reactions, they need to be free of inhibitors and contaminating activities that could interfere with either the cleavage analysis or the subsequent use of the cleaved DNA products for cloning, sequencing, etc. Two general classes of contaminating activities prevail: first, other endonuclease activities that could alter the number, the size, and the termini of fragments produced; second, exonuclease activities that could remove nucleotides from either the 3′ or 5′ ends of the resultant fragments and inhibit subsequent ligation and labeling experiments. Commercially available restriction enzymes are routinely characterized for and purified away from both types of nuclease contamination.

In addition to exonucleases that specifically degrade the 3′ and/or 5′ ends of double-stranded DNA, we have identified in several restriction endonuclease preparations a 3′ exonuclease activity specific for single-stranded DNA. Thus, DNA fragments with 3′ extended single-strand ends (e.g., *Hae*II and *Kpn*I) are readily degraded by this contaminating activity. Potential problems arising from contaminating exo- and endonucleases can be reduced by using the highest quality of restriction endonuclease available, the minimum quanity of enzyme required for complete digestion of the DNA, and the recommended assay conditions.

Troubleshooting Guide

In Table IX are listed a number of the common problems encountered when using restriction endonucleases, a probable cause, and a suggested

TABLE IX
TROUBLESHOOTING GUIDE

Problem	Probable causes	Suggested solutions
No cleavage	Inactive restriction enzyme	Check enzyme activity on unit substrate DNA.
	Presence of inhibitor, e.g., SDS, phenol, EDTA	Precipitate DNA twice with 1/2 volume of 7.5 *M* ammonium acetate plus 2 volumes of ethanol, or dialyze DNA sample.
	Nonoptimal reaction composition or temperature (thermophiles)	Prepare fresh buffer, check assay temperature.
	Inadequate gel separation of DNA substrate and fragments (e.g., *Sst*I cleavage of lambda DNA)	Lower the percentage of gel, or do double restriction enzyme digestion (e.g., *Eco*RI digested lambda DNA as substrate for *Sst*I).
	DNA methylation (e.g., *Eco*RII not cleaving pBR322 DNA)	Mix test DNA and unit substrate DNA, then cleave with selected enzyme; use isoschizomer insensitive to DNA methylation; replicate plasmid in mec^- dam^- *E. coli* host.
	DNA unmethylated (*Dpn*I requires methylated DNA)	Replicate plasmid in mec^+ dam^+ *E. coli* host. Use isoschizomer that cleaves unmethylated DNA (e.g., use *Sau*3AI rather than *Dpn*I)
	Other DNA modification	To identify, mix unit substrate DNA and test DNA, then cleave both with the selected restriction enzyme.
	Impure DNA	To detect, compare ability to cleave test DNA and unit substrate DNA. Remove impurities with an RPC-5 ANALOG column or by precipitating twice with 2 volumes of ethanol in the presence of 1/2 volume of 7.5 *M* ammonium acetate at −70° for 30 min.
	DNA has no recognition sequences for selected restriction enzyme.	Confirm restriction enzyme activity and lack of inhibitors as above. Do 10-fold excess units of enzyme. Cleave test DNA with several other restriction enzymes to ensure that impurities in the DNA or DNA methylation are not responsible for lack of cleavage.
Partial cleavage	Loss of restriction enzyme activity	Use 5- to 10-fold excess restriction enzyme. Check for conditions that cause enzyme activity loss, see below.

	Incorrectly diluted enzyme	Do unit titration assay or redilute from fresh enzyme stock.
	Presence of inhibitor(s), e.g., SDS, phenol, EDTA, or plasticizer from microfuge tubes	See above.
	Improper reaction conditions	Prepare fresh assay buffer, check assay temperature. Try overnight digestion; add 0.01% Triton X-100 to increase stability during incubation. Determine activity on unit substrate DNA.
	Test DNA requires more restriction enzyme for complete cleavage than unit substrate DNA (e.g., see Table V).	Do 5- to 10-fold excess restriction digest.
	DNA impure	See above.
	Methylation of only a subset of the recognition sequences (e.g., *Xba*I cleavage of lambda DNA)	See above.
	Other DNA modifications	See above.
	Portion of substrate DNA left unreacted on side of microfuge tube	Centrifuge 1–2 sec in Eppendorf centrifuge prior to incubation.
	Pipetting error (especially of viscous solutions	Use positive displacement pipettors; dilute so $>5 \mu l$ are pipetted.
	Annealed DNA ends (e.g., lambda DNA)	Heat DNA at 65° for 5 min prior to gel electrophoresis.
	Denaturation of restriction enzyme by assay reagents, temperature, and vortexing	Use recommended reaction conditions and temperature; avoid vigorous vortexing.
	Differences in nucleotide sequences adjacent to recognition site	Detect by cleaving a mixture of test DNA and unit substrate DNA; use 5- to 10-fold excess enzyme.
	Loss of restriction enzyme activity upon dilution	Dilute only into recommended storage buffer. Use immediately; diluted enzyme usually does not store well. Do 5- to 10-fold excess digest to obtain complete cleavage.
Persistent partial	Partial methylation (e.g., *Xba*I cleavage of lambda DNA)	See above
	Differences in nucleotide sequence adjacent to recognition site	See above
Difficulty cleaving supercoiled DNA	DNA structure	Relax DNA with topoisomerase I, then cleave; linearize plasmid first with another restriction enzyme; or use excess restriction enzyme.

(*continued*)

TABLE IX (*continued*)

Problem	Probable causes	Suggested solutions
Failure to obtain expected enzyme activity	Assayed on DNA different from that used to determine unit activity	Use DNA for unit assay for quantitation of enzyme activity.
	Pipetting errors (especially of viscous solutions)	Dilute so >5 μl can be added to reaction.
	Concentration of bovine serum albumin (BSA)	Use recommended BSA concentration in assay and storage buffers
	Loss of enzyme activity	See below.
More than expected number of DNA fragments	Restriction enzyme "star" activity	Detect by appearance of extra DNA fragments produced by cleavage of DNA used for unit determination; check assay conditions, especially glycerol concentration, or Mn^{2+} for Mg^{2+}; precipitate DNA twice with half volume of 7.5 *M* ammonium acetate and 2 volumes of ethanol at −70° for 30 min; minimize quantity of enzyme used.
	Presence of second restriction enzyme	Detect second activity by comparing restriction digest pattern to that expected for the DNA used for unit determination.
	Test DNA contaminated with another DNA	Detect by DNA minus enzyme on gel; assay other restriction enzymes on the same DNA substrate; purify test DNA from contaminant by either gel electrophoresis or RPC-5 ANALOG column chromatography.
No DNA observed	DNA quantitation in error (e.g., RNA contamination)	Treat DNA preparation with 100 μg/ml DNase-free RNase, phenol extract, then either dialyze or precipitate twice with ethanol.
	Nonspecific precipitation in reaction	Dialyze DNA or precipitate DNA twice with ethanol prior to assay.
Rapid loss of restriction enzyme activity upon storage	Improper storage temperature	Store enzymes in recommended storage buffer plus 50% glycerol at −20° in a non-frost-free freezer.

	Enzymes stored diluted.	Store restriction enzymes only in concentrated form.
	Incorrect storage buffer (e.g., *Eco*RI in Tris-HCl)	Use recommended storage buffer; check for pH changes with temperature.
	Low protein concentration	Store enzymes with 500 μg/ml nuclease-free BSA.
Diffuse DNA bands after gel electrophoresis	Protein binding to DNA	Heat cleaved DNA at 65° for 5 min in the presence 0.1% SDS prior to loading the gel.
	Exonuclease contamination	Detect by monitoring acid-soluble material after incubation of enzyme with radioactively labeled DNA; minimize the quantity of restruction enzyme used and/or incubation time.
Large precipitates after ethanol precipitation of DNA	$MgPO_4$ precipitation	Dialyze DNA after precipitation; use buffer with little or no phosphate; add excess EDTA to the DNA prior to precipitation to chelate the magnesium.
Poor litigation efficiency	High phosphate or salt carry-over from restriction digest	Dialyze DNA fragments after restriction digest; remove phosphate or salt with small molecular sieve column or multiple ethanol precipitations.
	Incomplete removal of restriction enzyme	Extract DNA after restriction digest with equal volumes of buffered phenol, chloroform, and ether, then precipitate with ethanol.
	Incomplete removal or inactivated bacterial alkaline phosphatase	Extract DNA after phosphatase treatment with buffered phenol, chloroform, and ether. Precipitate with ethanol.
	ATPase contamination	Extract the DNA with an equal volume of phenol, chloroform, ether, then precipitate with ethanol.
	Ligation of blunt ends	Use excess T4 DNA ligase (1–2 units per picomole of free ends).
	Exonuclease contamination	Detect as above; minimize amount of enzyme or incubation time.
	Unstable buffer components	Prepare fresh ligation buffer.
Poor kinase efficiency	Phosphate carry-over	Remove phosphate by either dialysis, molecular sieve, or RPC-5 ANALOG column.
	Self-annealing of GC-rich ends.	Heat DNA at 65° for 5 min prior to kinase reaction.

solution. This list is not necessarily complete, nor are the solutions unique, but the guide is intended to be a quick reference for effectively utilizing restriction endonucleases as tools for molecular biology.

Acknowledgments

We acknowledge Drs. H. Belle Isle, D. Appleby, A. Hu, and A. MarSchel of this laboratory for their contributions of unpublished data. We express our appreciation to P. Hammond for typing the manuscript and to Drs. D. Rabussay, J. George, J. A. Thompson, J. Kane, and D. Hendrick for their valuable comments and suggestions.

[2] Using T4 RNA Ligase with DNA Substrates

By CATHERINE A. BRENNAN, ALICE E. MANTHEY, and RICHARD I. GUMPORT

T4 RNA ligase forms a 3′ → 5′ phosphodiester bond between the 3′-hydroxyl and the 5′-phosphate of an oligoribonucleotide with the concomitant cleavage of ATP to AMP and PP_i.[1] See Uhlenbeck and Gumport[2] for a comprehensive review of the enzyme. In addition to the intramolecular circularization activity, the enzyme will also join oligonucleotides in an intermolecular reaction.[3–5] The oligonucleotide with the 3′-hydroxyl (the acceptor) must be three or more residues long, and it can be joined to an oligonucleotide bearing a 5′-phosphate (the donor) as short as a nucleoside 3′,5′-bisphosphate.[6,7] The product of the addition of a nucleoside 3′,5′-bisphosphate to an acceptor contains a 3′-phosphate and is therefore not a suitable substrate for addition of another donor. Removal of the 3′-phosphate and subsequent addition of another nucleoside 3′,5′-bisphosphate allows the stepwise synthesis of oligonucleotides.[7] The enzyme exhibits little specificity for the base portion of the nucleoside 3′,5′-bisphosphate, thereby facilitating the incorporation of nucleoside analogs into oligonucleotides.[7,8]

[1] R. Silber, V. G. Malathi, and J. Hurwitz, *Proc. Natl. Acad. Sci. U.S.A.* **69,** 3009 (1973).
[2] O. C. Uhlenbeck and R. I. Gumport, *in* "The Enzymes" (P. D. Boyer, ed.), 3rd ed., Vol. 15, p. 31. Academic Press, New York, 1982.
[3] G. Kaufman and U. Z. Littauer, *Proc. Natl. Acad. Sci. U.S.A.* **71,** 3741 (1974).
[4] G. Kaufman and N. R. Kallenbach, *Nature (London)* **254,** 452 (1975).
[5] G. C. Walker, O. C. Uhlenbeck, E. Bedows, and R. I. Gumport, *Proc. Natl. Acad. Sci. U.S.A.* **72,** 122 (1975).
[6] Y. K. Kikuchi, F. Hishinuma, and K. Sakaguchi, *Proc. Natl. Acad. Sci. U.S.A.* **75,** 1270 (1978).
[7] T. E. England and O. C. Uhlenbeck, *Biochemistry* **17,** 2069 (1978).
[8] J. R. Barrio, M. G. Barrio, N. J. Leonard, T. E. England, and O. C. Uhlenbeck, *Biochemistry* **17,** 2077 (1978).

ISBN 0-12-182000-9

The enzyme mechanism proceeds via the formation of two covalent intermediates. RNA ligase reacts with ATP to form an adenylylated enzyme with the release of PP_i.[9] In a second step, the enzyme transfers the adenylyl group to the 5′-phosphate of the donor to form the structure Ado-5′PP5′-X, where P5′-X represents the donor.[3,10] In the next step the 3′-hydroxyl of the acceptor displaces AMP and forms the phosphodiester bond.[10,11] The enzyme shows little specificity for the X structure of the second covalent intermediate; consequently, a wide variety of β-substituted ADP derivatives serve as substrates in an ATP-independent reaction.[12] The enzyme transfers the PX portion of Ado-5′PP-X to the 3′-hydroxyl of the oligonucleotides and releases AMP. This reaction allows oligonucleotides to be modified by the addition of nonnucleotide groups at their 3′ termini. These reactions have been extensively used in RNA syntheses and modifications. See Gumport and Uhlenbeck[13] for a review of the applications of the enzyme.

Although RNA ligase uses oligoribonucleotides much more efficiently than oligodeoxyribonucleotides, short DNA oligomers can be both circularized[14] and joined intermolecularly.[11] We have found conditions under which 2′-deoxyribonucleoside 3′,5′-bisphosphates can be added to DNA oligomers and single-strand DNA oligomers be joined in good yields.[15–18] We describe here how to purify the enzyme sufficiently free of nucleases that it can be used with DNA substrates, and, in addition, we specify the conditions necessary to perform these reactions.

Enzyme Purification

The preferred source of RNA ligase is from *Escherichia coli* infected with DNA negative (D*O*) mutants of T4 that allow the synthesis of enzyme but do not lyse the cells. The preparation of phage-infected cells has

[9] J. W. Cranston, R. Silber, V. G. Malathi, and J. Hurwitz, *J. Biol. Chem.* **249,** 7447 (1974).

[10] J. J. Sninsky, J. A. Last, and P. T. Gilham, *Nucleic Acids Res.* **3,** 3517 (1976).

[11] A. Sugino, T. J. Snopek, and N. R. Cozzarelli, *J. Biol. Chem.* **252,** 1732 (1977).

[12] T. E. England, R. I. Gumport, and O. C. Uhlenbeck, *Proc. Natl. Acad. Sci. U.S.A.* **74,** 4839 (1977).

[13] R. I. Gumport and O. C. Uhlenbeck, *in* "Gene Amplification and Analysis" Vol. 2: "Analysis of Nucleic Acid Structure by Enzymatic Methods" (J. G. Chirikjian and T. S. Papas, eds.), p. 313. 2, Elsevier/North-Holland, Amsterdam, 1981.

[14] T. J. Snopek, A. Sugino, K. L. Agarwal, and N. R. Cozzarelli, *Biochem. Biophys. Res. Commun.* **68,** 417 (1976).

[15] D. M. Hinton, J. A. Baez, and R. I. Gumport, *Biochemistry* **17,** 5091 (1978).

[16] D. M. Hinton and R. I. Gumport, *Nucleic Acids Res.* **7,** 453 (1979).

[17] M. M. McCoy and R. I. Gumport, *Biochemistry* **19,** 635 (1980).

[18] D. M. Hinton, C. A. Brennan, and R. I. Gumport, *Nucleic Acids Res.* **10,** 1877 (1982).

been described.[19] We have used cells infected with bacteriophage T4 bearing amber mutations in either gene *43* or *45* and have obtained satisfactory results with both.[19,20] The gene *45* mutant has the advantage of allowing purification of T4 DNA polymerase as well as T4 polynucleotide kinase and T4 DNA ligase from the cell extract.[21]

Assay

The enzyme can be assayed at all stages of purification by determining the circularization of $[5'\text{-}^{32}P]rA_n$ through the conversion of the label to a diester bond that is resistant to phosphomonoesterase.[1] The preparation of the labeled substrate[1] and the assay conditions have been described.[19] One unit of activity is defined as 1 nmol of $[5'\text{-}^{32}P](pA)_{11}$ converted to phosphatase resistance in 30 min at 37° at 1 μM oligomer termini concentraction.[19] It is difficult to quantify accurately the amount of enzyme with this assay (cf. Gumport and Uhlenbeck[13] for a discussion of this problem), but it can be used to follow the activity reliably through the purification. After initial purification, the adenylylation of the enzyme by ATP can serve as the basis for a direct stoichiometric assay by determining enzyme-AMP formation using labeled ATP[9,22] or for a catalytic assay of the ATP-PP_i exchange reaction using $[^{32}P]PP_i$.[9] The amount of pure protein can be determined by spectrophotometry. RNA ligase in 20 mM *N*-2-hydroxyethylpiperazine-*N'*-2-ethanesulfonic acid (HEPES-NaOH) (pH 7.5), 20 mM KCl, and 0.1 mM dithiothreitol (DTT) has an absorption maximum at 279 nm, a minimum at 251 nm, and a ratio of absorbances at 280 nm to 260 nm of 1.98. The molar extinction coefficient is 5.72×10^4 M^{-1} cm^{-1} at 280 nm using $M_r = 43{,}000$ for RNA ligase, and the $E_{1\,cm}$ is 1.3 at the same wavelength.[13] Because pure protein is available, we will express enzyme concentration in molar terms.

Several alternative procedures for the purification of RNA ligase have been described.[9,23,24] These protocols have been used almost exclusively

[19] M. M. McCoy, T. H. Lubben, and R. I. Gumport, *Biochim. Biophys. Acta* **562,** 149 (1979).

[20] R. I. Gumport, A. E. Manthey, J. A. Baez, M. M. McCoy, and D. M. Hinton, *in* "Nucleic Acids and Proteins" (Proc. Symp. Nucleic Acids and Proteins) (Z. Shen, ed.), p. 237. Science Press, Beijing, 1980.

[21] A. Panet, J. H. van de Sande, P. C. Loewen, H. G. Khorana, H. G. Raae, A. J. Lillehaug, and K. Kleppe, *Biochemistry* **12,** 5045 (1973).

[22] S. K. Vasilenko, A. G. Veniyaminova, V. I. Yamkovoy, and V. I. Maiyorov, *Biorg. Khim.* **5,** 621 (1979).

[23] J. A. Last and W. F. Anderson, *Arch. Biochem. Biophys.* **174,** 167 (1976).

[24] N. P. Higgins, A. P. Geballe, T. J. Snopek, A. Sugino, and N. R. Cozzarelli, *Nucleic Acids Res.* **4,** 3175 (1977).

PURIFICATION OF T4 RNA LIGASE FROM 149 g OF T4-INFECTED *Escherichia coli*

Fraction	Step	Protein (mg)	Specific activity (units/mg)	Activity recovery (%)
I	Crude extract	5393	46	100
II	Streptomycin supernatant	2207	90	81
III	DE-51 cellulose	196	489	38
IV	Affi-Gel Blue-I	57	1287	29
V	Matrex Gel Red A	36	1878	25
VI	Hydroxyapatite	14	—	—[a]
VII	Affi-Gel Blue-II	13.3	3020	16

[a] Not determined; however, the average activity recoveries from this step are greater than 80%.

to prepare enzyme for use with RNA substrates, and determinations of DNase activities at the protein concentrations we find necessary for efficient DNA joining have not been reported.

Procedure

The table shows a summary of a RNA ligase purification from 149 g of *E. coli* BB that have been infected with bacteriophage T4 *am* E10X3 (an amber mutant in gene *45* obtained from John Wiberg, University of Rochester, New York). Unless otherwise specified all operations with RNA ligase were done at 0–4° and solutions were centrifuged at 27,000 *g*. Additional details concerning some of the steps of the procedure may be found elsewhere.[19,20]

Crude Extract. Frozen infected cells were thawed in 500 ml of cold 50 m*M* tris(hydroxymethylaminomethane)-HCl (Tris-HCl) (pH 7.5), 2.5 m*M* DTT, 1.0 m*M* ethylenediaminetetraacetic acid (EDTA) (buffer I) and dispersed by stirring in a blender. The cells were disrupted in an Amicon French pressure cell (American Instrument Co.) at 1.6 kpsi. After two passages through the press, the suspension was centrifuged for 30 min; the supernatant was diluted with buffer I to A_{260} = 110 (fraction I, 1380 ml).

Streptomycin Sulfate Precipitation and Ammonium Sulfate Fractionation. A freshly prepared solution of 5% (w/v) streptomycin sulfate (Sigma Chemical Co.) was added dropwise to the stirring crude extract over a 15-min period to attain a final concentration of 0.8% (w/v). The mixture was stirred for 20 min and centrifuged for 20 min. T4 polynucleotide kinase can be purified from the pellet.[25,26] The supernatant (fraction II, 1635 ml) was adjusted to 35% saturation with $(NH_4)_2SO_4$ (0.209 g/ml), stirred for 20 min, and centrifuged for 20 min. The supernatant was decanted through Miracloth (Chicopee Mills, Inc., New York, New York) to remove floating debris. This step does not remove significant amounts of material but ensures that the next $(NH_4)_2SO_4$ pellet will sediment completely. The filtrate was brought to 50% saturation by the further addition of 0.094 g/ml of $(NH_4)_2SO_4$. The pellet was collected by centrifugation for 30 min and dissolved in a minimum volume (60–80 ml) of 20 m*M* HEPES-NaOH (pH 7.5), 40 m*M* NaCl, 1 m*M* DTT, 0.1 m*M* EDTA (buffer II). The sample was dialyzed three times in 1 liter of buffer II for a total of 2 hr.

DE-51 Cellulose Chromatography. The sample (1.02 g in 460 ml) was clarified by centrifugation and applied to a DE-51 cellulose (Whatman DE-51) column (3.9 cm diameter × 19.5 cm) that had been equilibrated in buffer II. The column was eluted with a 3-liter gradient from 0.04 to 0.1 *M* NaCl in buffer II at 200 ml/hr. T4 DNA ligase does not bind to the column and can be recovered by passing the loading effluent from the DE-51 column directly onto a DE-52 column equilibrated with buffer II. DNA ligase can be eluted from this column with a 0.05 to 0.4 *M* NaCl gradient in buffer II. The fractions from the DE-51 column containing RNA ligase (0.05–0.06 *M* NaCl) were concentrated by pressure filtration in an Amicon stirred cell apparatus using a PM-30 membrane, and the concentrate was dialyzed four times against 0.5 liter of 50 m*M* HEPES-NaOH (pH 7.5), 1 m*M* DTT, 10 m*M* $MgCl_2$ (buffer III) (fraction III, 51 ml).

Affi-Gel Blue Chromatography. The sample was clarified by centrifugation and applied to an Affi-Gel Blue column (1.9 cm diameter × 15 cm) as described.[19] After the absorbance of the effluent had returned to the baseline value, the column was eluted by applying one column volume of the equilibration buffer (buffer III) containing 0.2 *M* NaCl and 2 m*M* ATP and stopping the flow for 1 hr. Elution was continued at 30 ml/hr, and the fractions containing RNA ligase (5–6 column volumes) were concentrated by pressure filtration, as described in the preceding section, followed by dialysis twice in 1 liter buffer III containing 0.1 *M* NaCl for a total of 1.5 hr (fraction IV, 58 ml).

Matrex Gel Red A Chromatography. The sample was applied at 20 ml/hr to a Matrex Gel Red A (Amicon Corp., Lexington, Massachusetts) column (0.9 cm diameter × 30 cm) that had been equilibrated in buffer III containing 0.1 *M* NaCl. The column was eluted with a 250-ml gradient of 0.1 to 1.0 *M* NaCl in buffer III at 20 ml/hr. The fractions containing RNA

[25] V. Cameron and O. C. Uhlenbeck, *Biochemistry* **16,** 5120 (1977).
[26] D. A. Soltis and O. C. Uhlenbeck, *J. Biol. Chem.* **257,** 11332 (1982).

ligase were pooled, concentrated by pressure filtration, and dialyzed four times against 1 liter of 50 mM HEPES-NaOH (pH 7.5), 1 mM DTT, 0.1 M NaCl for a total of 2 hr to remove the $MgCl_2$. Dialysis was continued for 1 hr against 1 liter of 20 mM HEPES-NaOH (pH 7.5), 1 mM DTT, and 20 mM KP_i (pH 7.5) (buffer IV). Dialysis in buffer IV was repeated, and the sample was clarified by centrifugation (fraction V, 38 ml). Chromatography on Matrex Gel Red A resolves free enzyme from adenylylated enzyme, the former eluting at 0.6 M NaCl and the latter at 0.3 M NaCl.[20] Sometimes two peaks of RNA ligase activity are observed in the elution profile, and when this occurs they are combined and treated as described above.

Hydroxyapatite Chromatography. The sample was applied to and eluted from an hydroxyapatite (Bio-Rad HTP) column (0.9 cm diameter × 20 cm) as described.[19] RNA ligase elutes at approximately 60–80 mM KP_i. Fractions containing activity were partially concentrated by pressure filtration using a PM-10 membrane. Vacuum dialysis in a collodion bag (Schleicher & Schuell, 25,000 molecular weight cutoff) submerged in stirring 50 mM HEPES–NaOH (pH 7.5), 1 mM DTT further concentrated the enzyme. The dialysis medium was changed four times at 30-min intervals to remove P_i from the sample. The medium was then changed to buffer III twice for 30 min to return $MgCl_2$ to the solution (fraction VI, 11 ml). Phosphate must be removed from the solution before Mg(II) is added to avoid formation of a precipitate.

Affi-Gel Blue Chromatography II. The sample was applied to and eluted from an Affi-Gel Blue column (0.9 cm diameter × 15 cm) at 20 ml/hr as described in a previous section. The fractions containing RNA ligase (8–9 column volumes) were concentrated be pressure filtration with a PM-10 membrane and by vacuum dialysis in a collodion bag submerged in stirring 50 mM HEPES–NaOH (pH 7.5), 1 mM DTT, 25 mM NaCl (buffer V). The enzyme elutes from the Affi-Gel Blue column as a mixture of free and adenylylated enzyme because the elution buffer contains ATP and $MgCl_2$. To obtain free enzyme the sample must be dialyzed extensively (20–30 hr) against buffer V. Progress in removing adenylylates from the sample is monitored by determining the A_{280} to A_{260} ratio of the sample. When a value from 1.2 to 1.4 is attained and the enzyme has been concentrated to 6–10 mg/ml, the solution is clarified by centrifugation and an equal volume of glycerol is added (fraction VII, 7 ml). The enzyme is stored at −20°.

Comments

Enzyme purified by this procedure is sufficiently free of phosphomonoesterase, ATPase, DNase, and RNase to be used in the synthetic applications described here and elsewhere.[15–18,20,27] For many applications with RNA substrates, e.g., end-labeling RNA with [5′-^{32}P]prCp, fraction VI is satisfactory. Electrophoresis in polyacrylamide gels containing sodium dodecyl sulfate shows that greater than 95% of the protein migrates as the unadenylylated enzyme. A portion of the enzyme elutes from the final Affi-Gel Blue column in the adenylylated form. Although the last dialysis step does not remove all the adenylates from the solution as judged by spectrophotometry, the adenylylated enzyme must hydrolyze during the course of the dialysis because only free enzyme is observed upon polyacrylamide electrophoresis. It is important to dialyze long enough in the final step to lower the adenylate levels since the concentration of ATP in DNA joining reactions must be carefully regulated. In addition, concentrating the enzyme so that the final stock solution is 3–5 mg/ml is desirable in order to allow DNA joining reactions of small volumes to be assembled through direct addition of the stock enzyme solution. The yields of protein vary from 3 to 9 mg per 100 g of infected cells, and the activity varies from 7 to 15%. The enzyme in the storage buffer is stable for more than a year at −20° and loses less than 10% of its activity when left at room temperature for 1 week.

DNA Joining Reactions

Reaction conditions that differ from those usually used to join RNA molecules[28] are required to enable DNA acceptors to serve as substrates with RNA ligase. The essential features are that higher enzyme and oligonucleotide concentrations are required, longer reaction times are necessary, incubation at low temperature is imperative, ATP concentrations must be maintained at low values, and Mn(II) must be present. We will detail these conditions and describe how to prepare and analyze the reaction mixtures. Specific examples of the application of these methods to synthesize defined sequences of oligodeoxyribonucleotides have been reported.[18]

Reagents

HEPES–NaOH buffer, 0.5 *M* in water, pH 7.9 (all buffer pH values are measured at 0.05 *M* at room temperature)
Tris-HCl buffer, 1.0 *M* in water, pH 8.0
Dithiothreitol (DTT), 1.0 *M* in water, stored at −20°
$MnCl_2$, 0.1 *M* in water, stored at −20°

[27] R. I. Gumport, D. M. Hinton, V. S. Pyle, and R. W. Richardson, *Nucleic Acids Res. Symp. Ser. No.* **7**, 167 (1980).
[28] P. J. Romaniuk and O. C. Uhlenbeck, this volume [3].

$MgCl_2$, 0.1 *M* in water, stored at −20°
Spermine · 4 HCl, 10 m*M* in water, neutralized to pH 7.9 with NaOH, stored at −20°
Na_2 phosphocreatine, 10 m*M* in water, pH 9.0, stored at −20°
Na_3ATP, 1 m*M* in water, pH 7.5, stored at −20°
Glycerol (BRL, Inc., Redistilled Ultra Pure Reagent) stored neat at room temperature
Bovine serum albumin, 0.5 mg/ml in water, stored at −20°
Bacterial alkaline phosphatase (BAPC, Worthington Biochemical Corp.; dialyzed in 10 m*M* Tris-HCl, pH 8.0, 10 m*M* $MgCl_2$, 100 m*M* NaCl), 1 mg/ml stored at −20°
Creatine phosphokinase (Sigma, type I, rabbit muscle), 1750 units/ml in water, stored at −20°
Adenylate kinase (Sigma; myokinase Grade V, pig muscle, dialyzed in 50 m*M* HEPES–NaOH, pH 7.9, 1 m*M* DTT, 10 m*M* NaCl), 1700 units/ml in 50% glycerol, stored at −20°
RNase A (Sigma, type I-A, bovine pancreas), 37 mg/ml in 50 m*M* HEPES, pH 7.9, heated at 90° for 10 min and stored at −20°

Preparation of Nucleotide Substrates

The purity of all the nucleotide substrates is monitored by paper chromatography in 1-C_3H_7OH : conc. NH_3 : water (55 : 10 : 35) or by reverse-phase or anion-exchange high-performance liquid chromatography (HPLC). Sodium, ammonium, and triethylammonium salts of the nucleotides have all been used successfully. When high concentrations of nucleotides are used, it is probably better to use the sodium salts because inhibition of RNA ligase by ammonium ion has been observed. Nucleotides may be conveniently converted to the sodium form by chromatography on Sephadex-SP columns in the sodium form as described by Cartwright and Hutchinson.[29]

The Na^+ salts of the common 2′-deoxyribonucleoside 3′,5′-bisphosphates are available from PL Laboratories, Inc. The bisphosphate derivatives of other 2′-deoxynucleosides are prepared in a single step by reaction with pyrophosphoryl chloride.[8] Deoxyribonucleotides labeled in the 5′ position with ^{32}P are prepared from the 3′-monophosphate derivatives using [γ-^{32}P]ATP and the *pseT 1* mutant polynucleotide kinase that lacks an intrinsic 3′-phosphatase activity.[16] Oligonucleotide donors are prepared with 3′- and 5′-phosphate termini.[17] The 3′-phosphate acts as a blocking group to prevent circularization, self-addition, or multiple addi-

[29] I. L. Cartwright and D. W. Hutchinson, *Nucleic Acids Res.* **8,** 1675 (1980).

tions.[17,30] An oligodeoxyribonucleotide with 3′- and 5′-hydroxyl termini may be converted to a suitable donor by adding a 2′-deoxyribonucleoside 3′,5′-bisphosphate with RNA ligase followed by phosphorylation of the 5′-hydroxyl with ATP and *pseT 1* polynucleotide kinase.[17] The 3′-phosphate can be added to oligodeoxyribonucleotide without altering its sequence by adding a ribonucleoside 3′,5′-bisphosphate with RNA ligase and subsequently removing the 3′-phosphate with a phosphomonoesterase and the remaining ribonucleoside by periodate oxidation and β-elimination.[18] The 5′-phosphate is then added with polynucleotide kinase and ATP as described previously. The 5′-phosphate of the donor can be radiolabeled by using [γ-^{32}P]ATP and *pseT 1* polynucleotide kinase.[17] The phosphate penultimate to the 3′ terminus can be labeled by using [5′-^{32}P]pdNp and RNA ligase[16] and the labeled terminal 3′-phosphate similarly can be introduced by using [5′-^{32}P]prNp and the RNA ligase-periodate oxidation procedure.

Procedure

Two sets of reaction conditions that vary in their donor to acceptor to ATP concentration ratios are required to use DNA acceptors effectively with RNA ligase. One set of ratios of nucleotide components is used when the donor is a nucleoside 3′,5′-bisphosphate and another when it is an oligonucleotide. Both types of reactions contain the following components in common.

HEPES-NaOH, 50 m*M*, pH 7.9
DTT, 20 m*M*
Bovine serum albumin, 10 μg/ml
Creatine phosphokinase, 175 units/ml
Adenylate kinase, 170 units/ml
T4 RNA ligase, 15–50 μM

Single Nucleotide Donor Reactions.[15,16,18] To add a 2′-deoxyribonucleoside 3′-5′-bisphosphate donor to an oligodeoxyribonucleotide acceptor, the other components of the reaction mixture are as follows:

pdNp donor, 8 m*M*
dN(pdN)$_n$ acceptor, where $n \geq 2$, 1 m*M*
ATP, 0.4 m*M*
$MnCl_2$, 5 m*M*
Phosphocreatine, 1 or 40 m*M*
Spermine, 8 m*M*

High yields of conversion of acceptor to product are attained by driving the reaction with excess donor. The rate increases with increasing

[30] O. C. Uhlenbeck and V. Cameron, *Nucleic Acids Res.* **4,** 85 (1977).

substrate concentrations, and the donor concentration may be varied from 0.5 to 10 m*M* provided that the relative concentration of donor to acceptor to ATP remains 1 to 0.13 to 0.05. The acceptor to donor ratio is less critical than that of ATP to donor. Acceptor concentrations in excess of donor [5′-^{32}P]pdNp have been successfully used to maximize the incorporation of isotope into the product. A ratio of ATP to donor greater than 0.05 to 0.1 inhibits both the initial velocity and final yield of the reaction.[16] This inhibition is probably due to the conversion of the enzyme to the adenylylated form, which is inactive in the phosphodiester bond-forming step of the mechanism.[15] A low and constant concentration of ATP is maintained by a regeneration system composed of phosphocreatine, creatine phosphokinase, and adenylate kinase. With the sole exception of reactions in which pdGp is the donor, varying the level of phosphocreatine from 1 to 40 m*M* has little effect upon the yields of products. The purine donors tested were the bisphosphate derivatives of adenine, N^6-methyladenine, 2-aminopurine, and 2,6-diaminopurine.[16,31] When pdGp is the donor, the optimal concentration of phosphocreatine is 40 m*M*. The added stimulatory effect of the phosphagen on reactions using pdGp as the donor is independent of its role in generating ATP and is not understood.[16] Spermine may be varied from 2 to 8 m*M* with the higher concentrations being generally most effective.

Oligodeoxyribonucleotide Donor Reactions.[17,18] To join two oligodeoxyribonucleotides the reactions contain the following substances in addition to the common components:

pdN(pdN)$_n$pdNp donor, where $n \geq 1$, 1 m*M*
dN(pdN)$_n$ acceptor, where $n \geq 2$, 4 m*M*
ATP, 0.5 m*M*
$MnCl_2$, 10 m*M*
Phosphocreatine, 1 m*M*
Spermine, 2 m*M*

In contrast to the single nucleotide donor reaction, these reactions cannot be driven by excess donor because the oligonucleotide donor can bind in the acceptor site of the enzyme and thereby act as a competitive inhibitor of the reaction.[17] Acceptor and donor at equimolar concentrations can be joined,[17] but 1- to 10-fold molar excesses of acceptor over donor give faster rates and better yields. The rates of joining are proportional to the oligonucleotide concentrations, and reactions with 0.1–1 m*M* donor can be successfully performed provided that the ATP concentration is maintained from 0.1 to 0.9 that of the donor. The ATP concentration must be low in these reactions not only to prevent sequestering the enzyme in its adenylylated form, but also because high ATP levels lead to

[31] C. A. Brennan and R. I. Gumport, unpublished observations.

the modification of the 3′-phosphate of the donor.[17,18] Presumably, the donor binds to the adenylylated enzyme in the acceptor site and the adenylyl group is transferred to its 3′-phosphate to form an anhydride linkage yielding the structure pdN(pdN)$_n$pdN-3′pp-5′A.[18] Concentrations of 40 m*M* phosphocreatine lowered the product yield in all reactions, including those in which the 5′-terminal nucleotide of the donor was a dGMP residue.

Common Steps in the Protocol

For either set of conditions, the reactions are assembled by drying the combined nucleotides, oligonucleotides, spermine, and phosphocreatine under vacuum in 1.5-ml Eppendorf plastic tubes. These components are dissolved in a solution containing the remainder of the reaction ingredients with the exception of the RNA ligase. Reaction volumes (5–500 μl) are chosen to give the desired final concentrations of components. After the enzyme is added, the reactions are incubated at 17° for 1–10 days. Incubation is conveniently performed in a small heated water bath in a cold room or refrigerator. The tubes are wrapped in Parafilm and submerged as far as practicable to avoid concentration of the reaction by evaporation and condensation of water onto the walls of the tubes.

The manner of monitoring the progress of the reaction depends upon whether or not a labeled nucleotide substrate has been used. If a 5′-^{32}P-labeled donor was used, an aliquot can be treated with alkaline phosphatase and examined by paper chromatography or HPLC for radioactivity that is resistant to the enzyme and migrates as a larger oligonucleotide. The adenylylated donor (A-5′pp5′-donor) can also give rise to a phosphomonoesterase-resistant peak, which must not be confused with the authentic product. Similarly, if the radiolabel is in an internal nucleotide, chromatographic migration as a larger oligonucleotide indicates successful reaction.

Reaction mixtures without label can be assayed by anion-exchange HPLC on Zorbax-NH_2 columns (DuPont).[32] An aliquot containing 0.04–0.1 A_{260} of acceptor is deproteinized by dilution to ca. 500 μl with 50% (v/v) acetonitrile (MeCN) and passing it through an octadecyl-silica filter (Bondelut C_{18}, Analytichem International, Inc.). Alternatively, the aliquot is diluted twofold with MeCN, and the precipitate is removed by centrifugation at 12,000 *g* in an Eppendorf microcentrifuge. The resultant solutions are filtered through 0.45-μm Acrodisc filters (Gelman), evaporated under vacuum, and dissolved in 10 μl of water. The sample is injected into the HPLC column (4.6 cm × 25 cm) and eluted with 10%

[32] L. W. McLaughlin, F. Cramer, and M. Sprinzl, *Anal. Biochem.* **112,** 60 (1981).

methanol containing a gradient of 0.05 to 0.9 *M* KP_i, pH 4.5. The KH_2PO_4 used to form the gradients is purified on a Chelex-100 (Bio-Rad) column to remove contaminants that absorb UV light.[33] The formation of product is indicated by the appearance of a new peak of light-absorbing material eluting at higher salt concentrations that the acceptor and by the diminution of the peak due to the acceptor. The adenylylated donor (A-5′pp5′-donor) can also give rise to peak during these analyses. The activated donor can often be identified because it arises in reaction mixtures before any product is formed and decreases in amount as the reaction progresses. The maximum amount of activated donor that can possibly form is limited by the ATP concentration. Reaction mixtures that have been treated with alkaline phosphatase are also analyzed in the same chromatographic system. The removal of the 3′-phosphates from both the product and adenylylated donor reduces their retention times. The yields can be quantified by integrating the areas under the product and acceptor peaks and normalizing with calculated molar adsorption coefficients,[34] assuming the hypochromicity that occurs upon conversion of a nucleotide to a homopolymer.

The product can be isolated from the remainder of the reaction components by paper chromatography,[15,17] ion-exchange chromatography on DEAE-Sephadex[15] or Dowex AG1-X2,[35] or HPLC on ion-exchange (Zorbax-NH_2) or reverse-phase (C_8 or C_{18}) resins.[36,37] The method of choice will depend upon the availability of equipment and the size and amount of the oligonucleotide synthesized. As much as 5–10 A_{260} units of oligodeoxyoligonucleotides have been successfully purified on a 4.6 mm × 25 mm Zorbax-NH_2 column. The appropriate fractions are pooled, concentrated, and desalted on a Sephadex G-10 column (1.5 cm × 1 m) in 20% (v/v) ethanol, 30 m*M* triethylammonium bicarbonate, pH 8.0.[35] We have observed, however, that oligomers purified in this manner sometimes contain inhibitors of RNA ligase. Higher yields of products that are free of inhibitors may be obtained using DEAE-Sephadex columns.[15]

The product can be characterized by nearest-neighbor analysis if a 5′-^{32}P-labeled donor has been used.[38] The base composition of unlabeled product is determined by degrading it to deoxyribonucleosides with

[33] J. D. Karkas, J. Germershausen, and R. Liou, *J. Chromatogr.* **214,** 267 (1981).

[34] H. Büchi and H. G. Khorana, *J. Mol. Biol.* **72,** 251 (1972).

[35] G. R. Gough, C. K. Singleton, H. L. Weith, and P. T. Gilham, *Nucleic Acids Res.* **6,** 1557 (1979).

[36] M. E. Winkler, K. Mullis, J. Barnett, I. Stroynowski, and C. Yanofsky, *Proc. Natl. Acad. Sci. U.S.A.* **79,** 2181 (1982).

[37] H. J. Fritz, R. Belagaje, E. L. Brown, R. H. Fritz, R. A. Jones, R. G. Lees, and H. G. Khorana, *Biochemistry* **17,** 1257 (1978).

[38] R. D. Wells, T. M. Jacob, S. A. Narang, and H. G. Khorana, *J. Mol. Biol.* **27,** 237 (1967).

venom phosphodiesterase and alkaline phosphatase and quantifying the products by HPLC on a C_8 or C_{18}[39] resin using a methanol gradient in 50 m*M* NaP_i, pH 5.9 (personal communication, H. Coyer, PL Labs, Inc.). In addition, the sequence and purity of the product is determined by two-dimensional fingerprinting of 5′-^{32}P-labeled oligomer by standard methods.[40]

Comments

We have described two general sets of reaction conditions for using oligodeoxyribonucleotides as acceptors. One is for deoxyribonucleoside 3′,5′-bisphosphate donors, and the other for oligodeoxyribonucleotide donors. These conditions represent the best starting point for attempting a joining reaction. We have observed yields from 10 to >95% under these conditions, the single addition reactions being more efficient. From 1 to 200 nmol of acceptor have been used, and a total of ca 0.5 μmol of phosphodiester was formed with 1.7 mg of enzyme in these reactions. Particular donors and acceptors may warrant additional optimization studies if the yields are unsatisfactory. Increasing the substrate concentrations often helps because the apparent K_m for the acceptor oligonucleotide is in the millimolar range and the enzyme is therefore usually not saturated under the described conditions.[15,17] Varying the spermine and phosphocreatine concentrations may also be helpful. Duplex structure between the acceptor and donor is inhibitory, and the addition of RNase A can stimulate the reaction.[17,18] It presumably does so by acting as a DNA melting protein. However, RNase A can also stimulate the addition of 2′-deoxyribonucleoside 3′,5′-bisphosphate donors in reactions where duplex structures cannot form.[16] The addition of dimethyl sulfoxide from 10 to 20% (v/v) sometimes stimulates the oligomer joining reaction,[17] but not the addition of deoxyribonucleoside 3′,5′-bisphosphates.[16] The rates of both types of reactions increase with increasing enzyme concentration and in addition, the yield also increases in oligomer joining reactions. The dependence of yield upon enzyme concentration does not seem to be due to inactivation of the enzyme upon prolonged incubation.

There are two unexpected side reactions that may sometimes occur during these incubations. As mentioned, a minor side product may arise from the adenylylation of the 3′-phosphate of the donor to form a structure of the form pdN(pdN)$_n$pdN-3′pp5′A.[18] High ATP concentrations promote this reaction, and since low levels of ATP are used during synthesis this side-reaction has not caused serious difficulties. A second

[39] K. C. Kuo, R. A. McCune, and C. W. Gehrke, *Nucleic Acids Res.* **20,** 4763 (1980).
[40] C. D. Tu and R. Wu, this series, Vol. 65, p. 620.

problem may arise through the reversal of the reaction in an AMP-dependent removal of a pdNp residue from the donor or product. This reaction has been characterized and described by Krug and Uhlenbeck[41] and occurs more frequently with RNA substrates. When this reaction occurs, a new 3′-hydroxyl-terminated acceptor is formed; if it reacts with a donor, it may give rise to unexpected products. We have observed the reversal in one joining reaction with oligodeoxyribonucleotides and do not yet know how frequently it might present a difficulty.

Because the reactions of oligodeoxyribonucleotide acceptors are so slow compared to those of oligoribonucleotides, the large effects of acceptor base composition noted with the latter[7,28] are not observed. We have used acceptors up to nine nucleotides long. The limit is probably determined by the length at which it becomes impossible to attain millimolar concentrations. For example, we have been unable to perform the 2′-deoxyribonucleoside 3,5′-bisphosphate addition reaction on DNA restriction fragments at 1–5 μM. On the other hand, DNA restriction fragments do serve as efficient donors with oligoribonucleotide acceptors. Higgins *et al.*[42] showed that several different restriction fragments, either as duplex structures or after denaturation, would react efficiently with rA_n, rC_n, or rI_n acceptors. Oligodeoxyribonucleotides will not serve as acceptors with high molecular weight DNA donors. We have observed that oligodeoxyribonucleotides can be converted to reactive acceptors by adding a ribonucleoside residue to their 3′ termini.[20,43] Heteropolymeric acceptors of the form $(dN)_n rN$ can be synthesized by using a ribonucleoside 3′,5′-bisphosphate donor and RNA ligase under conditions described above for the nucleoside 3′,5′-bisphosphate addition reaction with oligodeoxyribonucleotides. We find the following conditions optimal for the addition of heteropolymeric acceptors to DNA donors.[43]

- HEPES-NaOH, 50 mM, pH 7.9
- DTT, 10 mM
- $MnCl_2$, 10 mM
- Bovine serum albumin, 50 μg/ml
- 5′-Phosphorylated DNA donor, 0.1–1.0 μM; e.g., a [5′-^{32}P]DNA restriction fragment 460 base pairs long isolated from a *Taq*I digest of plasmid DNA
- Acceptor, 50–200 μM; e.g., d(A-A-T-T)rC
- ATP, 10–100 μM
- T4 RNA ligase, 10–40 μM

[41] M. Krug and O. C. Uhlenbeck, *Biochemistry* **21,** 1858 (1982).

[42] N. P. Higgins, A. P. Gaballe, and N. R. Cozzarelli, *Nucleic Acids Res.* **6,** 1013 (1979).

[43] J. A. Baez and R. I. Gumport, unpublished observations.

The reactions are incubated for 2–5 hr at 17°. Successful reaction is monitored by determining the amount of radiolabel converted from phosphomonoesterase (PM) sensitivity to resistance. In addition, the product is susceptible to hydrolysis by spleen phosphodiesterase (SPD) because it contains a 5′-hydroxyl group. These criteria distinguish between the unreacted donor (PM sensitive, SPD resistant), the adenylylated donor (PM resistant, SPD resistant), and the product (PM resistant, SPD sensitive). Using these assays, yields of from 15 to >90% have been obtained with a variety of oligoribonucleotide and 3′-ribose-terminated oligodeoxyribonucleotide acceptors.[43] Acceptors must be in 50-fold or greater concentration excess over the donor. High ATP concentration (1 mM) inhibits the reaction. Enzyme concentrations greater than those of the donor are required for high yields. Heteropolymeric acceptors do not react successfully at the lower enzyme concentrations reported to be successful for oligoribonucleotides.[42]

Conclusion

There are two major difficulties involved in using DNA acceptors with RNA ligase. The first is that DNA is a much less efficient acceptor than is RNA. In one case in which a reaction was optimized for a dA_4 acceptor, rA_4 reacted at least 200 times faster.[17] The problem resides in the phosphodiester bond-forming step of the mechanism, since the adenylylated-donor intermediate forms rapidly and accumulates in DNA joining reactions.[15,17] The conditions we have described allow DNA acceptors to react provided that high enzyme concentrations and long reaction times (1–10 days) are used. The second major difficulty is in obtaining pure enzyme so that high concentrations can be incubated with the oligodeoxyribonucleotide substrates for the requisite time without degrading them. The purification described here surmounts this difficulty.

[3] Joining of RNA Molecules with RNA Ligase

By PAUL J. ROMANIUK and OLKE C. UHLENBECK

T4 RNA ligase catalyzes the ATP-dependent phosphodiester bond formation between a 5′-terminal phosphate (donor) and a 3′-terminal hydroxyl (acceptor). Although the enzyme can produce circular products with substrates of sufficient length that contain the required termini,[1,2]

[1] J. Leis, R. Silber, V. G. Malathi, and J. Hurwitz, *Adv. Biosci.* **8,** 117 (1972).
[2] R. Silber, V. G. Malathi, and J. Hurwitz, *Proc. Nat. Acad. Sci. U.S.A.* **69,** 3009 (1972).

METHODS IN ENZYMOLOGY, VOL. 100

ISBN 0-12-182000-9

RNA ligase is also a very effective intermolecular joining reagent for oligonucleotide synthesis.[3–8] In this chapter, we will concentrate on applications for oligoribonucleotide synthesis; the use of RNA ligase for the preparation of deoxy oligomers is discussed in this volume.[9] Comprehensive reviews of the T4 RNA ligase are also available.[10,11]

The intermolecular phosphodiester bond formation catalyzed by RNA ligase can be summarized as

$$\mathrm{E + ATP \rightleftharpoons E \sim AMP + PP_i} \quad (1)$$
$$\mathrm{E \sim AMP + pNpM_B \rightleftharpoons E + A5'ppNpM_B} \quad (2)$$
$$\mathrm{E + {}_{HO}XpYpZ_{OH} + A5'ppNpM_B \rightleftharpoons {}_{HO}XpYpZpNpM_B + AMP + E} \quad (3)$$
$$\text{Overall: } \mathrm{ATP + {}_{HO}XpYpZ_{OH} + pNpM_B \rightleftharpoons XpYpZpNpM_B + AMP + PP_i}$$

The 3′ terminus of the donor carries a blocking group B to ensure single addition to an acceptor. The minimum acceptor required by the enzyme is a trinucleoside diphosphate, and the minimum donor is a nucleoside 3′,5′-bisphosphate.[12] Oligomer substrates of certain nucleotide sequences are more reactive than others,[12,13] but most pairs of oligonucleotide substrates can be joined in good yield using the reaction conditions included here.

The versatility of RNA ligase as a reagent for oligoribonucleotide synthesis is illustrated by the following two examples. For preparative synthesis, the enzyme readily catalyzes the equimolar joining of two oligoribonucleotide blocks on a moderate scale (ca. 1 μmol). The enzyme can also be used to synthesize small amounts of internally ^{32}P-labeled oligomers of high specific activity.

[3] G. Kaufmann and Z. Littauer, *Proc. Natl. Acad. Sci. U.S.A.* **71,** 3741 (1974).
[4] G. C. Walker, O. C. Uhlenbeck, E. Bedows, and R. I. Gumport, *Proc. Natl. Acad. Sci. U.S.A.* **72,** 122 (1975).
[5] G. Kaufmann and N. R. Kallenbach, *Nature (London)* **254,** 452 (1975).
[6] E. Ohtsuka, S. Nishikawa, A. F. Markham, S. Tanaka, T. Miyake, T. Wakabayashi, M. Ikehara, and M. Sugiura, *Biochemistry* **17,** 4984 (1978).
[7] E. Ohtsuka, S. Nishikawa, R. Fukumoto, H. Uemura, T. Tanaka, E. Nakagawa, T. Miyake, and M. Ikehara, *Eur. J. Biochem.* **105,** 481 (1980).
[8] T. Neilson, E. C. Kofoid, and M. C. Ganoza, *Nucleic Acids Res. Symp. Ser.* **7,** 167.
[9] C. A. Brennan, A. E. Manthey, and R. I. Gumport, see this volume [2].
[10] R. I. Gumport, and O. C. Uhlenbeck, *in* "Gene Amplification and Analysis" Vol. 2: Analysis of Nucleic Acid Structure by Enzymatic Methods" (J. G. Chirikjian and T. S. Papas, eds.), p. 313. Elsevier/North-Holland, Amsterdam, 1981.
[11] O. C. Uhlenbeck, and R. I. Gumport, *in* "The Enzymes" (P. D. Boyer, ed.), 3rd ed., Vol. 15, p. 31. Academic Press, New York, 1982.
[12] T. E. England, and O. C. Uhlenbeck, *Biochemistry* **17,** 2069 (1978).
[13] E. Ohtsuka, D. Takefumi, H. Uemura, T. Taniyama, and M. Ikehara, *Nucleic Acids Res.* **8,** 3909 (1980).

Materials and Methods

Enzymes

T4 RNA ligase is availabe from a number of commercial sources, and can be purified from T4-infected *Escherichia coli*. The purification procedure we currently use is described in this volume.[9] For RNA joining reactions, the enzyme is stored at about 1 mg/ml in 20 m*M* HEPES, 1 m*M* DTT, 10 m*M* $MgCl_2$, pH 7.5, and 50% glycerol at −20°.

T4 polynucleotide kinase is used to prepare donor oligonucleotides by transfer of the γ-phosphate of ATP to a 5′-hydroxyl. The 3′-phosphatase activity of T4 polynucleotide kinase[14] makes the wild-type enzyme unsuitable when a 3′-phosphate is used as a donor blocking group. However, polynucleotide kinase isolated from the T4 *PseT 1* mutant can be used for this purpose, since it totally lacks 3′-phosphatase activity.[15] Both the wild-type[14] and mutant[15] enzyme can be purified from the same preparation of T4-infected cells used in RNA ligase isolation. Both types of enzymes are also commercially available. Polynucleotide kinase is generally stored at about 2000 units/ml (10 μg/ml) in 25 m*M* KH_2PO_4, 12.5 m*M* KCl, 2.5 m*M* DTT, 25 μ*M* ATP, pH 7.0, and 50% glycerol at −20°. The low concentration of ATP is required to prevent rapid inactivation of the enzyme.

Oligonucleotides

Oligoribonucleotides for RNA ligase reactions can be obtained in several different ways. Oligomers prepared by chemical synthesis have been used successfully in ligase joinings.[6–8] Primer-dependent polynucleotide phosphorylase[16] can be used to synthesize oligomers of the type $XpY(pZ)_n$ from commercial dinucleoside monophosphates and nucleoside 5′ diphosphates.[17] Although a mixture of products are obtained, conditions can be varied to obtain a specific oligomer in 10–25% yield. The addition of a specific ribonuclease to these polynucleotide phosphorylase reactions will allow the production of certain trimers in nearly quantitative yield.[17] Another important source of oligomers for RNA ligase reactions are fragments derived from partial or total ribonuclease digests of natural RNAs. By developing synthetic strategies that employ large portions of natural RNA molecules, the time devoted to synthesis is often reduced.

Oligonucleotides derived from these various sources can be easily

[14] V. Cameron, and O. C. Uhlenbeck, *Biochemistry* **16,** 5120 (1977).
[15] V. Cameron, D. Soltis, and O. C. Uhlenbeck, *Nucleic Acids Res.* **5,** 825 (1978).
[16] C. B. Klee, *Prog. Nucleic Acids Res.* **2,** 896 (1971).
[17] R. E. Thach, and P. Doty, *Science* **147,** 1310 (1965).

manipulated to a form suitable for use as an acceptor or donor. Oligomers from chemical synthesis and equilibrium polynucleotide phosphorylase reactions generally have free hydroxyls at both 5′ and 3′ termini, and they can be used directly as acceptors. Such oligonucleotides are converted to donors by first adding a suitable pNp to the 3′ terminus by RNA ligase, followed by 5′-phosphorylation with *PseT 1* polynucleotide kinase. Oligomers derived from nuclease digests, or prepared by nuclease-assisted polynucleotide phosphorylase reactions, usually have 5′-hydroxyl and 3′-phosphate termini. Treatment with ATP and *PseT 1* polynucleotide kinase yields a donor molecule, and treatment with a phosphomonoesterase yields an acceptor.

Before carrying out large-scale RNA ligase preparations, one frequently optimizes conditions and identifies products by doing trial reactions with radioactively labeled substrates. Donors are generally labeled on their 5′-terminal phosphate using [γ-^{32}P]ATP and polynucleotide kinase. Acceptors can be prepared with the 3′-terminal nucleotide labeled with ^{3}H by incubating a comparatively high concentration of a primer oligonucleotide, a low concentration of high specific activity ^{3}H-labeled nucleoside 5′-diphosphate and primer-dependent polynucleotide phosphorylase.[18] The combination of ^{3}H-labeled acceptor and ^{32}P-labeled donor permits rapid analysis of the products of trial reactions. In addition, hydrolysis of the intermolecular product of the reaction with mixed nucleases should give the predicted [^{3}H,^{32}P]nucleoside 3′- monophosphate.[19]

Separation Techniques

At the end of an RNA ligase reaction, a complex mixture of products, intermediates, and starting materials must be resolved. For products of chain length $\leq$12, descending paper chromatography on Whatman 3 MM is a very successful tool. The papers are developed in a buffer containing 1 *M* ammonium acetate, pH 7.5, and absolute ethanol. As chain length increases, the percentage volume of the aqueous component of the eluting buffer is also increased, as well as the total time of elution. A dodecamer product can readily be separated from heptamer acceptor and pentamer donor in 48 hr with a 70 : 30 (v/v) ammonium acetate–ethanol buffer. The appropriate band is desalted in ethanol and eluted with water to yield pure product that can be used directly for subsequent ligase reactions.[19]

Reaction mixtures can also be resolved by column chromatography. The elution of DEAE-Sephadex A-25 or DEAE-cellulose columns by gradients of triethylammonium bicarbonate for short oligomers and sodium chloride for longer oligomers has been used to purify RNA ligase

[18] O. C. Uhlenbeck, and V. Cameron, *Nucleic Acids Res.* **4,** 85 (1977).

[19] M. Krug, P. L. de Haseth, and O. C. Uhlenbeck, *Biochemistry* **21,** 4713 (1982).

reactions.[6,7] Chromatography on RPC-5 using potassium chloride gradients to purify oligodeoxynucleotide joining reactions has been used successfully.[20] Products eluted from columns are desalted by passage down a BioGel P2 or Sephadex G-10 column before being used in further ligase reactions.

Polyacrylamide gel electrophoresis is an excellent method for separating reaction products of longer chain lengths. The standard 20% polyacrylamide–7 *M* urea gel used for nucleic acid sequencing[21,22] has single nucleotide resolution to greater than 75 nucleotides. Oligomer products can be localized on the gel by autoradiography or ultraviolet shadowing and then eluted by a crush and soak procedure.[19]

High-performance liquid chromatography (HPLC) should be extremely useful for the purification of RNA ligase reactions. This technique is capable of readily purifying oligonucleotides of greater chain length than can be accomplished by paper chromatography and is a much quicker technique. Elution conditions have been reported that resolve all the starting substrates, intermediates, and products for RNA ligase reactions,[23] and conditions exist for the purification of preparative-scale reactions.[24] Recovery of product from HPLC columns is comparable to that of paper chromatography and conventional columns. Rapid analysis of trial reactions is also possible with HPLC.

Results

Moderate-Scale Joining Reactions

Conditions have been optimized for the effective joining of two oligoribonucleotides by RNA ligase.[10,11] The following conditions will give good yields for many donor–acceptor pairs with incubation for 10 hr at 17°.

HEPES, 50 m*M*, pH 8.3.
Dithiothreitol, 3 m*M*
Bovine serum albumin, 10 μg/ml
$MgCl_2$, 10 m*M*
Acceptor, 0.5 m*M*
Donor, 0.6 m*M*
ATP, 2 m*M*
RNA ligase, 180 μg/ml

[20] M. I. Moseman-McCoy, and R. I. Gumport, *Biochemistry* **19,** 635 (1980).
[21] H. Donis-Keller, A. Maxam, and W. Gilbert, *Nucleic Acids Res.* **4,** 2527 (1977).
[22] D. A. Peattie, and W. Gilbert, *Proc. Natl. Acad. Sci. U.S.A.* **77,** 4679 (1980).
[23] E. Romaniuk, L. W. McLaughlin, T. Neilson, and P. J. Romaniuk, *Eur. J. Biochem.* **125,** 639 (1982).
[24] L. W. McLaughlin, and E. Romaniuk, *Anal. Biochem.* **124,** 37 (1982).

Incubation is at the optimum 17°, although incubation temperatures from 0° to 37° have been used successfully. The product generally forms within the first 6 hr. Even if the reaction has not reached completion, little additional product forms by increasing the incubation time.[18] When very high yields are obtained using these conditions, it is often possible to reduce the amount of enzyme used. With a good acceptor–donor pair, almost quantitative conversion to product can be obtained with as little as 4 μg of RNA ligase per milliliter.[18]

In some cases, the conditions stated above may not give good yields. Since RNA ligase is sensitive to contaminants (such as ammonium or phosphate ions) that are commonly found in oligomers, purification by gel filtration chromatography on BioGel P2 or Sephadex G-10 can often improve the yield. However, poor yields in RNA ligase reactions are usually related to the nucleotide sequence or secondary structure of the donor or acceptor molecules.

Although systematic studies of reaction yields with a different combination of donors and acceptors are not available, some generalizations about the substrate specificity of RNA ligase can be made.[10] First, RNA ligase clearly acts more effectively with single-stranded RNA molecules. For example, the 5′-terminal phosphate of tRNA is inactive as a donor unless it is not base paired.[25,26] Second, acceptors with a high uridine content and donors with a high purine content are less active.[23] Thus, if several alternative synthetic pathways are available, a proper choice of joining reactions can often avoid reactions with low yields.

Some changes in the above reaction conditions can be made in an attempt to increase yield.

1. Although the conditions given contain relatively high enzyme concentrations (see, however, this volume[9]), more enzyme can be added to improve the yield.

2. The addition of 10–20% v/v dimethylsulfoxide (DMSO) to ligase reactions has led to dramatic increases in yield in some cases.[12,26,27] The enzyme is able to tolerate up to 40% v/v DMSO without an appreciable decrease in activity. The effectiveness of DMSO is not limited to cases where substrates may form secondary structures, but has been observed with simple acceptors as well.[23,26]

3. Although a threefold excess of ATP over donor is usually maintained to ensure that the equilibrium favors product, this ratio is sometimes detrimental. In reactions involving joining to a poor acceptor, the adenylated donor formed in the second step of the reaction will dissociate

[25] T. E. England, A. G. Bruce, and O. C. Uhlenbeck, this series, Vol. 65, p. 65.

[26] T. E. England, and O. C. Uhlenbeck, *Nature (London)* **275,** 560 (1978).

[27] A. G. Bruce, and O. C. Uhlenbeck, *Nucleic Acids Res.* **5,** 3665 (1978).

before product is formed. If the ATP concentration is too high, the free enzyme becomes adenylylated before it can rebind adenylylated donor. By using an ATP concentration stoichiometric with the donor, such "over adenylylation" cannot occur, and higher yields often result.

One additional problem that leads to poor yields is the possible reversal of the third step of the RNA ligase reaction. A detailed investigation of this phenomenon has been reported.[28] As illustrated in Scheme 1, reversal yields a mixture of three products, which can be identified during trial reactions by their different $^3H : ^{32}P$ ratios. Although reversal can occur in principle at any phosphodiester linkage, it occurs substantially only at 3′-terminal phosphates. Although measurable reversal is not observed for all combinations of donors and acceptors, its occurrence cannot always be predicted. At present, the best method to avoid reversal is not to use a phosphate as a donor blocking group. Suitable blocking groups include the photolabile *O*-nitrobenzyl[29] and acid-labile *O*-(α-methoxyethyl)[30] or ethoxymethylidine[6] groups.

Synthesis of Internally ^{32}P-Labeled RNA Fragments

If the objective is to obtain small amounts of radiochemically pure oligomer, the radiolabeling reaction with [γ-^{32}P]ATP and polynucleotide kinase and a synthesis reaction with RNA ligase can be combined as two sequential steps in the same reaction mixture.

Although [γ-^{32}P]ATP is commercially available, we prefer to synthesize it enzymically immediately before use. By modifying the procedure of Johnson and Walseth[31] by using HEPES buffer at pH 8.3, instead of Tris buffer at pH 9.0, the resulting [γ-^{32}P]ATP can be used directly in polynucleotide kinase and RNA ligase reactions. Since the five commercial enzymes used to prepare [γ-^{32}P]ATP have very low levels of ribonucleases, the ATP synthesis reaction can be terminated by heating to 90° for 2 min and used directly in a polynucleotide kinase reaction without further purification.

Preparation of the ^{32}P-labeled donor and subsequent ligation to an acceptor molecule are carried out sequentially in the same reaction tube, using, in a 10-μl final volume: 50 m*M* HEPES, pH 8.3; DTT, 3 m*M*; 10 μg of BSA per milliliter; 10 m*M* $MgCl_2$; 23 μ*M* donor; 20 μ*M* [γ-^{32}P]ATP (ca 400 Ci/mmol); and 100 U of *PseT 1* polynucleotide kinase per milliliter.

The reaction is initiated by the addition of enzyme and incubated at 37°

[28] M. Krug, and O. C. Uhlenbeck, *Biochemistry* **21,** 1858 (1982).

[29] E. Ohtsuka, H. Uemura, T. Doi, T. Miyake, S. Nishikawa, and M. Ikehara, *Nucleic Acids Res.* **8,** 601 (1979).

[30] J. J. Sninsky, J. A. Last, and P. T. Gilham, *Nucleic Acids Res.* **3,** 3157 (1976).

[31] R. A. Johsnon, and T. F. Walseth, *Adv. Cyclic Nucl. Res.* **10,** 135 (1979).

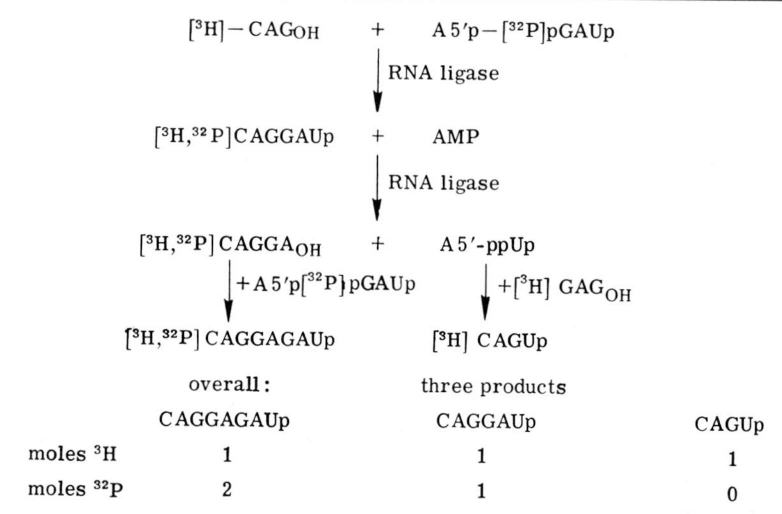

SCHEME 1. Products generated by reversal of the third step of the RNA ligase reaction.

for 2 hr. Although the reaction should be complete, a simple analysis by PEI-cellulose thin-layer chromatography developed in 0.8 *M* ammonium sulfate monitors transfer of ^{32}P from ATP ($R_p = 0.3$) to donor. The reaction is then heated at 90° for 2 min. Acceptor oligonucleotide, ATP, buffers, and RNA ligase are added to an aliquot (ca 5–9 μl) of the kinase reaction for the final joining in a 25-μl volume: 50 m*M* HEPES, pH 8.3; 3 m*M* DTT; 10 μg of BSA per milliliter; 10 m*M* $MgCl_2$; 8 μM acceptor; 6 μM ^{32}P-donor (from kinase reaction); 1 m*M* ATP; and 100 μg of RNA ligase per milliliter.

The reaction is then incubated at 16° for 4–12 hr. The product is purified by preparative gel electrophoresis using 20% polyacrylamide gels containing 7 *M* urea in 0.1 *M* Tris-borate, pH 8.1.[21] The gels are electrophoresed at 800 V for a period of time appropriate to the size of the expected product (ca 3 hr for a 21-mer). Product is located by autoradiography, and the gel slice is crushed and soaked in 1.5 ml of 1.0 *M* potassium acetate, pH 4.5, 1 m*M* EDTA overnight at 4°. After centrifugation to remove acrylamide fragments, $MgCl_2$ is added to 1 m*M* and the oligomer is precipitated at −60° with three volumes of ethanol. The product is collected by centrifugation and redissolved in distilled water.

This procedure has been used successfully in our laboratory to prepare radiolabeled amounts of biologically active oligoribonucleotides of sizes 14- to 21-mer.[19,32]

[32] O. C. Uhlenbeck, P. T. Lowary, and W. W. Wittenberg, *Nucleic Acids Res.* **10,** 3341 (1982).

[4] Exonuclease III: Use for DNA Sequence Analysis and in Specific Deletions of Nucleotides

By LI-HE GUO and RAY WU

Exonuclease III of *Escherichia coli* catalyzes the sequential hydrolysis of mononucleotides from the 3′ termini of duplex DNA molecules.[1] Using a high ratio of exonuclease III to DNA ends and moderate concentration of salt (90 m*M* KCl), digestion of DNA is relatively synchronous, removing approximately 10 nucleotides per minute from each 3′ terminus[2,3] at room temperature.

We have developed an improved enzymic method for DNA sequence analysis based on the partial digestion of duplex DNA with exonuclease III to produce DNA molecules with 3′ ends shortened to varying lengths. After exonuclease III treatment, 3′ ends are extended and labeled by repair synthesis[3] using the dideoxynucleotide chain termination method of Sanger *et al.*[4] The exonuclease III method (Procedures 8 and 9, and after Procedure 6 under DNA Sequencing) is the only sequencing method that can sequence both strands of a cloned DNA fragment over 1000 base pairs in length without prior gel fractionation of the fragments. We also describe a method for making deletions of nucleotides in DNA employing exonuclease III. Comparison is made between exonuclease III and *Bal*31 nuclease for making specific deletions.

In order to increase the ease and flexibility of cloning and sequencing any gene based on the exonuclease III method, we have constructed a family of pWR plasmids. They were derived from plasmid pUR222 of Rüther *et al.*[5] by deleting about 530 base pairs between the *amp* region and the *lac* region to produce pWR1. This deletion resulted in a fourfold increase in the copy number of the plasmid. A *Hin*dIII site was next inserted into the polylinker region of pWR1. The resultant pWR2 plasmid has eight unique restriction sites in the *lacZ′* gene, which can be used for cloning genes for sequencing by the exonuclease III method described here. Genes cloned in pWR2 can also be sequenced by other DNA sequencing methods[4,6] (procedures 7 and 10).

[1] C. C. Richardson, I. R. Lehman, and A. Kornberg, *J. Biol. Chem.* **239,** 251 (1964).
[2] R. Wu, G. Ruben, B. Siegel, E. Jay, P. Spielman, C. D. Tu, *Biochemistry* **15,** 734 (1976).
[3] L. Guo and R. Wu, *Nucleic Acids Res.* **10,** 2065 (1982).
[4] F. Sanger, S. Nicklen, and A. R. Coulson, *Proc. Natl. Acad. Sci. U.S.A.* **74,** 5463 (1977).
[5] U. Rüther, M. Koenen, K. Otto, and B. Müller-Hill, *Nucleic Acids Res.* **9,** 4087 (1981).
[6] A. M. Maxam and W. Gilbert, this series, Vol. 65, p. 499.

ISBN 0-12-182000-9

Properties of pUR222

The *lacZ'* gene encodes 59 amino acid residues of the α-peptide of β-galactosidase, a number that is sufficient for α-peptide activity.[8–11] Therefore bacteria harboring a plasmid that includes this region with its operator and promoter should make blue colonies on indicator plates containing isopropylthiogalactoside (IPTG) and 5-bromo-4-chloroindolyl-β-D-galactoside (X-gal). If an exogenous DNA fragment is inserted between the *lac* promoter and the *lacZ'* gene, or into the *lacZ'* gene, bacteria harboring this plasmid should give rise to white colonies. This makes the selection easy for colonies carrying inserts.[5,8–11]

Rüther *et al.*[5] constructed a multipurpose plasmid, pUR222, which contains six unique cloning sites (*Pst*I, *Sal*I, *Acc*I, *Hin*dII, *Bam*HI, and *Eco*RI) in a small region of its *lacZ'* gene. Bacteria harboring recombinant plasmids generally give rise to white colonies, whereas those containing only pUR222 form blue colonies on indicator plates. DNA cloned into this plasmid can be labeled and sequenced directly, after cutting with the proper restriction enzymes, using the procedure of Maxam and Gilbert[6]; it is not necessary to isolate the labeled fragment that is to be sequenced.

Construction and Properties of pWR Plasmids

Figure 1 shows the construction of pWR2, one of five related pWR plasmids. There are two *Pvu*II sites in pUR222, one located in the *lacZ'* gene and the other preceding the *lac* promoter. To remove the latter, plasmid pUR222 was first partially digested with restriction enzyme *Pvu*II. After partial digestion to cut only one of the two *Pvu*II sites, the linear pUR222 DNA was digested with *Bal*31 nuclease (Procedure 14) to delete about 530 base pairs. After ligation and transformation (Procedures 3 and 4), a number of blue colonies were obtained, and plasmid DNA was prepared from several of them. The DNA was analyzed by digestion with *Pvu*II and *Eco*RI and a colony with only one *Pvu*II site located in the *lacZ'* gene was selected.

For inserting a *Hin*dIII site[7] into the polylinker region, this plasmid was digested with *Eco*RI and *Bam*HI, and the resulting cohesive ends were filled in by using the large fragment of DNA polymerase I in the presence of four dNTPs (Procedure 10b). After electrophoresis (Proce-

[7] C. P. Bahl, K. J. Marians, R. Wu, J. Stawinsky, and S. A. Narang, *Gene* **1,** 81 (1976).

[8] B. Gronenborn and J. Messing, *Nature (London)* **272,** 375 (1978).

[9] J. Messing, B. Gronenborn, B. Müller-Hill, P. H. Hofschneider, *Proc. Natl. Acad. Sci. U.S.A.* **74,** 3642 (1977).

[10] U. Rüther, *Mol. Gen. Genet.* **178,** 475 (1980).

[11] A. Ullmann, F. Jacob, and J. Monod, *J. Mol. Biol.* **24,** 339 (1967).

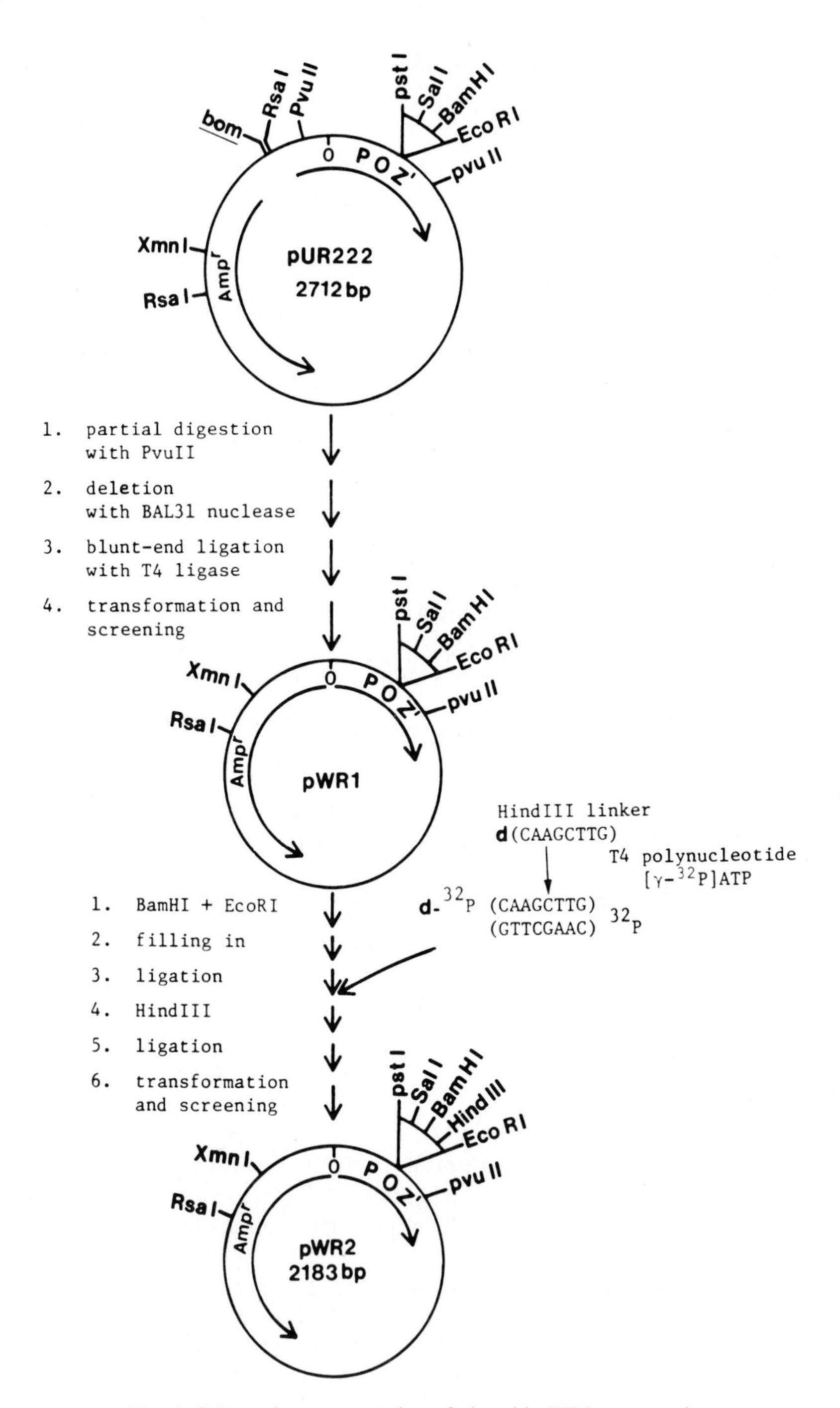

FIG. 1. Schematic representation of plasmid pWR2 construction.

dure 2), the plasmid DNA was ligated in the presence of a 5′-labeled *Hin*dIII adaptor, d(C-A-A-G-C-T-T-G), followed by *Hin*dIII digestion (Procedure 3). This adaptor has been chosen to restore the *Eco*RI and *Bam*HI sites. After electrophoresis to remove unused *Hin*dIII adaptor, the DNA was ligated to form a circle and was used to transform *E. coli* (Procedures 3 and 4). Transformants were screened for blue colonies; DNA was isolated from them and futher analyzed by separate digestion with *Eco*RI, *Hin*dIII, or *Bam*HI and electrophoresis. Five related plasmids were selected as being useful, and all of them gave very high copy number. One of them, pWR2, has the new features of a unique *Pvu*II site in the *lacZ′* gene, and a unique *Hin*dIII site added to the polylinker region. These sites are available for DNA cloning and sequencing in addition to sites in pUR222.[5]

Figure 2 shows the restriction map of pWR2. This plasmid is 2183 base pairs (bp) in length and consists of two functional regions. One region (355

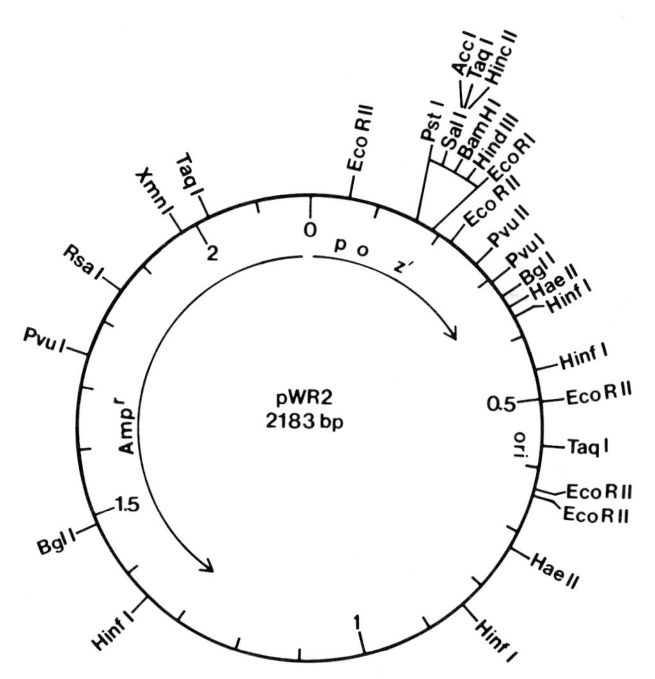

FIG. 2. Restriction map of pWR2. The positions of the *lac* region and the *Amp*ʳ region are indicated. Their junction is numbered as 0 (zero). The first base pair (bp) in the *lac* region (see Fig. 3) is designated nucleotide No. 1. There are 355 bp in the *lac* region, which is interrupted by the polylinker sequence, and 883 bp in the *Amp*ʳ region. Bacteria harboring recombinant plasmids with DNA inserts within the polylinker generally form white colonies, whereas those containing only pWR2 give rise to blue colonies on indicator plates.

bp) contains the *lac* promoter and operator and part of the *lacZ'* gene. The other region (1828 bp) contains the ampicillin-resistant gene (*amp* or *bla*) and the origin of replication (ori) of the plasmid. There are 10 unique restriction enzyme recognition sites in pWR2. Eight of them are located with the *lacZ'* gene, and two within the β-lactamase (*amp* or *bla*) gene. The locations of the restriction enzyme sites are tabulated in Table I. The

TABLE I
pWR2: LOCATIONS OF RESTRICTION RECOGNITION SEQUENCES

Enzymes[a]	Number of cleavage sites	Locations							
*Acc*I	1	173							
*Acy*I	1	1907							
*Alu*I	11	48	143	186	282	419	645	781	1038
		1559	1659	1722					
*Asu*I	5	301	1412	1491	1508	1730			
*Ava*II	2	1508	1730						
*Bam*HI	1	178							
*Bbv*I	12	165	260	333	382	400	819	884	887
		1093	1421	1610	1787				
*Bgl*I	2	342	1489						
*Dde*I	4	752	1161	1327	1867				
*Eco*RI	1	190							
*Eco*RII	5	65	232	503	624	637			
*Fnu*DII	4	524	1105	1435	1928				
*Fnu*4HI	18	166	261	334	383	401	404	522	677
		820	885	888	1094	1422	1611	1761	1788
		1883	2112						
*Gdi*II	1	1758							
*Hae*I	3	492	503	955					
*Hae*II	2	355	725						
*Hae*III	9	200	302	492	503	521	955	1413	1493
		1760							
*Hga*I	3	579	1157	1907					
*Hgi*AI	3	795	1956	2041					
*Hgi*CI	2	60	1318						
*Hgi*EII	1	1058							
*Hha*I	13	26		333	354	387	657	724	824
		998	1107	1500	1593	1930			
*Hin*cII	1	174							
*Hin*dIII	1	184							
*Hin*fI	4	377	452	848	1364				
*Hpa*II	10	88	684	831	857	1047	1451	1485	1552
		1622	1904						
*Hph*I	5	1221	1448	1844	2070	2085			
*Mbo*II	7	300	355	1127	1218	1973	2051	2160	

TABLE I (*continued*)

Enzymes[a]	Number of cleavage sites	Locations							
*Mnl*I	9	295	366	592	649	916	1316	1397	1527
		1733							
*Mst*I	2	332	1590						
*Pst*I	1	170							
*Pvu*I	2	313	1739						
*Pvu*II	1	282							
*Rru*I	1	1849							
*Rsa*I	1	1849							
*Sal*I	1	172							
*Sau*3AI	15	178	310	1043	1118	1129	1137	1215	1227
		1332	1673	1691	1737	1995	2012	2048	
*Taq*I	3	173	577	2020					
*Xmn*I	1	1970							

[a] The following enzymes do not digest pWR2. These sites are useful for sequencing the cloned gene if present.

*Ava*I, *Bal*I, *Bcl*I, *Bgl*II, *Bst*EII, *Cla*I, *Eco*RV, *Hpa*I, *Kpn*I, *Nco*I, *Nru*I, *Sac*I, *Sac*II, *Sma*I, *Sph*I, *Stu*I, *Tth*111I, *Xba*I, *Xho*I, *Xma*III

junction of the *lac* region and the *Amp*ʳ region is taken as the zero position on the physical map (see Figs. 2 and 3). The first base pair in the *lac* region is designated nucleotide No. 1.

In plasmid pWR2 seven unique restriction sites (*Pst*I, *Sal*I, *Acc*I, *Hinc*II, *Bam*HI, *Hin*dIII, and *Eco*RI) are located between codon No. 4 (thr) and No. 6 (ser) of the *lacZ'* gene (Fig. 4), and an eighth, *Pvu*II is in codon No. 35 (ser). DNA cloned into the *Bam*HI site of pWR2 can be directly sequenced by the chemical method. This is not possible with

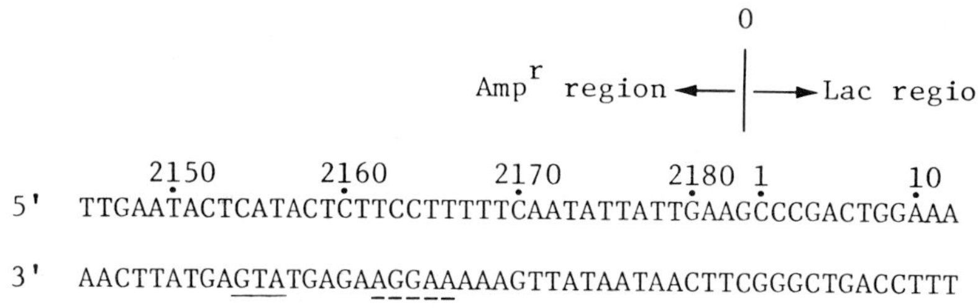

FIG. 3. DNA sequence of the intercistronic region between the *bla* gene (*Amp*ʳ region) and the *lac* region in pWR plasmids. ATG start codon for translation of β-lactamase is underlined. Sequences with homology to 16 S RNA are indicated by dashed underlines.

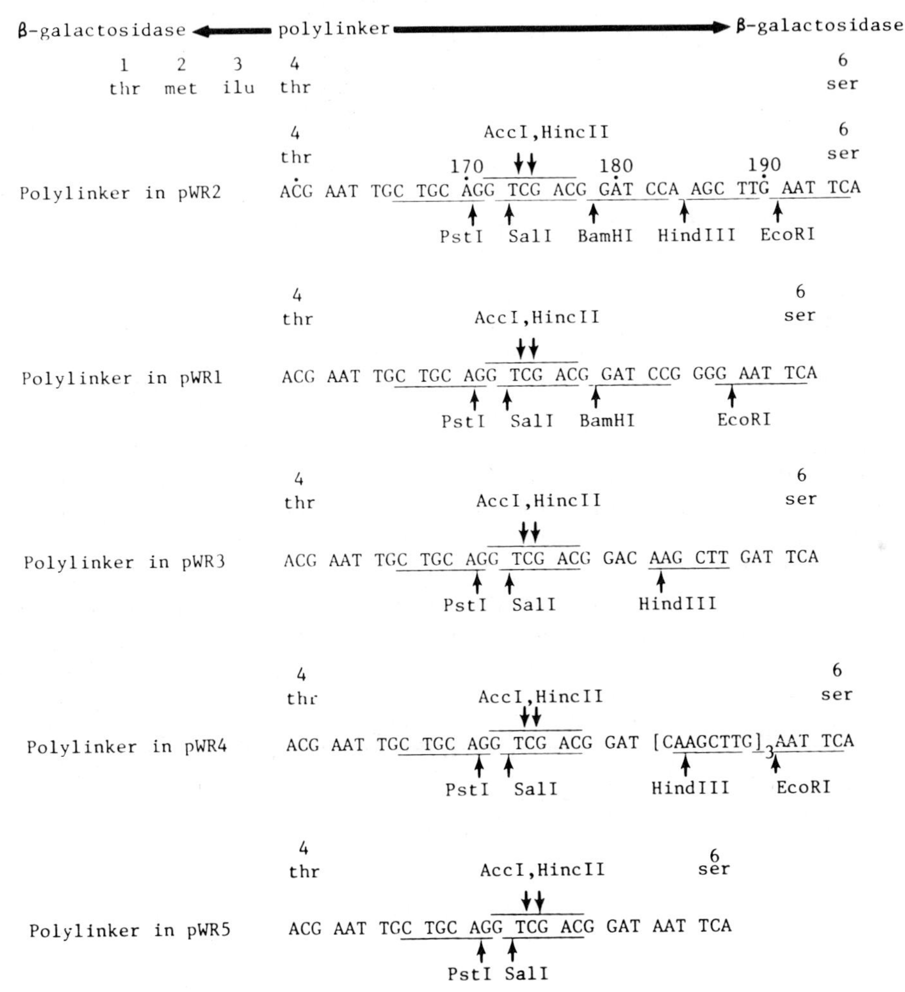

FIG. 4. The nucleotide sequences of polylinker inserted in the *lacZ'* gene region in pWR2 and its derivatives. The polylinkers are located between amino acid codon 4 (*thr*) and 6 (*ser*) of β-galactosidase gene.

pUR222.[5] The other pWR plasmids (Fig. 4) differ from pWR2 in the number of unique restriction sites in the polylinker region.

We have observed that bacteria carrying pWR2 without inserts in the polylinker region are capable of growing in either M9 minimal medium or L broth medium up to 1000 μg of ampicillin per milliliter. However, bacteria harboring recombinant plasmids with inserts in the polylinker region grow more slowly and can tolerate only up to 25 μg of ampicillin per milliliter. The reason for this decrease in resistance is not clear but could be related to the fact that 530 bp have been deleted between the *bla* promoter and the *lacZ'* gene in pWR2. In the *Amp*^{r} region of pWR2, there are only 27 base residues immediately preceding the start codon (ATG) of the β-lactamase gene (*bla*) (see Fig. 3, lower strand). This sequence is too short for the efficient functioning of the *bla* promoter. Efficient expression probably depends on another promoter within or downstream from the *lacZ'* gene, which is interrupted by insertion of a DNA fragment into the polylinker region.

Soberón *et al.*[12] showed that the deletion of the DNA region between the origin of replication and the *tet* gene of pBR322 did not affect the cloning capacity of the resulting plasmids. However, since a DNA region like the ColE1 relaxation site (*bom*) is deleted in these plasmids, they are superior EK2 vectors. Use of these vectors should permit lowering the degree of physical containment by at least one level.[13] The relaxation site (*bom*) in the pWR plasmids has also been deleted (Fig. 1), and therefore they are also superior vectors from the safety point of view.

Twigg and Sherratt[14] reported the construction of ColE derivatives in which deletion of a nonessential DNA region carrying the ColE1 relaxation site resulted in an increased copy number. In the construction of pWR plasmids, we also deleted this region from pUR222, and found that the copy number of pWR2 is at least four times higher than that of pUR222 or pBR322. The high yield of pWR2 and its derivatives allows the application of a rapid isolation procedure, yielding plasmid of sufficient purity and amount for DNA sequencing. The purity of the plasmid is high owing to a high ratio of plasmid DNA to chromosomal DNA.

Solutions, Buffers, Media, Gel Formulas, and Commercial Sources

Solutions

1. Chloroform-isoamyl alcohol (24 : 1, v/v), for extraction
2. Phenol saturated with 1 *M* Tris-HCl, pH 8, for extraction
3. 0.1 *M* EDTA–1.5 *M* NaOAc, for terminating enzyme reactions
4. 40% acrylamide stock solution for gel electrophoresis, 38% acrylamide–2% bisacrylamide in water, stable for at least half a year when stored at 4°

[12] X. Soberón, L. Covarrubias, and F. Bolivar, *Gene* **9,** 287 (1980).

[13] L. Covarrubias, L. Cervantes, A. Covarrubias, X. Soberón, I. Vichido, A. Blanco, Y. M. Kupersztoch-Portnoy, and F. Bolivar, *Gene* **13,** 25 (1981).

[14] A. J. Twigg and D. Sherratt, *Nature (London)* **283,** 216 (1980).

5. 5% ammonium persulfate for gel electrophoresis, stable for at least a year when stored at 4°
6. 0.3% each of bromophenol blue and xylene cyanole FF in 60% glycerol, for gel electrophoresis
7. 0.3% each of bromophenol blue and xylene cyanole FF in 10 *M* urea, for gel electrophoresis
8. 1 *N* NaOH–100 m*M* EDTA, for gel electrophoresis
9. 0.2 *N* NaOH–10 m*M* EDTA for gel electrophoresis
10. 1% agarose in TA or TB buffer plus 0.5 μg of ethidium bromide per milliliter

Buffers

11. 10 × T4 ligase buffer: 500 m*M* Tris-HCl, pH 7.6, 50 m*M* $MgCl_2$, 50 m*M* dithiothreitol, 5 m*M* ATP
12. 10 × TA buffer for gel electrophoresis: 400 m*M* Tris-HCl, pH 7.9, 30 m*M* NaOAc, 10 m*M* EDTA
13. 10 × TB buffer for gel electrophoresis: 500 m*M* Tris-borate, pH 8.3, 10 m*M* EDTA
14. TE buffer: 10 m*M* Tris-HCl, pH 8, 1 m*M* EDTA
15. 10 × RE buffer for restriction enzyme digestion and DNA polymerase repair synthesis: 500 m*M* Tris-HCl, pH 7.6, 500 m*M* KCl, 100 m*M* $MgCl_2$, 100 m*M* dithiothreitol
16. 10 × exonuclease III buffer: 660 m*M* Tris-HCl, pH 8, 770 m*M* NaCl, 50 m*M* $MgCl_2$, 100 m*M* dithiothreitol
17. 10 × λ exonuclease buffer: 500 m*M* Tris-HCl, pH 9.5, 20 m*M* $MgCl_2$, 30 m*M* dithiothreitol
18. 10 × terminal transferase buffer for tailing DNA: 1 *M* sodium cacodylate, pH 7, 10 m*M* $CoCl_2$, 2 m*M* dithiothreitol
19. 10 × S1 buffer for removal of single-stranded DNA: 0.5 *M* NaOAc, pH 4, 0.5 *M* NaCl, 60 m*M* $ZnSO_4$
20. 5 × *Bal*31 buffer for deletion of a segment of DNA with *Bal*31 nuclease: 100 m*M* Tris-HCl, pH 8, 60 m*M* $MgCl_2$, 60 m*M* $CaCl_2$, 3 *M* NaCl, 5 m*M* EDTA
21. Lysozyme solution for preparation of plasmid DNA: 2 mg/ml crystalline lysozyme, 50 m*M* glucose, 25 m*M* Tris-HCl, pH 8, 10 m*M* EDTA. Prepare daily from crystalline lysozyme, and stock solutions of other components; store at 4°
22. Lysozyme solution for mini preparation of plasmid DNA: 5 mg of crystalline lysozyme per milliliter in 25 m*M* Tris-HCl, pH 8
23. RNase A solution for preparation of plasmid DNA: 10 mg of

RNase A per milliliter in 50 mM NaOAc, pH 4.8, heated at 90° for 5 min prior to use

24. Triton solution for preparation of plasmid DNA: 0.3% Triton X-100, 150 mM Tris-HCl, pH 8, 200 mM EDTA
25. 30% polyethylene glycol (PEG) 6000–1.8 M NaCl for preparation of plasmid DNA

Media

26. YT broth contains (per liter): 8 g of (Bacto)tryptone, 5 g of (Bacto) yeast extract, 5 g of NaCl (if used in plates, add 15 g of agar), and deionized water. After autoclaving, if necessary, add ampicillin (10–20 μg/ml final concentration) after the solution has cooled to below 50°. 2 × YT broth (per liter): 16 g tryptone, 10 g yeast extract, and 5 g NaCl.
27. 20 × M9 salt: 140 g of Na_2HPO_4, 60 g of KH_2PO_4, 10 g of NaCl, 20 g of NH_4Cl, and H_2O to 1 liter
28. M9 medium containing IPTG and X-gal is made as follows: Agar, 15 g in 900 ml of water, is autoclaved. To the mixture add 50 ml of 20 × M9 salt, 20 ml of 20% casamino acids, 10 ml of 20% glucose, 5 ml of thiamin (1 mg/ml), 1 ml of 1 M $MgSO_4$, 0.1 ml of 1 M $CaCl_2$ (the components have been separately autocalved). After the mixture has cooled to below 50°, add ampicillin (20 μg/ml final concentration), IPTG (40 mg in 1.5 ml H_2O) and X-gal (40 mg in 1.5 ml of *N,N*-dimethylformamide). If using M9 medium for liquid culture, agar, IPTG, and X-gal should be omitted.

Polyacrylamide Gel Formulas

29. Volumes required for preparing polyacrylamide gels of various size and thickness:

Length (cm)	Width (cm)	Thickness (mm)	Volume required (ml)
40	35	3	500
40	20	3	260
40	35	2	400
40	20	2	220
40	35	1.5	300
40	35	0.6	100
40	35	0.4	70
80	35	0.4	150

30. Volumes of essential components needed to prepare polyacrylamide gels of different concentration:

Components	Percent concentration of polyacrylamide gel								
	3	4	5	6	8	10	12	15	20
	Volume (in ml) needed[a]								
40% acrylamide/ Bis[b] (19 : 1) mixture	15	20	25	30	40	50	60	75	100
10 × Tris-borate buffer	20	20	20	20	20	20	20	20	20
Distilled water	162.8	157.8	152.8	147.8	137.8	127.8	117.8	102.8	77.8
TEMED	0.2	0.2	0.2	0.2	0.2	0.2	0.2	0.2	0.2
5% $(NH_4)_2S_2O_4$	2	2	2	2	2	2	2	2	2

[a] Final volume = 200 ml.
[b] Bis = bisacrylamide.

31. Quantities of ingredients necessary for preparing denaturing (containing 8 *M* urea) acrylamide/bis mixture (19 : 1) of different polyacrylamide concentration:

Ingredients	Percent concentration			
	5	8	15	20
	Volume and weight required			
40% acrylamide/Bis (19 : 1) mixture (ml)	62.5	100	187.5	150
Urea, ultrapure (g)	240.24	240.24	240.24	126
10 × Tris-borate buffer (ml)	50	50	50	30
Distilled water (ml)	200	162.5	75	21.3
Total final volume (ml)	500	500	500	300
Final molarity of urea	8 *M*	8 *M*	8 *M*	7 *M*

32. Polynucleotide chain length corresponding to the migration rate of dye markers (bromophenol blue, and xylene cyanole FF) in different polyacrylamide concentrations of plain and 8 *M* urea-containing gel:

Percent concentration of plain and 8 M urea-containing polyacrylamide gel		Polynucleotide chain length, corresponding to the migration rate of dye	
		Bromophenol blue	Xylene cyanole FF
Plain			
Acrylamide	Acrylamide/Bis		
8%	29 : 1	20	125
5%	19 : 1	55	170
4%	19 : 1	83	400
4%	29 : 1	100	430
8 M Urea			
Acrylamide	Acrylamide/Bis		
20%	19 : 1	10	30
20%	29 : 1	12	35
15%	19 : 1	12	35
15%	29 : 1	14	48
12%	19 : 1	16	45
10%	19 : 1	18	55
8%	19 : 1	20	73
8%	29 : 1	23	90
5%	19 : 1	31	130–140
4%	19 : 1	50	250–300
3%	19 : 1	80	410

We thank Robert Yang for information on gels used for items 29–32.

Commerical Sources of Reagents, Isotopes, and Enzymes

[γ-^{32}P]ATP (>2000 Ci/mmol): The Radiochemical Centre, Amersham
Agarose: Bethesda Research Laboratories (BRL)
Acrylamide: Bio-Rad Laboratories
Adenosine triphosphate: P-L Biochemicals
Bisacrylamide: Bio-Rad Laboratories
Bromophenol blue: BDH Chemicals, Ltd.
5-Bromo-4-chloroindodyl-β-D-galactoside (X-gal): Sigma Chemical Co.
Bacto-tryptone: Difco Laboratories
Bacto-yeast extract: Difco Laboratories
Bovine pancreatic RNase A: Sigma Chemical Co.
Casamino acids: Difco Laboratories
*Bal*31 nuclease: BRL
[α-^{32}P]dNTPs (410 Ci/mmol): The Radiochemical Centre, Amersham
[^{35}S]Deoxyadenosine 5′-[α-thio]triphosphate (600 Ci/mmol): New England Nuclear

dNTPs: P-L Biochemicals
ddNTPs: P-L Biochemicals
Exonuclease III: BRL
λ exonuclease: New England BioLabs
DNA polymerase (Klenow fragment): New England BioLabs
*Hind*III linker (d-CAAGCTTG): Collaborative Research, Inc.
Isopropylthiogalactoside (IPTG): Sigma Chemical Company
MacConkey agar: Difco Laboratories
Polyethylene glycol 6000: J. T. Baker Chemical Corp.
Restriction enzymes: New England BioLabs and BRL
Reverse transcriptase: Life Science Incorporated
S1 nuclease: Sigma Chemical Co.
Sodium cacodylate: Fisher Scientific
T4 DNA ligase: New England BioLabs
T4 polynucleotide kinase: New England BioLabs
Terminal transferase: P-L Biochemicals
Thiamin hydrochloride (B_1): Calbiochem
Triton X-100: Sigma Chemical Co.
Xylene cyanole: McIB Manufacturing Chemists

Cloning into pWR Plasmids

The major steps underlying genetic engineering technology include isolation and specific cleavage of DNA, ligation of DNA fragments to a cloning vector, transformation and selection of the desired clone, confirming the cloned gene by physical mapping and DNA sequencing, and expression of the cloned gene. Usually several vectors are needed to serve these functions. However, it is most convenient if a single vector can serve all these functions. A plasmid has been constructed to serve all these functions. This plasmid, pWR2, has eight unique restriction enzyme sites in *lacZ'* gene (see Fig. 4) for convenient cloning of different DNA fragments. The desired clone can be readily selected by color change of clones from blue to colorless, and the insert DNA can be directly sequenced using different methods. The cloned gene can be expressed by using the *lac* promoter in this plasmid.

Plasmid pWR2 provides a wide selection of unique restriction sites for cloning exogenous DNA fragments. Table II lists these restriction sites and shows how these sites can be chosen for cloning of a large variety of restriction fragments and for sequencing the cloned DNA using either the exonuclease III method[3] or the chemical method.[6]

Sometimes there are problems with cloning of blunt-ended DNA into the *Hin*cII site in pWR2. We found that, when plasmid DNA digested

TABLE II
CLONING AND SEQUENCING IN PLASMID pWR2

Restriction fragments to be cloned	Cloning sites	Sequencing: Exo III method restriction enzyme, 1st cut	Exo III method restriction enzyme, 2nd cut	Chemical method restriction enzyme, 1st cut and labeling	Chemical method restriction enzyme, 2nd cut
*Eco*RI *Eco*RI*	*Eco*RI	*Pvu*II	*Pst*I	*Pvu*II	*Pvu*I or *Bgl*I
		*Hin*dIII	*Eco*RI	*Hin*dIII	*Bam*HI
*Hin*dIII	*Hin*dIII	*Eco*RI	*Pst*I	*Pvu*II	*Pvu*I or *Bgl*I
		*Sal*I	*Eco*RI	*Bam*HI	*Sal*I
*Bam*HI *Bgl*II	*Bam*HI	*Eco*RI	*Pst*I	*Hin*dIII	*Eco*RI
*Bcl*I *Sau*3A *Xho*II		*Sal*I	*Eco*RI	*Sal*I	*Pst*I
*Sal*I *Xho*I (*Ava*I)	*Sal*I	*Eco*RI	*Pst*I	*Hin*dIII	*Eco*RI
		*Pst*I	*Eco*RI		
*Acc*I *Asu*II *Cla*I	*Acc*I	*Eco*RI	*Pst*I	*Hin*dIII	*Eco*RI
*Hpa*II *Taq*I		*Pst*I	*Eco*RI		
*Pst*I	*Pst*I	*Eco*RI	*Pst*I	*Hin*dIII	*Eco*RI
		*Xmn*I	*Eco*RI		
Blunt-ended fragments	*Hin*cII	*Eco*RI	*Pst*I	*Hin*dIII	*Eco*RI
		*Pst*I	*Eco*RI		
	*Pvu*II	*Pvu*I	*Eco*RI	*Eco*RI	*Hin*dIII
		*Eco*Ri	*Pvu*I or *Bgl*I		

with *Hin*cII (Lot 11214, BRL) was blunt-end ligated to the *Hin*cII-digested DNA, there were a number of white clones that did not carry DNA inserts. The *lacZ'* gene had probably been inactivated by exonuclease contaminating the *Hin*cII. However, the white clones with DNA inserts can be distinguished from those without insert by the greater ampicillin resistance of the latter. Gardner *et al.*[15] also met with the same problems with blunt-ended cloning in M13mp7, but they were able to overcome this problem by using a particular batch of *Hin*cII (BRL, Lot 2651).

Procedures for Isolating, Digesting, Cloning, and Sequencing of DNA

Procedure 1. Restriction Enzyme Digestion of Plasmid DNA or Other DNA. To 0.5–3 μg of DNA, add 1–10 units of a restriction enzyme and

[15] R. C. Gardner, A. J. Howarth, P. Hahn, M. Brown-Lendi, R. J. Shepherd, and J. Messing, *Nucleic Acids Res.* **9**, 2871 (1981).

1 μl of 10 × RE buffer. Adjust the volume to 10 μl with distilled water and incubate at 37° for 30–60 min. For shotgun cloning, 0.5 μg of vector and 3 μg of DNA containing the desired insert fragment are digested together and then used directly for Procedure 3.

Procedure 2. Isolation of DNA Fragments by Gel Electrophoresis. To isolate a particular DNA fragment to be cloned, use Procedure 1 on 3 μg of the DNA carrying the desired fragment. Add 2 μl of dyes (0.3% bromophenol blue and xylene cyanole in 60% glycerol) to the reaction mixture. The sample is electrophoresed on a horizontal gel apparatus (10 × 8.2 cm) containing 1% low melting point agarose (BRL) plus 0.5 μg of ethidium bromide per milliliter in TB buffer at 60 V and 2 mA. (Warning: current should not be over 3 mA or 7.5 V/cm.) Cut out the band to be cloned with a razor blade and put it into an Eppendorf tube (1.5 ml). After melting the gel by heating at 70° for 2 min, measure the volume and add 0.1 volume of 5 *M* NaCl. Vortex and continue heating at 70° for 3 min. Extract the solution twice with an equal volume of phenol saturated with 1 *M* Tris-HCl (pH 8) and once with an equal volume of chloroform–isoamyl alcohol (24 : 1, v/v). Extract the aqueous phase several times with a large volume of *n*-butanol to remove ethidium bromide and to reduce the volume of the solution. Add 2.5 volumes of ethanol to the above solution, chill at −70° for 5 min, and centrifuge at 12,000 rpm (Eppendorf centrifuge) for 5 min. If there is a white salt precipitate at the bottom of the tube, add a little distilled water to dissolve it and repeat the ethanol precipitation step. Remove the supernatant and add 200–500 μl of ethanol to rinse the DNA pellet. Centrifuge the tube for 3 min and remove the supernatant. Dry the DNA pellet under vacuum for 5 min.

Procedure 3. Ligation of Insert into Vector. To vector plus linearized insert DNA (in 10 μl, from Procedure 1) or an isolated insert fragment plus linearized vector in 10 μl (from Procedure 1 plus 2), add 10 μl of 0.1 *M* EDTA–1.5 *M* NaOAc and 30 μl of TE buffer. Extract the mixture once with phenol saturated with 1 *M* Tris-HCl (pH 8) and once with chloroform–isoamyl alcohol (24 : 1, v/v). To the aqueous phase add 125 μl of ethanol. Chill the mixture at −70° for 5 min and centrifuge for 5 min. To the DNA pellet add 200 μl of ethanol and centrifuge for 3 min. Dry the plasmid DNA pellet under vacuum for 5 min. Resuspend the sample in 8 μl of H_2O and 1 μl of 10 × ligation buffer. After vortexing, add 1 μl of T_4 DNA ligase (3 units, New England BioLabs) to the mixture and incubate at 4° for 4–18 hr.

Procedure 4. Transformation

a. To the ligation mixture from Procedure 3, add 200 μl of *E. coli* $F^-Z^-\Delta M15recA$[5] or JM101,[8] made competent and stored frozen according to Morrison,[16] thawed on ice for 15 min. Incubate at 0° for 30 min. Heat at 42° for 2 min. Add 1 ml of 2× YT broth (no ampicillin) to the mixture and incubate at 37° for 60 min. Plate 100 μl of serially diluted samples in 10 m*M* NaCl on petri dishes containing (per milliliter) M9 agar medium plus 20 μg of ampicillin, 5 μg of thiamin, and 40 μg each of IPTG and X-gal or YT agar medium plus 20 μg/ml ampicillin, 40 μg/ml each of IPTG and X-gal. Incubate the plates at 37° for 15–24 hr, and select those bacteria that give white colonies.

b. To the ligation mixture from Procedure 3, add 40 μl of 0.1 *M* $CaCl_2$ and 200 μl of competent cells as in Procedure 4a. Incubate on ice for 60 min. Heat at 42° for 2 min, and transfer the mixture to 5 ml of YT broth. Shake at 37° for 4–5 hr. Transfer the culture into a 50-ml sterile centrifuge tube, and centrifuge at 8000 rpm for 5 min. Pour off the supernatant and resuspend the pellet in 1 ml of 10 m*M* NaCl. Plate 100 μl of serially diluted samples on petri dishes as described above. This slightly longer procedure yields two or three times more transformants.

Procedure 5. Large-Scale Preparation of Plasmid.[17] Inoculate a single colony in 20 ml of YT broth and shake overnight at 37°. Transfer the culture to 1 liter of M9 medium or YT medium plus 25 μg of ampicillin per milliliter for bacteria harboring plasmid pWR2, or 10 μg of ampicillin per milliliter for those harboring recombinant plasmids with inserted DNA in the polylinker region. Shake at 37° until A_{600} reaches 1.0. Add 150 mg of chloramphenicol and continue shaking at 37° overnight. Pour the culture into four 500-ml centrifuge bottles and incubate in ice for 15 min. Centrifuge at 8000 rpm for 10 min. Resuspend the pellets in 50 ml of lysozyme solution. Incubate on ice for 30 min. Add 1 ml of RNase A solution and 24 ml of Triton solution. Incubate on ice for 30 min, and centrifuge in a Beckman 50.2 Ti rotor at 3000 rpm for 1 hr at 4°. Transfer the supernatant into four 50-ml phenol-resistant plastic tubes with tight-fitting caps

[16] D. A. Morrison, this series, Vol. 68, p. 326.

[17] Plasmid DNA preparations may be contaminated with DNases or inhibitors of restriction enzymes. It is recommended that the following tests be carried out for each plasmid preparation. Three samples (1 μg of plasmid in each) are used for testing: sample (a), no incubation; sample (b), incubation with only restriction enzyme buffer at 37° for 3 hr; sample (c), incubation with a restriction enzyme and buffer at 37° for 3 hr (*Hin*dIII or *Pst*I are more sensitive to inhibitors than many other enzymes). After incubation, the samples are loaded on a mini-agarose gel (10 × 8.2 cm, see Procedures 2 and 12) for 1.5–2 hr. Sample (a) gives the percentage of form I plasmid DNA. Sharp bands in samples (b) and (c) indicate the lack of contaminating DNases, and smearing of DNA bands indicates the presence of DNases. Sample (c) serves to test whether an inhibitor of the restriction enzyme is present. Contaminated DNase may be removed by adding 3 *M* NaOAc to the plasmid DNA followed by ethanol precipitation. It may be necessary to carry out another phenol extraction and ethanol precipitation.

(Sarstedt, No. 60·547) and extract twice with phenol saturated with 1 *M* Tris-HCl (pH 8) and once with chloroform–isoamyl alcohol (24 : 1, v/v). Transfer the aqueous phase into a 500-ml centrifuge bottle. Add 8 ml of 3 *M* NaOAc and 200 ml of ethanol. The mixture is chilled at −70° for 20 min and centrifuged at 10,000 rpm for 30 min. Remove the supernatant and dry the DNA pellet under vacuum for 15 min. Resuspend the DNA pellet in 50 ml of TE buffer, and then add 20 ml of 30% PEG-6000–1.8 *M* NaCl. Incubate at 4° overnight. Centrifuge at 10,000 rpm for 20 min and remove the supernatant. Resuspend the plasmid DNA pellet in 5 ml of 0.3 *M* NaOAc and transfer into a 50-ml centrifuge tube. To the DNA solution add 13 ml of ethanol, chill at −70° for 15 min, and centrifuge at 10,000 rpm for 15 min. The DNA pellet is washed once with 20 ml of ethanol, dried, and resuspended in 2 ml of TE buffer. The final concentration of plasmid DNA is usually around 1 μg/μl. The total yield of plasmid pWR2 DNA using this chloramphenicol amplification procedure is very high, about 2 mg per liter of culture using pWR2 or its derivatives.

Procedure 6. Mini Preparation of Plasmid.[17] Transfer a single colony into 5 ml of YT broth or M9 medium with ampicillin as in Procedure 5 or streak on one-eighth of a petri dish containing YT broth agar medium plus 10 μg of ampicillin per milliliter. Incubate at 37° overnight. Pellet the bacteria at 10,000 rpm for 5 min or scrape up bacteria from the plate and transfer them into a 1.5-ml Eppendorf tube. Resuspend the cell pellet in 200 μl of cold TE buffer and 5 μl of 0.5 *M* EDTA (pH 8). Add 50 μl of lysozyme solution. Incubate on ice for 15 min; add 5 μl of RNase solution and 120 μl of Triton solution. Incubate on ice for 15 min, and centrifuge for 15 min. Extract the supernatant once with phenol and once with chloroform–isoamyl alcohol. To the aqueous phase add 30 μl of 3 *M* NaOAc and 700 μl of ethanol, chill at −70° for 10 min, and centrifuge for 5 min. Wash once with 1 ml of ethanol. The DNA pellet is dried and resuspended in 100 μl of TE buffer followed by addition of 40 μl of 30% polyethylene glycol 6000–1.8 *M* NaCl. Put on ice for 4 hr or at 4° overnight. Centrifuge for 15 min, and resuspend the plasmid DNA pellet in 40 μl of 0.3 *M* NaOAc followed by addition of 100 μl of ethanol. Centrifuge and wash the precipitate once with ethanol. Dry the plasmid DNA pellet and resuspend in 20 μl of TE buffer. The DNA concentration is usually around 1 μg/μl. Usually the DNA sample is pure enough for DNA sequencing using any one of three methods (procedures 7, 8, and 10).[3,4,6]

DNA Sequencing

There are three methods for sequencing a DNA fragment inserted in pWR2 without prior isolation and purification of the DNA fragment. They include the exonuclease III method of Guo and Wu,[3] the primer extension method of Sanger *et al.*,[4] and the chemical method of Maxam and Gilbert.[6]

One limitation with Sanger's method[4] is the requirement of single-stranded DNA as a template for hybridization with the added primer. One can overcome this limitation by cloning the DNA into a M13 phage[8] and isolate single-stranded DNA. One can use an alternative method[18] for sequencing double-stranded DNA by first linearizing a plasmid DNA with a restriction enzyme at a site far away from the DNA insert to be sequenced. The DNA is then heated at 100° for 3 min in the presence of a synthetic primer, followed by quenching the mixture to 0°. Subsequent steps are the same as the chain terminator procedure for DNA sequencing.[4]

Two universal primers (16-mers) have been designed and are being synthesized for sequencing any DNA fragments cloned into the polylinker region of pWR2. Primer I has the sequence of 5′ d(A-C-C-A-T-G-A-T-T-A-C-G-A-A-T-T), which can bind to the region between nucleotides 149 and 164 (see Fig. 8) and be extended rightward to sequence one strand of the cloned gene (see Fig. 6). Primer II has the sequence of 5′ d(C-A-C-G-A-C-G-T-T-G-T-A-A-A-A-C), which is complementary to nucleotides 221–206 and can bind to pWR and be extended leftward to sequence the other strand of the cloned gene. Primer II can be replaced by the 19-mer synthetic primer, 5′ d(T-T-G-T-A-A-A-A-C-G-A-C-G-G-C-C-A-G-T), for sequencing DNA cloned into M13mp2, mWJ22, or mWJ43.[19]

One potential problem with Maxam and Gilbert's method[6] is that a double-stranded DNA sometimes cannot be sequenced to give clean gel patterns because of breaks or gaps in the DNA. In these cases, it is best to separate the two strands of DNA prior to carrying out base-specific chemical cleavage reaction. Since strand separation procedure is time-consuming, we have made an improvement on the method of Rüther *et al.*[5] to overcome this difficulty. The 5′ ends of a plasmid linearized in the polylinker are first labeled with polynucleotide kinase and [γ-^{32}P]ATP followed by cutting at another restriction site in the polylinker region to produce a very short and a very long fragment, each labeled at the polylinker end. The single-end labeled fragments are digested with *E. coli* exonuclease III to destroy the very short piece and convert part of the long piece into single-stranded DNA before degradation by base-specific

[18] R. B. Wallace, M. J. Johnson, S. V. Suggs, Ken-ichi Miyoshi, R. Bhatt, and K. Itakura, *Gene* **16,** 21 (1981).

[19] R. Wu, L. Lau, H. Hsiung, W. Sung, R. Brousseau, and S. A. Narang, *Miami Winter Symp.* **17,** 419 (1980).

chemical reaction. This will give more reproducible sequencing results than the method employing double-stranded fragments.

Each of the above two methods can be used to sequence both strands of a short DNA fragment cloned in the polylinker region of pWR2, provided that the DNA is shorter than 400 bp. For sequence analysis of a longer DNA fragment the exonuclease III method[3] can be used, since it can determine the DNA sequences not only from both ends near the sites of cloning, but also from one or several restriction sites located within the cloned DNA (see Table II and Fig. 6). It is very likely that there are one or more restriction site(s) available in the middle of the DNA to be sequenced, and these sites can be used for the second digestion (e.g., X or Y in Fig. 6). The sequence can then be read from the second site within the cloned DNA. It should be pointed out that the second site within the cloned gene need not be a unique restriction site in the recombinant plasmid as long as other sites of the same enzyme are sufficiently distant so that no other labeled fragment falls into the size range of the labeled fragment to be sequenced.[3] If these sites are close to an identical site within the cloned gene to be utilized for the second digestion, Procedure 9 can be used to block one end of the DNA from exonuclease III digestion (see Fig. 7). If a site within the cloned gene is unique on the recombinant plasmid, the site can serve either as the second site or the first site.

All three sequencing methods can be used for sequence analysis of a gene cloned into pWR2. The primer method is the most simple and rapid for short DNA fragments (below 400 bp), but the exonuclease III method is the most useful, since it can determine sequences of both short and long DNA fragments (up to 2000 long) cloned into the plasmid. The major advantage is that a long DNA need not be fragmented as much for cloning and sequence analysis. For example, a 5000 bp DNA can be digested to give 3 or 4 fragments and each one cloned into pWR2 for complete sequence analysis by the exonuclease III method. In contrast, in the M13 method, a 5000 bp DNA needs to be digested to give about 15 fragments, and each of these needs to be cloned into M13mp8 or M13mp9 in both orientations to make a total of 30 clones. After clone selection and sequencing, additional effort is needed to line up the 15 fragments to give the linear sequence.

One requirement for the success of the exonuclease III method is the quality of restriction enzymes for the second digestion, which must be free of contaminating enzymes, so that the 5′ ends will remain intact. The restriction enzyme can be tested for contaminating enzymes according to the method described in Procedure 8. After *Pvu*II and exonuclease III digestion of pWR2 DNA followed by labeling with DNA polymerase I and second digestion with *Eco*RI, the reaction mixture is heated at 70° for

10 min to terminate the reaction. One portion of the sample will be saved as the control, the other portion will be incubated with the restriction enzyme to be tested at 37° for 30 min. After electrophoresis, if the gel pattern of the 92 bp long *Eco*RI-*Pvu*II fragment in the test sample is as clear as the control, the restriction enzyme is considered to be pure and useful for sequence analysis.

Procedure 7. Primer Method[4,18] *for DNA Sequencing.* A recombinant plasmid DNA can be first linearized using one of several sites (e.g., *Pvu*II) for sequencing the inserted DNA from primer I or primer II (see Fig. 6). The restriction enzyme digestion of plasmid DNA is the same as in Procedure 1, except that 0.4 pmol of DNA instead of 3 μg is used and incubated at 37° for 15–30 min. After incubation, add 1 μl of synthetic (16-mer) primer I or II (8–10 pmol). Transfer the mixture into a capillary, seal, heat at 100° for 3 min, and then quench at 0° for several minutes. Transfer the mixture into an Eppendorf tube containing [α-^{32}P]dATP (8 μCi, 410 Ci/mmol) that has been dried down. Add 3 μl of 1 × RE buffer into the tube. After vortexing and centrifuging, pipette aliquots of 3 μl into 4 Eppendorf tubes marked A, G, C, and T, respectively. To each tube add 1 μl of the appropriate ddNTP–dNTP mix (see Table III) and 1 μl of DNA polymerase (Klenow fragment 0.2–0.6 unit). Incubate at 23° for 10 min. To each reaction mixture, add 1 μl of 0.5 m*M* dATP (for chase) and incubate at 23° for an additional 10 min. Stop the reactions by 1 μl of 1 *N* NaOH–0.1 *M* EDTAand 4 μl of 10 *M* urea containing 0.3% bromophenol blue and xylene cyanole dyes. After standing at room temperature (23°) for 5 min, load the samples on a sequencing gel.[3,4,6] Alternatively, stop the reactions by ethanol precipitation as described in procedure 8(c).

Procedure 8. Exonuclease III Method for DNA Sequencing. There are two options for exonuclease III sequencing methods.[3] Here we describe only Method II using chain terminator. The principle of this method for

TABLE III
COMPOSITION OF ddNTP–dNTP MIXTURE[a]

Mix	ddATP (m*M*)	ddGTP (m*M*)	ddCTP (m*M*)	ddTTP (m*M*)	dATP (μ*M*)	dGTP (μ*M*)	dCTP (μ*M*)	dTTP (μ*M*)
ddATP–dNTP	0.6				2.5	500	500	500
ddGTP–dNTP		2			1	40	500	500
ddCTP–dNTP			2		1	500	40	500
ddTTP–dNTP				4	1	500	500	40

[a] Dideoxy- and deoxyribonucleoside triphosphate are dissolved in 1× RE buffer (50 m*M* Tris-HCl, pH 7.6, 50 m*M* KCl, 10 m*M* $MgCl_2$, 10 m*M* dithiothreitol). The mixtures can be stored at −20° and used for at least 3 months.

DNA sequencing is illustrated in Fig. 5. The restriction sites available in pWR2 for the exonuclease III sequencing method are tabulated in Table II.

To sequence a long DNA (e.g., 1000 bp) cloned in the polylinker region of pWR2, Fig. 6 gives a feasible strategy. Both X and Y sites are

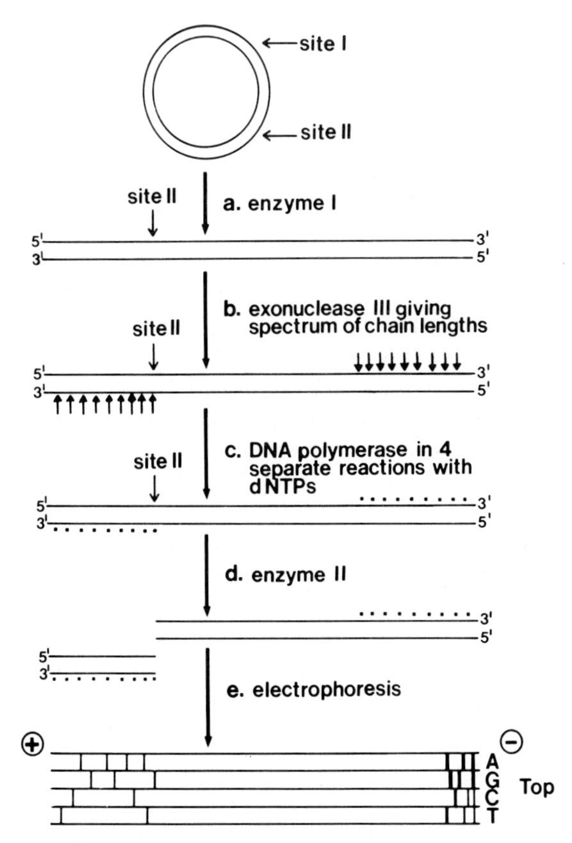

FIG. 5. The principle of the exonuclease III method for sequencing DNA. A fragment of DNA to be sequenced is cloned into a plasmid (represented by a double circle) between restriction sites I and II. In this example, the plasmid is first digested with restriction enzyme I to give a linear DNA. The digestion by exonuclease III (step b) gave a family of molecules with 3′ ends shortened. These molecules are an ideal template-primer system in which the shortened strands with 3′ ends serve as the primers. The digested DNA in step b is distributed in four tubes, and the 3′ ends of the DNA are extended (step C) using the chain termination method.[4] After digestion with a second restriction enzmye (step d), the DNA fragments are fractionated on a denaturing polyacrylamide gel (step e). The shorter fragments are well separated on the lower part of the gel (left-hand side of gel), and the longer fragments are retained near the top of the gel. Four lanes represent DNA fragments terminated with each of the four different ddNTPs.

available for sequencing both strands of DNA. If not all the requirements listed in the table for Fig. 6 can be met, one end of the restricted DNA can be blocked and the other end digested with exonuclease III, according to the strategy shown in Fig. 7 and described in Procedure 9. If X or Y site is unique in the recombinant plasmid, it may also serve for the first digestion to linearize the plasmid DNA and to obtain sequence information on both sides of these sites.

It is easy to determine the restriction sites in the cloned fragment by digesting it (either after gel separation or in the intact plasmid) with several restriction enzymes and running gel electrophoresis. The number of sites and the approximate size of fragments produced by the restriction enzymes are important, but an exact physical map is not necessary.

a. Digestion of plasmid DNA with a restriction enzyme. The mixture for the restriction enzyme digestion of plasmid DNA is the same as in Procedure 1, except that 1 pmol of DNA instead of 3 μg is used. If there is a need to remove 3′ protruding ends produced by restriction-enzyme digestion (3′ protruding ends cannot be efficiently cut by exonuclease III), 1–2 units of DNA polymerase (Klenow enzyme) are added to the same reaction mixture and incubated at 37° for 15 min after completion of the restriction enzyme digestion. The mixture is heated at 70° for 10 min.

b. Digestion of DNA with exonuclease III. To the above mixture add 4 μl of 10 × exonuclease III buffer, 26 μl of H_2O, and 1–4 μl of exonuclease III (see Table IV). It is advisable that the exonuclease III digestion be carried out beyond the second restriction site (i.e., the incubation time should be about 10 min longer than calculated). If the second site to be cut with a restriction enzyme is a unique site in the plasmid DNA, it is not necessary to strictly control the exonuclease III digestion. After completion of the exonuclease III digestion, add 10 μl 0.1 *M* EDTA–1.5 *M* NaOAc and extract the mixture once with 50 μl of phenol saturated with 1 *M* Tris-HCl (pH 8) and once with chloroform–isoamyl alcohol (24 : 1). Transfer the aqueous phase into another tube and precipitate the DNA by adding 130 μl of ethanol. Chill for 5 min at −70° and centrifuge at 0° for 5 min. Remove the supernatant and add 200 μl of ethanol to rinse the DNA pellet. Centrifuge for 3 min and remove the supernatant. Dry the DNA pellet under vacuum for 5 min.

c. Labeling of partially digested DNA and second restriction enzyme digestion. Resuspend the DNA pellet in 18 μl of H_2O and 2 μl of 10 × RE buffer, and heat at 70° for 5 min followed by a brief centrifugation (sometimes the heating may be omitted). Pipette 14 μl into a tube containing [α-^{32}P]dATP (8 μCi, 410 Ci/mmol) that has been dried down. Pipette aliquots of 3 μl into four tubes marked A, G, C, and T, respectively. Add to each tube 1 μl of the appropriate ddNTP–dNTP mix (see Table III) and

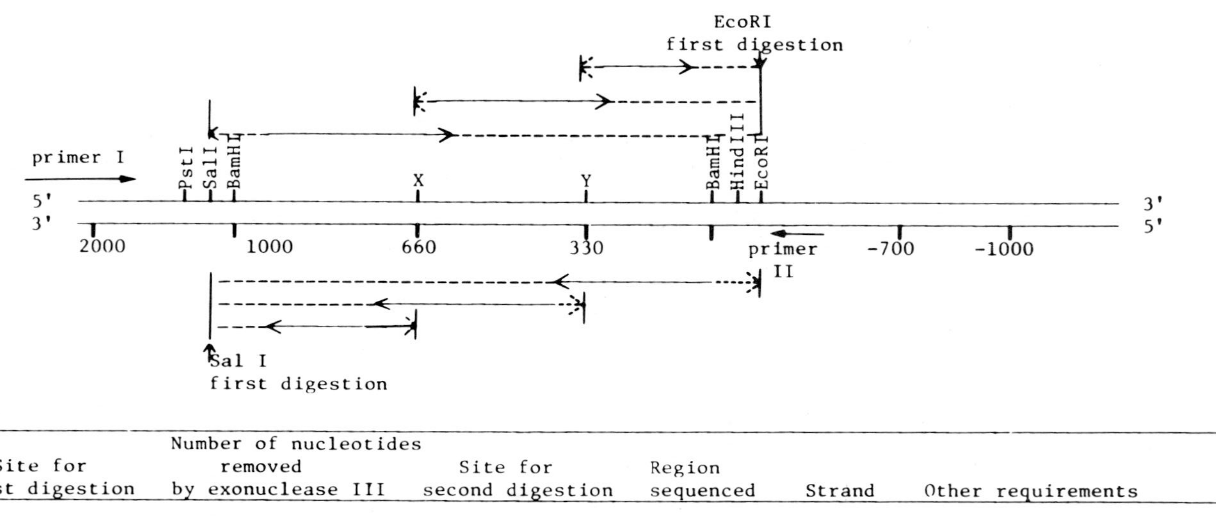

Site for first digestion	Number of nucleotides removed by exonuclease III	Site for second digestion	Region sequenced	Strand	Other requirements
EcoRI	1100	SalI	1000–600	upper	No EcoRI and SalI sites within the inserted DNA
EcoRI	700	X	600–300	upper	No other X site between nucleotides -1000 and +1000
EcoRI	400	Y	300–1	upper	No other Y site between nucleotide -700 and +700
SalI	1100	EcoRI	1–400	lower	No SalI and EcoRI sites within the inserted DNA
SalI	700	Y	350–700	lower	No other Y site between nucleotides +2100 and -100
SalI	400	X	600–1000	lower	No other X site between nucleotides +1800 and +200

TABLE IV
DIGESTION OF DNA WITH EXONUCLEASE III[a]

Number of nucleotides removed from each end of DNA	Incubation (min)	Exo III (U/pmol DNA)
100–250	10–25	15–20
250–500	25–50	20–25
500–750	50–75	25–30
750–1000	75–100	30–35
1000–1500	100–150	35–45

[a] To remove approximately 10 nucleotides per minute from each 3′ end of DNA, the concentration of *Escherichia coli* exonuclease III (ExoIII) needed is given in the table. A unit of exonuclease III is the amount of enzyme that liberates 1 nmol of mononucleotides from a sonicated DNA substrate in 30 min at 37°. BRL ExoIII, Lot 2429, was used.

1 μl of DNA polymerase (Klenow enzyme) (0.2–0.6 unit). Incubate at 37° or 23° for 10 min and then chase by adding 1 μl of 0.5 m*M* dATP and incubate for 5 min. To each tube add 1 μl of a restriction enzyme (1–2 units) and incubate at 37° for 10 min. Stop the reactions by addition of 1 μl of 1 *N* NaOH–0.1 *M* EDTA and 4 μl of 10 *M* urea containing 0.3% bromophenol blue and xylene cyanole dyes. Alternatively, stop the reactions by the addition of 1 μl of tRNA (10 μg/μl), 3 μl of 0.1 *M* EDTA–1.5 *M* NaOAc, and 25 μl of ethanol. After chilling and centrifugation, resuspend the labeled DNA pellets in 3 μl of 0.2 *N* NaOH–10 m*M* EDTA and 3 μl of 10 *M* urea containing 0.3% dyes. Let the preparation stand at room temperature (23°) for 5 min, then load the samples on a sequencing gel.[3,4,6]

Procedure 9. Methods to Digest and Sequence Only a Selected End of a Double-Stranded DNA Using Exonuclease III. Escherichia coli exonuclease III can digest a linear duplex DNA from the 3′ ends of both strands (see Fig. 5). DNA molecules with 3′ recessed or blunt ends are good substrates for exonuclease III digestion, but DNA with 3′ protruding ends are not efficiently digested. If the 3′ protruding end is long (e.g., >20

FIG. 6. A strategy for sequencing a long DNA fragment cloned in the polylinker region of pWR2 by using the exonuclease III method. The upper part of this figure shows a DNA fragment 1000 base pairs in length cloned in *Bam*HI site of the polylinker region. There are two restriction sites (X and Y) within the insert. If the restriction sites X and Y meet all the requirements listed in the table for this figure, both strands of the inserted DNA can be sequenced using the exonuclease III method without necessity of isolation and strand separation of DNA. Dashed arrows show directions of exonuclease III digestion of an inserted DNA; solid arrows show directions of DNA sequencing.

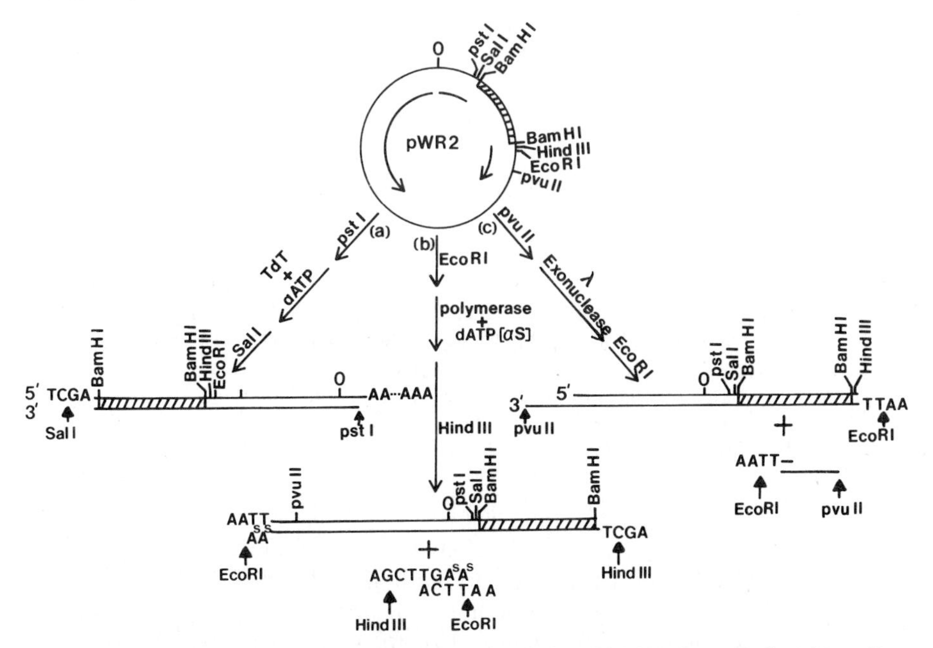

FIG. 7. Three methods for blocking one end of plasmid DNA from *Escherichia coli* exonuclease III digestion. For sequencing a DNA fragment, shown with hatched lines, which is cloned into the *Bam*HI site in the polylinker region of pWR2, one end of the linearized DNA can be blocked in one of three ways: (*a*) tailing one end of DNA with calf thymus terminal transferase (TdT); (*b*) introducing an α-thio nucleotide into one end of DNA with DNA polymerase I; (*c*) producing a long protruding 3′ end of DNA with λ exonuclease.

nucleotides), it would be resistant to exonuclease III digestion. Therefore a DNA with a recessed or blunt end at one end and a long 3′ single strand at the other end can be digested only at one end (e.g., an inserted DNA fragment shown with hatched lines in Fig. 7 can be digested with exonuclease III only from one 3′ end). Another strategy for blocking the 3′ end at one end of a DNA fragment is to introduce a 2′-deoxynucleoside 5′-*O*-(1-thiophosphate)[20] into one of the 3′ ends of DNA fragment using *E. coli* DNA polymerase I (Fig. 7). This modified 3′ end is resistant to exonuclease III digestion.[21] The application of these procedures to DNA sequencing is shown in Fig. 7. The procedures are also useful for making

[20] T. A. Kunkel, F. Eckstein, A. S. Mildvan, R. M. Koplitz, and L. A. Loeb, *Proc. Natl. Acad. Sci. U.S.A.* **78,** 6734 (1981).

[21] S. D. Putney, S. J. Benkovic, and P. R. Schimmel, *Proc. Natl. Acad. Sci. U.S.A.* **78,** 7350 (1981).

asymmetrical deletion of a stretch of DNA (see the section on application of plasmid pWR for expression of cloned genes).

Procedure 9a. Addition of a poly(dA) tail to 3′ ends of a DNA fragment (*Fig. 7*). Digestion of a plasmid DNA such as pWR2 with restriction enzyme *Pst*I is the same as in Procedure 1, except that 1 pmol of DNA instead of 3 μg is used. After the digestion, add 3 μl of 0.1 *M* EDTA–1.5 *M* NaOAc and 35 μl of ethanol to the reaction mixture, chill at −70° for 5 min, and centrifuge for 5 min. The DNA pellet is rinsed with 50 μl of ethanol, centrifuged for 3 min, and dried down under vacuum for 5 min. The DNA pellet is resuspended in 2 μl of 10× terminal transferase buffer, 1 μl of 1 m*M* dATP, terminal deoxynucleotidyltransferase (10 units), and distilled water is added to a final volume of 20 μl.[22] Incubate at 30° for 30 min, and add 10 μl of 0.1 *M* EDTA–1.5 *M* NaOAc and 30 μl of TE buffer. Extract the mixture once with phenol saturated with 1 *M* Tris-HCl (pH 8) and once with chloroform–isoamyl alcohol (24 : 1, v/v). After ethanol precipitation, washing, and drying, digest the tailed DNA with a restriction enzyme followed by exonuclease III. Extend and label the 3′ ends using the chain terminator procedure, and digest with a second restriction enzyme according to Procedure 8.

Procedure 9b. Introduction of an α-phosphorothioate nucleotide into one end of DNA to block that end from exonuclease III digestion (*Fig. 7*). Digestion of a plasmid DNA such as pWR2 with *Eco*RI is the same as in Procedure 1, except that 1 pmol instead of 3 μg of DNA is used. After digestion, add 12–24 pmol of dATP [α-S], 1 μl of 1 m*M* dTTP and 1 μl of DNA polymerase I (Klenow enzyme, 0.5–1 unit) for repair of *Eco*RI-digested ends. Reverse transcriptase may be used to replace Klenow enzyme for repair of restriction enzyme-digested ends. The final volume of the reaction mixture is 12–14 μl, and the incubation is carried out at 37° for 15 min. Add 5 μl of 0.1 *M* EDTA–1.5 *M* NaOAc. After ethanol precipitation, washing, and drying, digest the linear plasmid DNA with its modified 3′ ends with *Hin*dIII to produce two fragments, each with a modified 3′ end. This modified end is resistant to exonuclease III digestion, but the other 3′ end can be digested and then labeled using the chain-terminator procedure.

Procedure 9c. Producing long 3′ protruding ends with λ exonuclease (*Fig. 7*). Digestion of a plasmid DNA such as pWR2 with *Pvu*II enzyme is the same as in Procedure 1, except that 1 pmol instead of 3 μg of DNA is used. After *Pvu*II digestion, add 5 μl of 10× λ buffer, 35 μl of H_2O, and 2 units of λ exonuclease. Incubate at 6° or 23° for 30 min. Under these conditions, λ exonuclease removes 50–60 nucleotides (but no more) from

[22] G. Deng and R. Wu, *Nucleic Acids Res.* **9,** 4173 (1981).

each 5′ end of a DNA fragment. After digestion, add 10 μl of 0.1 *M* EDTA–1.5 *M* NaOAc to the reaction mixture and extract once with an equal volume of phenol and chloroform–isoamyl alcohol, respectively. After ethanol precipitation, washing, and drying, digest the DNA with long 3′ protruding tails with *Eco*RI or other restriction enzyme to produce two fragments, each with only one 5′ protruding or even end. This end alone can be digested by exonuclease III and labeled using the chain termination procedure.

Some Problems and Solutions for Use of the Exonuclease III Method

The rate of exonuclease III digestion can be controlled conveniently and averages about 10 nucleotides per minute at each end of DNA at 23° if the salt concentration in the digestion solution is 90 m*M* and the appropriate ratio of exonuclease III (units) to DNA (pmol) is used. The rate of exonuclease III digestion of DNA falls off with time; therefore, for removal of a large number of nucleotides the ratio of the enzyme to DNA must be increased for compensation (Table IV). In order to control the salt concentration during exonuclease III digestion, the salt used for the restriction enzyme cleavage of plasmid DNA must be taken into consideration. The rate of exonuclease III digestion was 10, 7.5, or 5 nucleotides per minute at 90 m*M*, 105 m*M*, or 125 m*M* salt, respectively.

We found that labeled oligonucleotides shorter than 15 residues usually give rise to very weak bands, even though exonuclease III digestion goes beyond the site to be cut by the second restriction enzyme. This may be due to the possibility that, after the digestion with the second enzyme, the short oligonucleotides dissociates from the template and is rapidly degraded.[23]

The purity of the restriction enzyme used for the second digestion is especially important because the common 5′ end of DNA fragments to be separated on the gel is produced by the enzyme. Contamination by exonucleases or other endonucleases would give extraneous bands in the gel pattern. Restriction enzymes that recognize six nucleotide sequences seem better than those that recognize four nucleotide sequences. Occasionally, there are some extraneous bands in one lane of gel pattern, especially in the A reaction lane when labeled dATP was used, but not in the other three lanes. The selection of another restriction enzyme for the second digestion may solve the problem.

It is recommended that the exonuclease III digestion be carried out beyond the site to be cut with the second restriction enzyme, by increas-

[23] A. J. H. Smith, this series, Vol. 65, p. 560.

ing the incubation time by about 10 min. Otherwise, some extraneous bands in the gel may appear, especially in the region for short oligonucleotides.

Other possible problems, causes, and solutions are tabulated in Table V. Any ambiguity in sequence analysis of one strand of DNA often can be resolved by determining the complementary sequence in the same region or by using different DNA sequencing methods described in this paper. As a rule, to be assured of absolute reliability, the sequence of both strands of a DNA molecule must be determined.

Procedure 10. Chemical Method for DNA Sequencing

a. Labeling plasmid DNA at the 5′ ends. The restriction sites available in pWR2 for chemical sequencing of DNA are tabulated in Table II. Plasmid DNA linearized according to Procedure 1 is labeled at the 5′ end by polynucleotide kinase and [γ-^{32}P]ATP as in the method of Maxam and Gilbert,[6] except that calf intestinal alkaline phosphatase (1 unit per 1 pmol of DNA) is used to remove the terminal phosphates and incubation is carried out at 60° for 30 min. After labeling the 5′ ends, the digestion by a second restriction enzyme produces two fragments each labeled only at one end (a very small and a very large fragment). It may be of advantage to digest the DNA by exonuclease III to produce single strands (Procedure 8a and b) before carrying out the base-specific chemical cleavage reactions.

b. Labeling plasmid DNA at 3′ recessive ends.[24] Plasmid DNA is linearized with a restriction enzyme according to Procedure 1. After digestion, transfer the reaction mixture into another Eppendorf tube containing [α-^{32}P]dNTP (5–10 μCi, 410 Ci/mmol) that has been dried down. After resuspension, add 1 μl of DNA polymerase (Klenow enzyme, 0.5–1 unit), and incubate at 23° for 10 min. For labeling *Eco*RI-digested ends of DNA, add 1 μl of 1 m*M* dATP and continue the incubation for 5 min. Heat the reaction mixture at 70° for 10 min to inactivate the polymerase, add 6–12 units of another restriction enzyme, and incubate at 37° for 30–60 min. Add 10 μl of 0.1 *M* EDTA–1.5 *M* NaOAc and 30 μl of TE buffer to stop the reaction. Extract DNA once with phenol saturated with 1 *M* Tris-HCl (pH 8) and once with chloroform–isoamyl alcohol (24:1, v/v). To the aqueous phase add 125 μl of ethanol, chill at −70° for 5 min and centrifuge for 5 min. Wash the DNA pellet with 200 μl of ethanol and dry under vacuum.

Base-specific chemical cleavage reactions are carried out according to the method of Maxam and Gilbert.[6]

[24] R. Wu, *J. Mol. Biol.* **51,** 501 (1970).

TABLE V
DIAGNOSIS AND CORRECTION OF PROBLEMS

Problem	Probable cause	Solution
A high background of gel bands	(a) Contamination of chromosomal DNA in plasmid DNA sample (b) A lot of nicks in plasmid DNA (a high concentration of form II DNA) (c) Plasmid DNA preparation is contaminated with DNase.	(a), (b), and (c) Further purify plasmid DNA by phenol and chloroform extraction, by adding NH_4OAc to 3 M and reprecipitate[17] with 2 volumes of EtOH, by agarose gel electrophoresis, or by CsCl banding.
	(d) The restriction enzyme for the first digestion of DNA is contaminated with DNase, especially if a large excess of enzyme is used.	(d) Digest plasmid DNA with a purer enzyme or a different restriction enzyme.
	(e) Labeled [α-^{32}P]dNTP is dirty.	(e) Get a clean [α-^{32}P]dNTP.
Band intensity in upper part of the gel is stronger than that in the lower part of the gel	(a) The incubation time of exonuclease digestion is too short.	(a) Carry out a longer incubation.
	(b) The salt concentration of exnuclease III digestion solution is too high.	(b) Dilute the salt concentration (K^+ and Na^+ ions) to 90 mM.
	(c) The specific activity of exonuclease III is lower than expected.	(c) Increase the amount of enzyme.
	(d) The ratio of ddNTP to dNTP is too low.	(d) Adjust the ratio of ddNTP to dNTP.
A band pattern in which every band appears as a doublet or triplet	(a) The second restriction enzyme has a contaminating exonuclease activity.	(a) The second restriction enzyme digestion should be carried out with a purer enzyme or incubation time should be shortened.
	(b) DNA polymerase had not been inactivated completely.	(b) After repair labeling, heat at 70° for a longer time prior to the second restriction enzyme digestion.

Extraneous gel bands or bands across all the lanes of the gel in a particular region of the sequence	(a) If the primers generated by exonuclease III digestion are capable of forming some internal secondary structure as a self-priming template, it may be labeled during the priming reaction and appears as extra bands after the second restruction enzyme cut. (b) The first or second restriction enzyme(s) may be contaminated with a double strand specific endonuclease. (c) The band(s) may be due to a pileup of DNA polymerase at a particular sequence.	(a) Heat the sample at 70° for 5 min prior to addition of cold and hot dNTPs mix and DNA polymerase; or the incubation of exonuclease III digestion should be carried out for a longer time. Using a different lot of DNA polymerase may help. (b) Use a new batch of restriction enzyme(s). (c) Use the chemical method of sequencing[6] for this portion of the sequence.
Band patterns obscure and superimposed upon one another	(a) The plasmid DNA to be sequenced may be contaminated with other DNA. (b) The base composition of the template DNA has a very asymmetric G : C ratio. (c) The rate of exonuclease III digestion of DNA is too rapid, so that the digestion is not synchronous. The primers generated in this way are not good for priming.	(a) Further purify the plasmid DNA. (b) Adjust the ratio of dideoxy- to deoxynucleoside triphosphates. (c) Lower the rate of ExoIII digestion by increasing the salt (K^+ or Na^+) concentration to 90 m*M* or decreasing the amount of ExoIII.
A sudden compression or increase in the spacing of bands, then returns to the normal band spacing	Due to the formation of secondary structure that moves abnormally during electrophoresis.	Run the gel at higher voltages or use dITP instead of dGTP for incorporation.
A large bubble across all the lanes in the upper part of the gel	The salt concentration of samples is too high.	Precipitate the DNA with ethanol prior to loading the gel.

Labeling a DNA Fragment Cloned in pWR2 as Probe for Molecular Hybridization

A DNA fragment cloned in pWR2 or its derivatives can serve as a probe for molecular hybridization. There are two methods for the synthesis of radioactive labeled probes with specific activities of at least 10^8 cpm per microgram of DNA. Method I uses primer I or primer II and extends it to get single-stranded labeled DNA (see Procedure 7). Single-stranded probes are free from interference by the complementary strand, and thus are more efficient in hybridizing to the target DNA. Method II makes double-stranded labeled DNA in which only one strand is labeled. The principle is the same as the exonuclease III method for DNA sequencing (see Fig. 5 and Procedure 8). These methods can produce radioactive probes with higher specific activity than those obtained by the method of nick translation.

Procedure 11. Method I for Labeling DNA. Linearize the recombinant plasmid DNA, add synthetic primer, heat, and quench the restriction enzyme digest as in Procedure 7. Transfer the DNA mixture into an Eppendorf tube containing [α-^{32}P]dNTP (50–200 μCi, 410 Ci/mmol) that has been dried. After resuspension, add 1 μl of 1 mM each of three other cold dNTPs and 1 μl of DNA polymerase I (Klenow enzyme, 0.5–1 unit). Incubate at 23° for 15 min and add 1 μl of 1 mM cold dNTP corresponding to the labeled dNTP; continue the incubation at 23° for 5 min. Stop the reaction by addition of 5 μl of 0.1 M EDTA–1.5 M NaOAc. After ethanol precipitation, washing, and drying, the labeled DNA can be used as a hybridization probe.

Procedure 12. Method II for Labeling DNA. Digest the recombinant plasmid DNA (1 μg) in RE buffer with 2–4 units of a restriction enzyme as in Procedure 1. Incubate at 37° for 30–60 min. To the mixture (10 μl), add 1 μl of 0.1 M KCl and the appropriate amount of exonuclease III, and incubate at 23° for appropriate time (see Table IV). Stop the reaction by heating the sample at 80° for 10 min. Transfer the sample into an Eppendorf tube containing [α-^{32}P]dNTP (50–200 μCi), which has been dried down. After resuspension, add 1 μl of 1 mM each of three other cold dNTPs and 1 μl of DNA polymerase I (Klenow enzyme, 0.5–1 unit). Incubate at 37° for 10 min, and add 1 μl of 1 mM cold dNTP corresponding to the radioactive nucleotide and 1 μl of a second restriction enzyme (2–4 units). Incubate at 37° for 15 min, and stop reaction by addition of 5 μl of 0.1 M EDTA–1.5 M NaOAc. After ethanol precipitation, washing and drying, the labeled DNA fragment (one strand labeled) is electrophoresized in 1% low melting point agarose as in Procedure 2 or regular agarose. If using regular agarose for isolation of labeled DNA frag-

TABLE VI
EXPRESSION OF A GENE CLONED IN pWR2

Number of base pairs preceding the start codon within a gene to be cloned in pWR2	Sites in which a gene may be cloned for expression
$3n$	Filled-in *Acc*I
	S1 nuclease-treated *Eco*RI
	S1 nuclease-treated *Hin*dIII
	S1 nuclease-treated *Bam*HI
	S1 nuclease-treated *Sal*I
	S1 nuclease-treated *Pst*I
$3n + 1$	*Hin*cII
	*Pvu*II
$3n + 2$	S1 nuclease-treated *Acc*I
	Filled-in *Eco*RI
	Filled-in *Hin*dIII
	Filled-in *Bam*HI
	Filled-in *Sal*I

ment, TA buffer and 60 V (10 mA) are used. The labeled DNA fragment is eluted from the agarose gel by electrophoresis.[25]

Application of Plasmid pWR for Expression of Cloned Genes

If a DNA fragment carrying a protein-coding sequence is inserted into one of the unique restriction sites in the *lacZ'* gene of pWR and the resultant reading frame is in phase with the *lacZ'* gene, a protein fused with β-galactosidase can be produced in bacteria. Owing to the high copy number of the pWR plasmid and the few proteins encoded by the plasmid, the fusion protein can be readily identified by gel electrophoresis. If the inserted gene is not more than about 300 nucleotides long and the reading frame is in phase with the *lacZ'* gene, the colonies in question are light blue, showing that some fused α-peptide of β-galactosidase has been produced. It is easy to recognize and select these colonies in which the cloned gene is expressed as a protein fused to the α-peptide. The sites in β-galactosidase gene of pWR2 for insertion of blunt-ended DNA fragments to produce fused proteins are shown in Table VI.

For a gene cloned in the polylinker region, if the product of a single protein rather than a fused protein is desired, the stretch of nucleotides between the Shine–Dalgaro (SD) sequence of α-peptide and the polylinker (see Fig. 8), or between the SD sequence and the start codon (ATG)

[25] R. C.-A. Yang, J. Lis, and R. Wu, this series, Vol. 68, p. 176.

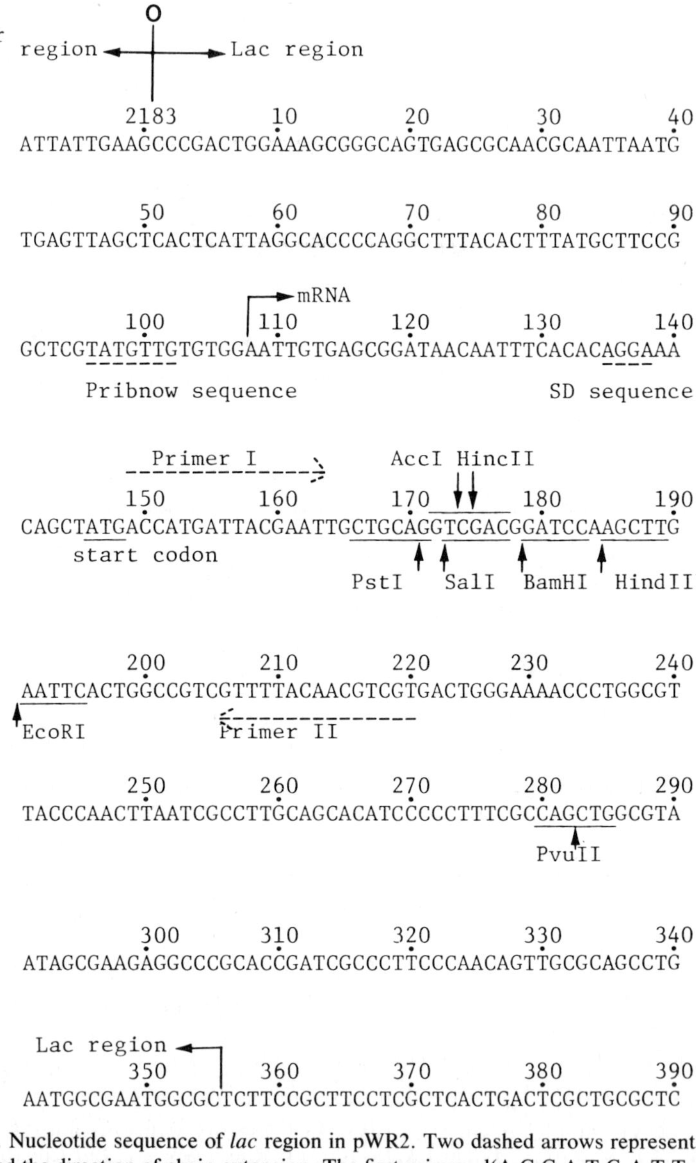

FIG. 8. Nucleotide sequence of *lac* region in pWR2. Two dashed arrows represent two primers and the direction of chain extension. The first primer, d(A-C-C-A-T-G-A-T-T-A-C-G-A-A-T-T), can bind to the region between nucleotides 149 and 164, and the second primer, d(C-A-C-G-A-C-G-T-T-G-T-A-A-A-A-C), is complementary to nucleotides 221–206. Each primer can be used to determine the sequence of a specific strand of a DNA fragment cloned in the polylinker region.

of the cloned gene, may be deleted by use of Procedure 13 or 14. For the deletion of a stretch of nucleotides between the SD region and the polylinker (for example, a gene cloned in the *Eco*RI site in pWR2), the recombinant plasmid DNA is linearized by *Bam*HI digestion, followed by blocking these ends from exonuclease III digestion using Procedure 9b (introduction of α-phosphorothioate nucleotides into these ends). Then *Pst*I digestion followed by exonuclease III treatment is carried out by using Procedures 8a and 8b to remove the stretch of nucleotides between positions 140 and 170 (Fig. 8). The resultant plasmid DNA is digested with *Hin*dIII and treated with S_1 nuclease using Procedure 13. The plasmid DNA is next self-ligated using Procedure 3. For the deletion of a stretch of nucleotides between the SD region and the start codon of the cloned gene, the recombinant plasmid DNA is first linearized with a restriction enzyme (*Pst*I or *Sal*I or *Bam*HI or *Hin*dIII) followed by the use of Procedure 14 (*Bal*31).

Deletion of a Specific Region of DNA

Deletion of a stretch of nucleotides from a plasmid DNA with exonuclease III and S_1 nuclease using Procedures 8a, 8b, and 13 usually is more easily controlled than with *Bal*31 nuclease (Procedure 14). *Escherichia coli* exonuclease III digests DNA from the 3′ ends with almost no base specificity. Under appropriate conditions of salt and enzyme concentration, the rate of digestion is about 10 nucleotides per minute at each end of DNA at 23°.[2,3] Exonuclease III digestion followed by S_1 nuclease[2] is useful for making deletions if the objective is to get a family of deletions covering all sizes. On the other hand, the rate of *Bal*31 nuclease digestion depends to a great extent upon the DNA sequence and the type of the terminus. For example, Fig. 9a shows that the patterns of *Bal*31 digestions of *Sma*I- and *Ava*I-treated pYT2 DNA[3] are different. The rate of *Bal*31 digestion of *Ava*I ends is faster than that of *Sma*I ends, although their base composition is almost the same. Figure 9b shows *Bal*31 digestion of *Hha*I-treated pYT2 DNA. The rate of digestion of *Hha*I ends is faster than that of *Ava*I ends. Degradation by *Bal*31 of the G-rich region in the 5′ → 3′ strand of pYT2 DNA is slow, so that G bands on the gel pattern in Fig. 9b are much more intense. The advantage of *Bal*31 is that the rate of digestion is relatively fast, so that it is convenient for deleting a very large stretch of DNA (such as 2000 base pairs). The disadvantage is that not all sizes of DNA are represented in the final digestion mixture. The rate of digestion of an A:T-rich region can be as much as 100-fold greater than that of a G-rich region. The rate of digestion is also sequence specific. We found that among the G-containing dinucleotide sequences, the rate follows the pattern of TG > GT = CG > AG > GC = GA > GG,

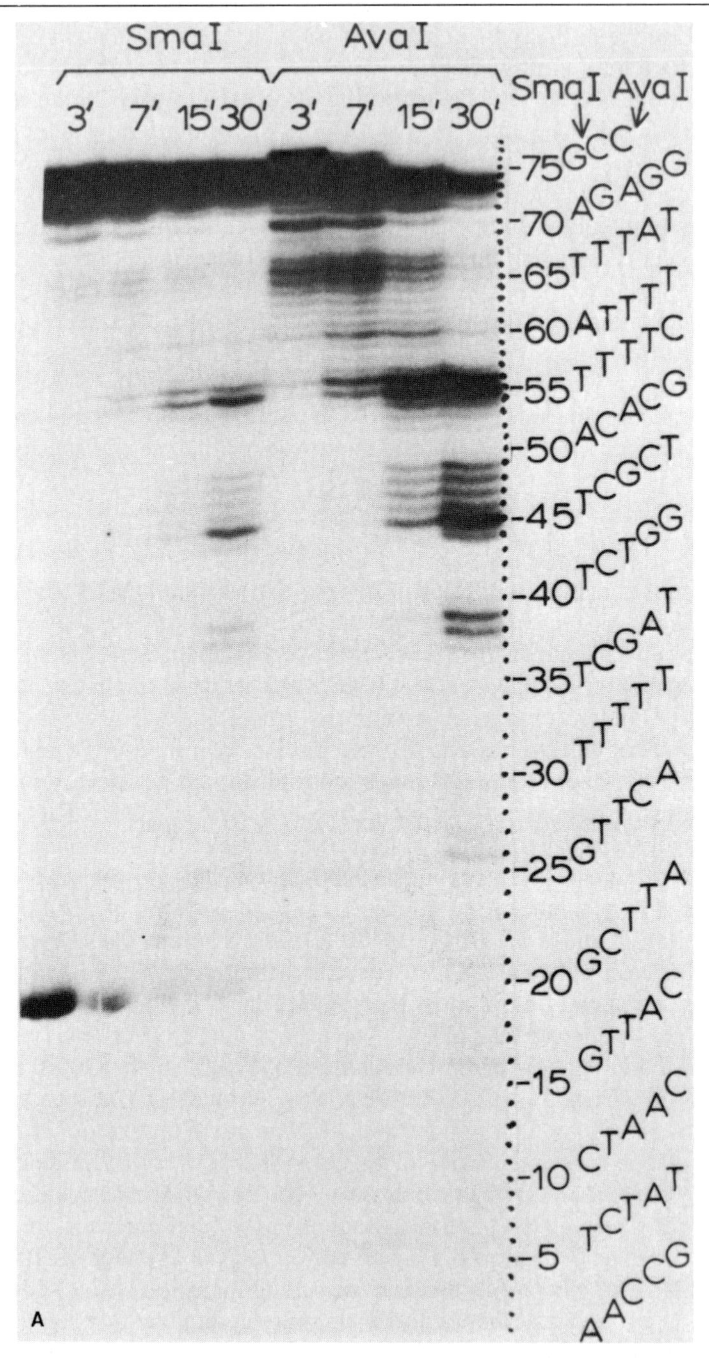

FIG. 9. (A) autoradiogram of *Bal*31 nuclease digestions of *Sma*I- and *Ava*I-treated pYT2 DNA. (B) Autoradiogram of *Bal*31 nuclease digestion of *Hha*I-treated pYT2 DNA.

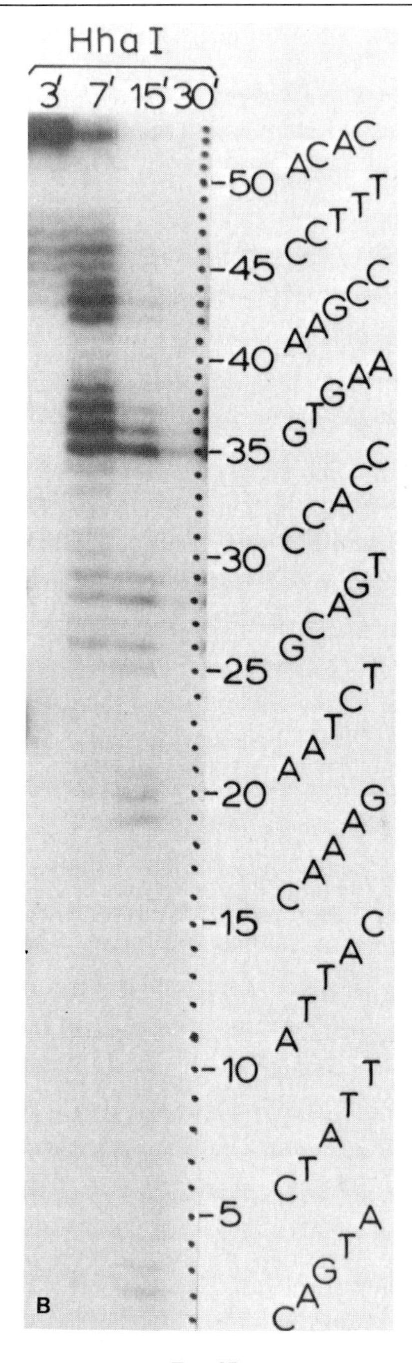

HhaI
3' 7' 15' 30'
-50
-45
-40
-35
-30
-25
-20
-15
-10
-5
ACAC
CCTTT
AAGCC
GTGAA
CCACC
GCAGT
AATCT
CAAAG
ATTAC
CTATT
CAGTA
B

FIG. 9B

and the majority of the DNA termini generated by *Bal*31 are G-C base pairs.

Procedure 13. Deleting a Stretch of Nucleotides with Exonuclease III and S_1 Nuclease. Plasmid DNA is linearized with a restriction enzyme using Procedure 1 and then digested with exonuclease III using Procedures 8a and 8b, but not extracted with phenol and chloroform–isoamyl alcohol. After completion of exonuclease III digestion, immediately add 4.5 μl of 10 × S_1 nuclease buffer (2) and S_1 nuclease (20 units per microgram of DNA). Incubate at 23° for 15 min. Add 10 μl of 0.1 *M* EDTA–1.5 *M* NaOAc and 25 μl of TE buffer, extract the mixture once with 50 μl of phenol and once with 50 μl of chloroform–isoamyl alcohol (24 : 1, v/v). Precipitate the DNA in the aqueous phase with ethanol, rinse DNA pellet once with ethanol and dry under vacuum. The DNA is then self-ligated using Procedure 3.

Procedure 14. Deleting a Stretch of Nucleotides with Bal31 Nuclease. Plasmid DNA is linearized with a restriction enzyme using Procedure 1. To this, add 8 μl of 5 × *Bal*31 buffer, 20 μl of water, *Bal*31 enzyme (1 unit of enzyme per microgram of DNA), and incubate at 23°. Between 5 and 10 bp are removed per minute from each end of *Sma*I-cut DNA or over 200 bp of *Hin*dIII-cut or *Eco*RI-cut DNA. After completion of the *Bal*31 digestion, add 2 μl of 0.5 *M* EDTA to stop the reaction. Extract the mixture with 50 μl of phenol and 50 μl of chloroform–isoamyl alcohol. Precipitate the DNA from the aqueous phase with ethanol, rinse the DNA pellet once with ethanol, and dry under vacuum.

Acknowledgments

We thank Ullrich Rüther for the pUR222 plasmid and the *E. coli* $F^-Z^-\Delta M15recA$. We are grateful to J. Yun Tso and Robert Yang for valuable help and discussion. This work was supported by Research Grant GM29179 from the National Institutes of Health.

[5] Terminal Transferase: Use in the Tailing of DNA and for *in Vitro* Mutagenesis

By GUO-REN DENG and RAY WU

Terminal deoxynucleotidyltransferase catalyzes the addition of deoxynucleotides to the 3′ termini of DNA. Bollum first purified this enzyme from calf thymus, named it calf thymus DNA polymerase,[1] and

[1] F. J. Bollum, *Fed. Proc. Fed. Soc. Exp. Biol. Med.* **17,** 193 (1958).

ISBN 0-12-182000-9

used it to add homopolymer deoxynucleotide tails to denatured DNA.[2] In 1962, this enzyme was recognized as being different from DNA polymerase, and the name was changed to terminal transferase.[3] Lobban and Kaiser[4] and Jackson *et al.*[5] made use of terminal transferase to add homopolymer tails to DNA in preparation for cloning P22 DNA and SV40 DNA fragments, respectively. These are the first examples of the construction of recombinant DNA *in vitro*. By use of this tailing method, any double-stranded DNA fragment can be joined to a cloning vehicle. Since each DNA fragment carries the same type of tail, it cannot hybridize with another molecule of the same species. Thus, after cloning, each transformant should represent the desired recombinant DNA. These advantages make this technique particularly useful for the construction of recombinant DNA molecules for cloning.[6–10]

In this chapter we present an improved procedure for the efficient addition of homopolymer tails to DNA. We also describe a new procedure for the addition of a single nucleotide to the 3′ end of a DNA as a method for *in vitro* mutagenesis, which should prove to be useful in investigating the control and expression of genes.

Materials and Reagents

dATP, dTTP, dCTP, dGTP, ATP, UTP, CTP, and GTP were obtained from P-L Biochemicals, Inc.; [α-^{32}P]dATP, [α-^{32}P]dTTP, [α-^{32}P]dCTP, [α-^{32}P]dGTP (specific activities: 410 Ci/mmol), and [γ-^{32}P]ATP (2000–3000 Ci/mmol) were purchased from Amersham Corporation; [^{35}S]deoxyadenosine 5′-(α-thio)triphosphate (600 Ci/mmol) was from New England Nuclear.

Deoxynucleotide decamer d(A-A-C-T-T-G-A-C-C-C) was synthesized in the laboratory of S. A. Narang, and generously made available to us.

[2] F. J. Bollum, *J. Biol. Chem.* **235,** PC18 (1960).
[3] J. S. Krakow, C. Contsogeorgopoulos, and E. S. Canellakis, *Biochim. Biophys. Acta* **55,** 639 (1962).
[4] P. E. Lobban and A. D. Kaiser, *J. Mol. Biol.* **78,** 453 (1973).
[5] D. A. Jackson, R. H. Symons, and P. Berg, *Proc. Natl. Acad. Sci. U.S.A.* **69,** 2904 (1972).
[6] S. Nakanishi, A. Inoue, T. Kita, S. Numa, A. C. Y. Chang, S. N. Cohen, J. Nunberg, and R. T. Schimke, *Proc. Natl. Acad. Sci. U.S.A.* **75,** 6021 (1978).
[7] L. Villa-Komaroff, A. Efstratiadis, S. Broome, P. Lomedico, R. Tizard, S. P. Naber, W. L. Chiek, and W. Gilbert, *Proc. Natl. Acad. Sci. U.S.A.* **75,** 3727 (1978).
[8] J. G. Seidman, M. H. Edgell, and P. Leder, *Nature (London)* **271,** 582 (1978).
[9] J. T. Wilson, L. B. Wilson, J. K. deRiel, L. Villa-Komaroff, A. E. Efstratiadis, B. G. Forget, and S. M. Weissman, *Nucleic Acids Res.* **5,** 563 (1978).
[10] H. Land, M. Grez, H. Hauser, W. Lindenmaier, and G. Schütz, *Nucleic Acids Res.* **9,** 2251 (1981).

Mercury-bound glass beads were a gift from K. J. O'Brien of New England Nuclear. Sodium cacodylate was purchased from Fisher Scientific Company.

Terminal deoxynucleotidyltransferase was obtained from P-L Biochemicals, Inc. (Milwaukee, Wisconsin). Restriction endonuclease *Bgl*I, *Pvu*II, *Eco*RI, *Hin*dIII, *Bam*HI, *Sal*I, *Hae*III, and *Hin*fI were purchased from New England BioLabs, Inc.; *Pst*I, *Hinc*II, exonuclease III and bacterial alkaline phosphatase were from Bethesda Research Laboratories, Inc. T4 polynucleotide kinase was obtained from Biogenics, Inc. Avian myeloblastosis virus reverse transcriptase was the gift of Dr. J. Beard of Life Science, Inc.

Addition of Deoxynucleotide Homopolymers to 3′ Ends of DNA

DNA molecules with even ends or 3′ recessive ends are inefficient as primers for the terminal transferase-catalyzed tailing reaction. But the efficiency of tailing is increased appreciably when the DNA molecules are treated with λ exonuclease to remove a short stretch of nucleotides from the 5′ end.[4,5] Since the λ-exonuclease reaction was difficult to control, Roychoudhury *et al.*[11] adopted a Co^{2+}-containing buffer that allowed terminal transferase to add tails to all types of 3′ termini of duplex DNA (even end, 3′ protruding end, and 3′ recessive end) without prior λ exonuclease treatment. Brutlag *et al.*[12] and Humphries *et al.*[13] chose a low ionic strength buffer containing Mg^{2+} to add tails. These different conditions for tailing have been reviewed.[14,15]

Using a gel electrophoresis method, we analyzed the efficiency of adding homopolymer tails to duplex DNA with different types of ends and found conditions for adding dG or dC tails of 15–40 nucleotides in length, and dA or dT tails of 30–80 nucleotides to DNA fragments, which are the lengths most suitable for DNA hybridization and cloning.[16] The two different buffer systems used for adding homopolymer tails are the following: (*a*) $CoCl_2$ buffer: 100 m*M* sodium cacodylate (pH 7.0), 1 m*M* $CoCl_2$, 0.2 m*M* dithiothreitol; (*b*) $MnCl_2$ buffer: 100 m*M* sodium cacodylate (pH 7.1), 2 m*M* $MnCl_2$, 0.1 m*M* dithiothreitol. In our experiments, double-stranded DNA fragments with 3′ protruding ends are obtained from *Bgl*I

[11] R. Roychoudhury, E. Jay, and R. Wu, *Nucleic Acids Res.* **3,** 863 (1976).

[12] D. Brutlag, K. Fry, T. Nelson, and P. Hung, *Cell* **10,** 509 (1977).

[13] P. Humphries, R. Old, L. W. Coggins, T. McShane, C. Watson, and J. Paul, *Nucleic Acids Res.* **5,** 905 (1977).

[14] T. Nelson and D. Brutlag, this series, Vol. 68, p. 41.

[15] R. Roychoudhury and R. Wu, this series, Vol. 65, p. 42.

[16] G. Deng and R. Wu, *Nucleic Acids Res.* **9,** 4173 (1981).

digestion of pBR322, fragments with blunt ends from *Hin*cII digestion, and fragments with 3′ recessive ends from *Bam*HI digestion. In order to analyze the tailing efficiency by gel electrophoresis, the fragments were kinase-labeled at the 5′ ends.[17]

Efficiency of Tailing DNA with Different Types of 3′ Ends

Terminal transferase requires single-stranded DNA as primer.[18,19] Thus, the incorporation efficiency is the highest for DNA with 3′ protruding ends.[4,5]

The DNA fragments with different types of 3′ ends are shown below.

The 3′ ends of the upper strands are 3′-protruding, even or 3′-recessive, respectively. The 5′ ends of the upper strands were ^{32}P-labeled beforehand. Although the 3′ ends of both upper and lower strands can accept tails, only the ^{32}P-labeled upper strands are detected after electrophoresis and autoradiography. The results in Table I show that in most cases the longest tail length and the highest percentage of DNA with tails are obtained in DNA with 3′-protruding ends (Fig. 1). We expected even-ended DNA to be more efficiently tailed (higher percentage of molecules with tails) than DNA with 3′-recessive ends. However, this is not true in certain cases, such as the addition of dC tails to *Hin*cII fragments as compared with *Bam*HI fragments in both kinds of buffers. The results also show that for DNA with either even ends or 3′-recessive ends, $CoCl_2$ buffer gives higher efficiencies in dA and dT tailing, thus $CoCl_2$ buffer is recommended. Since $MnCl_2$ buffer gives higher efficiencies in dC or dG tailing and more uniform length of tails (Table I and Figs. 1 and 2), it is recommended for dC and dG tailing.

[17] G. Chaconas and J. H. van de Sande, this series, Vol. 65, p. 75.

[18] K. Kato, J. M. Gonçalves, G. E. Houts, and F. J. Bollum, *J. Biol. Chem.* **242,** 2780 (1967).

[19] F. J. Bollum, this series, Vol. 10, p. 145.

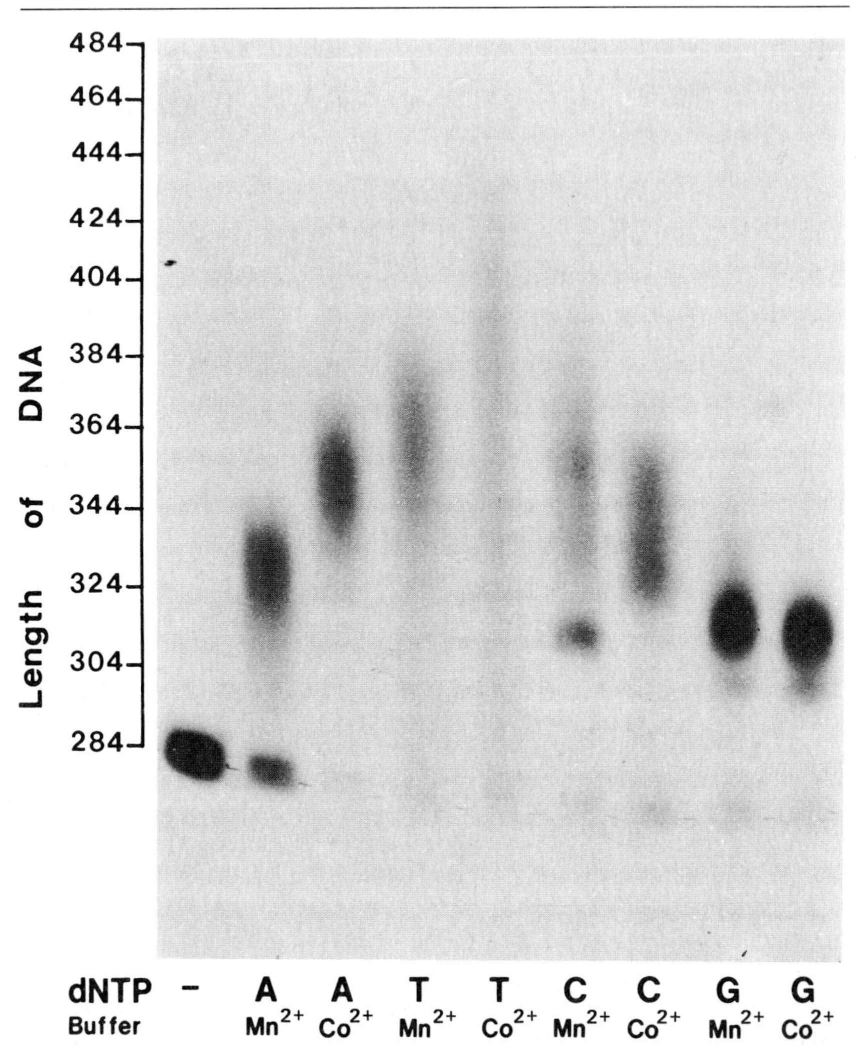

FIG. 1. Addition of homopolymer tails to 3′ ends of DNA fragments cut by *Bgl*I. Homopolymer tails were added to the 3′ ends of 5′-^{32}P-labeled pBR 322-*Bgl*I DNA fragments (final concentration: 60 pmol of ends per milliliter of DNA). The reaction mixture contained either the $MnCl_2$ buffer or the $CoCl_2$ buffer, one unlabeled dNTP (dTTP, dCTP, dATP, or dGTP, 100-fold excess over the 3′ end of DNA), and terminal transferase (500 units/ml). Incubations were carried out at 30° for 60 min. The labeled DNA was precipitated by ethanol, denatured in 0.3 *N* NaOH, and loaded on a 5% polyacrylamide gel in 8 *M* urea. Only the short restriction fragment of each experiment is shown in the figure.

TABLE I
EFFICIENCY OF TAILING REACTIONS USING DNA WITH DIFFERENT TYPES OF 3′ ENDS[a]

dA Tails	^{32}P-pBR322 DNA cut with	Types of ends	$MnCl_2$-containing buffer		$CoCl_2$-containing buffer	
			Average tail length	% of DNA with tails	Average tail length	% of DNA with tails
dA	*Bgl*I	3′-protruding	48 (30–65)	72	76 (58–95)	91
	*Hinc*II	even	—	>1	141	17
	*Bam*HI	3′-recessive	40	2	40	4
dT	*Bgl*I	3′-protruding	78 (50–106)	89	100	90
	*Hinc*II	even	42	15	167	27
	*Bam*HI	3′-recessive	50	14	109	28
dC	*Bgl*I	3′-protruding	74 (40–108)	91	72 (51–92)	89
	*Hinc*II	even	23 (15–30)	30	43	20
	*Bam*HI	3′-recessive	35 (28–42)	79	45	47
dG	*Bgl*I	3′-protruding	46 (36–56)	96	42 (32–52)	92
	*Hinc*II	even	25 (14–33)	95	19	64
	*Bam*HI	3′-recessive	42 (36–48)	72	21	15

[a] Homopolymer tails were added to 5′-labeled pBR322 DNA (60 pmol of 3′ ends per milliliter, cut by *Bgl*I, *Hinc*II, or *Bam*HI) with unlabeled dATP, dTTP, dCTP, or dGTP (100-fold excess over 3′ ends of DNA). The transferase reactions were carried out at 30° for 60 min. Average tail length and percentage of DNA molecules with tails were estimated from gel electrophoresis analysis.[16] There are two cuts in pBR322 DNA by *Hinc*II digestion; data of only the shorter *Hinc*II fragment are shown. The numbers in parentheses are the range of the length for over 90% of the tailed molecules.

Addition of Tails with Different Levels of dNTPs

At concentration of dNTPs below the K_m for terminal transferase, the rate of the tailing reaction is dependent on the levels of dNTPs. With this consideration, one can control the reaction by varying the amounts of dNTPs added to the mixture. As the amounts of dNTPs are increased from 20-fold excess over DNA ends to 200-fold excess, both the tail length and the percentage of DNA molecules with tails increase. For DNA with 3′-protruding ends, a ratio of dNTP to DNA of about 20 : 1 can be used to give homopolymer tails 15–40 nucleotides in length.[16] In other experiments (data not shown) using DNA with even ends and 3′-recessive ends, a ratio of dNTP to DNA of about 100 : 1 was the best. It is clear from Fig. 2 that a higher percentage of tailed DNA molecules cannot be obtained by increasing the amounts of dCTP or dGTP from 100- to 200-fold excess for tailing reaction of even ends.

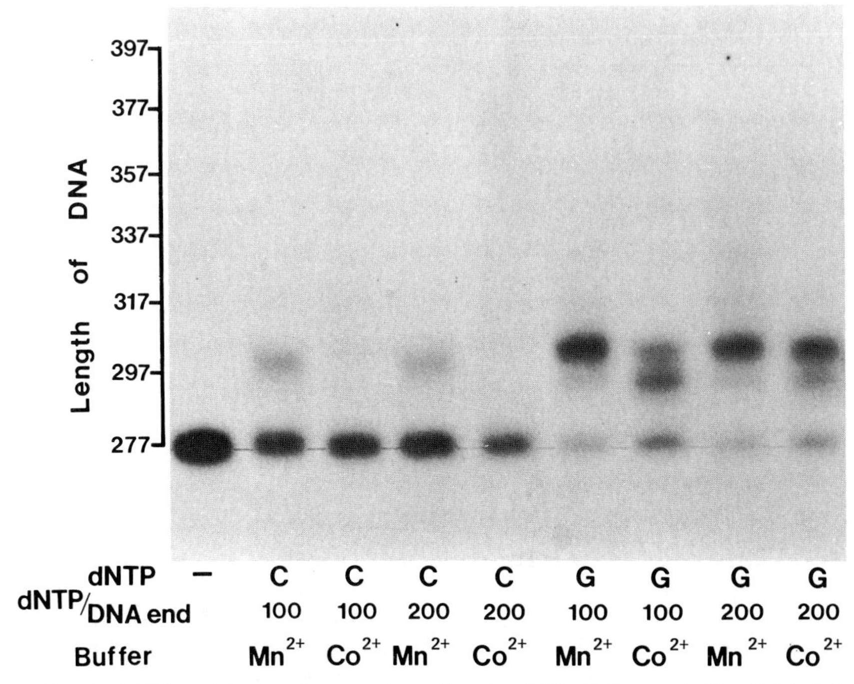

FIG. 2. Addition of homopolymer tails to 3′ ends of *Hinc*II fragment. The 5′ labeled pBR322-*Hinc*II fragment (60 pmol of ends per milliliter) was incubated with dCTP or dGTP (100- or 200-fold excess over 3′ ends of DNA), and terminal transferase (500 units/ml) at 30° for 60 min. Reactions were carried out either in $MnCl_2$ or $CoCl_2$ buffer. Electrophoresis was performed as described in the legend of Fig. 1 except that a 3.5% polyacrylamide gel was used.

Restoration of the Recognition Sequences of Restriction Enzymes

To facilitate excision of inserted DNA fragments from recombinant vehicles after cloning, one can restore the restriction sequence at the ends of the DNA fragments by the proper choice of homopolymer tail. For example, if a dG tail is added to the 3′ end of a *Pst*I-cut fragment, the *Pst*I sequence will be restored after annealing with a DNA fragment carrying a dC tail and cloning.[6,7,10,20–23] Other restriction sequences (with 5′ protruding ends) also can be restored by repair synthesis to produce even ends before addition of tails.[16]

[20] A. Dugaiczyk, cited in F. Bolivar *et al.*, *Gene* **2,** 95 (1977).

[21] M. B. Mann, R. N. Rao, and H. O. Smith, *Gene* **3,** 97 (1978).

[22] F. Bolivar, *Gene* **4,** 121 (1978).

[23] A. Otsuka, *Gene* **13,** 339 (1981).

For cloning a DNA fragment into the *Pst*I site of pBR322, dC tails and dG tails were added to the DNA fragment to be cloned and to *Pst*I-cut pBR322 respectively, according to the conditions mentioned above. These two DNA fragments with complementary tails were annealed,[24] and the product was used to transform competent cells of *E. coli* 5346.[25] Tetracycline (10 μg/ml) and ampicillin (50 μg/ml) in L-broth agar plates were used to screen the transformants. The clones that survive on tetracycline plates but not on ampicillin plates were supposed to be those carrying a recombinant DNA. We found that 80% of the tetracycline-resistant clones were ampicillin sensitive. This means that 80% of the vector molecules had received insertions. (It is known that the *Pst*I-cut pBR322 fragment without a tail can ligate back to itself to form a circle at its cohesive ends during annealing. If the dG tailing efficiency of the vector is low, after annealing with the dC tailed DNA fragments to be cloned, most of the transformants will contain the vector only.) Several selected clones were sequenced, and it was found that the inserted DNA fragments were present in most clones. These recombinant plasmids retained the *Pst*I sequence at both sides of the insertion, and the dC : dG tail was 15–20 base pairs long.

Addition of a Single Nucleotide to 3′ Ends of DNA for *in Vitro* Mutagenesis

Terminal transferase needs the 3′ end of single-stranded DNA as primer, but does not need a template. Thus, a mismatched nucleotide can be added by using the transferase and a specific dNTP that does not complement the one in the template strand. Since the creation of site-specific mutation is very useful in studying gene control and expression, we wanted to establish optimal conditions under which terminal transferase can add single mismatched nucleotides to DNA to obtain specific mutants.

The general scheme for creating a site-specific mutation is as follows. An example is given for generating a specific base change at the *Sal*I site of a DNA molecule (Fig. 3). The DNA is digested with *Sal*I to give a linear DNA. To one-half of the DNA sample, we add a mismatched d$\overset{*}{A}$ using terminal transferase, followed by repair synthesis to fill in the cohesive end. To the other half of the DNA sample, we add S1 nuclease to digest away the single-stranded cohesive end. Each fragment is digested with a

[24] S. L. Peacock, C. M. McIver, and J. J. Monahan, *Biochim. Biophys. Acta* **655,** 243 (1981).
[25] D. A. Morrison, this series, Vol. 68, p. 326.

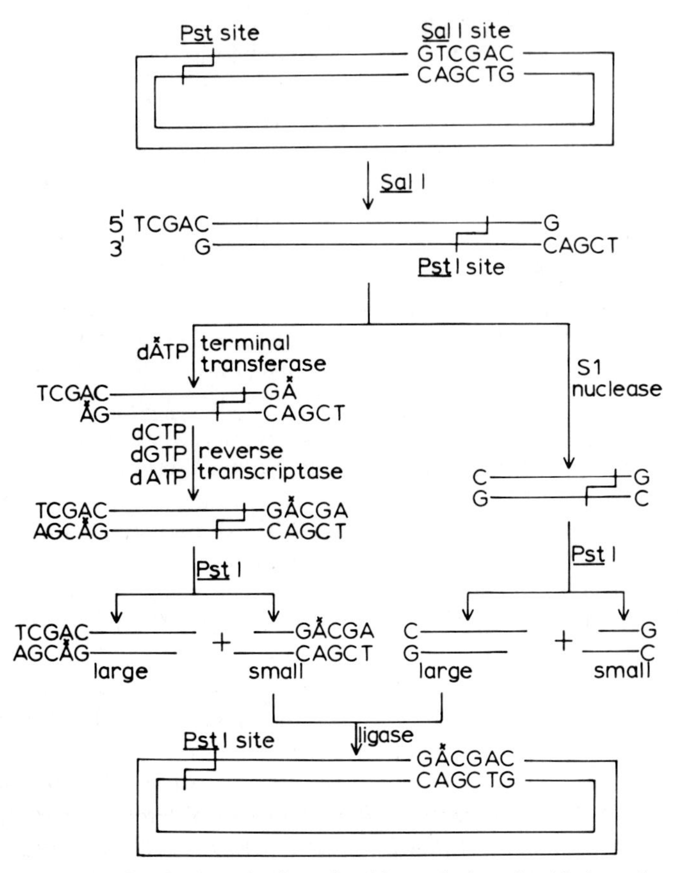

FIG. 3. A scheme for the introduction of a mismatched nucleotide by using terminal transferase. The mismatched nucleotide is shown with an asterisk on top. After transformation, 50% of the daughter plasmid will become the mutant (with one base pair altered), which leads to loss of the *Sal*I site, and the other 50% of the plasmid will be wild type.

second enzyme (such as *Pst*I) to give a large fragment and a small one. The mutated small fragment is ligated to the S1 nuclease-digested large fragment by blunt-end ligation. After transforming host cells with this DNA mixture, one can digest the DNA samples with *Sal*I and select the cloned DNA which has lost the *Sal*I site. Since the base change introduced by this *in vitro* method is present in one strand only, during replication of this altered plasmid, one daughter plasmid will become the mutant with an altered base pair and the other daughter plasmid will be the same as the wild type.

TABLE II
ADDITION OF dNTPs TO 3′ END OF DECAMER[a]

Buffer systems	dNTPs added	Fraction of each product (%)				
		Unchanged	1 added	2 added	3 or more	Total
$CoCl_2$	dATP	65	25	6	4	100
	dTTP	69	23	6	2	100
	dCTP	51	32	13	4	100
	dGTP	54	32	9	5	100
$MnCl_2$	dATP	91	7	1	1	100
	dTTP	78	20	2	0	100
	dCTP	58	30	9	3	100
	dGTP	58	13	3	26	100

[a] Conditions were the same as described in the legend of Fig. 4 except that only those results from 30 min of incubation are included and both the $CoCl_2$ and the $MnCl_2$ buffer systems were used.

Addition of a Single Nucleotide to a Decamer

First we used a deoxynucleotide decamer, 5′-d(A-A-C-T-T-G-A-C-C-C), as a primer for testing the conditions for the addition of a single nucleotide. This decamer was incubated in $CoCl_2$ buffer with one of the four unlabeled deoxynucleotides (twofold excess over decamer) and terminal transferase at 30° for 1–30 min. The sample was fractionated on a 15% polyacrylamide gel; the autoradiograph is shown in Fig. 4. By scintillation counting of the gel slices containing different length of oligomer, the efficiencies of single addition of dNMPs were calculated (Table II). Under favorable conditions, about 30% of the decamers received a single nucleotide. When the same experiments are carried out in the presence of $MnCl_2$ in place of $CoCl_2$, the results are less favorable (Table II). In experiments in which the dNTPs are replaced by the NTPs (2-, 5-, and 10-fold excess over the decamer primer), the results of single NMP addition[26,27] are less favorable for AMP and UMP incorporation; however, those of CMP and GMP are about the same as the incorporation of dCMP and dGMP (data not shown).

Addition of a Single Nucleotide to a Restriction Site of DNA

In the next experiment, we examined whether a single nucleotide could be incorporated to the 3′ recessive ends of duplex DNA. For this

[26] R. Roychoudhury and H. Kössel, *Eur. J. Biochem.* **22,** 310 (1971).

[27] R. Roychoudhury, E. Jay, and R. Wu, *Nucleic Acids Res.* **3,** 863 (1976).

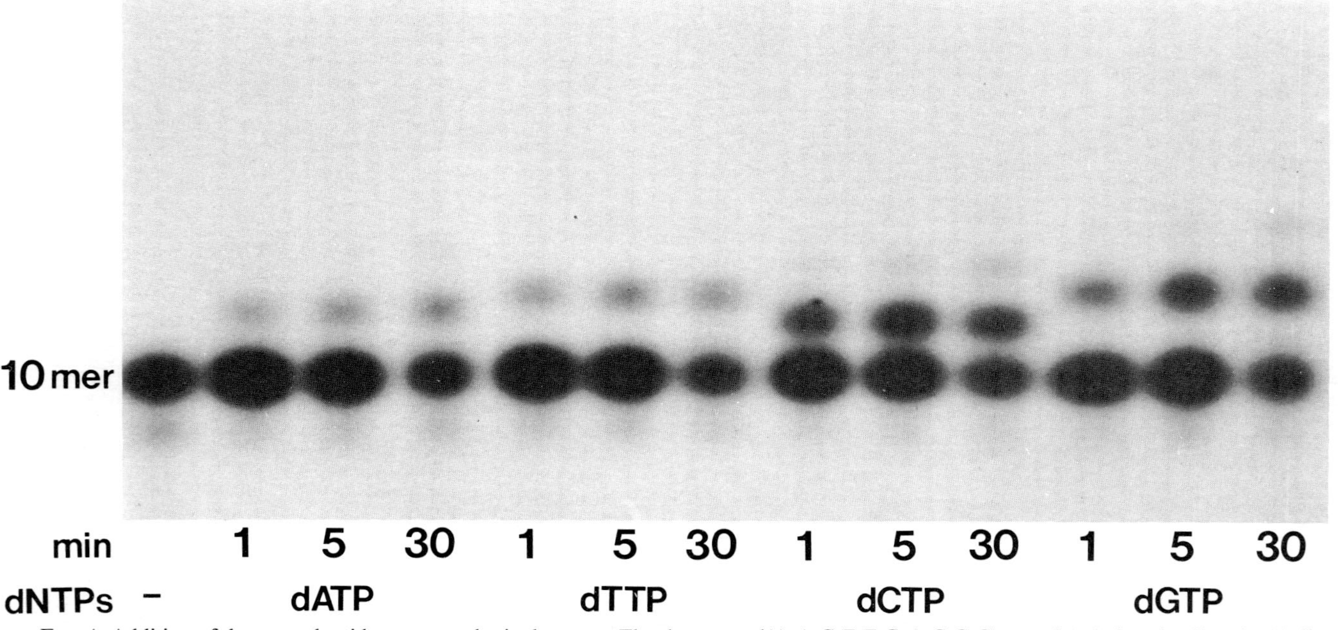

FIG. 4. Addition of deoxynucleotides to a synthetic decamer. The decamer, d(A-A-C-T-T-G-A-C-C-C), was labeled at the 5′ end with ^{32}P using T4 polynucleotide kinase: 50 pmol of the decamer, 100 pmol of [γ-^{32}P]ATP, and 10 units of T4 kinase (Biogenics, Lot 570) were added to 100 μl of solution containing 50 mM Tris-HCl (pH 7.6), 10 mM $MgCl_2$, and 5 mM dithiothreitol. The mixture was incubated at 37° for 1 hr, and the labeled decamer was purified on a small DE-52 column. Deoxynucleotides (100 pmol/ml) were added to the 3′ end of 5′-labeled decamer (50 pmol/ml) in the presence of terminal transferase (500 units/ml) and $CoCl_2$ buffer. The mixture was incubated at 30° for 1, 5, and 30 min. The samples were loaded onto a 15% polyacrylamide gel (350 × 400 × 1.5 mm) containing 50 mM Tris-borate (pH 8.1), 1 mM EDTA, and 8 M urea. The gel was run at 13 mA (200–400 V) for 16 hrs.

purpose, pBR322 was digested with *Eco*RI, kinase-labeled at 5′ ends of both *Eco*RI sites, and cut again with *Hin*dIII. Two fragments with different length (one is 31 base pairs; the other is 4331 base pairs) were obtained, and each one was labeled at the 5′ end of one strand only. These fragments were incubated with one of the four dNTPs, terminal transferase in $CoCl_2$ buffer (for adding dATP or dTTP) or $MnCl_2$ buffer (for dCTP or dGTP). After running the samples on a 15% polyacrylamide gel, the tail length and percentage of molecules with tails were calculated (Table III). We found that one dAMP or dTMP and 1–3 dCMP or dGMP could be added to the 3′-recessive ends of duplex DNA if 5- to 10-fold excess of dNTPs (compared with DNA 3′ ends) were used.

Stability of the Mismatched Nucleotide

The next question to be answered is whether after one mismatched nucleotide is incorporated, more nucleotides can be added to its 3′ end by using DNA polymerase or AMV reverse transcriptase according to the instructions of the template. In other words, during repair synthesis is the mismatched nucleotide stable without being removed by the 3′ → 5′ exonuclease activity (editing function) of the enzyme?

To test the stability of the mismatched nucleotide at the 3′ end of a DNA, pBR322 was digested by *Sal*I to give the linear form. The *Sal*I-cut pBR322 was incubated with [α-^{32}P]dTTP and reverse transcriptase to introduce a single labeled dT to give DNA 1 shown below. A mismatched

TABLE III
ADDITION OF dNTPs TO 3′-RECESSIVE END OF DUPLEX DNA[a]

Buffer	dNTPs added	dNTP/DNA end	Tail length	Percent of molecules with tails (%)
$CoCl_2$	dATP	5	1	20
	dATP	10	1	25
	dTTP	5	1	15
	dTTP	10	1	20
$MnCl_2$	dCTP	5	2	50
	dCTP	10	3	60
	dGTP	5	1	75
	dGTP	10	3	80

[a] 5′-Labeled DNA fragments (50 pmol of ends per milliliter) were incubated with one of the four dNTPs (5- or 10-fold excess over DNA ends), terminal transferase (1000 units/ml) in $CoCl_2$ or $MnCl_2$ buffer at 37° for 30 min. Results were calculated from gel electrophoresis.

nucleotide was then added by terminal transferase, and the product (DNA 2) was divided in three aliquots for repair synthesis: (*a*) dGTP was added to the DNA fragment; (*b*) dATP; (*c*) both dGTP and dATP. The conditions for repair synthesis were the same as described above except that a 200-fold excess of triphosphates over the DNA ends were used. The procedures are summarized below.

$$\begin{array}{l}\text{-G}\\ \text{-CAGCT}\end{array} \xrightarrow{[^{32}\text{P}]\text{dTTP}} \underset{(1)}{\begin{array}{l}\text{-G}\dot{\text{T}}\\ \text{-CAGCT}\end{array}} \xrightarrow{\text{dTTP}} \underset{(2)}{\begin{array}{l}\text{-G}\dot{\text{T}}\text{T}\\ \text{-CAGCT}\end{array}}$$

$$\text{(a)} \xrightarrow{\text{dGTP}} \begin{array}{l}\text{-G}\dot{\text{T}}\text{TG}\\ \text{-CAGCT}\end{array} \quad (3)$$

$$\text{(b)} \xrightarrow{\text{dATP}} \begin{array}{l}\text{-G}\dot{\text{T}}\text{T}\\ \text{-CAGCT}\end{array} \quad (4)$$

$$\text{(c)} \xrightarrow[\text{dATP}]{\text{dGTP}} \begin{array}{l}\text{-G}\dot{\text{T}}\text{TGA}\\ \text{-CAGCT}\end{array} \quad (5)$$

All the products in each step (1–5) were cut by *Hin*fI to give two labeled fragments with different length, 20 and 201 base pairs, respectively, which were easily separated on a 15% polyacrylamide gel (Fig. 5). By counting the gel slices containing different bands, it was estimated that 20% of the molecules received one unlabeled dTMP and less than 5% recieved two or more (lane 2 of Fig. 5). In lane 3, almost all the molecules that received one unlabeled dTMP were extended further by adding one dGMP; in lane 5, almost all those molecules which received one unlabeled dTMP were extended to give blunt-ended DNA by adding one dGMP and one dAMP. The product in lane 4 is similar to that in lane 2, since there is no dGTP in the reaction mixture for the addition of the next nucleotide. The fact that in lanes 3, 4, and 5 the longer fragments at G, T, A position are not less than the longer fragment at the T position in lane 2 indicates that after single addition, during the repair synthesis, the mismatched nucleotide is not removed by the reverse transcriptase. This result agrees with the report by Battula and Loeb[28] that AMV reverse transcriptase lacks the proofreading function. We conclude from these experiments that it is possible to direct the incorporation of a specific mismatched nucleotide to a specific location on a duplex DNA at a restriction site. A similar approach was communicated to us by Douglas R. Smith, who proposed to use RNA ligase (instead of terminal transferase) to add a mismatched nucleotide.

Addition of a Single Nucleotide to an Exonuclease III-Digested DNA

We asked the question whether it is possible to produce an *in vitro* mutation at any place on the DNA outside the restriction site. Since

[28] N. Battula and L. A. Loeb, *J. Biol. Chem.* **251,** 982 (1976).

[29] L. Guo and R. Wu, *Nucleic Acids Res.* **10,** 2065 (1982).

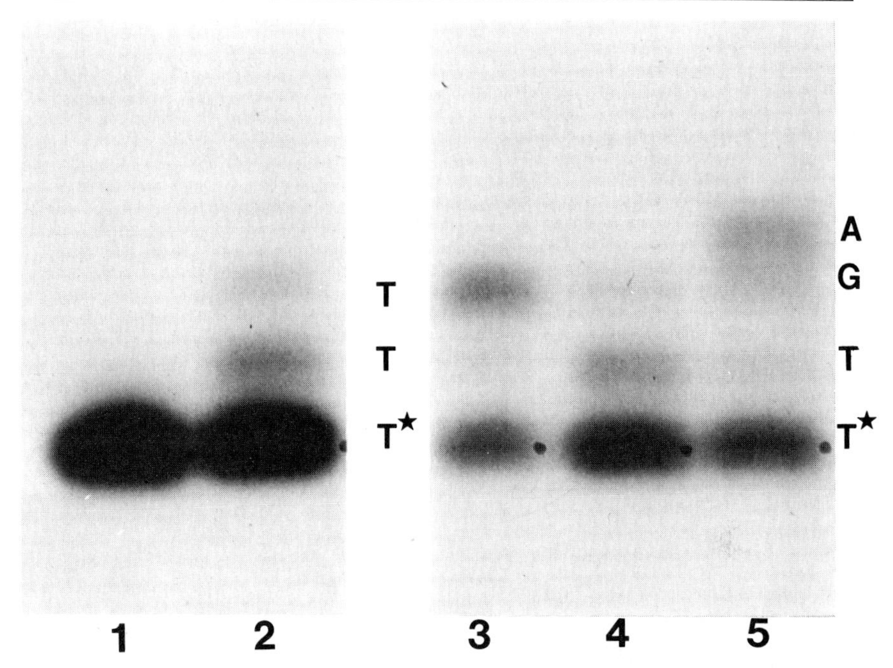

FIG. 5. Single addition and repair synthesis at the *Sal*I site. *Sal*I cut pBR322 (50 pmol/ml) was labeled with [α-^{32}P]dTTP (5-fold excess over DNA ends), 50 m*M* Tris-HCl (pH 7.6), 50 m*M* KCl, 5 m*M* $MgCl_2$, 5 m*M* dithiothreitol, and AMV reverse transcriptase (500 units/ml) at 37° for 30 min to label the 3′ end of the DNA. After extraction with neutralized phenol, chloroform : isoamyl alcohol (24 : 1) and precipitation twice with ethanol, the labeled pBR322-*Sal*I fragment (100 pmol of ends per milliliter) was incubated with dTTP (500 pmol/ml) and terminal transferase (1000 units/ml) in $CoCl_2$ buffer at 37° for 30 min, extracted with phenol, chloroform, and precipitated twice with ethanol. The remaining cohesive end was filled in by repair synthesis with dGTP and dATP (see the text). After digestion with *Hin*fI, the products were loaded onto a 15% polyacrylamide gel. Lanes 1–5 represent the products 1–5. T*, T, G, and A represent the positions of the DNA fragment receiving [^{32}P]dTMP, dTMP, dGMP, and dAMP, respectively. Only the short fragments from *Sal*I–*Hin*fI digestion are shown.

exonuclease III can remove nucleotides from the 3′ ends of a DNA duplex, and the extent of digestion is easily controlled, we used this enzyme to remove a number of nucleotides from the 3′ ends and have the area of interest exposed as a single-stranded region.[29] Thus, a DNA fragment with one mismatched nucleotide at almost any position can be made. To test this approach experimentally, pBR322 was digested by *Sal*I and then treated with exonuclease III to remove about 20 nucleotides (see the legend of Fig. 6). The digested DNA fragments were divided into six aliquots and incubated with [α-^{32}P]dATP and either reverse transcriptase

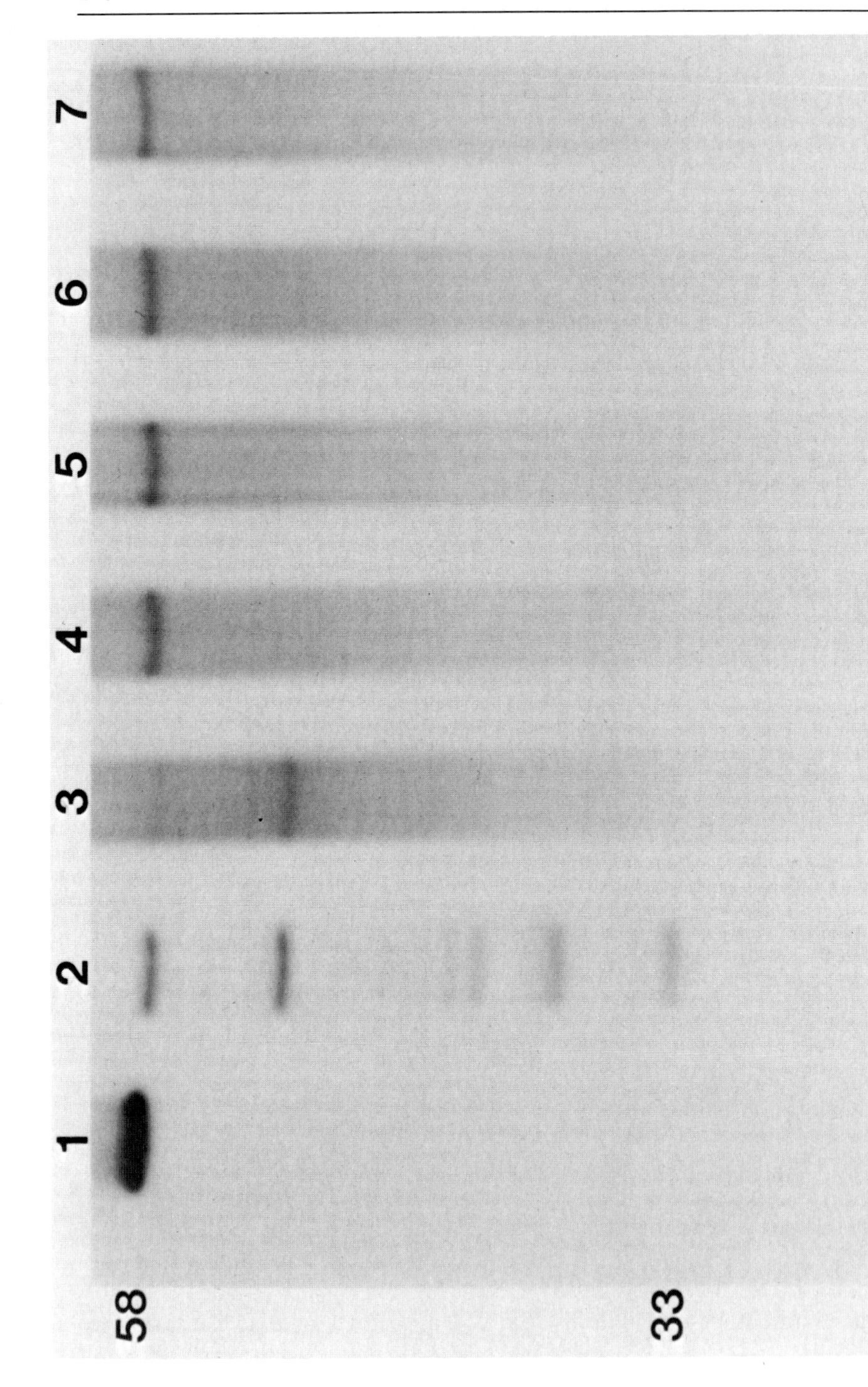
1
2
3
4
5
6
7
58
33

to add a labeled dAMP to match the dTMP in the complementary strand (Fig. 6, lane 2) or terminal transferase to add one (or two) mismatched dAMP to the 3′ ends of the DNA molecules of different lengths (lanes 3–7). The DNA of lanes 4–7 was next subjected to repair synthesis with reverse transcriptase using all four unlabeled dNTPs. After digestion with *Hae*III, all samples were fractionated by gel electrophoresis. The result in lane 2 shows that about 80% of the DNA molecules lost between 5 and 25 nucleotides after exonuclease III digestion. The pattern in lane 3 indicates that [^{32}P]dAMP was added to the 3′ ends of the DNA molecules of different lengths. The upper bands in lanes 4–7 represent full-length DNA fragment (compare to lane 1) resulting from repair synthesis. Since in the reaction of repair synthesis there were no ^{32}P-labeled dNTPs, this 58 base pair-long DNA must come from the DNA fragments that first received a [^{32}P]dAMP using the terminal transferase and then was lengthened by repair synthesis with four dNTPs. Results from different experiments show that 30–70% of the fragment that had received the [^{32}P]dAMP could be repaired to the full length.

We conclude from these experiments that in the overall reaction of single nucleotide addition, the production of mismatched base pairs is only around 20% of the DNA molecules. That means most of the DNA molecules do not receive a mismatched nucleotide. After transformation with this mixture of DNA molecules, the screening of the transformants for the mutants may be difficult unless there is an efficient method to select for the mutants. Selection by direct DNA sequencing would be too time consuming.

Addition of a Single Mismatched dAMP [αS]

We next tested the following approach to overcome the above problem. For incorporating the mismatched nucleotide by terminal trans-

FIG. 6. Single addition of mismatched nucleotide followed by repair synthesis of exonuclease III-treated duplex DNA. Lane 1: pBR322 cut with *Sal*I and labeled with dTTP, dCTP, dGTP, and [^{32}P]dATP by repair synthesis. Lanes 2–7: pBR322 digested with *Sal*I and then with exonuclease III for 2 min. Approximately 20 nucleotides were removed from each 3′ end. After extraction with phenol and chloroform and precipitation with ethanol, the DNA was handled in six different ways: in lane 2, it was repair labeled with [^{32}P]dATP; in lane 3, it was extended with [^{32}P]dATP by terminal transferase; in lanes 4–7, it was labeled as in lane 3, the DNA was extracted with chloroform, passed through a DE-52 column, and precipitated by ethanol to remove unreacted [^{32}P]dATP; the DNA was then repaired with all four dNTPs by reverse transcriptase (in lanes 4 and 5, the repair buffer contained 50 m*M* KCl; in lanes 6 and 7, 100 m*M* KCl. The concentrations of reverse transcriptase were 1200 units/ml in lanes 4 and 6 and 2400 units/ml in lanes 5 and 7). All the DNA samples (lanes 1–7) were digested with *Hae*III before loading on a 15% polyacrylamide gel. Only the short *Hae*III fragments are shown on this figure.

ferase, we replaced dNTP with deoxynucleoside 5′-[α-thio]triphosphate. After incorporation, the DNA molecules received a sulfur nucleotide that can be selectively adsorbed on a mercury glass-bead affinity column, whereas those DNA molecules without a sulfur nucleotide pass through the column and can be eliminated. In this way, only the sulfur-containing DNA molecules, which carry the mismatched nucleotide, are selected and purified before ligation to the vector and transformation of host cells.

To test the incorporation efficiency, pBR322 was cut with *Sal*I, and incubated with [^{35}S]deoxyadenosine 5′-[α-thio]triphosphate[30] and terminal transferase. The results in Table IV indicate that the rates of incorporation are at least twice as high in $MnCl_2$ buffer as in $CoCl_2$ buffer, and more incorporation can be obtained when more dATP[αS] is added.

We next determined the percentage of the DNA molecules that received one or more dAMP[αS] molecules. *Sal*I cut pBR322 was first labeled at the 3′ end with [α-^{32}P]dTTP in the presence of AMV reverse transcriptase, and the labeled DNA was incubated with dATP[α^{35}S] and terminal transferase. After digestion with *Hin*fI, the samples were fractionated by gel electrophoresis. From Fig. 7 and Table V, we find that 8–9% of the DNA molecules receive one dAMP[αS] when the amounts of dATP[αS] are varied from 2-fold to 10-fold excess over DNA ends. With increasing concentration of dATP[αS], more DNA molecules obtain two, three, or more dAMP[αS], whereas the number of the molecules receiving one dAMP[αS] remain the same. Thus, a ratio of dATP[αS] : DNA end of 2 is recommended.

Purification of Sulfur-Labeled DNA by Mercury Glass Beads

The next step is to purify the sulfur-labeled DNA from unlabeled DNA. Since thiol compounds can bind to mercury atoms on glass beads, the sulfur-labeled DNA fragments will be adsorbed on the mercury affinity column, whereas the unlabeled DNA fragment will pass through. When the column is eluted with a buffer containing a high concentration of 2-mercaptoethanol, the sulfur-labeled DNA can be eluted from the column and recovered.[31]

Mercury glass beads (200 mg) were soaked overnight in 10 ml of 50 m*M* Tris-HCl (pH 7.6), 100 m*M* NaCl. After degassing the slurry of glass beads in a small flask connected with a water aspirator for 5 min, 25 mg of

[30] T. A. Kunkel, F. Eckstein, A. S. Mildvan, R. M. Koplitz, and L. A. Loeb, *Proc. Natl. Acad. Sci. U.S.A.* **78,** 6743 (1981).

[31] A. E. Reeve, M. M. Smith, V. Pigiet, and R. C. C. Huang, *Biochemistry* **16,** 4464 (1977).

TABLE IV
INCORPORATION OF dAMP[αS] TO DNA IN DIFFERENT BUFFER SYSTEMS[a]

dATP[αS]/DNA end	pmol dAMP[αS] incorporated/100 pmol DNA ends	
	$CoCl_2$ buffer	$MnCl_2$ buffer
2	4	18
5	10	32
10	20	43

[a] *Sal*I cut pBR322 (50 pmol of ends per milliliter) was incubated with dATP[αS] (2-, 5-, 10-fold excess over DNA ends) and terminal transferase (1000 units/ml) at 37° for 30 min. The reaction was performed in either $CoCl_2$ or $MnCl_2$ buffer to compare their efficiencies for supporting the incorporation. Then, the mixtures were loaded on DE-52 columns, the unreacted triphosphates eluted with 0.35 *M* ammonium formate, and the DNA eluted with 2 *M* NaCl. A portion of each eluate was counted in a scintillation counter after addition of the scintillator Aquasol-2 (from New England Nuclear). Results were calculated from counting different eluates after DE-52 chromatography. Chemical synthesis of all four unlabeled or labeled dNTP(αS) can be carried out according to the method of Kunkel *et al.*[30]

the glass beads (based on dry weight) were placed in a plastic pipette tip plugged with sterilized cotton for small-scale assay. The column was washed with: (*a*) 2 ml of 50 m*M* Tris-HCl (pH 7.6), 100 m*M* NaCl; (*b*) 1 ml of 50 m*M* Tris-HCl (pH 7.6), 100 m*M* NaCl, 14 m*M* 2-mercaptoethanol;

TABLE V
INCORPORATION OF dAMP[αS] WITH DIFFERENT AMOUNTS OF dATP[αS][a]

dATP[αS]/DNA end	Fraction of products (%)				
	Unchanged	1 added	2 added	3 or more	Total
2	87	8	4	1	100
5	86	8	5	1	100
10	83	9	6	2	100

[a] Conditions of this experiment are the same as those set forth in legend of Fig. 7. Results were calculated from counting the gel slices after electrophoresis.

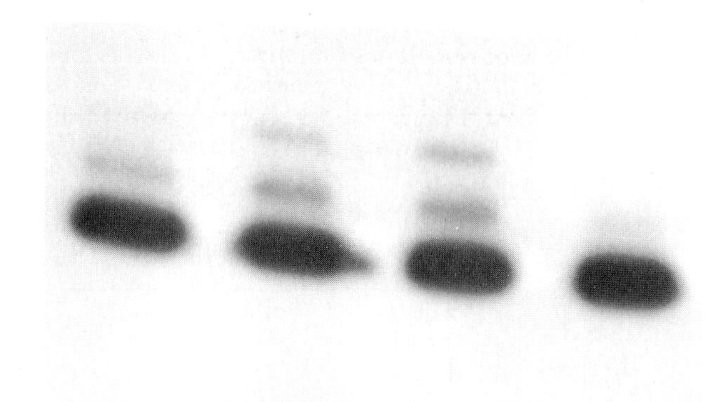

FIG. 7. Incorporation of dAMP[αS] to DNA. *Sal*I-cut pBR322 (50 pmol of ends per milliliter), was labeled with ^{32}P, and the mixture was passed through a DE-52 column to remove the unreacted triphosphate. The ^{32}P-labeled DNA was incubated with different amounts of dATP[αS] (2-, 5-, 10-fold excess over DNA ends) in the presence of terminal transferase (1000 units/ml) and $MnCl_2$ buffer. Reaction was carried out at 37° for 30 min. After digestion with *Hin*fI, samples were loaded on a 15% polyacrylamide gel (350 × 400 × 0.3 mm). After electrophoresis, the gel was exposed to an X-ray film. Since the radiation of ^{35}S is weak, the bands from the gel are attributable to ^{32}P radiation.

(*c*) 1 ml of 50 m*M* Tris-HCl (pH 7.6), 100 m*M* NaCl, 10 m*M* $Hg(OAc)_2$; (*d*) 10 ml of 50 m*M* Tris-HCl (pH 7.6), 100 m*M* NaCl.

When kept in solutions, the mercury glass beads trap many small air bubbles, which will lead to the nonspecific adsorption of DNA (e.g., ^{32}P-labeled DNA without sulfur). Thus, in order to lower the nonspecific adsorption, it is important to degas the slurry of glass beads thoroughly before packing the column and not to allow the air bubbles to be trapped in the column during the whole procedure of washing and elution.

To test the ability of this column to specifically adsorb sulfur-labeled DNA, two kinds of DNA fragments were chosen. *Sal*I-cut pBR322 labeled with [α-^{32}P]dTTP by repair synthesis, which represented DNA without sulfur nucleotide; and *Sal*I-cut pBR322 labeled with [^{35}S]deoxyadenosine 5′-[α-thio]triphosphate and terminal transferase, which represented DNA with sulfur nucleotide. The DNA samples were extracted with neutralized phenol and chloroform and precipitated with ethanol to remove the unreacted triphosphates, the enzymes, and the 2-mercaptoethanol or dithiothreitol that were present in the labeling buffer.

The two types of labeled DNA were each resuspended in 100 μl of 50 m*M* Tris-HCl (pH 7.6), 100 m*M* NaCl, and applied separately onto the mercury glass bead columns. The nonadsorbed DNA was washed through the columns (peak I DNA) with 50 mM Tris-HCl (pH 7.6), 100 mM NaCl, 0.4 mM 2-mercaptoethanol, and the sulfur-labeled DNA was eluted (peak II DNA) with 50 mM Tris-HCl (pH 7.6), 100 mM NaCl, 14 mM 2-mercaptoethanol. Table VI gives the percentage of [^{35}S]DNA or [^{32}P]DNA that appeared in the two peaks. From counting the fractions, about 73% of the ^{35}S-labeled DNA was found to be adsorbed to the column and eluted in

TABLE VI
ELUTIONS OF ^{35}S-LABELED DNA AND ^{32}P-LABELED DNA FROM THE MERCURY AFFINITY COLUMNS[a]

Type of DNA applied on column	Percent DNA found in	
	Peak I	Peak II
[^{32}S]DNA	27	73
[^{32}S]DNA	99	1

[a] Labeled DNA samples (0.25 pmol of each, containing about 30,000 counts of either ^{35}S or ^{32}P) were ethanol precipitated and resuspended in 100 μl of 50 mM Tris-HCl (pH 7.6), 100 mM NaCl, and loaded separately onto mercury glass bead columns (passed through column by gravity at room temperature). The columns (with 80 μl of wet glass beads) were washed with 400 μl of 50 mM Tris-HCl (pH 7.6), 100 mM NaCl, 0.4 mM 2-mercaptoethanol (the concentration is critical) (peak I DNA), and then eluted three times with 100-μl portions of 50 mM Tris-HCl, 100 mM NaCl, 14 mM 2-mercaptoethanol (the combined eluates are called peak II DNA). Each fraction was counted either with the addition of a liquid scintillator for ^{35}S-labeled DNA or by Cerenkov counting for ^{32}P-labeled DNA.

The mercury column can be washed and regenerated after use with the same procedure for washing as described above. When not in use, the column should be stored in a refrigerator. Before reusing, it is necessary to add 400 μl of 50 mM Tris-HCl (pH 7.6) and 100 mM NaCl and to mix the glass beads with the solution using a siliconized Pasteur pipette and degas the slurry in the plastic tip.

peak II; in contrast, only 1% of the ^{32}P-labeled DNA (without ^{35}S) was nonspecifically adsorbed and appeared in peak II. Since about 13% of the DNA molecules receive dAMP[αS] (Table V) after the terminal transferase reaction, out of the 87% of unlabeled molecules only 0.9% would appear in the second peak after passing through the mercury glass bead column. Out of the 13% of the ^{35}S-labeled DNA fragment, 9.5% would appear in the second peak. Thus, after mercury affinity column, the ratio of labeled DNA to the unlabeled DNA will increase from 13 : 87 to 9.5 : 0.9. In other words, about 90% of the DNA molecules found in peak II are labeled with ^{35}S-labeled nucleotide, which represent those with one or two mismatched nucleotides. This enrichment of molecules with mismatched nucleotides will allow the selection of mutant clones by DNA sequencing after transformation of host cells. This method of cloning can be applied to any DNA molecule without the requirement for a specific screening procedure.

Acknowledgments

This work was supported by research Grants DAR 79-17310 from the National Science Foundation and GM 27365 from the National Institutes of Health, United States Public Health Service. We thank Kennedy J. O'Brien for providing us with mercury-derivatized glass beads and a protocol for its use.

[6] Generation of Overlapping Labeled DNA Fragments for DNA Sequencing: A Mercury–Thiol Affinity System

By JAMES L. HARTLEY and JOHN E. DONELSON

The two techniques for determining nucleotide sequences in DNA restriction fragments are the chemical modification method of Maxam and Gilbert[1] and the dideoxynucleotide chain termination method of Sanger *et al.*[2] Both methods are quick, reliable and easy to learn using well-documented manuals describing the procedures.[3,4] Therefore, much of the challenge in a DNA sequencing project nowadays is not in performing the actual sequencing steps themselves, but in devising a strategy for generat-

[1] A. M. Maxam and W. Gilbert, *Proc. Natl. Acad. Sci. U.S.A.* **74,** 560 (1977).

[2] F. Sanger, S. Nicklen, and A. R. Coulson, *Proc. Natl. Acad. Sci. U.S.A.* **74,** 5463 (1977).

[3] A. M. Maxam and W. Gilbert, this series, Vol. 65, p. 499.

[4] Bethesda Research Labs User Manual for M13 Sequencing. Available from Bethesda Research Laboratories, Inc., Gaithersburg, Maryland 20760.

ISBN 0-12-182000-9

ing a family of overlapping restriction fragments whose combined sequences will provide the complete sequence of the region to be determined.

Typically, a DNA sequencing project involves the determination of a partial restriction enzyme map of the region to be sequenced,[5] generation of some sequence data, and computer search of the partial sequence information for additional restriction sites; then the whole procedure is repeated again and again until all the sequence gaps are filled.[6] Alternatively, a large number of small restriction fragments are sequenced blindly, and overlapping sequences are aligned with the aid of a computer program until the complete sequence is determined.[7] These "shotgun" sequencing strategies[5–9] require increasing effort as the number of nucleotides needed to fill the remaining gaps decreases. Thus the time needed to obtain the final 10% of a sequence may be equal to that needed for the initial 90%. Subcloning of some restriction fragments overcomes some of this difficulty but itself involves time-consuming fragment isolations, cloning manipulations, and plasmid or phage purifications.

The *in vitro* procedure described here was devised to generate quickly a large number of overlapping restriction fragments that are labeled at one end for immediate Maxam–Gilbert sequencing. The method includes mercuration of plasmid DNA, thiol agarose affinity purification, and end-labeling by DNA polymerase repair. One preparative agarose gel electrophoresis step yields many terminally labeled fragments that are immediately ready for the sequencing reactions without strand separation or secondary cutting. Furthermore, the map positions of all the labeled ends are determined by inspection of the relative migrations of the labeled fragments on the preparative gel.

General Strategy of the Procedure

Figure 1 demonstrates the application of the procedure to the sequence determination of a fragment cloned in a plasmid. The plasmid DNA is opened at a restriction enzyme site near the region to be sequenced, and about 200 nucleotides are removed from both 3′ ends by T4 DNA polymerase in the absence of deoxynucleoside triphosphates. Then a mixture of mercurated and nonmercurated triphosphates are added. In

[5] U. Rüther, M. Koenen, K. Otto, and B. Müller-Hill, *Nucleic Acids Res.* **9**, 4087 (1981).
[6] J. Messing, R. Crea, and P. Seeburg, *Nucleic Acid Res.* **9**, 309 (1981).
[7] J. G. Sutcliffe, *Cold Spring Harbor Symp. Quant. Biol.* **43**, 77 (1979).
[8] A. M. Frischauf, H. Garoff, and H. Lehrach, *Nucleic Acids Res.* **8**, 5541 (1980).
[9] W. M. Barnes, *in* "Genetic Engineering" (J. K. Setlow and A. Hollaender, eds.), Vol. 2, p. 185. Plenum, New York, 1980.

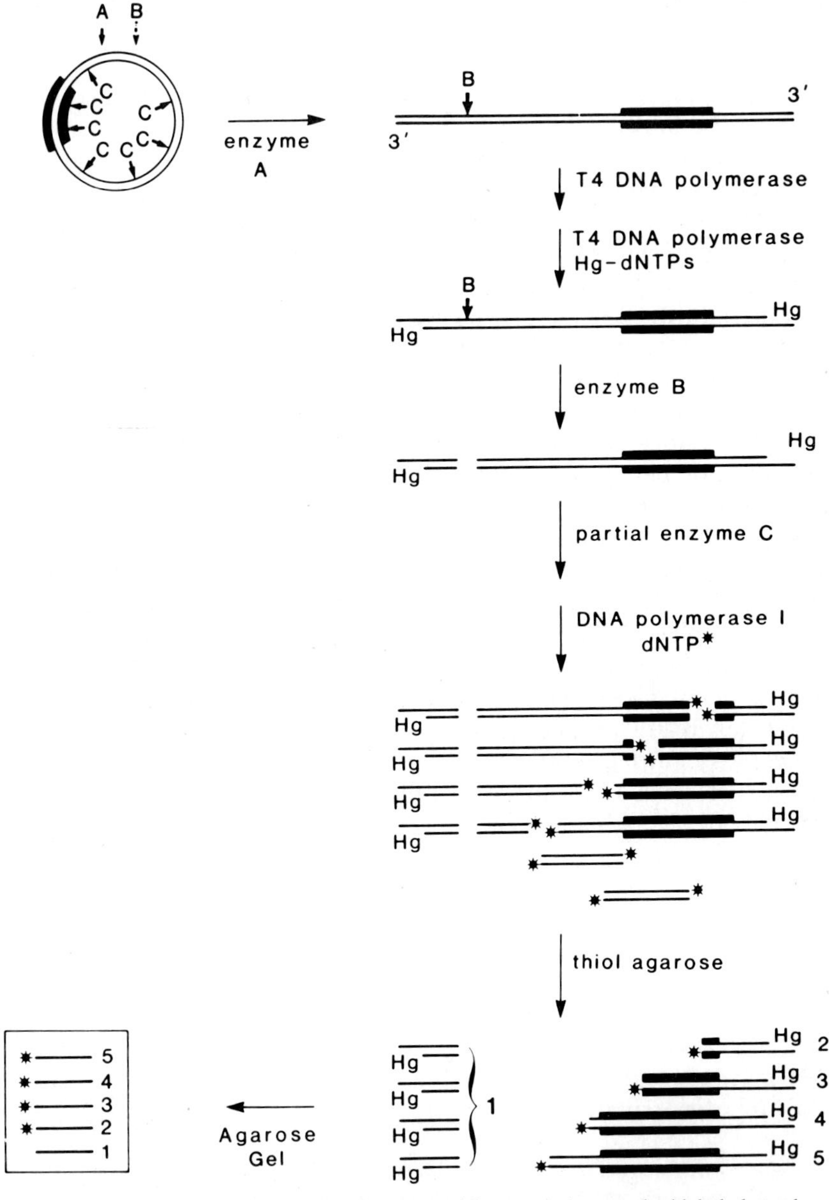

FIG. 1. Schematic representation of the protocol for generating terminal labeled overlapping DNA fragments for Maxam–Gilbert sequence analysis. In general the region to be sequenced is in a plasmid (shown as a double circle). First the plasmid is linearized with restriction enzyme A whose recognition site occurs only once in the plasmid. Then about

the presence of triphosphates, the 5′ → 3′-polymerase activity overwhelms the 3′ → 5′-exonuclease activity,[10] resulting in repaired, flush ends containing mercury atoms in the major grooves.[11] The extent of mercuration is thus a function of the extent of the initial 3′ → 5′-exonuclease digestion. After removal of the DNA polymerase and unincorporated triphosphates by phenol extraction and gel filtration, respectively, the labeled end of the DNA farthest from the region to be sequenced is removed by digestion with a second restriction enzyme (enzyme B in Fig. 1). This prevents a ladder of partial restriction fragments from arising from this labeled end that would complicate the subsequent agarose gel step at the end of the procedure.

The mixture of these two mercurated fragments is then divided into aliquots and partially digested with restriction enzymes that generate protruding 5′ ends (enzyme C in Fig. 1). In practice, enzyme C is usually (but not always) one or more of the restriction enzymes that recognize cleavage sites four or five nucleotides long, since these restriction enzymes produce more partial fragments than those that recognize six-nucleotide sequences. The ends generated by these partial digestions are labeled with *E. coli* DNA polymerase I and [α-^{32}P]deoxynucleotide triphosphates under conditions in which the mercurated, flush ends are not affected. The sequential incubations with enzyme B, enzyme C, and DNA polymerase I are conducted in the same buffer when possible. The products of these reactions are applied directly to small columns of thiol agarose (one for each restriction enzyme used in the partial digests). The unmercurated partial digestion products are washed off the column, and the bound mercurated fragments are eluted with 2-mercaptoethanol and applied directly to a preparative gel. Fragments of interest are eluted from the gel for sequencing by the Maxam and Gilbert technique.

[10] P. H. O'Farrell, E. Kutter, and M. Nakanishi, *Mol. Gen. Genet.* **179,** 421 (1980).
[11] R. M. K. Dale and D. C. Ward, *Biochemistry* **14,** 2458 (1975).

200 nucleotides are removed from each 3′ end using the 3′-exonuclease activity of T4 DNA polymerase. The four deoxynucleoside triphosphates (dNTPs), two of which are mercurated, are added to the reaction, and the polymerase activity of T4 DNA polymerase repairs the 3′ ends. One mercurated end is removed with a second restriction enzyme, and partial digestions with restriction enzyme C (and additional restriction enzymes if necessary) are conducted. After terminal labeling with [α-^{32}P]dNTPs using *Escherichia coli* DNA polymerase I, the fragments are passed through a thiol agarose column. The bound mercurated fragments are eluted with 2-mercaptoethanol and applied immediately to a preparative agarose gel for separation on the basis of size. Fragments of interest are eluted and subjected to the cleavage reactions for sequence determination.

Materials and Reagents

In addition to the normal materials and reagents used in DNA sequencing projects, two items are required—mercurated deoxynucleoside triphosphates and agarose to which thiol groups are attached. We have used 5-mercuri-2′-deoxycytidine 5′-triphosphate (Hg-dCTP), 5-mercuri-2′-deoxyuridine 5′-triphosphate (Hg-dUTP), and thiol agarose (AgThiol), which are available from P-L Biochemicals. Other conventional enzymes and reagents and their commercial sources are restriction enzymes (Bethesda Research Laboratories and New England BioLabs), T4 DNA polymerase and *E. coli* DNA polymerase I (New England BioLabs), unlabeled deoxynucleoside triphosphates (Sigma), [α-^{32}P]dCTP and [α-^{32}P]dATP, 2000–3000 Ci/mmol (Amersham), BioGel A-15m (Bio-Rad), and agarose (Seakem ME from Marine Colloids Division of FMC).

Application of the Method

The sequence determination of the cDNA insert in plasmid pcB 1.1 will be used below to illustrate the method (Fig. 2). This cDNA is inserted into the *Pst*I site of pBR322 via the oligo(dG) : oligo(dC) tailing technique and contains the coding sequence for the variable surface glycoprotein of *Trypanosoma brucei* clone Iltat 1.1. Plasmid pcB 1.1 is maintained in *E. coli* strain HB101, and its construction, identification, and initial characterization are described by Majiwa *et al.*[12] In practice the method can be applied to sequence analysis of DNA fragments inserted at any site within pBR322.

Step 1. Mercuration of the DNA. Plasmid pcB 1.1 DNA was prepared by cesium chloride–ethidium bromide gradient centrifugation.[13] Linear pcB 1.1 DNA molecules were generated by digesting 300 μg (86 pmol) of intact plasmid at either its single *Eco*RI site (751 bp on one side of the cDNA insert) or its single *Ava*I site (2187 bp on the other side of the cDNA insert) as shown in Fig. 2. Sequences of the inside strand are determined from the *Eco*RI-opened plasmid, and sequences on the outside strand are determined from the *Ava*I-opened plasmid. After ethanol precipitation, the DNAs were redissolved in 67 m*M* potassium acetate, 33 m*M* Tris acetate pH 7.8, 10 m*M* magnesium acetate, 10 m*M* 2-mercaptoethanol, 100 μg of bovine serum albumin per milliliter to a final concentration of 500 μg of DNA per milliliter. The reducing agent is required for

[12] P. A. O. Majiwa, J. Young, P. T. Englund, S. Z. Shapiro, and R. O. Williams, *Nature (London)* **297,** 514 (1982).

[13] D. B. Clewell and D. R. Helinski, *J. Bacteriol.* **110,** 1135 (1972).

incorporation of mercurated nucleotides.[14] T4 DNA polymerase (0.2 unit/μg DNA) was added and incubated (without added triphosphates) for 60 min at 37°, then transferred to an ice bath. Stock solutions (10 m*M*) of dGTP, Hg-dCTP, Hg-dUTP, and [α-^{32}P]dATP (0.1 Ci/mmol) were added to final concentrations of 100 μM each, and the reaction was incubated another 10 min at 37°. Tubes were placed on ice, and aliquots were precipitated with trichloroacetic acid to determine the extent of mercuration. The reaction mixtures were extracted with phenol and applied to small (bed volumes of 10–20 times the reaction volumes) columns of BioGel A-15m equilibrated and eluted in 10 m*M* Tris-HCl (pH 7.4), 150 m*M* NaCl, 10 m*M* $MgCl_2$. Void volume fractions were located by incorporated ^{32}P and pooled. Flushness of the mercurated ends was tested by adding 1 μl (10 μCi) each of aqueous [α-^{32}P]dATP and [α-^{32}P]dCTP (3000 Ci/mmol) to a 25-μl aliquot of the pooled void volume fractions, adding 5 units of *E. coli* DNA polymerase I, and incubating on ice for 5 min. The mercurated DNA was judged acceptable if less than 5% of the ends were labeled with one nucleotide.

Step 2. Secondary Cut and Partial Restriction Enzyme Digestions. All the enzyme reactions described prior to the affinity chromatography were conducted in the high-salt buffer used for the BioGel A-15m gel filtration, i.e., 10 m*M* Tris-HCl (pH 7.4), 150 m*M* NaCl, 10 m*M* $MgCl_2$. The two DNAs mercurated at either the *Eco*RI or *Ava*I site were cut to completion at 37° with *Bam*HI (enzyme B of Fig. 1). Aliquots of each DNA were removed and partially digested with *Msp*I, *Taq*I, *Hin*fI, or *Hin*dIII (enzyme C of Fig. 1), which were chosen on the basis of preliminary restriction mapping of the cDNA insert.[14a] The amount of DNA and the degree of cutting desired were determined by the number of sites between the mercurated end and the region of interest for each enzyme. Five to 10 pmol of plasmid were cut for each restriction site between the mercurated end and the farthest site of interest. For example, four *Msp*I fragments were sought from the *Eco*RI-opened plasmid, extending counterclockwise from the mercurated *Eco*RI end to the *Msp*I sites at map positions 4880, 4638, and 3548 in the vector portion (equivalent to *Msp*I sites at positions 3900, 3658, and 3548 in pBR322), and at position 4229 in the cDNA insert (see Fig. 2). As a consequence, 40 pmol of the *Eco*RI-cut, mercurated plasmid were cut to 25% completion with *Msp*I, as judged by analytical gel electrophoresis or by a DNA polymerase I labeling assay

[14] R. M. K. Dale, D. C. Livingston, and D. C. Ward, *Proc. Natl. Acad. Sci. U.S.A.* **70,** 2238 (1973).

[14a] H. O. Smith and M. Birnstiel, *Nucleic Acids Res.* **3,** 2837 (1976).

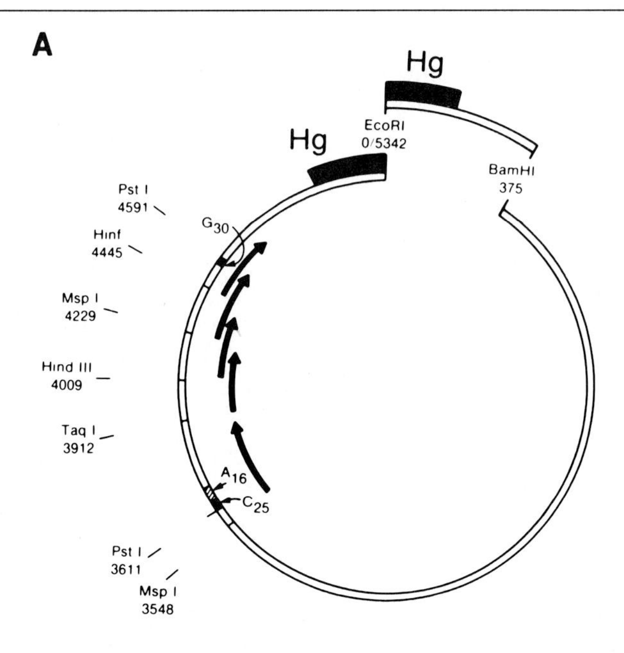

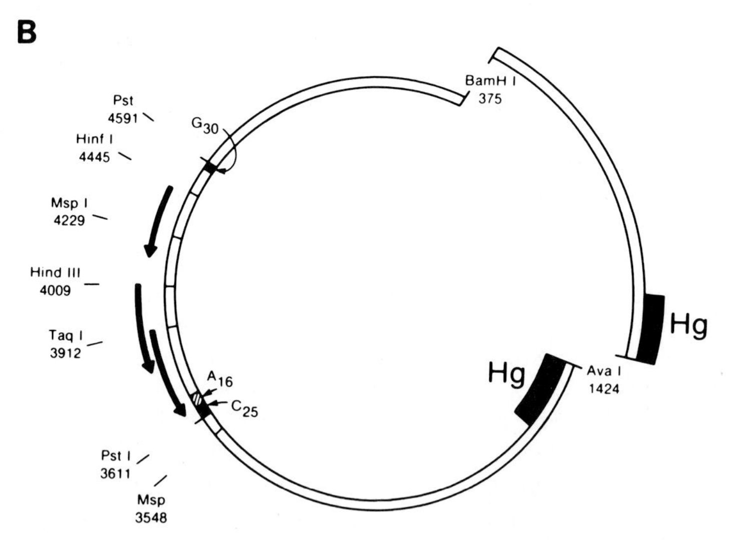

FIG. 2. The strategy used to sequence the cDNA insert of plasmid pcB 1.1, which is cloned at the *Pst*I site of plasmid pBR322, using the oligo(dG) : oligo(dC) tailing procedure.[12] The plasmid was linearized at either the *Eco*RI site (diagram A) or the *Ava*I site (diagram B) in pBR322, mercurated nucleotides were introduced at these ends, as shown by the solid

(see below). Other digestions of the *Eco*RI-cut, mercurated plasmid were (*a*) a 50% *Taq*I digestion of 10 pmol of DNA; (*b*) a total *Hin*dIII digestion of 5 pmol of DNA; and (*c*) a total *Hin*fI digestion of 5 pmol of DNA. For the *Ava*I-cut, mercurated plasmid the digestions were (*a*) a 50% *Taq*I cut of 20 pmol, (*b*) a 14% *Hin*fI cut of 70 pmol; and (*c*) a 100% *Hin*dIII cut of 5 pmol.

The extent of partial digestion was judged by analytical 0.8% agarose gels, or by a quantitative assay as follows. A small aliquot of DNA was cut to completion with each enzyme to be used. Five microliters of either the partial digest or the complete digest containing about 1 μg of DNA were added to 50 μl of freshly prepared partial digest assay mix (2 μM each of dGTP and dTTP, 2.5 μCi (100 pmol) each of [α-^{32}P]dATP and [α-^{32}P]dCTP, and 5 units of *E. coli* DNA polymerase I in 70 m*M* NaCl, 70 m*M* Tris-HCl, pH 7.4, 10 m*M* $MgCl_2$). After incubating 4° for 10 min, the trichloroacetic acid precipitable counts were compared for the partial and complete digest. The ratio indicated the extent of the partial digestions. The actual digestions were kept on ice during the assay or the analytical gel, and reincubated as necessary.

Step 3. Labeling the Nonmercurated Ends of the Partial Restriction Fragments. The DNA termini created by the restriction enzymes used to generate the partial restriction fragments could be labeled with either [^{32}P]dCMP (*Msp*I or *Taq*I ends) or [^{32}P]dAMP (*Hin*fI or *Hin*dIII ends) using the DNA repair reaction of *E. coli* DNA polymerase I. Either [α-^{32}P]dATP or [α-^{32}P]dCTP, 3000 Ci/mmol, was added to appropriate partial digests at a concentration of 0.5–2 pmol label per picomole of "labelable" termini, and DNA polymerase I was added to 100 units/ml. The reaction buffer remained the same as used in step 2 (10 m*M* Tris-HCl, pH 7.4, 150 m*M* NaCl, 10 m*M* $MgCl_2$). After 10 min of incubation on ice, a mixture of dGTP, dCTP, dATP, and dTTP was added to a final concentration of 25 μM each and incubated for 5 min on ice to eliminate possible heterogeneity of label incorporation. One-tenth volume of 0.5 *M* EDTA, pH 8.5, was added to stop the reactions.

---

rectangular boxes and the symbol Hg, and *Bam*HI was used for the secondary cut (enzyme B of Fig. 1). The other restriction enzymes (enzyme C of Fig. 1) shown in the figure were used to obtain partial restriction fragments for sequence analysis. Numbers indicate the locations of the restriction sites in base pairs using the single *Eco*RI site as the reference position 1. Arrows indicate the locations and lengths of the sequenced regions on both strands. G_{30} and C_{25} show the locations and sizes of the two oligo(dG) : oligo(dC) boundaries. A_{16} shows the location of a 16 nucleotide oligo(dA) region that corresponds to the poly(rA) at the 3′ end of the mRNA from which this cDNA was synthesized. Note that mercuration at the *Eco*RI site permits sequence determination of one strand, whereas mercuration at the *Ava*I site permits sequence determination of the other strand.

One of the concerns in developing this general procedure was that the mercurated ends might also be labeled with the high-specific-activity radioactive nucleotides during this step. This would occur if the mercurated nucleotides in step 1 had not been incorporated all the way to the 3′ termini to generate flush double-stranded ends. In practice, labeling of the mercurated ends was not a problem if caution was exercised to ensure that the incorporation of mercurated nucleotides was complete before the mercuration reaction was terminated (see step 1).

Step 4. Affinity Chromatography of the Mercurated, Labeled Fragments. The thiol agarose was prepared for affinity chromatography by stirring 1-ml aliquots twice for 10 min in 10 ml of DTT buffer (0.1 *M* Tris HCl, pH 7.4, 0.3 *M* NaCl, 1 m*M* EDTA, 20 m*M* dithiothreitol, freshly prepared), collected by centrifugation, and washed twice in T100N buffer (10 m*M* Tris HCl, pH 7.4, 100 m*M* NaCl). The centrifuged pellets were suspended in 0.5 ml of T100N and transferred to plugged 1 ml (blue) disposable plastic pipette tips to make columns of about 0.7 ml bed volume. Gel beds were washed with at least 5 bed volumes of T100N. Labeling mixtures were applied directly to the thiol agarose columns, and the nonadsorbed material was reapplied to the columns twice. After washing with 5 bed volumes of T100N, adsorbed DNAs were eluted with T100N containing 0.1 *M* 2-mercaptoethanol. Fractions (100 μl) were counted without addition of scintillation fluid by Cerenkov radiation, and peak fractions were pooled and applied directly to a preparative 1.2% agarose gel. Because earlier fractions contained the larger adsorbed DNAs, these were selected over later fractions.

The analytical gel shown in Fig. 3 shows an example of three sets of mercurated partial-digestion products that were applied to the thiol agarose column (lane 1) and the fragments that bound (lane 2) or did not bind (lane 3) to the column. The nonmercurated fragments produced by the partial restriction enzyme digestion did not adsorb to the thiol agarose at all, while the adsorbed mercurated fragments bound to varying extents and were eluted with 2-mercaptoethanol. The bands in Fig. 3 vary in intensity owing to the different sensitivities of the restriction sites during the partial digestion. The adsorbed bands were precisely those predicted from the maps of the cDNA insert and the vector. The mercury substituents of the DNAs did not alter the appearance of the bands. Nearly 100% of the mercurated DNAs smaller than 1 kb adsorbed to the thiol column whereas 30–50% of the DNAs of 3–4 kb were bound. The use of an affinity medium with more thiol groups, such as a reported sulfhydryl-cellulose,[15] may improve the binding of the large fragments, but, in prac-

[15] P. C. Feist and K. J. Danna, *Biochemistry* **20,** 4243 (1981).

tice, the binding of only 30–50% of the larger fragments was not a problem, since the number of counts in each partial fragment band was greater than that needed for the cleavage reactions.

Step 5. Preparative Agarose Gel and Chemical Cleavage Reactions

The mixture of mercurated fragments of each partial restriction digest (that had been eluted with T100N and 0.1 *M* 2-mercaptoethanol) were pooled and applied directly to a 1.2% agarose gel run at 4° for about 15 hr at 100 V in a buffer of 100 m*M* Tris borate, 1 m*M* EDTA, pH 8.1. Bands of interest were electroeluted within dialysis bags as described.[16,17] The labeled DNA was concentrated from the elution buffer by adsorbing to small (200 μl bed volume) columns of DEAE-cellulose (Whatman DE-52) equilibrated in 10 m*M* Tris-HCl, pH 7.4, 1 m*M* EDTA, and then eluting with 0.5 *M* Tris base, 1.5 *M* sodium acetate. Eluted DNAs were precipitated with 2.5 volumes of ethanol and used directly for the sequencing reactions.

Figure 4 shows the preparative gel of the same three mercurated partially digested DNAs discussed in Fig. 3. The bands indicated by arrows are the same fragments indicated by arrows in Fig. 3 and were eluted for sequence determination.

The actual sequence determinations of the isolated mercurated, ^{32}P-labeled fragments were carried out by the Maxam and Gilbert procedure[3] using the A > G, G, C, and C + T reactions. The cleavage products were fractionated on thin 6% polyacrylamide gels (0.4 mm × 25 cm × 87 cm) containing 7 *M* urea. Three applications of each set of cleavage products were made, spaced about 10 hr apart. The first 100 nucleotides could be read from the last application, the next 100 from the second, and another 100 from the first, for a total of about 300 nucleotides.

Additional Remarks and Observations

The 980 bp cDNA sequence of pcB 1.1 was determined using this procedure as summarized in Figs. 2–4 and has been reported elsewhere.[18,19] Several features of the mercuration procedure described here influence its application to a variety of sequencing projects.

1. All steps are *in vitro*. The starting material is the intact plasmid. No strand separation or secondary restriction enzyme digests are required if

[16] H. O. Smith, this series, Vol. 65, p. 371.
[17] M. W. McDonnell, M. N. Simon, and F. W. Studier, *J. Mol. Biol.* **110,** 119 (1977).
[18] A. Rice-Ficht, K. K. Chen, and J. E. Donelson, *Nature (London)* **294,** 53 (1981).
[19] J. L. Hartley, K. K. Chen, and J. E. Donelson, *Nucleic Acids Res.* **10,** 4009 (1982).

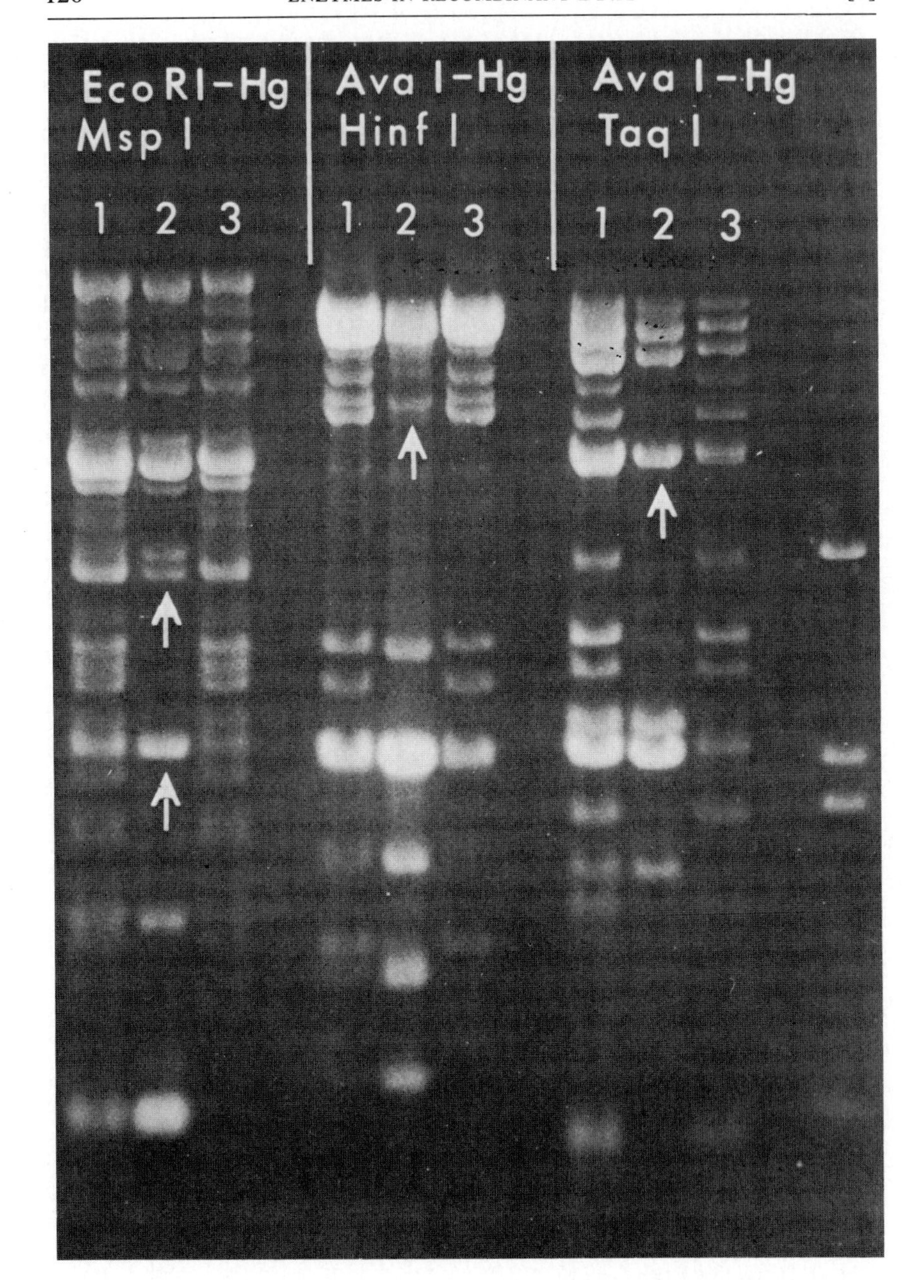
EcoRI-Hg
Msp I
1 2 3
Ava I-Hg
Hinf I
1 2 3
Ava I-Hg
Taq I
1 2 3

labeling is done by DNA polymerase I repair. These features allow the rapid generation of an entire set of end-labeled fragments. Because the largest portion of time is devoted to confirming the various steps (completeness of the restriction cuts, extent of mercuration, etc.), experience permits all the steps through the affinity chromatography to be conducted in 1 or 2 days, followed by overnight preparative gel electrophoresis and recovery of individual fragments. Therefore, many terminal-labeled overlapping fragments ready for sequencing can be generated quickly.

2. A number of different partial restriction digests can be performed on the same mercurated DNA substrate and run on the same preparative gel after the labeling and affinity steps. Fragments can then be selected from the gel on the basis of their spacing. Thus, of the *Taq*I, *Hin*fI, etc., fragments produced, only those 200–300 bp different in length, and not necessarily from the same digest, need be sequenced to get overlapping information. Other fragments can be (*a*) saved in case the expected overlaps are not achieved or (*b*) discarded if they appear to be mixtures of two or more fragments because of closely spaced sites for the same restriction enzyme. The result is much faster and less labor-intensive than "shotgun" sequencing, since each fragment is chosen on the basis of its migration relative to others on the preparative gel.

3. All the sequence obtained from a particular mercurated DNA will be from the same strand. To obtain the sequence of the opposite strand, the plasmid must either be opened on the other side of the insert (e.g., the *Ava*I site vs the *Eco*RI site of pcB 1.1 as shown, Fig. 2), or the insert must be cloned in both orientations. Alternatively, it should also be possible to label the 3′ ends of the partial digest with DNA polymerase I, and the 5′ ends with T4 polynucleotide kinase. In the latter instance, the label on the mercurated end would have to be cut off at some point, perhaps with a restriction enzyme that cleaves near the end. For example, in the case of pcB 1.1, it would be possible to open the plasmid at its single *Cla*I site (position 25), mercurate, secondary cut with *Bam*HI, partially restrict, 5′-end label, and then remove the mercurated label by cleavage with *Eco*RI

FIG. 3. An analytical 0.85% agarose gel stained with ethidium bromide showing aliquots of the mercurated, partially restricted, terminal-labeled plasmid DNA that was applied to the thiol agarose column (lanes 1), the partial restriction fragments that bound to the column (lanes 2), and the partial restriction fragments that passed through the column (lanes 3). The first three lanes show the fragments obtained from the plasmid that was mercurated at the *Eco*RI site and partially digested with *Msp*I; the middle three lanes and the right-hand three lanes show the fragments obtained from plasmid mercurated at the *Ava*I site and partially digested with *Hin*fI or *Taq*I, respectively. Arrows show fragments that were eluted from a preparative agarose gel (Fig. 4) for DNA sequence determination. Bands on the far right are standard length markers.

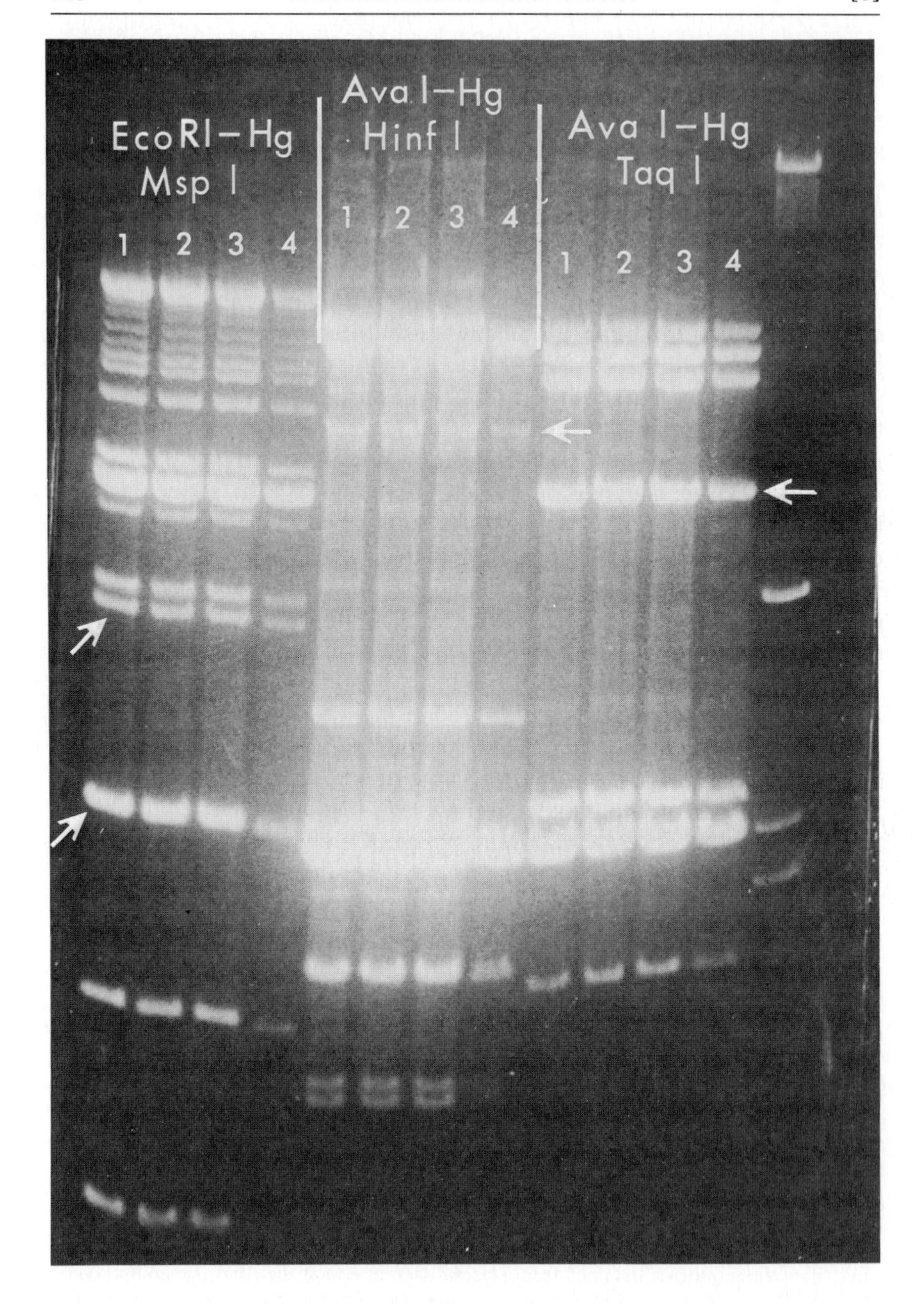
EcoRI–Hg
Msp I
1 2 3 4
AvaI–Hg
HinfI
1 2 3 4
AvaI–Hg
TaqI
1 2 3 4

at position 1 without removing so many mercurated nucleotides that the fragments would not bind to the thiol column. The sequence of both strands could then be determined from different aliquots of the same set of fragments.

4. Although over 30 commercially available restriction enzymes produce recessed 3′ ends that can be labeled by DNA polymerase repair, those which have short recognition sequences are the most likely candidates for use in generating the partial digestion products because they cut frequently. Among these are restriction enzymes with 4 or 5 long recognition sites such as *Taq*I, *Msp*I (*Hpa*II), *Sau*3AI, *Hin*fI, *Dde*I, *Eco*RII or *Bst*NI, *Sau*96I, and *Ava*II or *Sin*I. Since we have sequenced DNA molecules of over 6000 nucleotides using DNA polymerase repair labeling exclusively,[20] combinations of these enzymes should suffice for most requirements.

5. Preliminary mapping of the insert by the method of Smith and Birnstiel[14a] is rapid and useful. The amount of DNA and the degree of completion of the partials needed can be judged, and the enzymes for the partial digestions can be judiciously chosen.

6. The DNA region to be sequenced must not contain sites for either of the enzymes used to open the plasmid or to cut off one end after mercuration. Since pBR322 contains at least 16 unique restriction sites (see Appendix of New England BioLabs catalog), this requirement should not generally pose a problem. The varying sensitivities of restriction sites to the partial digestions, and the fact that each plasmid molecule yields only one (mercurated) "sequenceable" fragment, dictate that 5–10 pmol of plasmid be used for each fragment to be prepared. As a consequence, the plasmid is ideally opened quite close to the insert, so that relatively few sites occur between the mercurated end and the DNA to be sequenced.

Finally, the preparative gel shown in Fig. 4 demonstrates that "sequenceable" fragments of 3000 bp and more can be obtained by the approach described above. Because the labor involved is independent of the number of fragments until the final step of the protocol (elution of the labeled DNAs from the preparative gel), the time required is virtually the

[20] J. L. Hartley and J. E. Donelson, *Nature (London)* **286,** 860 (1980).

FIG. 4. Ethidium bromide stain of a preparative 0.85% agarose gel of three sets of mercurated, labeled, partial restriction fragments that bound to the thiol agarose column (see the three lanes 2 in Fig. 3). Lanes 1–4 in this figure are multiple applications of the same three samples described in Fig. 3. Arrows show the fragments that were eluted for sequencing and point to the same fragments as do the arrows in Fig. 3. Bands on the far right are standard-length DNA fragments.

same whether the number of fragments sought is 4 or 40. Given suitable restriction sites in the vector and sufficient sites in the insert, the method presented here permits rapid sequencing of the cloned DNAs of several thousand base pairs or more with little data handling and a minimum of sequencing runs. These characteristics suggest the approach has wide applicability to projects requiring the determination of DNA sequences. In addition the procedure may have uses in other projects that require the isolation of partial restriction fragments—for example, the isolation of DNA fragments of increasing size but containing a common sequence.

Acknowledgments

We thank Kenneth K. Chen and David Dorfman for doing the Maxam–Gilbert DNA sequence determinations and John Young, Phelix Majiwa, and Richard Williams of ILRAD, Nairobi, Kenya, for providing plasmid pcB 1.1. The research was supported by USPHS Grants AM25295 and AI 16950 and NSF Grant PCM 76-1341.

Section II

Enzymes Affecting the Gross Morphology of DNA

A. Topoisomerases Type I
Articles 7 and 8

B. Topoisomerases Type II
Articles 9 through 11

[7] HeLa Toposiomerase I[1]

By LEROY F. LIU

Similar to the prokaryotic systems, there are two types of DNA topoisomerases in animal cells.[2-6] Both type I and type II DNA topoisomerases have been purified to homogeneity from HeLa cells.[7,8] The type I DNA topoisomerase from HeLa cells (HeLa topoisomerase I) similar to all other eukaryotic type I DNA topoisomerases, introduces transient single-stranded breaks on duplex DNA,[7] resulting in the relaxation of a superhelical DNA or other types of topological rearrangements of DNA.[2-5] HeLa topoisomerase I can also spontaneously break down single-stranded DNA, generating active enzyme-linked fragments of single-stranded DNA.[9,10] The exact biological functions of animal topoisomerase I are still unknown. Processes such as DNA sequence rearrangements,[9-11] DNA replication, RNA transcription, and genetic recombination have been suggested to require such an activity.[2,3]

Assay Method

Principle. The most convenient assay for HeLa topoisomerase I is the relaxation of superhelical DNA.

Reagents

Reaction buffer: 40 m*M* Tris-HCl, pH 7.5, 120 m*M* KCl, 10 m*M* $MgCl_2$, 0.1 m*M* EDTA, 0.1 m*M* dithiothreitol, 30 μg of bovine serum albumin and 20 μg of superhelical DNA per milliliter, and the enzyme.

[1] This work was supported by an NIH grant and the Chicago Community Trust/Searle Scholars Program.

[2] J. C. Wang and L. F. Liu, *in* "Molecular Genetics" (J. H. Taylor, ed.), Part 3, pp. 65–88. Academic Press, New York, 1979.

[3] J. J. Champoux, *Annu. Rev. Biochem.* **47,** 449 (1978).

[4] N. R. Cozzarelli, *Cell* **22,** 327 (1980).

[5] M. Gellert, *Annu. Rev. Biochem.* **50,** 879 (1981).

[6] L. F. Liu, C. C. Liu, and B. M. Alberts, *Cell* **19,** 697 (1980).

[7] L. F. Liu and K. G. Miller, *Proc. Natl. Acad. Sci. U.S.A.* **78,** 3487 (1981).

[8] K. G. Miller, L. F. Liu, and P. T. Englund, *J. Biol. Chem.* **256,** 9334 (1981).

[9] M. D. Been and J. J. Champoux, *Proc. Natl. Acad. Sci. U.S.A.* **78,** 2883 (1981).

[10] B. D. Halligan, J. L. Davis, K. E. Edwards, and L. F. Liu, *J. Biol. Chem.* **257,** 3995 (1982).

[11] J. C. Wang, R. I. Gumport, K. Javaherian, K. Kirkegaard, L. Klevan, M. L. Kotewicz, and Y. C. Tse, *in* "Mechanistic Studies of DNA Replication and Genetic Recombination" (B. M. Alberts, ed.), pp. 769–784. Academic Press, New York, 1980.

ISBN 0-12-182000-9

Stop solution: 5% (w/v) Sarkosyl, 50 mM EDTA, pH 8.0, 25% (w/v) sucrose, 0.25 mg of bromophenol blue per milliliter.

TBE buffer for agarose gel electrophoresis: 0.09 M Tris-borate, pH 8.3, 2.5 mM EDTA.

Agarose gel, 0.7%

Procedure. Reactions (20 μl) were incubated at 30° for 30 min and stopped by 5 μl of the stop solution. The samples were then analyzed by electrophoresis on a 0.7% agarose gel. Because of the sensitivity of the assay, activities in the crude extracts can be easily monitored by serial dilutions.

Definition of Enzyme Unit. One unit of the enzyme is defined by 50% relaxation of the plasmid DNA (0.4 μg) under our assay conditions.

Enzyme Purification

Cell Growth and Isolation of Nuclei. HeLa cells (S-3) were grown in suspension in MEM (minimal essential medium) supplemented with 5% horse serum, and 100 units of penicillin and 0.1 mg of streptomycin sulfate per milliliter. Fifteen liters of culture (5 to 6 × 10^5 cells/ml) were processed for each enzyme purification. All procedures were carried out between 0 and 4°. The cells (about 20 g wet weight) were pelleted and washed twice with PBS (0.01 M sodium phosphate, pH 7.5, 0.15 M NaCl) and then resuspended in 125 ml of extraction buffer (5 mM potassium phosphate, pH 7.5, 2 mM $MgCl_2$, 1 mM phenylmethylsulfonyl fluoride (PMSF) (added as a 0.1 M solution in isopropanol), 1 mM mercaptoethanol, 0.5 mM dithiothreitol (DTT), 0.1 mM EDTA. After 30 min, the swollen cells were homogenized with a Dounce homogenizer (loose pestle). Cell disruption was monitored by a phase microscope and usually required at least 20 strokes. The nuclei were collected by centrifugation (2000 g for 10 min) and then washed once with 5 mM potassium phosphate, pH 7.5, 1 mM PMSF, 1 mM mercaptoethanol, 1 mM DTT.

Lysis Nuclei and PEG Precipitation. The washed nuclei were resuspended in total of 125 ml of the nuclei wash buffer. EDTA was then added to a final concentration of 4 mM. After 1 hr on ice, nuclei were lysed by the slow addition of an equal volume of 2 M NaCl, 100 mM Tris-HCl, pH 7.5, 10 mM mercaptoethanol, and 1 mM PMSF. DNA in the gelatinous nuclear extract was precipitated by the slow addition of 125 ml of 18% (w/v) polyethylene glycol (PEG) (in 1 M NaCl, 50 mM Tris-HCl, pH 7.5) with constant stirring and then removed by centrifugation at 15,000 g for 25 min.

Hydroxyapatite Chromatography. The PEG supernatant was loaded directly onto a hydroxyapatite column (2 × 10 cm) equilibrated with 1 M NaCl, 50 mM Tris, pH 7.5, 6% (w/v) PEG, 10 mM mercaptoethanol,

1 mM PMSF. After washing the column with 100 ml of 0.2 M potassium phosphate, pH 7.0, 10% (w/v) glycerol, 10 mM mercaptoethanol, and 1 mM PMSF, the column was developed with 200 ml of a linear salt gradient (0.2 to 0.7 M potassium phosphate in the same column buffer). The fractions containing the topoisomerase activity (eluted between 0.4 and 0.5 M) were pooled.

Phosphocellulose Chromatography. The pool of the hydroxyapatite fractions was diluted with an equal volume of 10% (w/v) glycerol, 5 mM potassium phosphate, pH 7.0, 10 mM mercaptoethanol, 1 mM PMSF, and 0.1 mM EDTA before loading onto a phosphocellulose (P-11) column (1 × 0.5 cm) equilibrated with 0.2 M potassium phosphate, pH 7.0 in solution A (10% glycerol, 10 mM mercaptoethanol, 1 mM PMSF, and 0.1 mM EDTA). After washing the column with 5 ml of the column buffer, the enzyme was eluted with 15 ml of a linear salt gradient (0.2 to 0.7 M potassium phosphate, pH 7.0, in solution A). The fractions containing topoisomerase activity (between 0.5 and 0.6 M) were pooled.

Single-Stranded DNA Cellulose Chromatography. The pool of the phosphocellulose fractions was dialyzed (twice, for 2 hr) against 0.1 M KCl in solution B (40 mM Tris-HCl, pH 7.5, 10% glycerol, 10 mM mercaptoethanol, 1 mM PMSF, and 0.1 mM EDTA) and loaded onto a single-stranded DNA cellulose column (1 × 0.3 cm). The column was then washed with 4-ml aliquots of 0.1 M, 0.2 M, 0.3 M, 0.4 M, and 1.0 M KCl in solution B. The activity was eluted in the 0.4 M wash. The pooled enzyme fractions were then dialyzed against 50% (w/v) glycerol, 70 mM potassium phosphate, pH 7.0, 0.5 mM DTT, and 0.1 mM EDTA, and stored at −20°. The enzyme has been stored for 2 years without loss of activity.

A typical purification is summarized in the table.

Purification of HeLa Topoisomerase I

Step	Volume (ml)	Total protein (mg)	Specific activity (units/ml)	Yield[a] (%)
1. PEG supernatant	375	262	8.8×10^5	(100)
2. Hydroxyapatite chromatography	35	4.4	3.4×10^7	65
3. Phosphocellulose chromatography	2	0.7	2.1×10^8	65
4. Single-stranded DNA cellulose chromatography	1	0.5	2.2×10^8	48

[a] The enzyme activity was measured by serial dilution of the enzyme using an enzyme diluent containing 6% (w/v) polyethylene glycol (PEG) in the reaction buffer. The inclusion of PEG in the enzyme diluent stimulates the topoisomerase activity fivefold.

Properties of the Purified Enzyme

Molecular Weight. The native enzyme is a monomeric protein of 100,000 daltons. The previously identified nicking-closing enzyme from rat liver (about 66,000 daltons) is most likely a fully active proteolytic fragment of topoisomerase I.[7] HeLa topoisomerase I, as well as other eukaryotic type I topoisomerases, is very sensitive to proteolysis. The highest molecular weight form of topoisomerase I from other eukaryotic tissues ranges from 100,000 to 130,000.[11,12] HeLa topoisomerase I is highly rich in lysine and is strongly inhibited by heparin.[13] More than 90% of the enzyme activity resides in the nucleus.

Enzyme Activity. HeLa topoisomerase I remains active in the absence of Mg(II) ion, but the activity is 5- to 50-fold reduced. In the absence of Mg(II) ion, the presence of EDTA does not further reduce the activity. Topoisomerase I binds tightly to histone H1 and nonhistone HMG proteins and is stimulated by these proteins 5- to 50-fold depending on the assay conditions.[13] Heparin strongly inhibits topoisomerase I.[13] Topoisomerase I binds tightly to both DNA and RNA. Studies have shown that topoisomerase I also binds tightly to nucleosomes and ribonucleoproteins (RNP).[13]

Topological Reactions. Topoisomerase I can relax both positive twists and negative twists of DNA to completion. In addition, topoisomerase I promotes a very inefficient reaction that catenates nicked circular DNA.[14,15] The reverse reaction (decatenation) has not been demonstrated.[16] Topoisomerase I can also promote the complete renaturation of two complementary single-stranded DNA rings.[17,18]

Covalent Strand Transfer Reactions. Topoisomerase I spontaneously breaks down single-stranded DNA to smaller fragments.[9,10] Each cleavage generates a free 5′-OH end and an enzyme-linked 3′-phosphoryl end. The enzyme-linked single strands are enzymatically potent. They can react with the 5′-OH groups of a variety of DNA, resulting in the intra- or intermolecular covalent joining of two pieces of DNA and release of the enzyme.[9,10]

Specificity. The cleavage sites of topoisomerase I on both double- and single-stranded DNA have been determined by the nucleotide sequencing

[12] W. S. Dynan, J. J. Jendrisak, D. A. Hager, and R. R. Burgess, *J. Biol. Chem.* **256,** 5860 (1981).

[13] K. Javaherian and L. F. Liu, *Nucleic Acids Res.* (in press).

[14] Y. C. Tse and J. C. Wang, *Cell* **22,** 269 (1980).

[15] P. O. Brown and N. R. Cozzarelli, *Proc. Natl. Acad. Sci. U.S.A.* **78,** 843 (1980).

[16] L. F. Liu, J. L. Davis, and R. Calendar, *Nucleic Acids Res.* **9,** 3979 (1981).

[17] K. Kirkegaard and J. C. Wang, *Nucleic Acids Res.* **5,** 3811 (1978).

[18] J. J. Champoux, *Proc. Natl. Acad. Sci. U.S.A.* **74,** 3800 (1977).

method.[19] Topoisomerase I shows rather weak sequence specificity as defined by its cleavage reaction. The majority of the cleavage sites map at the exact nucleotide positions, whether single-stranded DNA or its corresponding double-stranded DNA were used.[19] The nucleotides that are linked to the enzymes at the cleavage sites are predominantly pyrimidine nucleotides.[19]

[19] K. E. Edwards, B. D. Halligan, J. L. Davis, N. L. Nivera, and L. F. Liu, *Nucleic Acids Res.* **10,** 2565 (1982).

[8] Multiple Forms of Rat Liver Type I Topoisomerase

By SUSAN R. MARTIN, WILLIAM K. MCCOUBREY, JR., BETTY L. MCCONAUGHY, LISA S. YOUNG, MICHAEL D. BEEN, BONITA J. BREWER, and JAMES J. CHAMPOUX

The type I topoisomerase from rat liver nuclei was originally described as a single-subunit enzyme with a molecular weight of 66,000.[1] In subsequent experiments we noticed that the addition of the protease inhibitor phenylmethylsulfonyl fluoride (PMSF) altered the elution profile from phosphocellulose and resulted in the isolation of higher molecular weight forms of the enzyme. Liu and Miller[2] have reported the isolation of a 100,000 molecular weight topoisomerase as well as a 67,000 molecular weight species from HeLa cells. They presented evidence showing that the smaller form of the enzyme was probably derived from the larger form by proteolytic cleavage either *in vivo* or *in vitro*. Following these leads, we have reexamined the purification of the rat liver topoisomerase in the presence of PMSF. Under these conditions we can identify an additional two, possibly three, forms of the topoisomerase that are separable by chromatography on phosphocellulose. Here we describe the assay and partial characterization of these rat liver type I topoisomerases.

Materials and Reagents

The sources for most of the materials have been given previously.[1] PMSF from Sigma was dissolved in absolute ethanol at 100 m*M* and added to buffers just prior to use. Column buffer is 1 m*M* EDTA, 0.5 m*M* dithiothreitol, 10% glycerol containing the indicated concentration of KPO_4, pH 7.4.

[1] J. J. Champoux and B. L. McConaughy, *Biochemistry* **15,** 4638 (1976).
[2] L. F. Liu and K. G. Miller, *Proc. Natl. Acad. Sci. U.S.A.* **78,** 3487 (1981).

ISBN 0-12-182000-9

Methods

Agarose Gel Assay

Reaction mixtures (final volume 20 μl) contain 20 m*M* Tris-HCl, pH 7.5, 150 m*M* KCl, 1 m*M* EDTA, SV40 DNA at 5 μg/ml, and 2 μl of enzyme diluted in column buffer with 20 m*M* KPO_4, pH 7.4. These conditions are specific for the type I enzyme since the type II enzymes require Mg^{2+} and ATP for activity.[3-5] The reactions are incubated for 10 min at 37° and stopped by the addition of 10 μl of 2.5% sodium dodecyl sulfate (SDS), 25 m*M* EDTA, 25% Ficoll, 0.03% bromophenol blue. The samples are subjected to electrophoresis for 16–20 hr at 0.8 V/cm in a 1% agarose gel containing 90 m*M* Tris-borate, pH 8.3, 2.5 m*M* EDTA, to separate the superhelical (form I) and relaxed (form Ir) forms of the closed circular DNA.[6] The gel is stained with 0.1 μg of ethidium bromide per milliliter in the electrophoresis buffer and photographed with UV illumination.

Under these conditions relaxed closed circles and nicked circles (form II) migrate close together. Occasionally crude extracts contain an endonuclease that exhibits some activity under these assay conditions. In such a case it is important to distinguish between the nuclease and topoisomerase activities. Separation of form Ir and form II DNAs can be conveniently accomplished by subjecting the stained gel to an additional 3 hr of electrophoresis at 2 V/cm in the presence of 0.1 μg of ethidium bromide per milliliter. The ethidium bromide increases the mobility of the form Ir relative to the form II and allows one to quantitate all three species in the same experiment (Fig. 1).

Quantitation of enzyme activity is achieved by running a set of reactions containing serial twofold dilutions of the fraction to be assayed (Fig. 1). The amount of enzyme that gives 50% conversion of the superhelical substrate to relaxed product is defined as one unit and can usually be estimated simply from a visual inspection of the gel photograph. Units determined in this manner are identical to the units determined by the filter assay described previously.[1] Although this method is laborious and time consuming, it remains our method of choice to quantitate the enzyme activity in crude extracts containing extraneous nucleic acids. After removal of nucleic acids one can easily and rapidly quantitate the topoisomerase by a fluorometric assay.

[3] T. S. Hsieh and D. Brutlag, *Cell* **21,** 115 (1980).

[4] M. I. Baldi, P. Benedetti, E. Mattoccia, and G. P. Tocchini-Valentine, *Cell* **20,** 461 (1980).

[5] L. F. Liu, *Mechanistic Studies of DNA Replication and Genetic Recombination, ICN-UCLA Symp. Mol. Cell. Biol.* **19,** 817 (1980).

[6] W. Keller, *Proc. Natl. Acad. Sci. U.S.A.* **72,** 2550 (1975).

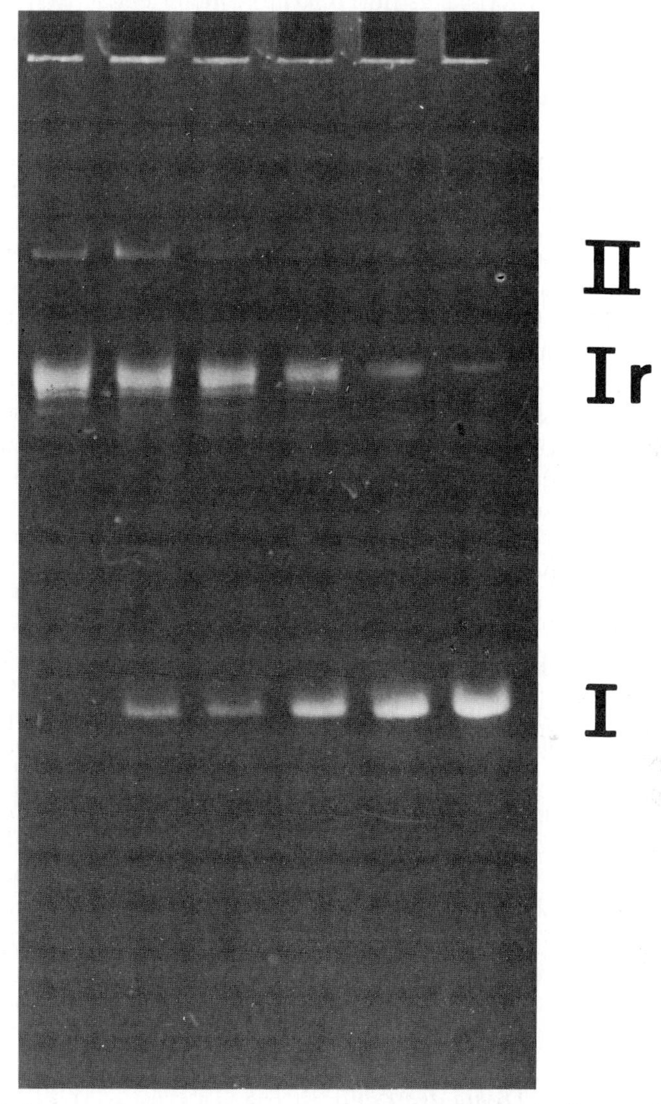

FIG. 1. Agarose gel assay. A crude extract from human fibroblast cells grown in culture was prepared by lysis of nuclei with 1 M NaCl as described.[1] The extract was diluted 1/10, and a 2-μl aliquot was assayed (lane a). Successive twofold dilutions were similarly assayed for lanes b through f. A vertical 1% agarose gel was run for 20 hr at 0.8 V/cm, stained with ethidium bromide, run an additional 3 hr at 2 V/cm, and photographed (see text). The positions of nicked (form II), relaxed closed circles (form Ir), and superhelical (form I) SV40 DNA are indicated.

Fluorometric Assay

Reaction mixtures (final volume 0.30 ml) contain 20 m*M* Tris-HCl, pH 7.5, 1 m*M* EDTA, 170 m*M* NaCl, 4 μg of ethidium bromide per milliliter, and 4 μg of SV40 DNA per milliliter. Under these conditions, sufficient dye is bound to the DNA to introduce positive superhelical turns that can be relaxed by the topoisomerase.[7] Relaxation of the positive turns results in approximately a 30% increase in bound ethidium.[8] By carrying out the reaction in the cuvette of a fluorometer, one can measure the increase in fluorescence that accompanies the increase in binding. The reaction is initiated by the addition of 2–10 μl of the fraction to be assayed directly to the reaction mix in the cuvette. The sample is mixed rapidly by inversion and placed in the fluorometer; the time course of the fluorescent increase is monitored by a recorder attached to the fluorometer. The number of fluorometric units in the reaction mixture is defined as the reciprocal of the time in minutes required for half-maximal reaction at room temperature. We have found that, for halftimes in the range from 0.5 to 5 min, the units defined in this way are directly proportional to enzyme concentration in the reaction.[1]

Phosphocellulose Chromatography

The procedures for the high-salt lysis of rat liver nuclei and the removal of DNA by polyethylene glycol precipitation have been described.[1] We now include PMSF at 1 m*M* in all buffers from the time of tissue homogenization. The polyethylene glycol supernatant (100–150 ml) is dialyzed against column buffer containing 0.25 *M* KPO_4 and loaded onto a phosphocellulose column (1.5 $\times$ 4 cm) at a flow rate of 30 ml/hr. The column is washed with 100 ml of the same buffer and eluted with a linear gradient from 0.25 to 0.85 *M* KPO_4 in 240 ml of column buffer. The fractions are assayed for topoisomerase activity with the fluorometric assay, and the salt gradient is determined from the refractive index.

In the absence of PMSF, all the topoisomerase activity elutes in a peak centered at 0.36–0.37 *M* KPO_4.[1] In the presence of PMSF (Fig. 2) the first peak of activity appears at approximately 0.39 *M* KPO_4, followed by a second peak at 0.44 *M* and a third broader peak at 0.48 *M* KPO_4.

Sephadex G-150 Chromatography

Samples to be chromatographed are concentrated to 1.2–1.5 ml, and glycerol is added to a final concentration of 25%. The sample is layered

[7] J. J. Champoux and R. Dulbecco, *Proc. Natl. Acad. Sci. U.S.A.* **69,** 143 (1972).
[8] W. Bauer and J. Vinograd, *J. Mol. Biol.* **54,** 281 (1970).

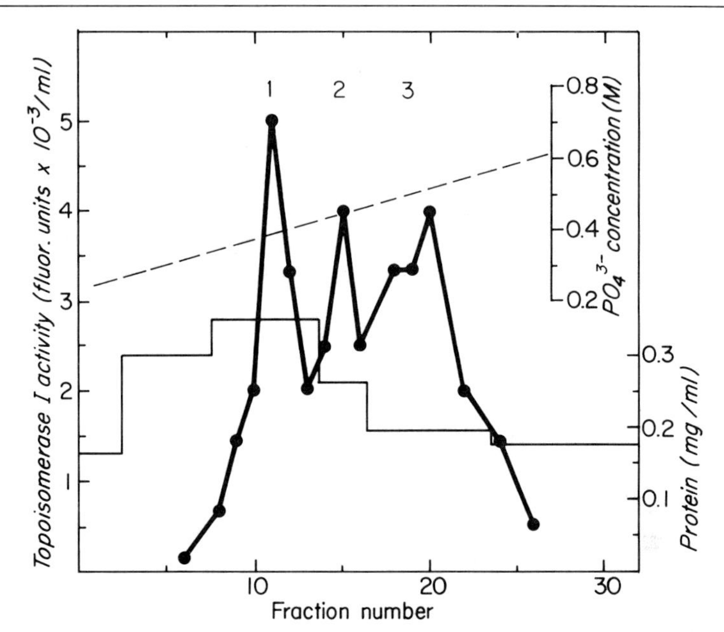

FIG. 2. Phosphocellulose chromatography. Rat liver topoisomerase activity (heavy line, filled circles) was assayed by the fluorometric assay. Protein concentrations for the indicated regions are indicated by the thin solid line. Phosphate concentration (dashed line) was measured with a refractometer. The peaks of topoisomerase activity are numbered in the order in which they elute from phosphocellulose.

onto a Sephadex G-150 column (2.6 × 80 cm) equilibrated with column buffer containing 70 m*M* KPO_4. The column is eluted with the same buffer at a flow rate of 10 ml/hr, and the activity is determined fluorometrically.

Characterization of Topoisomerase Species

The first peak of activity from the phosphocellulose column (0.39 *M* KPO_4) shown in Fig. 2 was pooled, dialyzed, and chromatographed on carboxymethyl–Sephadex (CM-50) as described.[1] Analysis of this material by Sephadex G-150 gel filtration yielded a single peak of activity eluting between 180 and 200 ml (Fig. 3, open circles). For comparison, the elution profile is shown for the 66,000 molecular weight form prepared in the absence of PMSF and analyzed in a separate experiment on the same column (Fig. 3, triangles). These results indicate that the first peak from the phosphocellulose column shown in Fig. 2 is significantly larger than the original 66,000 molecular weight species. Chromatography of this phosphocellulose fraction on phenyl Sepharose followed by SDS–acryl-

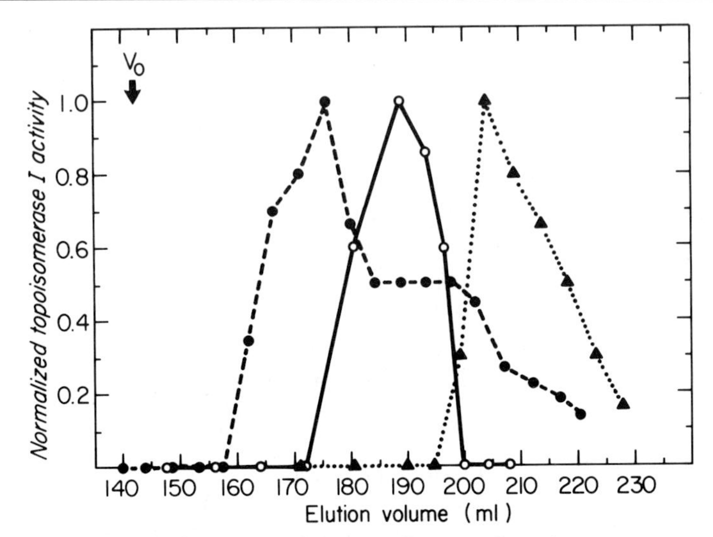

FIG. 3. Sephadex G-150 chromatography. Rat liver type I topoisomerase was purified in the absence of phenylmethylsulfonyl fluoride and chromatographed on G-150 (triangles). This species has been shown previously to cochromatograph with bovine serum albumin and has been assigned a native molecular weight of 66,000.[1] In separate experiments the first peak eluting from phosphocellulose (open circles) and the third peak from phosphocellulose (filled circles) were analyzed on the same column.

amide gel analysis[9] revealed a major band with a molecular weight of 77,000 that could be shown by the method of Hager and Burgess[10] to possess topoisomerase activity. Thus it appears that little or no 66,000 molecular weight enzyme is detectable when the isolation is carried out in the presence of PMSF.

The third peak from the column shown in Fig. 2 was pooled, concentrated, and analyzed directly by Sephadex G-150 chromatography. The results of this analysis are also presented in Fig. 3 (filled circles). Most of the activity eluted ahead of the 77,000 molecular weight species at a position consistent with a molecular weight of approximately 100,000. SDS-acrylamide gel analysis of this peak after phenyl Sepharose chromatography confirmed the presence of a 100,000 molecular weight species that could be shown to have topoisomerase activity.[9] The presence of this species is consistent with the observations of Liu and Miller[2] on the HeLa cell enzyme.

[9] B. L. McConaughy and J. J. Champoux, unpublished observations.

[10] D. A. Hager and R. R. Burgess, *Anal. Biochem.* **109,** 76 (1980).

The amount of the second peak observed during phosphocellulose chromatography was variable, and we have not characterized this species any further.

Comments

We have not established in a rigorous way the relationship between these various species of the type I topoisomerase from rat liver cells. However, there is reason to suspect that phosphocellulose peak 1 and possibly peak 2 are derived either *in vivo* or *in vitro* from peak 3 by proteolytic cleavage, similar to the situation reported for HeLa cells.[2] The observation that the total units of activity recovered at the phosphocellulose stage is the same whether or not we include PMSF suggests that, in the absence of PMSF, all the forms are converted to a common species by proteolysis in the crude extract. In addition, prolonged storage of the 77,000 molecular weight form in the absence of PMSF results in a breakdown of the species to the 66,000 molecular weight form. It is noteworthy that none of the 66,000 form of the enzyme is detectable in the presence of PMSF. Presumably this form of the enzyme is not present *in vivo*. We have not yet been able to determine whether the phosphocellulose peak 1 and possibly peak 2 detected in the presence of PMSF preexist in the cell or are generated after the cells are broken open. If they are produced by proteolysis of the large form (peak 3) in the crude extracts, then it is possible that a protease different from the one that generates the 66,000 form is involved.

The existence of a protease-resistant domain that retains nicking–closing activity suggests that the protease-sensitive portion of the molecule may have other functions, such as substrate recognition or subcellular compartment recognition. The presence of multiple species in extracts prepared in the presence of PMSF raises the interesting possibility that different-sized species may have different physiological roles in the cell.

We have previously shown that the processivity of the 66,000 molecular weight form of the enzyme decreases as the monovalent cation concentration is raised from 50 mM to 150 mM.[11] Similar experiments have shown that the largest form of the enzyume (100,000 molecular weight) is more processive than the small form at 150 mM salt.[9] Thus, it appears that the DNA binding of the large form is somewhat tighter than that of the

[11] B. L. McConaughy, L. S. Young, and J. J. Champoux, *Biochim. Biophys. Acta* **655,** 1 (1981).

small form. This observation may be related to the elution of the large form at the higher salt concentrations from phosphocellulose. Attempts to discern other differences between the largest and the smallest forms have failed.

The purified eukaryotic type I topoisomerase is useful in studying the effect of ligands on the secondary and tertiary structure of DNA and in preparing a variety of different topological variants of circular DNAs. The enzyme is active over a wide range of monovalent salt concentrations below 0.2 *M*, does not require a divalent cation, and can completely relax positive as well as negative superhelical turns. In addition, the rat liver enzyme will break single-stranded DNA to yield a covalent DNA–enzyme complex.[12] The attached enzyme remains active and can attach a free single-strand to the end of the broken strand.[13] We are currently investigating the possibility of exploiting this reaction for the cloning of single-stranded DNA. For these and similar studies the topoisomerase prepared from rat liver nuclei in the absence of PMSF appears to be adequate and, unlike the larger forms described here, can easily be obtained as a stable, homogeneous species in a reasonable yield.

Acknowledgments

This work has been supported by National Institutes of Health Research Grant GM23224. W. K. M. and L. S. Y. were predoctoral trainees of the National Institutes of Health Grant 1-T32-GM07270, and B. J. B. was supported by National Institutes of Health postdoctoral fellowship 5-F32-GM07234.

[12] M. D. Been and J. J. Champoux, *Nucleic Acids Res.* **8,** 6129 (1980).
[13] M. D. Been and J. J. Champoux, *Proc. Natl. Acad. Sci. U.S.A.* **78,** 2883 (1981).

[9] *Escherichia coli* Phage T4 Topoisomerase

By KENNETH N. KREUZER and CORNELIS VICTOR JONGENEEL

Topoisomerases are enzymes that alter DNA topology by changing the linking number of circular duplex DNA molecules and by interconverting topologically knotted or catenated DNA forms. The so-called type II topoisomerases act by a mechanism involving the passage of a duplex segment of DNA through a transient double-strand break in another segment of DNA.[1] A novel ATP-dependent type II topoisomerase with

[1] Reviewed by N. R. Cozzarelli, *Science* **207,** 953 (1980) and M. Gellert, *Annu. Rev. Biochem.* **50,** 879 (1981).

ISBN 0-12-182000-9

DNA-dependent ATPase activity has been isolated from extracts of bacteriophage T4-infected *E. coli* cells.[2,3] The enzyme is composed of three subunits, coded for by phage genes *39, 52,* and *60.* It has a high specific activity for topoisomerization reactions and can be easily purified to near homogeneity in milligram amounts (see below).

A role for the T4 topoisomerase in the initiation of T4 DNA replication is suggested by physiological studies of topoisomerase-deficient mutants. Conditional mutations in genes *39, 52,* or *60* result in the "DNA-delay" phenotype, in which the onset of DNA synthesis is delayed and the cumulative total of DNA synthesized is reduced under restrictive conditions.[4] Autoradiographic studies have revealed that the rate of replication fork movement is normal, but the number of forks is reduced, as expected if the defect is confined to the initiation of replication forks.[5] The residual replication observed in these mutants appears to be dependent on the host DNA gyrase.[6] This observation, together with the fact that T4 topoisomerase cannot supercoil various *E. coli* plasmid DNAs *in vitro,* led to the proposal that the topoisomerase is a supercoiling enzyme specific for the T4 replication origin.[2] We are currently studying the interaction of T4 topoisomerase with DNA in the origin region of the T4 chromosome and attempting to reconstitute the initiation reaction *in vitro* in order to test this and other models for the involvement of the enzyme in the DNA replication process.

Purification

The following purification procedure differs significantly from previously published methods.[2,3] The present protocol gives both better yields (up to 5 mg per 100 g of cells) and a higher final purity.

Buffers

Buffer A: 20 m*M* NaCl, 40 m*M* Tris-HCl (pH 8.1), 10 m*M* $MgCl_2$, 2 m*M* $CaCl_2$, 1 m*M* Na_3EDTA, 1 m*M* 2-mercaptoethanol; 1 m*M* phenylmethylsulfonyl fluoride (PMSF) and 10 m*M* benzamidine-HCl are added immediately before use.

[2] L. F. Liu, C.-C. Liu, and B. M. Alberts, *Nature (London)* **281,** 456 (1979).

[3] G. L. Stetler, G. J. King, and W. M. Huang, *Proc. Natl. Acad. Sci. U.S.A.* **76,** 3737 (1979).

[4] R. H. Epstein, A. Bolle, C. M. Steinberg, E. Kellenberger, E. Boy de la Tour, R. Chevalley, R. S. Edgar, M. Susman, G. H. Denhardt, and A. Lielausis, *Cold Spring Harbor Symp. Quant. Biol.* **28,** 375 (1964).

[5] D. McCarthy, C. Minner, H. Bernstein, and C. Bernstein, *J. Mol. Biol.* **106,** 963 (1976).

[6] D. McCarthy, *J. Mol. Biol.* **127,** 265 (1979).

Buffer B: 100 mM NaCl, 20 mM Tris-HCl (pH 8.1), 5 mM Na_3EDTA, 1 mM 2-mercaptoethanol; 1 mM PMSF is added immediately before use.

Buffer C: 20 mM Tris-HCl (pH 8.1), 1 mM Na_3EDTA, 1 mM 2-mercaptoethanol, 10% (v/v) glycerol; in addition, buffer C1 contains 0.15 M NaCl, buffer C2 contains 0.25 M NaCl, buffer C3 contains 0.6 M NaCl, and buffer C4 contains 2 M NaCl.

Buffer D: Equimolar amounts of KH_2PO_4 and K_2HPO_4 (pH ~6.8), 10 mM 2-mercaptoethanol, 10% (v/v) glycerol. The potassium phosphate concentrations are: buffer D0, 0.02 M; D1, 0.1 M; D2, 0.25 M; D3, 0.4 M; D4, 0.55 M; D5, 1 M.

Buffer E1: 40 mM Tris-HCl (pH 7.8), 20 mM NaCl, 0.5 mM Na_3EDTA, 1 mM 2-mercaptoethanol, 10% (v/v) glycerol. Buffer E2 is E1 supplemented with 25% (w/v) $(NH_4)_2SO_4$.

Buffer F: 30 mM potassium phosphate (pH 7.2), 10 mM 2-mercaptoethanol, 0.5 mM Na_3EDTA, containing 10% or 50% (v/v) glycerol.

Bacterial and Bacteriophage Strains

Escherichia coli B_E is used as the host; it is infected with the bacteriophage T4 double mutant *amN134 amBL292* (genes 33^- 55^-). This mutant cannot shift from early to late gene expression and overproduces early gene products. It was obtained from Dr Junko Hosoda (University of California, Berkeley).

Purification Procedure

Step 1. Growth of T4-Infected Cells. Escherichia coli B_E is grown at 37° in a 220-liter fermentor in H broth (Difco). At a cell density of 6×10^8/ml, T4 bacteriophage is added at a multiplicity of infection of 5 to 10. Two and a half hours after infection, the cells are collected in a Sharples centrifuge at room temperature. The cell paste is placed on disposable plastic trays, wrapped in heat-sealable plastic bags, and then quickly frozen in liquid nitrogen and stored at −70°. This procedure typically yields about 600 g of cell paste per 220-liter fermentor run.

Step 2. Lysis of the Cells. Cell paste (250 g) is added to 750 ml of buffer A and blended at high speed in a Waring blender until the mixture becomes a homogeneous suspension without any visible ice crystals. After a short centrifugation (5 min at 5000 rpm in the Sorvall GSA rotor) to remove trapped air bubbles, the resulting liquid is divided into two equal aliquots in stainless steel beakers and kept on ice. Each aliquot is sonicated with the large tip of a Branson Sonifier at maximum power, keeping the temperature of the sample below 10°. Sonication is considered complete when the turbidity of the sample, monitored by its absorption at 600 nm, no longer decreases. This usually requires 5-8 min of sonication. The two aliquots are combined, 20 mg of pancreatic DNase I (Worthington) are added, and the lysate is incubated for 15–20 min at 15°. The viscosity should decrease dramatically during this last step.

Step 3. Centrifugation and Dialysis. The lysate is clarified by two rounds of centrifugation, being first spun for 90 min at 14,000 *g* (9000 rpm in the Sorvall GSA rotor). The resulting supernatant is then centrifuged for 4 hr at 140,000 *g* (35,000 rpm in the Beckman type 35 rotor). The second supernatant is dialyzed against three 20-liter changes of buffer B, allowing at least 6 hr for each change. The lysate must be dialyzed extensively at this point, as residual Mg^{2+} will activate the previously added DNase I, resulting in degradation of the DNA-cellulose column. The use of a large number of small dialysis bags helps make this step more effective. At the end of dialysis, prechilled glycerol is added to a final concentration of 10% (v/v).

Step 4. Single-Stranded DNA-Cellulose Chromatography. The cleared, dialyzed lysate is applied at a flow rate of ~100 ml/hr to a 2.5 × 30 cm (~150 ml packed volume) DNA-cellulose column containing about 1 mg of single-stranded calf thymus DNA (Worthington Biochemicals) per milliliter of bed volume.[7] The column is washed with buffer C1 until the effluent is free of protein, and bound proteins are eluted stepwise at the same flow rate with about three column volumes each of buffers C2, C3, and C4. The topoisomerase elutes with buffer C2 (0.25 *M* NaCl). It can be detected with a topoisomerase assay (see below), but we find it more convenient to analyze fractions by sodium dodecyl sulfate–polyacrylamide gel electrophoresis (SDS–PAGE), as the topoisomerase subunits are easily detectable on a gel at this stage (see Fig. 1, lane c). The fraction eluted with buffer C4 (2 *M* NaCl) can also be saved, as it provides an excellent source of the T4 gene *32* protein (helix destabilizing protein) and the T4 gene *61* protein (a component of the T4 primase).

Step 5. Hydroxyapatite Chromatography. A 2.5 × 20 cm column of Bio-Rad HTP hydroxyapatite (100 ml packed volume), containing 20% (w/w) Whatman CF-11 cellulose to improve its flow characteristics, is equilibrated with buffer D1. The pH of the D series buffers is kept near 6.8. (The final yield is dramatically reduced, and the topoisomerase elutes at a higher phosphate concentration, when the column is run at a higher pH.) The DNA-cellulose eluate is loaded directly onto the hydroxyapatite column, which is then rinsed with buffer D1, followed by stepwise washes with buffers D2 through D5, all at a flow rate of 120 ml/hr. The topoiso-

[7] B. M. Alberts and G. Herrick, this series, Vol. 21, p. 198.

a
b
c
d
e

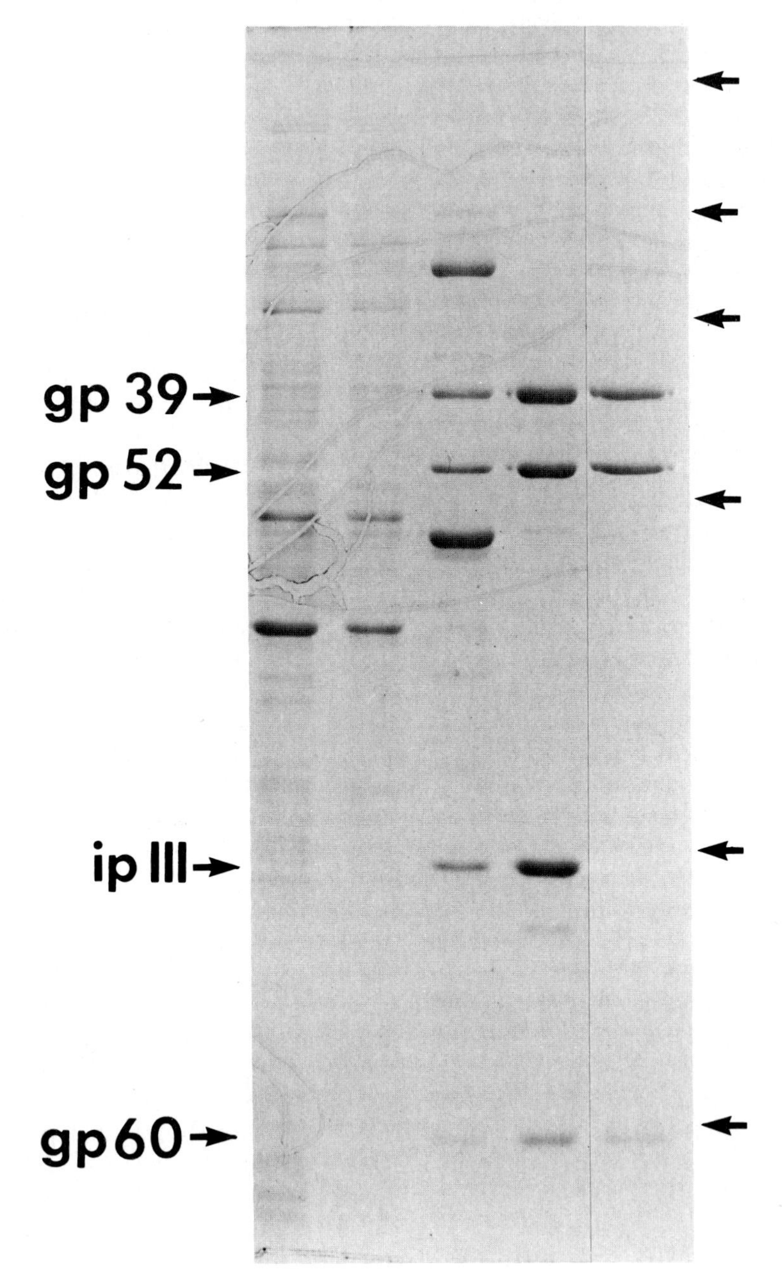
gp 39
gp 52
ip III
gp60

merase elutes with buffer D3 (0.4 *M* potassium phosphate), and is most easily detected by SDS–PAGE.

The pooled hydroxyapatite column fractions are generally too dilute for loading onto the gel filtration column or for storage. They are therefore usually concentrated to a volume of 3–5 ml by passage over a norleucine-Sepharose[8,9] column in the following manner. Solid $(NH_4)_2SO_4$ is added to the pool to a final concentration of 25% (w/v), and the enzyme is loaded at a flow rate of 2 column volumes per hour on a small (1–2 ml bed volume per 10 mg of protein) norleucine-Sepharose column equilibrated with buffer E2. The column is washed with buffer E2, and then stripped of protein with buffer E1.

The enzyme recovered at this stage is usually free of contaminating nuclease activities, but it contains a variable amount of ipIII (a T4-encoded protein present inside the mature phage head). Since ipIII does not interfere with the topoisomerase activity, the purification can be stopped at this point. In this case, the enzyme is dialyzed twice against buffer F containing 10% glycerol and then once against this buffer containing 50% glycerol, and stored at −20°.

Step 6. Gel Filtration. The concentrated hydroxyapatite eluate is adjusted to 20–25% glycerol and layered on top of a 2.5 × 50 cm column of Sephacryl S-300 Superfine (Pharmacia). Buffer D1 is used for elution at a flow rate of about 25 ml/hr (5 ml cm^{-2} hr^{-1}). Topoisomerase elutes at about 1.4 times the excluded volume (V_0 = 90 ml), while ipIII and another minor contaminant come out at more than two times V_0. After gel filtration, the topoisomerase is essentially homogeneous (>90% pure), as judged by Coomassie Blue staining of an SDS gel containing the protein (see Fig. 1, lane e). Occasionally, as exemplified in the table, some ipIII will remain as a minor contaminant, but judicious pooling of fractions can

[8] C. F. Morris, H. Hama-Inaba, D. Mace, N. K. Sinha, and B. M. Alberts, *J. Biol. Chem.* **254,** 6787 (1979).

[9] R. A. Rimerman and G. W. Hatfield, *Science* **182,** 1268 (1973).

FIG. 1. Sodium dodecyl sulphate–polyacrylamide gel electrophoresis of fractions from a topoisomerase purification. T4 topoisomerase was purified using the protocol described in the text. Samples from the various fractions generated in the purification procedure were analyzed by electrophoresis through a 12.5% polyacrylamide gel [U. K. Laemmli, *Nature (London)* **227,** 680 (1970)]. The protein bands were detected by staining with Coomassie Brilliant Blue. Lane a: crude cell lysate; lane b: fraction I (cleared, dialyzed lysate); lane c: fraction II (single-stranded DNA-cellulose pool); lane d: fraction III (hydroxyapatite pool); lane e: fraction V (Sephacryl S-300 pool). The molecular weight (M_r) standards (Bethesda Research Laboratories) indicated on the right of the figure are, in descending order: myosin heavy chain, M_r 200,000; phosphorylase *b*, M_r 92,500; bovine serum albumin, M_r 68,000; chicken ovalbumin, M_r 45,000; α-chymotrypsinogen, M_r 25,700; β-lactoglobulin, M_r 18,400.

PURIFICATION OF T4 TOPOISOMERASE

Fraction	Step	Volume (ml)	Protein mg/ml	Protein Total (mg)	Purity[a] (%)	Yield[a] (%)	Purification (fold)
I	Cleared lysate	810	6.75	5470	ND[b]	(100)	—
II	Single-stranded DNA cellulose	110	1.31	144	12	>90	38
III	Hydroxyapatite	210	0.22	46.2	41	>90	119
IV	Norleucine-Sepharose	11	3.1	34.1	41	75	119
V	Sephacryl S-300	44	0.16	7.04	93	39	269

[a] Purity and yield were determined by scanning Coomassie Blue-stained sodium dodecyl sulfate–polyacrylamide gels with an integrating laser scanning densitometer.
[b] ND, not determined.

minimize this contamination. The enzyme can be reconcentrated, if necessary, by loading it at a flow rate of 1 column volume per hour on a small (1 ml bed volume per 20 mg of protein) phosphocellulose column equilibrated with buffer D0, washing the column with the same buffer, and stripping it with buffer D3. The enzyme is finally dialyzed into buffer F (10% glycerol twice, then 50% glycerol) and stored at −20°. Under these conditions, purified topoisomerase has lost no detectable activity after 2 years.

Assay Methods

Relaxation of Superhelical DNA

The standard assay mixture (20 μl) for measuring T4 topoisomerase-catalyzed relaxation of superhelical DNA contains 40 m*M* Tris-HCl (pH 7.8), 60 m*M* KCl, 10 m*M* $MgCl_2$, 0.5 m*M* dithiothreitol, 0.5 m*M* Na_3EDTA, 30 μg of nuclease-free albumin per milliliter, 0.5 m*M* ATP, and 0.3 μg of supercoiled pBR322 plasmid DNA. Fresh serial dilutions of the enzyme in a buffer containing 50% glycerol, 30 m*M* potassium phosphate (pH 7.2), 10 m*M* 2-mercaptoethanol, 0.5 m*M* Na_3EDTA, and 50 μg of nuclease-free albumin per milliliter are tested for activity. After incubation for 30 min at 30°, the reaction is terminated by the addition of 5 μl of 5% SDS, 20% Ficoll, and 0.1% each bromophenol blue and xylene cyanole. The reaction products are visualized by ethidium bromide staining after electrophoresis through a 1% agarose gel in TBE running buffer

(89 m*M* Tris base, 89 m*M* boric acid, and 2.5 m*M* Na_3EDTA). One unit is defined as the amount of enzyme that catalyzes one-half relaxation of the 0.3 μg of DNA in the 20-μl assay mixture.

Relaxation activity due to the topoisomerase can be detected even in crude extracts of T4-infected *E. coli,* but only after extensive dilution.[2] Extracts from cells infected with topoisomerase-deficient mutants show lower levels of a different relaxation activity resulting from the sequential action of T4-induced nucleases and DNA ligase.[10]

Formation and Resolution of Knotted and Catenated Circular Duplex DNAs

The conditions for preparing knotted pBR322 DNA are as follows: Samples containing 50 m*M* Tris-HCl (pH 7.5), 60 m*M* KCl, 10 m*M* $MgCl_2$, 0.5 m*M* Na_3EDTA, 0.5 m*M* dithiothreitol, 30 μg of nuclease-free albumin per milliliter, 20 μg of pBR322 DNA per milliliter, and 33 μg of T4 topoisomerase per milliliter are incubated at 30° for 3 min, and the reaction is terminated by extracting with neutralized phenol.[11] Knotted DNA forms can be detected by agarose gel electrophoresis as above; resolution is enhanced by nicking of the reaction products with DNase I.[11]

In order to produce DNA catenanes, the standard relaxation reaction conditions are used, with the addition of a DNA-condensing agent, such as histone H1 at a weight ratio of histone to DNA of 0.2.[12] The catenated forms of circular DNA produced are detected by agarose gel electrophoresis. Under the standard relaxation reaction conditions both unknotting and catenane resolution occur and can be detected by agarose gel electrophoresis. Since neither of these reactions is carried out by combinations of nucleases and ligases, they provide a more specific assay for the topoisomerase than does the relaxation assay. Liu and Davis[13] described an especially convenient substrate for unknotting assays: phage P4 DNA, which is naturally knotted when ligated immediately after extraction from phage heads.

Hydrolysis of ATP

We have used three different methods to measure the hydrolysis of ATP to ADP and inorganic phosphate, only one of which will be discussed in detail.

[10] L. F. Liu, personal communication.

[11] L. F. Liu, C.-C. Liu, and B. M. Alberts, *Cell* **19,** 697 (1980).

[12] L. F. Liu, *in* "Mechanistic Studies of DNA Replication and Genetic Recombination" (B. M. Alberts, ed.), p. 817. Academic Press, New York, 1980.

[13] L. F. Liu and J. L. Davis, *Nucleic Acids Res.* **9,** 3979 (1981).

Method 1. The reaction products can be separated from each other by thin-layer chromatography on polyethyleneimine-cellulose.[14] The best eluent for the separation of free phosphate from ATP is 1 *M* HCOOH with 0.5 *M* LiCl; 0.75 *M* NaH_2PO_4 (pH 3.5) is used for optimal separation of ADP from ATP. Using [γ-^{32}P]ATP (freshly purified over DEAE-Sephadex)[15] as a substrate, results are quantitated by cutting out the $^{32}PO_4^{3-}$ spot and counting it in a liquid scintillation counter. This method is recommended when a high sensitivity is required.

Method 2. The $^{32}PO_4^{3-}$ released from [γ-^{32}P]ATP can be measured by the differential charcoal absorption method of Zimmerman and Kornberg.[16] Briefly, this method takes advantage of the fact that ATP is adsorbed by charcoal under acidic conditions, whereas inorganic phosphate is not. The charcoal absorption method is convenient for a semiquantitative measurement of the ATPase activity of a large number of samples.

Method 3. As our standard ATPase assay, we have modified a spectrophotometric method described by Panuska and Goldthwait.[17] The hydrolysis of ATP to ADP is linked to the oxidation of NADH to NAD^+ by the combined action of pyruvate kinase and L-lactate dehydrogenase. Assay conditions are the same as for the relaxation of pBR322 DNA, with the following additions: 2 m*M* phosphoenolpyruvate, 0.15 m*M* NADH, and 20 U/ml each of L-lactate dehydrogenase and pyruvate kinase (both type II from Sigma Chemical Co., freshly resuspended in assay buffer). To attain maximal rates, 5 μg of double-stranded DNA per milliliter should be present. The initial velocity of the reaction at 30° is determined from the rate of change in absorbance at 340 nm. Using a microprocessor-controlled Hewlett–Packard HP8450 spectrophotometer, we can determine accurate initial rates from the first 2 min of reaction. The rate of ATP hydrolysis is easily calculated from the equation:

$$\text{Rate of hydrolysis (micromoles min}^{-1}\text{ ml}^{-1}) = -dA/dt\ (\text{min}^{-1})/\varepsilon_{340}^{\text{m}M}\ (\text{NADH})$$

or, in numerical terms:

$$\text{Rate of hydrolysis} = -dA/dt \times 0.16.$$

The spectrophotometric method has the advantage of being rapid and very accurate. Its major disadvantage is that it requires relatively large amounts of enzyme for reliable rate measurements, as the specific activity of the topoisomerase is only 1–1.5 U/mg under optimal conditions (one

[14] K. Randerath and E. Randerath, this series, Vol. 12A, p. 323.

[15] W. E. Wehrli, D. L. M. Verheyden, and J. G. Moffatt, *J. Am. Chem. Soc.* **87,** 2265 (1965).

[16] S. Zimmerman and A. Kornberg, *J. Biol. Chem.* **236,** 1480 (1961).

[17] J. R. Panuska and D. A. Goldthwait, *J. Biol. Chem.* **255,** 5208 (1980).

unit of ATPase is defined as the amount of enzyme required to hydrolyze 1 μmol of ATP in 1 min at 30°).

It should be noted that the ATPase assay is not a good way to detect the enzyme in crude extracts, as there are a number of other DNA-dependent ATPases in T4-infected *E. coli* cells.[18] In particular, the product of the *dda* gene has a specific activity more than 100-fold higher than the topoisomerase.

DNA Cleavage Reactions

Efficient cleavage of double-stranded DNA is induced by the topoisomerase under the standard relaxation conditions, provided that 500 μg of oxolinic acid are added per milliliter.[19] Unique length ^{32}P-labeled linear DNA substrates are used to allow detection of infrequent cleavage events by gel electrophoresis, and treatment of the reaction products with SDS is required to reveal the covalent topoisomerase–DNA complex. Cleavage of single-stranded DNA is observed under the same conditions without oxolinic acid, but SDS treatment is not required.[20] The products of both types of cleavage reactions are analyzed by gel electrophoresis after treatment with proteinase K (EM Biochemicals) at 100 μg/ml for 30 min at 37°.

We have developed a protocol for specifically purifying cleaved duplex DNA molecules with the topoisomerase covalently attached.[19] This filter-binding procedure is a modification of the glass fiber (Whatman GF/C) filter method of Thomas *et al.*[21] and Coombs and Pearson.[22] Topoisomerase reactions with ^{32}P-labeled duplex DNA substrates are incubated as above and terminated by the addition of SDS to 0.2%. The detergent is removed by a rapid gel filtration method,[23] and one-fifth final volume of 5× binding buffer is added to the eluate [binding buffer contains 50 m*M* Tris-HCl (pH 7.8), 200 m*M* KCl, 10 m*M* $MgCl_2$, and 0.5 m*M* Na_3EDTA]. Covalent protein–DNA complexes are then collected on GF/C filters as follows: Two GF/C filters 7 mm in diameter (one on top of the other) are placed onto a larger GF/A filter; 150 μl of binding buffer are delivered to the top disk slowly enough so that liquid does not spill over; the buffer will soak through by capillary action. After this prewash, the filter pair is moved to a dry area of the GF/A filter, and the gel filtration

[18] K. Ebisuzaki and S. B. Jellie, *J. Virol.* **37,** 893 (1981).

[19] K. N. Kreuzer and B. M. Alberts, *J. Biol. Chem.* (submitted).

[20] K. N. Kreuzer, *J. Biol. Chem.* (submitted).

[21] C. A. Thomas, Jr., K. Saigo, E. McLeod, and J. Ito, *Anal. Biochem.* **93,** 158 (1979).

[22] D. H. Coombs and G. D. Pearson, *Proc. Natl. Acad. Sci. U.S.A.* **75,** 5291 (1978).

[23] For a detailed description of the rapid gel filtration, see Kreuzer and Alberts.[19] The method was modified from the original method of M. W. Neal and J. R. Florini, *Anal. Biochem.* **55,** 328 (1973).

eluate (up to 100 μl) is applied slowly, followed immediately by 150 μl of binding buffer. The filter pair is then washed four more times with 150 μl of binding buffer, moving the filter pair to a dry area of the GF/A filter before each wash. The top filter, which contains most of the protein–DNA complexes, is eluted twice with 20 μl of 10 m*M* Tris-HCl (pH 7.8) and 0.1% SDS, and the eluate is treated with proteinase K as above to eliminate the covalently bound protein. After another rapid gel filtration step, the labeled DNA fragments obtained are analyzed by gel electrophoresis. This procedure is sufficiently sensitive to detect topoisomerase-induced double-stranded DNA cleavage events in crude cell extracts supplemented with oxolinic acid, or cleavage by the purified enzyme even in the absence of the inhibitor.

Properties of the Topoisomerase

Physical Properties

T4 topoisomerase contains three different subunits, coded for by the phage genes *39, 52,* and *60.*[2,3] The molecular weights of the subunits were originally estimated to be 63,000, 52,000, and 16,000, respectively.[2] More recent measurements of their migration rates in SDS–PAGE, compared to calibrated sets of standards, yield molecular weight (M_r) values of 57,000 for the gene *39* protein, 48,000 for the gene *52* protein, and 18,000 for the gene *60* protein (see Fig. 1). The M_r 23,000 contaminant described by Liu *et al.*[2] is ipIII, and our purified topoisomerase does not contain the M_r 110,000 protein reported by Stetler *et al.*[3] In our purest preparations, the three subunits are present in equimolar amounts. Cruder fractions often contain a slight excess of the gene *52* protein. This suggests that there may be a "free" pool of this subunit, possibly with an *in vivo* function of its own. The number of copies of each subunit making up the functional form of the enzyme has not yet been established; its behavior on gel filtration columns is consistent with it being a hexamer containing two copies of each subunit.[24]

The products of genes *39* and *52* have been reported to be tightly membrane-associated.[25,26] However, it seems unlikely that the enzyme is an integral membrane protein, as it can be solubilized quantitatively by as little as 0.2 *M* NaCl. Also, topoisomerase shows none of the solubility problems usually associated with highly hydrophobic proteins. On the other hand, these properties do not preclude a secondary association

[24] M. Munn, unpublished results.

[25] W. M. Huang, *Virology* **66,** 508 (1975).

[26] B. J. Tacaks and J. P. Rosenbusch, *J. Biol. Chem.* **250,** 2339 (1975).

between the enzyme and the cell membrane; whether this is functional or adventitious remains to be established. Topoisomerase tends to stick to dialysis membranes; it is also unusually soluble in ammonium sulfate: concentrations as high as 2.3 *M* (70% saturation at 0°) fail to precipitate it.

A velocity sedimentation analysis of a crude T4-infected cell extract on a preparative sucrose gradient reveals that topoisomerase activity sediments as a 15–20 S complex whose functional significance in unclear.[2] The complex is disrupted by increasing the salt concentration or by ribonuclease treatment. Affinity chromatography of a T4-infected cell extract over a column of topoisomerase covalently bound to Bio-Rad Affi-Gel 10 reveals an M_r 30,000 T4 protein specifically binding to the enzyme.[27] So far, nothing is known about the genetic origin or other properties of this protein.

Enzymic Properties

Topoisomerization Reactions. Most of the topological interconversions catalyzed by the T4 topoisomerase are ATP-dependent; they include the relaxation of both positively and negatively supercoiled DNA and the knotting, unknotting, catenation, and decatenation of circular duplex DNA. As depicted schematically in Fig. 2, all these reactions can be explained by the double-strand passage mechanism of the type II topoisomerases.[1] The hydrolysis of ATP is probably required only for turnover of the enzyme, since a nonhydrolyzable analog of ATP will apparently support one round of topoisomerization.[2] A very low level of knotting and relaxation can be detected in the absence of exogenously added ATP or ATP analog.[11] This may reflect a small proportion of enzyme molecules that are purified in an "ATP-charged" form or, alternatively, may indicate that the enzyme can, albeit rarely, proceed through a reaction cycle totally uncoupled from ATP binding or hydrolysis. However, there is generally a tight coupling between ATP hydrolysis and strand-passage reactions (see below).

The relaxation of both positively and negatively supercoiled DNA is catalyzed by the T4 enzyme with about equal efficiency.[2] As expected from the double-strand passage model, supercoils are removed in steps of two rather than one.[11] The reaction requires ATP, Mg^{2+}, and K^+, and under the optimized conditions described above, a specific activity of about 10^7 U per milligram of protein is measured.

Whether catenated and knotted DNA forms are destroyed or created by the topoisomerase depends on the influence of the reaction conditions on the structure of the DNA. Decatenation and unknotting are catalyzed

[27] C. V. Jongeneel, unpublished results.

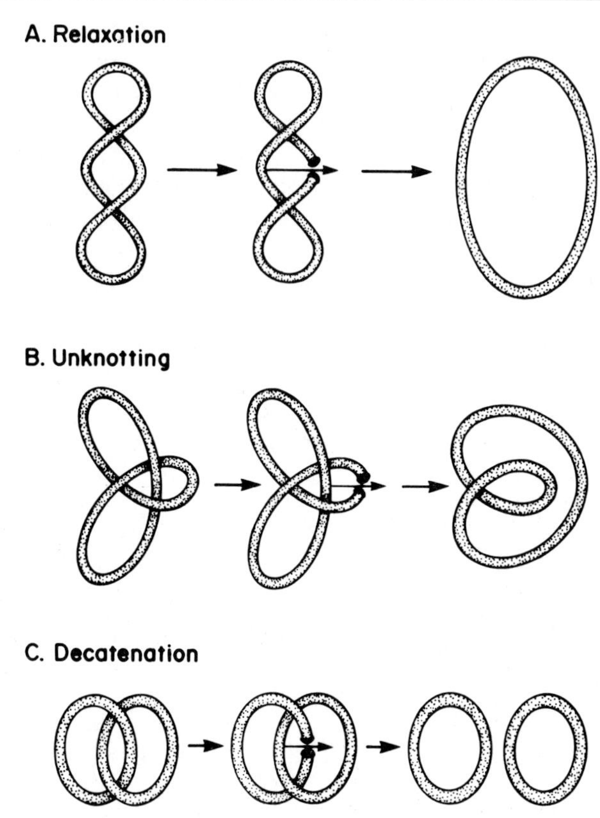

FIG. 2. Deoxyribonucleic acid topoisomerization reactions catalyzed by the T4 topoisomerase.

by the T4 topoisomerase under the standard conditions for relaxation, since catenanes and knotted DNA are energetically unfavorable compared to simple monomeric circles at low DNA concentrations in the absence of condensing agents. Catenation and knotting of circular duplex DNA are favored by DNA-condensing agents, such as histone H1, spermidine, and polymin P. In the case of DNA gyrase, it has been shown convincingly that production of catenanes occurs only under those ionic conditions where the DNA-condensing agents induce multimolecular aggregation of the DNA.[28] We presume that the same holds for the T4 topoisomerase. Liu *et al.*[11] demonstrated that high concentrations of T4 topoisomerase in the absence of either ATP or any of the above-mentioned condensing agents resulted in the conversion of about 50% of the

[28] M. A. Krasnow and N. R. Cozzarelli, *J. Biol. Chem.* **257**, 2687 (1982).

substrate pBR322 into a knotted form. While the mechanistic details of this knotting reaction have yet to be explained fully, the presumption is that the superhelical structure of the substrate and the topoisomerase binding act to reduce the effective statistical segment length of the DNA without causing aggregation and thus favor unimolecular knot formation. The trace amount of topoisomerase activity detectable in the absence of ATP or ATP analog is presumed to be responsible for the actual strand-passage reaction. While the addition of ATP would be expected to greatly increase strand-passage activity, it would also result in the relaxation of the initially supercoiled DNA and therefore increase the effective statistical segment length, disfavoring the knotted structure.

Hydrolysis of ATP. The T4 topoisomerase has an ATPase activity that is strongly DNA-dependent, producing ADP and inorganic phosphate.[2] As mentioned above, the hydrolysis of ATP is tightly coupled to the strand-passage reaction and is presumed to be required to regenerate the original form of the enzyme at the end of the reaction cycle. ATP can be replaced by dATP,[29] but not by any of the other ribo- or deoxyribonucleoside triphosphates, as a cofactor for the topoisomerase reaction.[2] Both strand-passage and ATPase reactions have an absolute requirement for Mg^{2+}. The topoisomerase reaction requires K^+ ions, and cannot proceed with equivalent amounts of Na^+. ATPase activity is also severely reduced in the absence of K^+.

The rate of ATP hydrolysis in the presence of double-stranded DNA (5 μg/ml) is 1–1.5 μmol mg^{-1} min^{-1}; the rate in the absence of DNA is more than 20-fold lower. The apparent K_m for double-stranded DNA is about 1.5 μg/ml (4.5 μM nucleotides). Denatured DNAs or the single-stranded genomic DNAs from phages ϕX174 or fd stimulate the ATPase, but they are severalfold less effective than double-stranded DNA. These DNA cofactor requirements are similar to those of DNA gyrase,[30] and are consistent with the DNA substrate specificity of type II topoisomerases. A rough calculation of the stoichiometry of ATP hydrolysis shows that 1–2 molecules of ATP are hydrolyzed for each strand-passage event. This ratio is maintained at suboptimal K^+ concentrations and most likely reflects a tight coupling between the two reactions.

The T4 topoisomerase-catalyzed relaxation of positively or negatively supercoiled DNA poses a paradox, in that it requires ATP hydrolysis in a reaction that is thermodynamically favored. It is likely that *in vivo* ATP hydrolysis is coupled to an energy-requiring reaction that has yet to be defined. Liu *et al.*[2] speculated that topoisomerase could introduce local

[29] C. Zetina-Rosales, unpublished results.

[30] A. Sugino and N. R. Cozzarelli, *J. Biol. Chem.* **255,** 6299 (1980).

supercoiling at the primary replication origin(s) of T4 DNA by recognizing two oriented DNA sequences bracketing the origin. The finding that ATP dramatically affects the efficiency of topoisomerase-induced cleavage of glucosylated hydroxymethylcytosine-containing T4 DNA (see below) adds credence to this hypothesis. Another likely possibility, however, is that there are cofactors yet to be discovered that will allow the enzyme to use the energy of ATP hydrolysis effectively and thereby express its full *in vivo* activity.

DNA Cleavage Reactions. An intermediate in the DNA strand-passage reaction of type II topoisomerases consists of duplex DNA broken in both strands with protein covalently attached to each of the two newly formed 5′ ends.[1] This reaction intermediate of the T4 topoisomerase was originally detected at a low frequency after addition of detergent to the standard relaxation reaction.[2] Subsequently, it was found that the addition of the DNA gyrase inhibitor, oxolinic acid, at 500 μg/ml greatly increases the yield of the detergent-induced covalent complex, causing nearly every enzyme molecule to cleave the duplex DNA (assuming that the functional form of the enzyme is a hexamer).[19] Oxolinic acid also inhibits the relaxation activity of the T4 topoisomerase, with half-maximal inhibition being observed at a drug concentration of roughly 250 μg/ml. Thus, although 25-fold higher drug levels are required for the T4 enzyme, oxolinic acid seems to act identically on the T4 topoisomerase and the *E. coli* DNA gyrase.[1]

The sites of DNA cleavage induced by the T4 topoisomerase have been used as markers of the preferential sites of action of the enzyme on T4 DNA.[19] The filter-binding assay we developed (see above) enabled us to purify those substrate DNA molecules that were cleaved by the topoisomerase away from all uncleaved DNA molecules. Using this assay, it was found that the glucose moieties attached to the hydroxymethylcytosine residues of native T4 DNA greatly increase the specificity with which the topoisomerase recognizes T4 DNA in the absence of ATP; in addition, the efficiency of cleavage is increased severalfold in the presence of ATP.[19] This finding may be relevant to the physiological action of the topoisomerase, since T4 mutants deficient in DNA glucosylation show disturbances in replication (at least in certain *E. coli* hosts) and in recombination.[31]

The T4 topoisomerase also cleaves single-stranded DNA in a site-specific fashion, but this cleavage is mechanistically distinct from the double-strand DNA cleavage discussed above.[20] Single-strand cleavage occurs under standard reaction conditions, is inhibited rather than en-

[31] Reviewed in Kreuzer and Alberts.[19]

hanced by oxolinic acid, and results in virtually every enzyme molecule being covalently attached to a broken DNA end. The sites of single-stranded DNA cleavage are distinct from those of oxolinic acid-induced double-stranded DNA cleavage, and many of the single-strand cleavage sites correlate with regions of secondary structure in the DNAs.

Comparison with Other Type II Topoisomerases

All of the known type II topoisomerases catalyze the relaxation of supercoiled DNA (changing the linking number in steps of 2) and the production and resolution (depending on the conditions) of both knotted and catenated covalently closed duplex DNA circles.[1] The two best-studied type II enzymes, *E. coli* DNA gyrase and the T4 topoisomerase, share several additional properties. Both enzymes are multisubunit proteins with DNA-dependent ATPase activity and generally show a tight coupling between ATP hydrolysis and catalytic strand-passage reactions. Nonhydrolyzable ATP analogs allow a single round of topoisomerization, and it therefore appears that ATP hydrolysis is required only for turnover of both enzymes.[2,32] Thus, the binding of ATP (or the nonhydrolyzable analog) is thought to induce a conformational change that allows either enzyme to proceed through a reaction cycle. Originally, oxolinic acid was shown to be a specific inhibitor of *E. coli* DNA gyrase and to interfere with resealing of the cleaved-DNA reaction intermediate.[1] It is now clear that the drug acts similarly on the T4 topoisomerase, except that higher drug levels are required for both inhibition of relaxation and induction of DNA cleavage (see above). The rules of DNA sequence recognition by gyrase and T4 topoisomerase must be related, since a substantial subset of sites in duplex ϕX174 DNA are cleaved in common by both enzymes.[19]

While these results indicate a functional or evolutionary relatedness of the *E. coli* DNA gyrase and the T4 DNA topoisomerase, the differences between the two enzymes may turn out to be more interesting than their similarities in deducing their biological roles. While gyrase can use its ATP hydrolysis to drive the energetically unfavorable reaction of introducing negative supercoils into all circular duplex DNA substrates tested, no such activity has yet been detected for the T4 topoisomerase. Mechanistically related to this difference is the finding that DNA is wrapped around DNA gyrase as part of its supercoiling cycle;[33,34] it appears that no

[32] A. Sugino, N. P. Higgins, P. O. Brown, C. L. Peebles, and N. R. Cozzarelli, *Proc. Natl. Acad. Sci. U.S.A.* **75,** 4838 (1978).

[33] L. F. Liu and J. C. Wang, *Cell* **15,** 979 (1978).

[34] C. L. Peebles, N. P. Higgins, K. N. Kreuzer, A. Morrison, P. O. Brown, A. Sugino, and N. R. Cozzarelli, *Cold Spring Harbor Symp. Quant. Biol.* **43,** 41 (1979).

such wrapping occurs with the T4 topoisomerase, since DNA circles are relaxed to completion even in the presence of a large excess of this enzyme.[10]

A second difference between the T4 topoisomerase and DNA gyrase relates to the recognition of secondary structure in DNA. As discussed above, T4 topoisomerase cleaves single-stranded DNA, and many of the cleavage sites correlate to regions of secondary structure in the DNA.[20] Single-stranded DNA is an effective inhibitor of relaxation by the T4 enzyme, presumably because most or all of the enzyme becomes covalently linked to the newly cleaved ends. DNA gyrase is not inhibited by single-stranded DNA, nor does it cleave this substrate. Possible relationships between secondary structure recognition by the T4 enzyme and its role in the formation of replication bubbles are discussed by Kreuzer.[20]

Eukaryotic type II DNA topoisomerases have thus far been isolated from *Drosophila* embryos, *Xenopus* eggs, and cultured CHO and HeLa cells.[11,35,36] The eukaryotic type II enzymes resemble the T4 topoisomerase more closely than *E. coli* DNA gyrase with respect to the distinguishing characteristics discussed above. Thus, the eukaryotic enzymes examined to date require ATP to relax DNA, but they are not capable of supercoiling circular DNA substrates *in vitro*. Site-specific supercoiling by these enzymes has been postulated.[12] Although cleavage of single-stranded DNA has not yet been tested, the eukaryotic enzymes bind tightly to single-stranded DNA,[10] and so it seems likely that they will also turn out to recognize DNA secondary structure. Thus, it is reasonable to view the T4 topoisomerase as the best prokaryotic model for eukaryotic type II topoisomerases. Perhaps, therefore, the initiation of T4 replication bubbles, with its apparent topoisomerase involvement, will resemble initiation in eukaryotic cells.

Acknowledgments

We thank Bruce Alberts for many fruitful discussions and critical reading of this manuscript. Experiments performed by the authors were supported by Grant GM 24020 from the National Institute of General Medical Sciences to Bruce Alberts. K. N. K. was supported by postdoctoral fellowships from the Anna Fuller Fund and NIH, and C. V. J. is a fellow of the Leukemia Society of America.

[35] T. Hsieh and D. Brutlag, *Cell* **21,** 115 (1980).
[36] M. I. Baldi, P. Benedetti, E. Mattoccia, and G. P. Tocchini-Valentini, *Cell* **20,** 461 (1980).

[10] Purification and Properties of Type II DNA Topoisomerase from Embryos of *Drosophila melanogaster*

By TAO-SHIH HSIEH

DNA topoisomerases are enzymes that can transiently break the DNA backbone bonds and thereby interconvert DNA topological isomers.[1-3] These enzymes are classified into two types according to their mechanisms of action.[4] Type I DNA topoisomerases work by making reversible single-stranded breaks, and type II enzymes work by passing a segment of DNA through a transient double-stranded break. Both types of topoisomerases have been isolated from a wide variety of sources. Bacterial DNA gyrases[5-7] and bacteriophage T4-induced topoisomerase[8,9] are very well characterized prokaryotic type II enzymes. They have also been isolated from various eukaryotic sources including *Xenopus* oocytes,[10] *Drosophila* embryos,[11] and HeLa cell nuclei.[12] The type II DNA topoisomerases were first discovered in eukaryotes because of their ability to catenate and decatenate DNA duplex rings and to unknot the circular DNA with topological knots. These enzymes are also capable of relaxing both the positively and negatively supercoiled DNA with a unique character that they will change the linking number of DNA in steps of two. All the above reactions are in complete accord with the notion that these enzymes are type II topoisomerases. Despite the fact that all the topoisomerization reactions they catalyze require the presence of ATP, they have not been shown to introduce superhelical turns into a relaxed DNA molecule, which marks a clear distinction between these enzymes and

[1] J. C. Wang and L. F. Liu, *in* "Molecular Genetics" (J. H. Taylor, ed), Part 3, p. 65. Academic Press, New York, 1979.

[2] N. R. Cozzarelli, *Science* **207,** 953 (1980).

[3] M. Gellert, *Annu. Rev. Biochem.* **50,** 879 (1981).

[4] L. F. Liu, C.-C. Liu, and B. M. Alberts, *Cell* **19,** 697 (1980).

[5] M. Gellert, K. Mizuuchi, M. H. O'Dea, and H. A. Nash, *Proc. Natl. Acad. Sci. U.S.A.* **73,** 3872 (1976).

[6] A. Sugino, C. L. Peebles, K. N. Kreuzer, and N. R. Cozzarelli, *Proc. Natl. Acad. Sci. U.S.A.* **74,** 4767 (1977).

[7] L. F. Liu and J. C. Wang, *Proc. Natl. Acad. Sci. U.S.A.* **75,** 2098 (1978).

[8] L. F. Liu, C.-C. Liu, and B. M. Alberts, *Nature (London)* **281,** 456 (1979).

[9] G. L. Stetler, G. J. King, and W. M. Huang, *Proc. Natl. Acad. Sci. U.S.A.* **76,** 3737 (1979).

[10] M. I. Baldi, P. Benedetti, E. Mattoccia, and G. P. Tocchini-Valentini, *Cell* **20,** 461 (1980).

[11] T. Hsieh and D. Brutlag, *Cell* **21,** 115 (1980).

[12] K. G. Miller, L. F. Liu, and P. T. Englund, *J. Biol. Chem.* **256,** 9334 (1981).

ISBN 0-12-182000-9

bacterial DNA gyrase. A preliminary account on the purification of type II topoisomerase from *Drosophila* early embryos has been presented earlier.[11] I will describe here a procedure for purifying this enzyme to homogeneity and some of its further characterization.

Assay Method

Principle. The ATP-dependent topoisomerization reactions that are useful in assaying *Drosophila* type II topoisomerase are relaxation of supercoiled DNA, catenation, decatenation, and unknotting of double-stranded circular DNA (Fig. 1). It should be emphasized that the ATP-dependence of these reactions offers a valuable distinction for the reactions catalyzed by type II topoisomerase from those catalyzed by the type I enzyme (compare lanes 1–4 with lanes 5–8, Fig. 1). The preparation of substrates and the reaction conditions for assaying the enzyme by the relaxation, catenation, and decatenation reactions have already been described in detail.[11] Liu *et al.*[13] showed that the unknotting of knotted DNA isolated from the bacteriophage P4 capsids can be used in assaying eukaryotic type II topoisomerases. The purified *Drosophila* type I topoisomerase cannot efficiently catalyze the catenation, decatenation, and unknotting reactions of the circular duplex DNA, being either in the intact or nicked form (lanes 9–12, Fig. 1; and unpublished results). It is, however, more advantageous to employ the unknotting reaction as the assay for type II topoisomerase than the catenation or decatenation reaction, since one can obtain a large amount of knotted P4 DNA readily and there is no need for the presence of a DNA condensing reagent as for the catenation reaction, which will introduce one extra variable in the reaction condition. Furthermore, the amount of activity required for the unknotting of the knotted DNA is roughly equivalent to that required for relaxing the same amount of supercoiled DNA.

Procedure

Assay solution: 10 m*M* Tris-HCl (pH 7.9), 50 m*M* KCl, 100 m*M* NaCl, 10 m*M* $MgCl_2$, 1.25 m*M* ATP, and 50 μg of bovine serum albumin (BSA; from Miles) per milliliter.

Enzyme diluent: 1 mg of BSA per milliliter, 15 m*M* sodium phosphate (pH 7.2), 50 m*M* NaCl, 10% glycerol, and 0.1 m*M* dithiothreitol

DNA substrate: Knotted bacteriophage P4 DNA is prepared as described[13] except that the differential centrifugation steps are performed before the CsCl isopycnic banding of the P4 capsids.

[13] L. F. Liu, J. L. Davis, and R. Calendar, *Nucleic Acids Res.* **9**, 3979 (1981).

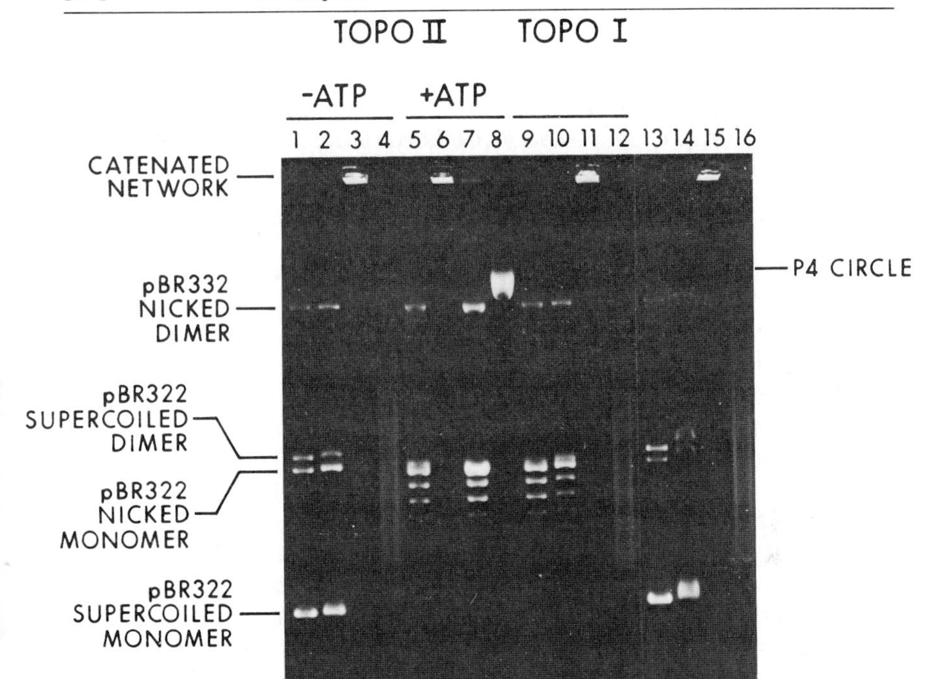

FIG. 1. Analysis of DNA products from topoisomerization reactions by agarose gel electrophoresis. Lanes 1, 5, and 9 are for assaying relaxation of superhelical turns; 2, 6, and 10 are for catenation reaction; 3, 7, and 11 are for decatenation reaction; and 4, 8, and 12 are for unknotting reaction. Lanes 1–8 show the reaction catalyzed by type II topoisomerase (10 units of fraction VI enzyme for each reaction), among which lanes 1–4 are like lanes 5–8 except that there is no ATP present in the reaction mixture. For comparison, lanes 9–12 show the reactions by *Drosophila* type I topoisomerase (10 units per reaction) under the same reaction conditions and without the presence of ATP, and lanes 13–16 are the controls without any enzyme added. Plasmid DNA pBR322 is used in lanes 1, 2, 5, 6, 9, 10, 13, and 14. Approximately 1 μg of histone H1 isolated from *Drosophila* tissue culture cells is added to each of the reactions shown in lanes 2, 6, 10, and 14 to mediate the catenation reaction. The purified, catenated pBR322 DNA and the knotted bacteriophage P4 DNA is used in reactions shown in lanes 3, 7, 11, and 15 and in lanes 4, 8, 12, and 16, respectively. See the text for details on the reaction conditions and gel electrophoresis.

Unknotting reaction: Usually 1 μl of the sample to be assayed is added to an assay solution of 20 μl containing 0.3 μg of knotted P4 DNA. The incubation is continued at 30° for 15 min, and the reaction is stopped by adding 3 μl of a mixture containing 50 μg of proteinase K (Boehringer Mannheim) per milliliter, 0.67% sodium dodecyl sulfate (SDS), 67 m*M* Na_3EDTA, 26.7% sucrose, 0.067% in bromophenol blue and xylene cyanole FF. The mixture is incubated at 50° for 30 min and then loaded on an agarose gel to analyze the products of the topoisomerization reaction.

Agarose gel electrophoresis: The DNA is routinely analyzed by the electrophoresis in a 1% agarose horizontal slab gel immersed in a buffer of 36 m*M* Tris base, 30 m*M* NaH_2PO_4 and 1 m*M* Na_3EDTA, pH 7.7, which is run at a voltage gradient of about 2 V/cm for 16 hr. The gel is stained in water with 1 μg/ml of ethidium bromide, and the DNA is visualized by the fluorescence from illuminating the gel with UV light. The knotted P4 DNA appears as a smear between the positions of pBR322 supercoiled monomer and pBR322 nicked dimer (lanes 4, 12, and 16, Fig. 1), while the unknotted P4 circles run as a distinct band slightly above the position of pBR322 nicked dimer (lane 8, Fig. 1).

Definition of Units. One unit of enzyme is defined as the amount of enzyme required to unknot 0.3 μg of knotted P4 DNA under the reaction condition described above. The concentration of enzyme in a sample is assessed by assaying the serially diluted samples, from which one can obtain a semiquantitative estimate of the enzyme activity. The specific activities are expressed as units per milligram of protein. Protein concentration is determined by the Coomassie blue dye binding method[14] using BSA as a standard.

Purification Procedures

Reagents and Materials

Buffer G: 30 m*M* Tris-HCl (pH 7.9), 10 m*M* Na_3EDTA, 1 *M* NaCl, 10 m*M* $NaHSO_3$, and 1 m*M* PMSF (phenylmethylsulfonyl fluoride, Sigma). The protease inhibitors $NaHSO_3$ and PMSF are added from a stock solution of 1 *M* $NaHSO_3$ (pH adjusted to 6.9) and 100 m*M* PMSF (in isopropanol) immediately before use.

Buffer A: 50 m*M* Tris-HCl (pH 7.9), 1 m*M* Na_3EDTA, 10 m*M* $NaHSO_3$, and 1 m*M* PMSF

Buffer P: 15 m*M* sodium phosphate (pH 7.2), 0.1 m*M* Na_3EDTA, 10% glycerol, 0.1 m*M* dithiothreitol, 10 m*M* $NaHSO_3$, 1 m*M* PMSF

Buffer K: 10% glycerol, 0.1 m*M* Na_3EDTA, 0.1 m*M* dithiothreitol, 10 m*M* $NaHSO_3$, 1 m*M* PMSF, and the indicated concentration of potassium phosphate (pH 6.8)

Buffer S: 50 m*M* potassium phosphate (pH 7.2), 200 m*M* NaCl, 0.1 m*M* Na_3EDTA, 0.1 m*M* dithiothreitol, 10 m*M* $NaHSO_3$, 1 m*M* PMSF

[14] M. M. Bradford, *Anal. Biochem.* **72,** 248 (1976).

Saturated (100%) ammonium sulfate solution: 767 g of ammonium sulfate (Fisher) are dissolved in 1 liter of water and the pH is adjusted to 7.0 with NaOH solution. The concentration of ammonium sulfate is approximately 4.1 *M*.

Polymin P (polyethyleneimine) solution 10% (v/v) is made by diluting the polymin P stock (Bethesda Research Laboratories) with water and adjusting the pH to 7.4 with HCl solution.

Phosphocellulose: P-11 cellulose phosphate, 5.3 mEq/g (Whatman)

Hydroxyapatite (BioGel HTP, Bio-Rad Laboratories)

DNA agarose, prepared by covalently linking denatured salmon sperm DNA to agarose BioGel A-1.5m according to the procedure of Arndt-Jovin *et al.*[15]

Phosphocellulose, hydroxyapatite, and DNA agarose are washed and equilibrated in buffer P.

Collection of Drosophila Embryos. Adult fruit flies of *Drosophila melanogaster* (Oregon R, P2) are raised in an incubator (Percival) with automatic controls on temperature (25°), humidity (70%), and light–dark cycle. The embryos are collected for a period of 16 hr on the corn syrup–agar plates. The chorions of the embryos are removed by immersing them in 1% sodium hypochlorite for 2 min and washed thoroughly with distilled water. The embryos are quickly frozen and stored at −80°.

Purification Steps

Unless otherwise noted, the following steps are carried out at 0–4° and the centrifugations are performed at 12,000 rpm (23,000 g) with a GSA rotor in a Sorvall RC 5B centrifuge.

Step 1. Preparation of Extract. Frozen embryos (250 g) are thawed and mixed with 1 liter of buffer G in a Waring blender. The mixture is blended for 2 min, and the complete disruption of embryos can be readily checked with a dissecting microscope. The homogenate is immediately centrifuged for 30 min, and the supernatant is filtered through two layers of Miracloth (fraction I).

Step 2. Polymin P and Ammonium Sulfate Precipitation. Fraction I is added with one-tenth of its volume of 5 *M* NaCl solution and mixed with stirring of 10% polymin P solution (30 ml per liter of fraction I). The stirring is continued for another 20 min, and the mixture is centrifuged for 20 min. To the supernatant is added two volumes of saturated ammonium

[15] D. J. Arndt-Jovin, T. M. Jovin, W. Baehr, A.-M. Frischauf, and M. Marquardt, *Eur. J. Biochem.* **54,** 411 (1975).

sulfate solution, and the mixture is stirred for 20 min. The pellet is collected by centrifugation for 20 min and resuspended in 1 liter of 50% ammonium sulfate in buffer A. The suspension is stirred for 20 min, and the pellet is collected by centrifugation. The pellet is then suspended in 1 liter of 10% ammonium sulfate in buffer A, and the suspension is cleared by centrifugation for 20 min. The supernatant is made 60% in ammonium sulfate by slowly adding, with stirring, solid salt (325 g per liter of supernatant). After the salt is dissolved, the mixture is stirred gently for 20 min and the pellet from a centrifugation for 20 min is suspended in 1500 ml of buffer P (fraction II).

Step 3. Phosphocellulose Chromatography. Fraction II is loaded at a flow rate of 280 ml/hr to a 600-ml phosphocellulose column (6.4 cm in diameter and 19 cm in length) that has been equilibrated in buffer P with 50 m*M* NaCl. The column is then washed with 2 liters of buffer P with 0.2 *M* NaCl and eluted with a 6-liter gradient of 0.2 to 1.5 *M* NaCl in buffer P. The enzyme begins to elute at about 0.45 *M* NaCl. The active fractions are pooled (fraction III).

Step 4. Hydroxyapatite Chromatography. Fraction III is loaded to a 100-ml hydroxyapatite column (4 cm in diameter and 8 cm in length) equilibrated in buffer P with 50 m*M* NaCl. The column is washed with 750 ml of buffer K with 0.1 *M* potassium phosphate and eluted with a gradient of 1.5 liters from 0.1–0.7 *M* potassium phosphate in buffer K. The enzyme is eluted from 0.4 *M* potassium phosphate. The active fractions are pooled and dialyzed into buffer P with 50 m*M* NaCl (fraction IV).

Step 5. DNA-Agarose Chromatography. Fraction IV is loaded to a 12 ml DNA-agarose column (1.5 cm in diameter and 7 cm in length) equilibrated with 50 m*M* NaCl in buffer P. The column is washed with 80 ml of 0.1 *M* NaCl in buffer P, then eluted with a 200-ml gradient of 0.1–1.0 *M* NaCl in buffer P. The enzyme is eluted starting from 0.3 *M* NaCl. The active fractions are pooled (fraction V).

Step 6. Glycerol Gradient Sedimentation. Fraction V is concentrated with a 0.5-ml column of hydroxyapatite and then dialyzed into 200 m*M* NaCl in buffer P. About 0.25 ml of this dialyzed and concentrated fraction V enzyme is carefully layered on each 4.5 ml of 10–30% glycerol gradient in buffer S. The gradient is centrifuged in a Sorvall AH 650 rotor at 44,000 rpm (180,000 g) for 15 hr at 4°. Fractions of the gradient are collected from the bottom in 6 drops per fraction. The active fractions are pooled (fraction VI). For long-term storage of the purified enzyme, fraction VI can be dialyzed into 50% glycerol, 15 m*M* sodium phosphate, pH 7.2, 100 m*M* NaCl, 0.1 m*M* Na_3EDTA, 0.1 m*M* dithiothreitol, and 0.1 m*M* PMSF, and stored at −20°.

PURIFICATION OF TYPE II DNA TOPOISOMERASE FROM *Drosophila melanogaster* EMBRYOS[a]

Fraction and step	Volume (ml)	Protein (mg/ml)	Activity (units/μl)	Specific activity (units/mg)	Yield (%)
I. Homogenate	1200	14.8	70	4.7×10^3	(100)
II. Polymin P and $(NH_4)_2SO_4$ precipitations	1700	6.0	60	1.0×10^4	120
III. Phosphocellulose	1600	0.3	20	6.7×10^4	38
IV. Hydroxyapatite	350	0.17	30	1.8×10^5	13
V. DNA-agarose	20	1.0	250	2.5×10^5	6
VI. Glycerol gradient sedimentation	3	0.17	500	2.9×10^6	1.8

[a] From 250-g embryos.

Notes on the Purification

The table summarizes the purification steps and shows the specific activity and yield at each step. The enzyme is purified from the total embryo extract. We have attempted to fractionate the disrupted embryos into nuclear and cytoplasmic fractions and found about equal distribution of the enzyme among these fractions. Since the early embryos were shown to have a large amount of topoisomerases in the cytoplasm fraction,[11] it is likely that some of the enzyme in the cytoplasmic fraction may be stored there during oogenesis. We have also noticed that the embryos collected over a longer period (16 hr) is a better source for isolating topoisomerase than the early embryos (collected in 2 hr), since not only can it be collected in a larger quantity, but also it offers at least 5 times more topoisomerase per embryo as assayed either biochemically or immunochemically.[16]

There are both types of topoisomerases in the embryo extract,[11] and most of the type I enzyme is fractionated into soluble fraction of the 50% ammonium sulfate fractionation. The rest of the type I enzyme is further separated from the type II enzyme by phosphocellulose chromatography.

Physical Properties

The physical properties of fraction VI enzyme were analyzed by gel filtration chromatography and glycerol gradient sedimentation. By using

[16] Unpublished result.

proper calibration from the markers with known physical properties, the sedimentation velocity[17] and the Stokes' radius[18] of the enzyme were determined to be 10 S and 65 Å, respectively. From combining these two physical parameters, we therefore estimate that the molecular weight (M_r) of the enzyme is 3.2×10^5 and its friction ratio is 1.7.[19] In a denaturing gel electrophoresis system,[20] the fraction VI enzyme gives four major polypeptides with the molecular weights (M_r) of 170,000, 151,000, 141,000, and 132,000 (Fig. 2). We believe that the intact enzyme is a dimer of the polypeptide chains of M_r 170,000, and the smaller polypeptides are the proteolytic cleavage products of the large one and are still active in the topoisomerization reactions. These conclusions are based on the following three lines of experiments.[16] First, the isolated M_r 170,000 and 132,000 polypeptides from SDS gel electrophoresis gave almost identical proteolytic cleavage patterns when they were treated with protease V8 and analyzed by the SDS gel electrophoresis according to the procedure of Cleveland *et al.*[21] Second, the structural relatedness of these polypeptides can also be established by an immunochemical approach. The antibodies prepared as directed against either M_r 170,000 and 132,000 polypeptides can clearly cross-react with each other as demonstrated by the electrophoretic transfer and blotting experiment according to the procedure of Towbin *et al.*[22] Finally, all four polypeptides can form a covalent complex with radioactively labeled DNA, which indicates that the degree of proteolysis in our enzyme preparation will not make these polypeptides inactive in the topoisomerization reactions. However, we do not yet know all the effects of the mild proteolysis on the enzyme in catalyzing the topoisomerization reaction. It is apparent the protease inhibitors we used throughout the purification steps are not capable of inactivating all the proteolytic enzymes in the *Drosophila* embryo extract.

Enzymatic Properties

The fraction VI enzyme can catalyze the following reactions in an ATP-dependent fashion: relaxation of superhelical turns, catenation, decatenation, and unknotting of circular duplex DNAs (see Fig. 1). It can alter the linking number of DNA only in steps of two. We have also

[17] R. G. Martin and B. N. Ames, *J. Biol. Chem.* **236,** 1372 (1961).
[18] J. Porath, *Pure Appl. Chem.* **6,** 233 (1963).
[19] L. M. Siegel and K. J. Monty, *Biochim. Biophys. Acta* **112,** 346 (1966).
[20] U. K. Laemmli, *Nature (London)* **227,** 680 (1970).
[21] D. W. Cleveland, S. G. Fischer, M. W. Kirschner, and U. K. Laemmli, *J. Biol. Chem.* **252,** 1102 (1977).
[22] H. Towbin, T. Staehelin, and J. Gordon, *Proc. Natl. Acad. Sci. U.S.A.* **76,** 4350 (1979).

FIG. 2. Sodium dodecyl sulfate gel electrophoresis of purified *Drosophila* type II topoisomerase. About 2 μg of fraction VI enzyme are analyzed by electrophoresis in 8% polyacrylamide gel. The gel is stained with 0.2% Coomassie Brilliant Blue dye, destained by diffusion, and photographed. The size of the polypeptides is calculated using *Escherichia coli* RNA polymerase holoenzyme, *E. coli* DNA polymerase I, and bovine serum albumin as markers. The molecular weights are given in the text and are also abbreviated at the left-hand side of the figure (170K, 170,000; 132K, 132,000).

learned that the *Drosophila* type II topoisomerase at a high concentration can knot circular duplex DNA molecules.[16] All these properties clearly show that this is a type II DNA topoisomerase and works by passing a segment of DNA through a reversible double-stranded break. These properties are also shared by a less purified preparation of the enzyme from either *Xenopus* oocytes[10] or *Drosophila* early embryos,[11] as well as by a homogeneous preparation of the enzyme from HeLa cell nuclei.[12] It is interesting to note the similarity of the physical properties between the type II enzymes from *Drosophila* embryos and HeLa cell nuclei. The HeLa enzyme is also a dimer of polypeptide chains of M_r 172,000.[12]

The double-stranded breakage and rejoining mechanism used by the *Drosphila* type II enzyme can be inferred from the structure of a reaction intermediate as well, which is a DNA molecule with a double-stranded break and covalent linkage of the enzyme to the broken ends. We can trap approximately 30% of the DNA in the reaction mixture in this intermediate form if we stop the reaction with a denaturant such as SDS. The topoisomerase-cleavage site has 3′-hydroxyl ends and 4-nucleotide long, 5′-protruding ends that are covalently linked to the enzyme molecules.[16] A similar intermediate has also been observed in reactions with bacterial DNA gyrase.[6,23] As discussed in the preceding section, the covalent enzyme–DNA complex formation reaction can be useful in demonstrating which polypeptides in the enzyme preparation are involved in the breaking and rejoining of the DNA molecule during the topoisomerization reaction.

Acknowledgment

This work is supported by a grant GM29006 from National Institute of General Medical Science. I wish to thank Dr. Leroy Liu for providing bacterial and phage strains used in the preparation of phage P4 knotted DNA.

[23] M. Gellert, K. Mizuuchi, M. H. O'Dea, T. Itoh, and J.-I. Tomizawa, *Proc. Natl. Acad. Sci. U.S.A.* **74,** 4772 (1977).

[11] *Escherichia coli* DNA Gyrase

By RICHARD OTTER and NICHOLAS R. COZZARELLI

Deoxyribonucleic acid gyrase is a prokaryotic DNA topoisomerase essential for cell growth that is responsible for maintaining the negatively supercoiled state of DNA.[1] Gyrase changes the topological linking number of closed circular DNA in steps of two by transiently creating a double-strand break and passing another DNA segment through the break.[2] When ATP is present, the result of strand passage is the introduction of negative supercoils (Fig. 1A). In the absence of ATP, gyrase removes negative supercoils, albeit inefficiently (Fig. 1A). High local concentrations of DNA caused by DNA aggregation favor intermolecular strand passage or catenation (Fig. 1B). Decatenation occurs if the DNA is dispersed (Fig. 1B). Removal of knots from DNA is the intramolecular analog to decatenation (Fig. 1C). The concerted breakage and reunion of DNA that is the hallmark of topoisomerases can be uncoupled by sequential addition of nalidixic acid and a protein denaturant.[3] Cleavage at specific DNA sequences results from covalent attachment of a gyrase subunit (Fig. 1D). Gyrase binds DNA in a unique way (Fig. 1E); the DNA is coiled in a positive sense around the enzyme, which causes compensating negative supercoils to be segregated elsewhere in the DNA.[4] Lastly, gyrase is a DNA-dependent ATPase (Fig. 1F).[5]

Gyrase's varied activities *in vitro* probably reflect a varied physiological role. Besides its documented place in DNA supercoiling, gyrase may also be involved in resolving catenanes and knots in the bacterial cell.[1] The *Escherichia coli* gyrase, which we will emphasize, has been best characterized, but the *Micrococcus luteus* enzyme is strikingly similar in spite of the evolutionary divergence of these bacteria.[6]

[1] N. R. Cozzarelli, *Science* **207,** 953 (1980); M. Gellert, *Annu. Rev. Biochem.* **50,** 879 (1981).
[2] P. O. Brown and N. R. Cozzarelli, *Science* **206,** 1081 (1979); K. Mizuuchi, L. M. Fisher, M. H. O'Dea, and M. Gellert, *Proc. Natl. Acad. Sci. U.S.A.* **77,** 1847 (1980).
[3] M. Gellert, K. Mizuuchi, M. H. O'Dea, T. Itoh, and J. Tomizawa, *Proc. Natl. Acad. Sci. U.S.A.* **74,** 4772 (1977); A. Morrison and N. R. Cozzarelli, *Cell* **17,** 175 (1979); A. Sugino, N. P. Higgins, and N. R. Cozzarelli, *Nucleic Acids Res.* **8,** 3865 (1980).
[4] L. F. Liu and J. C. Wang, *Cell* **15,** 979 (1978).
[5] K. Mizuuchi, M. H. O'Dea, and M. Gellert, *Proc. Natl. Acad. Sci. U.S.A.* **75,** 5960 (1978); A. Sugino, N. P. Higgins, P. O. Brown, C. L. Peebles, and N. R. Cozzarelli, *Proc. Natl. Acad. Sci. U.S.A.* **75,** 4838 (1978). A. Sugino and N. R. Cozzarelli, *J. Biol. Chem.* **255,** 6299 (1980).
[6] L. F. Liu and J. C. Wang, *Proc. Natl. Acad. Sci. U.S.A.* **75,** 2908 (1978).

METHODS IN ENZYMOLOGY, VOL. 100

ISBN 0-12-182000-9

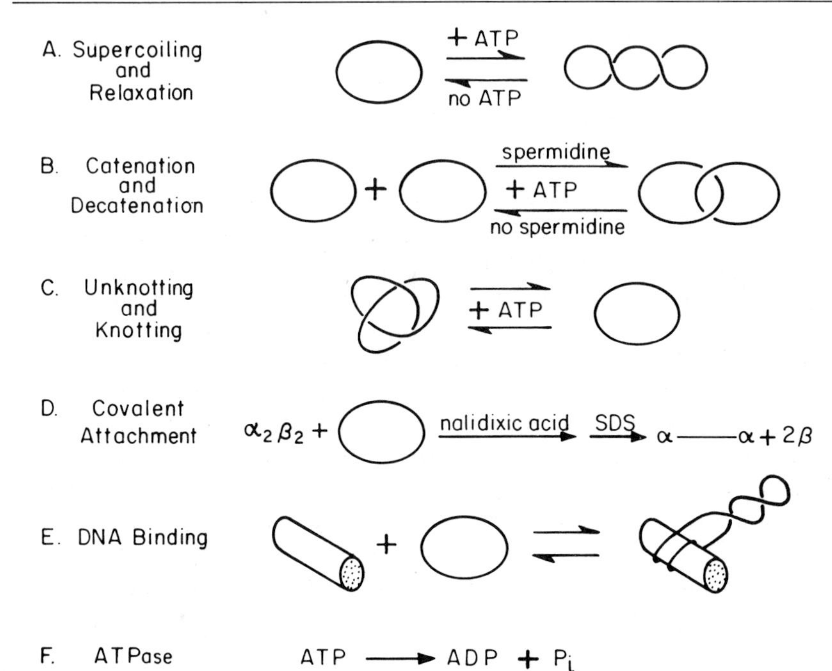

FIG. 1. Activities of DNA gyrase. (A) Gyrase, in the presence of ATP, will introduce negative supercoils into relaxed covalently closed circular DNA; in the absence of ATP, gyrase relaxes supercoils. (B) Gyrase catenates DNA rings that are aggregated by a condensing agent such as spermidine; catenated substrates are decatenated under nonaggregating conditions. (C) Knotted molecules are quickly untangled by gyrase. (D) Gyrase cleaves DNA after incubation with nalidixic acid and treatment with sodium dodecyl sulfate (SDS) by forming a covalent complex between the 5′ ends of the DNA and the α protomers encoded by the *gyrA* gene. (E) DNA binds to gyrase (represented by the cylinder) by wrapping in a positive sense creating compensatory negative supercoils elsewhere in the DNA. (F) In the presence of DNA, gyrase hydrolyzes ATP to ADP and P_i.

Gyrase is a tetramer consisting of two protomers of 105,000 daltons encoded by the *gyrA* gene and two encoded by *gyrB* of 95,000 daltons. The A subunit contains the active site for DNA breakage and reunion, and the B subunit contains the ATP binding site. The two subunits can be selectively inactivated by drugs or mutations. The novobiocin family of antimicrobials, including coumermycin and chlorobiocin, interferes competitively with ATP binding to the B subunit and therefore with energy transduction.[7] The nalidixic acid family of drugs inhibits DNA breakage

[7] A. Sugino and N. R. Cozzarelli, *J. Biol. Chem.* **255,** 6299 (1980); A. Sugino, N. P. Higgins, P. O. Brown, C. L. Peebles, and N. R. Cozzarelli, *Proc. Natl. Acad. Sci. U.S.A.* **75,** 4838 (1978).

and reunion but not DNA binding or ATPase; most resistant mutants are in *gyrA*. Clinically promising nalidixic acid analogs have been synthetized that have a 100-fold increased potency.[8] The availability of selective inhibitors and mutants[9] has allowed extensive physiological studies of gyrase's role *in vivo*[10] and determination of the particular function of each subunit.

This report will cover the assay, purification, and reactions of DNA gyrase. The enzyme is a valuable reagent for preparing negatively and positively supercoiled DNA, forming or resolving catenanes and knots, and analyzing the structure of complex forms of DNA.[11]

Assays

The preferred gyrase assay measures supercoiling of DNA as analyzed by gel electrophoresis. The supercoiling reaction is unique to gyrase, and the assay is sensitive and relatively quick. All plasmid DNAs that have been tested are active in the assay. The unit of gyrase activity is defined as the amount sufficient to convert 23 fmol[12] of relaxed ColE1 DNA to the fully supercoiled form in 30 min at 30° under the conditions detailed below.

Gyrase is usually purified as two separate subunits that are inactive by themselves (except for an ATPase activity detectable in concentrated solutions of B subunit). The holoenzyme is reconstituted simply by mixing the subunits together. Various factors affect the rate of reconstitution, but in the presence of DNA the enzyme is re-formed in the order of seconds. One subunit is usually added in excess, so that a more precise estimate of the activity of the limiting subunit can be made.

The substrate for the supercoiling assay is a relaxed plasmid DNA that is conveniently prepared using the topoisomerase in a crude rat liver

[8] A. Ito, K. Hirai, M. Inoue, H. Koga, S. Suzue, T. Irikura, and S. Mitsuhashi, *Antimicrob. Agents Chemother.* **17,** 103 (1980).

[9] K. N. Kreuzer, K. McEntee, A. P. Geballe, and N. R. Cozzarelli, *Mol. Gen. Genet.* **167,** 129 (1978); M. Gellert, K. Mizuuchi, M. H. O'Dea, T. Itoh, and J. Tomizawa, *Proc. Natl. Acad. Sci. U.S.A.* **74,** 4772 (1977); E. Orr, N. F. Fairweather, I. B. Holland, and R. H. Pritchard, *Mol. Gen. Genet.* **177,** 103 (1979); M. Gellert, M. H. O'Dea, T. Itoh, and J. Tomizawa, *Proc. Natl. Acad. Sci. U.S.A.* **73,** 4474 (1976).

[10] E. C. Engle, S. H. Manes, K. Drlica, *J. Bacteriol.* **149,** 92 (1982); N. F. Fairweather, E. Orr, and I. B. Holland, *J. Bacteriol.* **142,** 153 (1980); M. Filutowicz, *Mol. Gen. Genet.* **177,** 301 (1980).

[11] J. C. Marini, K. G. Miller, and P. T. Englund, *J. Biol. Chem.* **255,** 4976 (1980).

[12] DNA concentrations are expressed in terms of moles of plasmid.

nuclear extract with EDTA added to inhibit nucleases.[13] Topoisomerase prepared in a few hours from only 10 g of liver can relax 10 nmol of plasmid DNA. The 1.0-ml reactions contain 20 m*M* Tris-HCl (pH 7.6), 5 m*M* EDTA, 0.2 *M* KCl, 50 μg of serum albumin per milliliter, 35 pmol of DNA, and relaxing enzyme. After 1 hr at 37°, the product is extracted with phenol and precipitated with ethanol. If there is more than about 20% contamination with nicked or linear DNA, the relaxed DNA should be purified by equilibrium density gradient centrifugation in the presence of ethidium bromide.[14]

The standard supercoiling assay contains, in 20 μl, 23 fmol of relaxed ColE1 DNA, 50 m*M* Tris-HCl (pH 7.6), 20 m*M* KCl, 10 m*M* $MgCl_2$, 2 m*M* dithiothreitol, 1.5 m*M* ATP, 5 m*M* spermidine · Cl_3, 50 μg of bovine serum albumin per milliliter, and gyrase. Enzyme is diluted in 50 m*M* Tris-HCl (pH 7.6), 20 m*M* KCl, 2 m*M* dithiothreitol, and 50 μg of albumin per milliliter. After 30 min at 30°, the reactions are stopped with 7 μl of 2% sodium dodecyl sulfate, 20% Ficoll, and 400 μg of bromophenol blue per milliliter. The samples are loaded on a 1% agarose gel (13 × 15 × 0.4 cm) and electrophoresed at 4 V/cm for at least 5 hr for clear resolution. For routine assays where wide separation of topoisomer bands is not required, results can be obtained more rapidly if electrophoresis is at a higher voltage or if mini gels are used. The gels are stained with ethidium bromide (0.5 μg/ml) for at least 15 min and photographed without destaining under ultraviolet light. Electrophoretic resolution of supercoiled and relaxed DNA is no more difficult than other types of electrophoretic separations. A standard buffer, TBE (89 m*M* Tris base, 89 m*M* boric acid, 2.5 m*M* EDTA), and a medium EEO agarose such as Sigma type II are satisfactory. Assays of crude enzyme preparations should have a control for the amount of nicking because nicked circular DNA is not resolved from the relaxed closed circular DNA substrate. Contaminating endonucleases can sometimes be differentially inhibited by 500 μg of tRNA per milliliter, 10 m*M* spermidine, or 25 m*M* potassium phosphate, pH 7.6. The assay is nonlinear with respect to appearance of fully supercoiled DNA, and activity estimates should be based on dilutions that give between 50% and 100% supercoiling. Gyrase is active between 15 m*M* and 100 m*M* salt, has a broad pH optimum, but is relatively temperature sensitive—the optimum is 30°. Several nonelectrophoretic methods can be used to measure supercoiling. Negative supercoiling increases the affinity of DNA for ethidium bromide, whose bound form can be quantitated fluorometri-

[13] J. J. Champoux and B. L. McConaughy, *Biochemistry* **15,** 4638 (1976); W. S. Dynan, J. J. Jendrisak, D. A. Hager, and R. R. Burgess, *J. Biol. Chem.* **256,** 5860 (1981); D. E. Pulleyblank and M. J. Ellison, *Biochemistry* **21,** 1155 (1982).

[14] D. B. Clewell and D. R. Helinski, *Proc. Natl. Acad. Sci. U.S.A.* **62,** 1159 (1969).

cally.[15] Two other topoisomerase assays are based on the selective retention of supercoiled DNA by nitrocellulose filters under certain conditions.[16,17]

Gyrase can also be assayed by its catenating activity. With a relaxed DNA substrate, huge interlinked DNA networks are formed. The networks can be separated quickly from starting material by bench-top centrifugation[18] or filtration through nitrocellulose filters[19]; use of a labeled substrate makes these rapid, quantitative assays. Catenation does, however, require more gyrase than supercoiling and the assays are less useful for crude preparations where large DNA products are produced by several other mechanisms.

The cleavage of DNA and formation of a covalent complex between DNA and DNA gyrase has been used to monitor gyrase activity during its purification. Labeled DNA is incubated with DNA gyrase in the presence, per milliliter, of 20 μg of oxolinic acid or 200 μg of nalidixic acid for 30 min at 30° under supercoiling reaction conditions. Addition of 0.2% sodium dodecyl sulfate (SDS) causes approximately 40% of the gyrase in the reaction to be bound covalently to the DNA, and this complex is retained by a nitrocellulose filter.[3] The blanks in this reaction are very low because the SDS decreases nonspecific binding to the filter. The assay can be made specific for gyrase by controls in the absence of oxolinic acid or by demanding dependence on added complementary gyrase subunit if only one subunit is being assayed. The assay need not be done with circular DNA, and it is not confounded by relatively high endonuclease levels. It is less sensitive, however, than the supercoiling assay.

Gyrase Purification

DNA gyrase has been purified from *E. coli*,[20] *M. luteus*,[6] *Bacillus subtilis*,[21] and *Pseudomonas*.[22] Several laboratory strains of *E. coli* K12

[15] A. R. Morgan, D. H. Evans, J. S. Lee, and D. E. Pulleyblank, *Nucleic Acids Res.* **7,** 571 (1979).

[16] T. C. Rowe, J. R. Rusche, M. J. Brougham, and W. K. Holloman, *J. Biol. Chem.* **256,** 10354 (1981).

[17] J. J. Champoux and B. L. McConaughy, *Biochemistry* **15,** 4638 (1976).

[18] M. A. Krasnow and N. R. Cozzarelli, *J. Biol. Chem.* **257,** 2687 (1982).

[19] R. Low and A. Kornberg, personal communication.

[20] M. Gellert, K. Mizuuchi, M. H. O'Dea, and H. A. Nash, *Proc. Natl. Acad. Sci. U.S.A.* **73,** 3872 (1976); A. Sugino, C. L. Peebles, K. N. Kreuzer, and N. R. Cozzarelli, *Proc. Natl. Acad. Sci. U.S.A.* **74,** 4767 (1977); N. P. Higgins, C. L. Peebles, A. Sugino, and N. R. Cozzarelli, *Proc. Natl. Acad. Sci. U.S.A.* **75,** 1773 (1978).

[21] A. Sugino and K. F. Bott, *J. Bacteriol.* **141,** 1331 (1980); E. Orr and W. L. Staudenbauer, *J. Bacteriol.* **151,** 524 (1982).

[22] Unpublished results of the authors.

and *E. coli* B have been assayed for gyrase activity; all were similar except that strain H560 had about 5 times more gyrase than the others (about 2 × 10^6 units per gram of cell paste). H560 is also convenient because it lacks endonuclease I activity. Both *E. coli gyrA* and *gyrB* genes have been cloned into multicopy plasmids; the *gyrB* gene has been placed under transcriptional control of the bacteriophage λ P_L promoter.[22a] Strains containing these plasmids make excellent gyrase subunit overproducers. Since wild-type *E. coli* strains produce approximately 10 times as much gyrase A subunit as B subunit, a strain that overproduces the B subunit is particularly valuable.

The *E. coli* and *M. luteus* gyrases are usually purified as their separated subunits. The difference in DNA binding properties of the subunits and the holoenzyme has been taken advantage of in two purification schemes. Klevan and Wang[23] removed trace contaminants from highly purified subunits by adsorbing reconstituted gyrase to DNA cellulose. Neither subunit alone binds DNA or the DNA analog heparin. In the modification of the Staudenbauer and Orr[24] purification of subunit B that follows, gyrase holoenzyme is first adsorbed to and eluted from heparin–agarose. In a later step, separated subunit B is passed through this column whereas contaminants remain adsorbed.

Escherichia coli cells are grown in a vigorously aerated fermentor at 30° until late exponential phase. A yield of 10 g wet weight of cell paste per liter of medium is typical. A defined medium (containing per liter 1 g of $(NH_4)_2SO_4$, 3 g of KH_2PO_4, 5.25 g of K_2HPO_4, 0.5 g of sodium citrate, 0.12 g of $MgSO_4$) supplemented with 1% casein hydrolyzate and 1% glucose yields a higher level of gyrase than a rich medium such as LB. The cells are harvested and resuspended in an equal volume of a solution containing 20% sucrose and 0.1 *M* Tris-HCl, pH 7.6. The cells may now be frozen in liquid nitrogen for long-term storage at −70°; no loss of gyrase activity is observed for at least a year.

The frozen cells are thawed in a beaker placed in a 4° water bath. The cells are occasionally stirred until they reach a temperature of 1°. Then 0.5 *M* dithiothreitol is added to 2 m*M*; 0.5 *M* EDTA to 20 m*M*; 0.2 *M* phenylmethylsulfonyl fluoride in ethanol to 1 m*M*; and egg white lysozyme to 200 μg/ml. The solution is transferred to Beckman type 35 ultracentrifuge bottles and incubated at 0° for 20 min. The cells are lysed by addition of 10% Brij-58 and 4 *M* KCl to final concentrations of 0.1% and 75 m*M*. The lysate is carefully stirred and then centrifuged at 140,000 *g* for 90 min at

[22a] M. Kodaira and A. Kornberg, personal communication.

[23] L. Klevan and J. C. Wang, *Biochemistry* **19,** 5229 (1980).

[24] W. L. Staudenbauer and E. Orr, *Nucleic Acids Res.* **9,** 3589 (1981).

2°. The supernatant is removed, dialyzed against buffer A (25 m*M* Tris-HCl (pH 7.6), 2 m*M* dithiothreitol, 10% glycerol) plus 50 m*M* KCl, and then diluted by half with the same buffer to lower the viscosity. Extracts of strain H560 have about 2 units of gyrase per microliter at this stage.

The diluted lysate is applied to a heparin–agarose affinity column (1 ml column volume per 10 g of cell paste) equilibrated with buffer A + 50 m*M* KCl. The column is washed with two column volumes of buffer A + 50 m*M* KCl and eluted with a five column volume 50 to 500 m*M* KCl gradient in buffer A. Free subunit A (or subunit B with a B overproducing strain) passes through, whereas gyrase activity peaks broadly at about 0.1 *M* KCl. The holoenzyme pool is loaded directly onto a novobiocin–agarose affinity column[24] equilibrated with buffer A + 0.15 *M* KCl (1 ml column volume for 100 g of cell paste). The column is washed extensively with buffer A + 0.15 *M* KCl. Two column volumes of buffer A plus 0.8 *M* KCl dissociates the gyrase subunits and elutes subunit A. The B subunit is then eluted with 5 *M* urea in buffer A and dialyzed against buffer A. Any precipitate is removed by centrifugation. The dialysate is applied to a second heparin–agarose column (1 ml column volume per 100 g of cell paste) equilibrated with buffer A + 50 m*M* KCl. The gyrase B subunit does not bind heparin in the absence of A subunit and therefore passes through. The pass-through is concentrated by adsorption to a small hydroxyapatite column (1 ml column volume per 500 g of cell paste) equilibrated with 50 m*M* potassium phosphate (pH 6.8), 2 m*M* dithiothreitol, and 10% glycerol and then eluted with the same buffer with 400 m*M* phosphate. This pool is then dialyzed against enzyme storage buffer (50% glycerol, 50 m*M* potassium phosphate (pH 7.6), 0.2 m*M* EDTA, 1 m*M* dithiothreitol) and maintained at −20°. The activity of this physically homogeneous preparation is lost with a half-life of about a year.

Pure gyrase subunit A can be obtained by the published procedure[20,25] starting with strain H560 or a subunit A overproducing strain. Smaller amounts, equivalent to the amount of subunit B, can be prepared as a by-product of the B purification. The A subunit eluted from the novobiocin–agarose column is dialyzed against buffer A + 50 m*M* KCl and applied to a heparin–agarose column (1 ml column volume per 100 g of cell paste). Subunit A will pass through the column. It may be concentrated and stored as with the B subunit, or it can be further purified by phosphocellulose chromatography as described.[25] The A subunit is quite stable, and no decrease in activity is observed after a year.

[25] A. Sugino, C. L. Peebles, K. N. Kreuzer, and N. R. Cozzarelli, *Proc. Natl. Acad. Sci. U.S.A.* **74,** 4767 (1977).

DNA Gyrase Reactions and Uses

The formation of negatively supercoiled DNA by gyrase has been described above. It should be noted that the *in vitro* products are more negatively supercoiled than native DNA.[26] Additional alterations of DNA tertiary and quaternary structure carried out by gyrase include relaxation, catenation, decatenation, unknotting, and positive supercoiling.

Relaxation. Relaxation of supercoiled DNA by gyrase occurs only in the absence of ATP or when novobiocin blocks the ATPase site; it has the same ionic requirements as supercoiling. This reaction is about 15 times slower than supercoiling and quite easily observed if a sensitive system is used to show the change in supercoiling. We use a small plasmid, the 1.7 kb pA03,[27] that has few supercoils so that each supercoil has a large effect on electrophoretic mobility. Removal of 20 supercoils from native ColE1 DNA does not alter its electrophoretic mobility significantly, whereas relaxation of pAO3 by just 1 supercoil can be easily detected. The composite acrylamide–agarose gel of Peacock and Dingman[28] gives excellent resolution of pAO3 topoisomers. The gel components are in TBE buffer: 0.5% agarose, 1.9% acrylamide, 0.1% bisacrylamide, 0.02% TEMED, and 0.05% ammonium persulfate. The agarose and buffer are boiled until solution is complete, and the lost water is replaced. The acrylamide and bisacrylamide are added, and the solution is slowly mixed on a magnetic stirrer until the temperature falls to 40°. The TEMED and ammonium persulfate solution are added, and the gel is poured. A slight layering at the top of vertical gels signals completion of polymerization. Horizontal gels must be poured and polymerized under a nitrogen atmosphere. The gel is run at 5 V/cm for about 12 hr. Plasmids larger than about 3 kb are not well resolved in this gel system. Figure 2 shows the pAO3 products of a relaxation reaction run on a composite gel.[29] Partially relaxed pAO3 displayed on a composite gel is also an excellent source of single DNA topoisomers for determining topoisomerase type.

Catenation. Under suitable conditions, gyrase will efficiently catenate DNA rings.[18] With ColE1 DNA there are two products depending on the substrate supertwist density. Supercoiled ColE1 DNA treated with gyrase yields mainly dimers, trimers, and tetramers. With relaxed ColE1 DNA,

[26] M. Gellert, K. Mizuuchi, M. H. O'Dea, and H. A. Nash, *Proc. Natl. Acad. Sci. U.S.A.* **73,** 3872 (1976).

[27] A. Oka, N. Nomura, M. Morita, H. Sugisaki, K. Sugimoto, and M. Takanami, *Mol. Gen. Genet.* **172,** 151 (1979).

[28] A. C. Peacock and C. W. Dingman, *Biochemistry* **7,** 668 (1968).

[29] F. Dean, M. A. Krasnow, R. Otter, M. M. Matzuk, S. J. Spengler, and N. R. Cozzarelli, *Cold Spring Harbor Symp. Quant. Biol.* **47,** in press (1983).

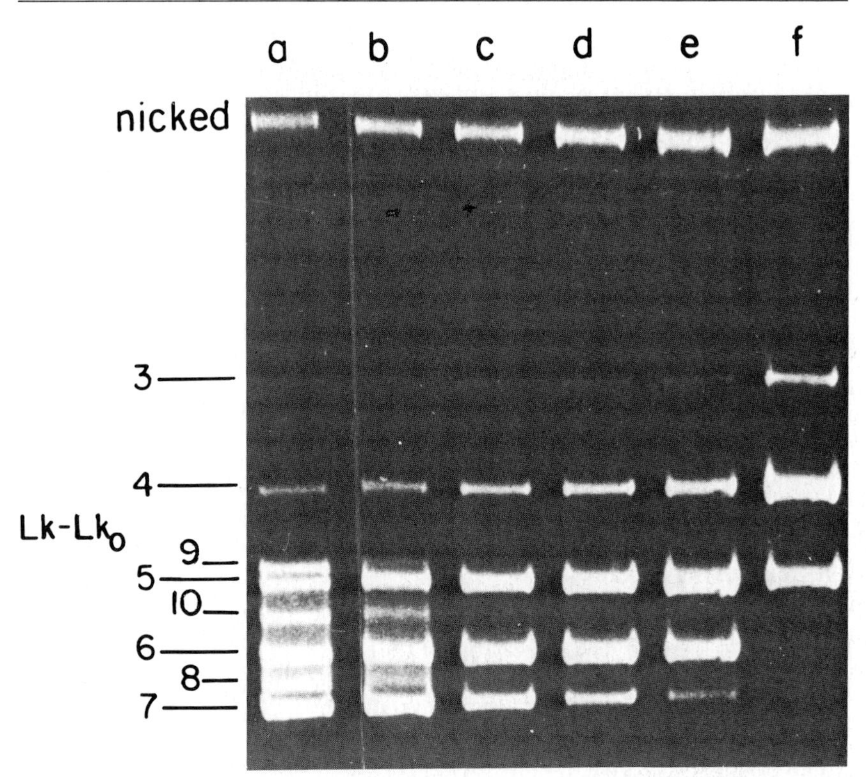

FIG. 2. Progressive relaxation of plasmid pAO3. Native plasmid pAO3 (lane a) was progressively relaxed (lanes b through f), and the DNA was displayed by electrophoresis through a composite acrylamide agarose gel as described in the text. The scale at the left indicates the number of supercoils in each topoisomer. Note that the mobility of pAO3 DNA increases with negative supercoiling until a maximum is reached, whereupon further supercoiling retards the DNA.[29]

however, the products are networks of thousands of interlinked circles. The small plasmid pAO3, even when relaxed, gives only dimeric, trimeric, and tetrameric products.

Typical catenation conditions are 20 m*M* Tris-HCl (pH 7.6), 2 m*M* $MgSO_4$, 2 m*M* dithiothreitol, 50 μg of serum albumin per milliliter, 15 m*M* KCl, 0.5 m*M* ATP, and 5 m*M* spermidine. Five supercoiling units of enzyme will catalyze formation of the equilibrium distribution of products starting with 23 fmol of DNA in a 20-μl reaction in 30 min at 30°. Catenation results because the cations neutralize greater than 90% of the DNA electrostatic charge[30] causing DNA aggregation. The high DNA concen-

[30] R. W. Wilson and V. A. Bloomfield, *Biochemistry* **18,** 2192 (1979).

tration in the reaction environment favors the intermolecular reaction of catenation. A simple rule for ensuring aggregating ionic conditions is to have the monovalent and divalent cation concentration less than 10 times the spermidine concentration.

Decatenation and Unknotting. These reactions are carried out under supercoiling conditions at subaggregating levels of spermidine.[18] Nicked substrates are untangled less efficiently than covalently closed circles. The conversion by gyrase of a mutimeric form of DNA to monomers is very good evidence for a catenated substrate structure.

Introduction of Positive Supercoils. The best way to prepare positively supercoiled DNA is through the stoichiometric binding of DNA gyrase to DNA. Gyrase is either bound to singly nicked circular DNA, and DNA ligase subsequently added to seal the nick,[31] or bound to covalently closed DNA in the presence of a relaxing topoisomerase, such as eukaryotic topoisomerase I.[4] In both cases, ATP must be absent or gyrase will remove all positive twists introduced. When the bound gyrase is removed, the positive supercoils around the protein will be converted to free positive supercoils.

Acknowledgments

This work was supported by NIH Research Grants GM-21397 and GM-22729 from the National Institute of General Medical Sciences.

[31] R. E. Depew and J. C. Wang, *Proc. Natl. Acad. Sci. U.S.A.* **72,** 4275 (1975).

Section III

Proteins with Specialized Functions Acting at Specific Loci

[12] The Bacteriophage Lambda Terminase Enzyme

By M. Gold, A. Becker, and W. Parris

Terminase plays a central role in the maturation of λ DNA and the morphogenesis of the phage head. Reviews[1,2] deal with these processes in detail and should be consulted for the genetic and biological mechanisms involved. *In vitro,* terminase is a multifunctional enzyme that can be used to package λ DNA and its derivatives and thus has proved to be extremely useful as a method in recombinant DNA technology. With highly purified preparations, packaging efficiencies of over 10^9 plaques per microgram of DNA can be readily achieved. The principles of *in vitro* packaging were established by Kaiser and Masuda[3] and a procedure developed by Becker and Gold,[4] which in its various modification[5-7] has been generally adopted for use in many laboratories engaged in gene cloning and amplification. These methods rely on the use of crude extracts and/or some partially purified terminase. The purification of terminase to near homogeneity is here described.

Assay of Terminase

Packaging. A complete and detailed description of this method is given elsewhere in this series.[8] Mature DNA isolated from phage particles can be used as substrate. All terminase fractions to be assayed are diluted into the sonic extract of induced NS428, the prohead donor. Further refinements and details of this method have been described[9,10] but are not necessary for monitoring the purification steps. A unit of terminase activ-

[1] H. Murialdo and A. Becker, *Microbiol. Rev.* **42,** 529 (1978).

[2] M. Feiss and A. Becker, *in* "Lambda II" (R. W. Hendrix, R. J. Roberts, F. W. Stahl, and R. A. Weisberg, eds.) Cold Spring Harbor Laboratory, Cold Spring Harbor, New York, in press, 1983.

[3] A. D. Kaiser and T. Masuda, *Proc. Natl. Acad. Sci. U.S.A* **70,** 260 (1973).

[4] A. Becker and M. Gold, *Proc. Natl. Acad. Sci. U.S.A.* **72,** 581 (1975).

[5] N. Sternberg, D. Tiemeier, and L. Enquist, *Gene* **1,** 255 (1977).

[6] H. E. Faber, D. O. Kiefer, and F. R. Blattner, *in* Application to the National Institutes of Health for EK2 certification of a host-vector system for DNA cloning. Supplement X (1978).

[7] T. Maniatis, E. F. Fritsch, and J. Sambrook, *in* "Molecular Cloning—A Laboratory Manual," p. 256. Cold Spring Harbor Laboratory, Cold Spring Harbor, New York, 1982.

[8] L. Enquist and N. Sternberg, this series, Vol. 68, p. 281.

[9] A. Becker, H. Murialdo, and M. Gold, *Virology* **78,** 277 (1977).

[10] S. Benchimol, A. Becker, H. Murialdo, and M. Gold, *Virology* **91,** 205 (1978).

ISBN 0-12-182000-9

ity is defined as that amount yielding 10^6 phages under the conditions of the assay.[4]

Cos-Cleavage. Terminase can form a complex with either mature or immature DNA,[11] and in the latter case the cohesive end sites (*cos*) are cut[12] endonucleolytically to regenerate the 5′-protruding, 12 nucleotide-long termini, which are characteristic of the lambdoid phage. This cleavage reaction is independent of proheads and packaging and can be carried out in several ways. One assay[12] involves the generation and detection of new restriction-endonuclease fragments from immature concatameric λ DNA after being cleaved at *cos* by terminase. However, simpler procedures have been devised that are based on the fact that the *cos* site and its flanking sequences can be inserted into small, well characterized plasmids to form "cosmids[13]," which are good substrates for terminase *in vitro*.[14,15]

Principle of Assay

The assay is based on appearance of two distinct fragments of DNA when cosmids that have been prelinearized at a unique restriction site are cleaved at *cos* by terminase. Any cosmid can be used, and the restriction site should be chosen so that two pieces of unequal length will result from terminase action. The fragments are resolved by electrophoresis in agarose gels.

Reagents

Buffer A: 0.02 *M* Tris-HCl, pH 8.0, 3 m*M* $MgCl_2$, 1 m*M* EDTA, 7 m*M* 2-mercaptoethanol

Supercoiled plasmid C25 DNA. This cosmid was constructed by Dr. L. Moran. It contains a 1850 bp *cos*-containing fragment from λ Charon 1[16] inserted into the *Bam*HI site of plasmid pBR322. A unique *Pst*I site is located 1150 bp from *cos*. However, any similar cosmid can be utilized.

Restriction endonuclease *Pst*I

Wrap mix: 6 m*M* Tris-HCl, pH 7.4, 15 m*M* ATP, 18 m*M* $MgCl_2$, 60 m*M* spermidine-HCl, 30 m*M* 2-mercaptoethanol, 0.5 *M* KCl

[11] A. Becker, M. Marko, and M. Gold, *Virology* **78,** 291 (1977).

[12] A. Becker and M. Gold, *Proc. Natl. Acad. Sci. U.S.A.* **75,** 4199 (1978).

[13] J. Collins, this series, Vol. 68, p. 309.

[14] G. Miller and M. Feiss, *Virology* **109,** 379 (1981).

[15] M. Gold and W. Parris, manuscript in preparation.

[16] F. R. Blattner, B. S. Williams, A. E. Blechl, K. Denniston-Thompson, H. E. Faber, L. A. Furlong, D. J. Grunwald, D. O. Kiefer, D. D. Moore, J. W. Schumm, E. L. Sheldon, and O. Smithies, *Science* **196,** 161 (1977).

Terminase. Depending on the preparation, the early fractions of the purification are unsuitable for this assay. Very pure fractions require a host factor (see below).
EDTA, 0.5 *M*
Glycerol bromophenol mixture: 50% Glycerol, 5% Sarkosyl, 0.025% bromophenol blue
Sodium dodecyl sulfate (SDS) 5%
Agarose gel, 1.4%
Acetate buffer: 0.04 *M* Tris-HCl, pH 7.7, 0.005 *M* sodium acetate, 0.001 *M* EDTA
Ethidium bromide, 10 mg/ml

Linearization of C25 Cosmid DNA

To 25 μl of buffer A, add 1.5 μg of C25 DNA and 1 unit of *Pst*I enzyme. This is incubated at 37° for 3 hr.

Cleavage of Linearized C25 with Terminase

To the above reaction add 3 μl of 0.5 *M* KCl, 3 μl of Wrap mix, and 3 μl of terminase. Mix and incubate at 22° for 40 min. The reaction is stopped by the addition of 5 μl of 0.5 *M* EDTA, 5 μl of the glycerol bromophenol mixture, and 5 μl of 5% SDS.

The samples are heated for 5 min at 65°, then loaded on an agarose gel and electrophoresed at 25 mA for 16 hr. Ethidium bromide is added after running to a concentration of 2 μg/ml to visualize the DNA bands, which are photographed. Negatives can be scanned with a densitometer to quantitate the extent of cleavage.

Purification of Terminase[17]

Bacterial Strain

Terminase is prepared from induced 594/TI (λd$gal_{805}cI_{857}Sam_7$) cells. This is a T1 phage-resistant, galactose-transducing, thermoinducible, lysis-defective lysogen with a λ deletion-substitution between the middle of gene *W* and *att* so that the only left-arm bacteriophage genes expressed are *Nu1* and *A*.

Methods

Growth of Bacteria. Bacteria are grown in a 200-liter Fermacell fermentor (New Brunswick Scientific Co., Inc.) using LB broth neutralized

[17] M. Gold and A. Becker, *J. Biol. Chem.*, in press.

to pH 7.3 with NaOH. Usually the inoculum consists of 3 liters of overnight cultures grown at 32°. The fermentor is maintained at 32°, and growth is continued until $A_{650} = 0.6$ (2×10^8 cells/ml) is reached. Air flow during culture is kept at 4 ft^3/min, and agitation at 200 rpm; Antifoam (diluted 1 : 100) is added to control foaming. At this point the temperature of the culture was raised to 45° over a period of 10 min and maintained at that temperature for 13 min in order to achieve complete prophage induction. The culture is then cooled to 37° using the Rapid Chiller; this process requires about 5 min. The induced culture is aerated with agitation at 37° for 1 hr then chilled to 15° over a period of approximately 10 min. The cells are harvested by centrifugation in a refrigerated continuous-flow Sharples centrifuge at 15,000 rpm and stored at −20°. About 600 g (wet weight) of induced bacteria are routinely harvested in this manner and full enzyme activity can still be recovered after over 2 years of storage.

Subsequently, all operations are carried out at 0–5°, and all centrifugations involving large volumes are for 40 min at 15,000 *g* in either the GS-3 or GSA rotor of a Sorvall centrifuge. Where smaller volumes are involved, the SS-34 rotor can be used with centrifugation at 25,000 *g* for 30 min. All buffers contain 10 m*M* 2-mercaptoethanol and 1 μM phenylmethylsulfonyl fluoride (PMSF).

A typical work-up is described in the following subsections.

Preparation of Extract. Frozen cells (4600 g) were broken into small pieces manually and suspended in 16 liters of 0.02 *M* Tris, pH 8.0, containing 0.01 *M* $MgCl_2$ and 0.001 *M* EDTA. The suspension was disrupted with a Polytron type TP 45/2 homogenizer (Brinkmann Instruments) at a setting of 8; the temperature was maintained at between 5 to 10° by external cooling. When the mixture reached a uniform consistency that did not change with further homogenization, extraction was completed by treatment with a Biosonik III ultrasonic generator using a 3/4-inch probe operating at full power. The extract temperature did not exceed 12° during this process, which continued until centrifugation of a small sample indicated that cell breakage was essentially complete. The crude extract (20 liters) was taken to the next step without centrifugation.

Polymin P Precipitation: Ammonium Sulfate I. Polyethyleneimine (Polymin P, Miles Laboratories) was diluted 5-fold in water and brought to pH 8.0 with HCl; 2500 ml of this solution was added slowly with stirring to the crude extract over a period of 1 hr. After a further 1 hr of stirring, the heavy suspension was centrifuged and the supernatant discarded. The pellets were pooled and homogenized to a uniform suspension in 6 liters of 0.05 *M* ammonium succinate, pH 6.0, using the Polytron. After stirring for 30 min and centrifugation, the supernatant was discarded and the pellets were again extracted in the same way with 5 liters of 0.1 *M* ammonium succinate, pH 6.0. Finally the pellets were eluted with two successive extractions of 5 liters of 0.2 *M* ammonium succinate, pH 6.0, each, these supernatants were retained and combined. Solid ammonium sulfate (1640 g) was added slowly with stirring, and after 2 hr the suspension was centrifuged and the pellets were discarded. Solid ammonium sulfate (2354 g) was added to the supernatant slowing with stirring; after 14 hr the suspension was centrifuged. The pellets were dissolved in a minimum volume of 0.02 *M* Tris, pH 8.0, and dialyzed against 12 liters of this buffer for 16 hr two successive times (ammonium sulfate I). The final volume of the ammonium sulfate fraction was 2250 ml. At this stage of purification the terminase fraction could be kept frozen in liquid N_2 for several months without any loss of activity. When mixed with an equal volume of glycerol, some preparations have retained full activity for over 2 years when stored at $-15°$.

Hydroxyapatite Chromatography: Ammonium Sulfate II. The ammonium sulfate I fraction was loaded on a column of hydroxyapatite (Clarkson Chemical Co.) that had been equilibrated with 0.01 *M* potassium phosphate buffer, pH 6.5. The column was 15 cm in diameter and 20 cm high, and the flow rate was maintained at 1.5 liters/hr. The column was washed with 4 liters of buffer and then developed with a linear gradient of potassium phosphate buffer, pH 6.5 (0 to 0.5 *M*). The total volume of the gradient was 24 liters and fractions of 1200 ml were collected. Each fraction was brought to 65% saturation with solid ammonium sulfate; after standing overnight the suspensions were centrifuged and the pellets were dissolved in a minimum volume of 0.02 *M* Tris, pH 8.0, and 1 m*M* EDTA and dialyzed for 16 hr against 6 liters of the same buffer. Terminase activity eluted as a peak between 0.2 and 0.3 *M* potassium phosphate, and these fractions were pooled (ASII, 307 ml).

Heparin-Agarose Column Chromatography: Ammonium Sulfate III. A column (10.5×22 cm) of heparin-agarose was prepared[18] and equilibrated with 4 liters of 20 m*M* Tris-HCl, pH 8.0, containing 1 m*M* EDTA; 300 ml of the AsII fraction were applied to the column and washed with 4 liters of equilibration buffer. The column was then developed with a linear gradient of KCl (0 to 0.5 *M*) in the same buffer. The total volume of the gradient was 12 liters. Fractions of 600 ml were collected. Each fraction was precipitated by the addition of solid ammonium sulfate to a final saturation of 70%; after standing overnight, the precipitates were collected by centrifugation. The precipitates from the individual fractions were redissolved in 20 m*M* Tris, pH 8.0, containing 1 m*M* EDTA and were dialyzed against 10^3 volumes of this buffer for 15 hr. The final volume of each of the fractions was about 20 ml. The fractions were assayed

[18] T. A. Bickle, V. Pirrotta, and R. Imber, *Nucleic Acids Res.* **4,** 2561 (1977).

in the packaging reaction, then frozen and kept in liquid nitrogen. Terminase activity was recovered in four fractions at a position between 0.14 and 0.22 *M* KCl on the gradient. This step gave an apparent recovery of 35% of the units applied with an enrichment factor or about 2.

BioRex 70 Column Chromatography: Ammonium Sulfate IV. A column (5.5 × 25 cm) of BioRex 70 (Bio-Rad) was prepared and equilibrated with 0.01 *M* sodium phosphate buffer, pH 6.7, containing 0.01 *M* NaCl and 1 m*M* EDTA. The active ASIII fractions were thawed, pooled (80 ml), and applied to the column. The column was then washed with 2 liters of equilibrating buffer and developed with a linear gradient of NaCl (0.01 to 0.4 *M*) in equilibration buffer. The total volume of the gradient was 4 liters, and fractions of 200 ml were collected. Again, each fraction was made 70% in ammonium sulfate and left to precipitate overnight. The precipitates were then collected by centrifugation, redissolved in equilibration buffer, and dialyzed against 5 liters of the same buffer. After dialysis, the individual fractions (range of volumes, 2–2.5 ml) were assayed for biological activity. Activity was recovered in six fractions corresponding to 0.18 to 0.26 *M* NaCl on the gradient, these fractions were pooled and kept in liquid nitrogen (ASIV). The BioRex step gave a recovery of 65% of the input activity and an overall 2.7-fold purification. Considerable contaminating nuclease activity is removed in this step.

ATP-Agarose Column Chromatography. ATP-agarose type 4 (PL) was used to construct a column (0.9 × 16 cm) that was equilibrated with 5 volumes of 0.02 *M* Tris buffer, pH 8.0. The ASIV active fractions were thawed, pooled (14.5 ml), and applied to the column over a period of 2.5 hr. The column was then washed with 36 ml of buffer followed by four successive 36-ml washes of (*a*) 10 m*M* sodium pyrophosphate; (*b*) 10 m*M* adenosine monophosphate and 10 m*M* $MgCl_2$; (*c*) 15 m*M* adenosine triphosphate and 5 m*M* $MgCl_2$; and (*d*) 0.5 *M* NaCl. All the preceding were in equilibration buffer. Fractions of 6 ml were collected, assayed directly for biological activity, and frozen in liquid nitrogen. Over 90% of the activity seen on this column was found in the fractions eluted by ATP; 60% of the total input activity was recovered, and the purification was 5- to 6-fold. The ATP-agarose fractions were stable in liquid nitrogen without prior concentration.

DEAE-Sephadex A-25 Chromatography and Concentration. The pooled, active ATP-agarose fractions were further purified and concentrated on a column (0.9 × 3 cm) of DEAE-Sephadex A-25 (Pharmacia). The column was first equilibrated with 10 m*M* Tris-HCl, pH 7.1, containing 100 m*M* KCl and 5 m*M* dithiothreitol. The enzyme was applied, and the column was washed with 10 ml of equilibration buffer. A linear KCl gradient (0.1 to 0.5 *M*) in equilibration buffer was applied to the column.

PURIFICATION OF λ TERMINASE

Fraction	Volume (ml)	Total protein (mg)	Total activity (units)	Specific activity (units/mg)
I. Extract	2.00×10^4	1.7×10^6	1.6×10^6	0.96
II. Polymin P eluate (ASI)	2.25×10^3	1.0×10^5	1.4×10^6	14
III. Hydroxyapatite (ASII)	3.07×10^2	7.0×10^3	5.4×10^5	78
IV. Heparin-agarose (ASIII)	88	1.3×10^3	1.9×10^5	142
V. BioRex 70 (ASIV)	15	3.2×10^2	1.2×10^5	384
VI. ATP-agarose	30	36	7.2×10^4	2,087
VII. DEAE-Sephadex A-25	9	0.4	2.3×10^4	57,500

The total gradient was 30 ml, and 1.5-ml fractions were collected, dialyzed against 0.02 *M* Tris, pH 8.0, containing 10% glycerol, and assayed. Activity was detected in two peaks, over a total of six fractions. The two peaks correspond to 0.22 *M* KCl and 0.26 *M* KCl, respectively, on the gradient. This double-peaking is characteristic of this purification step. The active fractions were stored in liquid nitrogen.

A summary of the purification is given in the table.

Properties of Terminase

ASI fractions of terminase kept in liquid N_2 for longer than 2 years have retained full activity, and the most highly purified fractions are stable for at least a year when stored in this way.

Terminase has a native molecular weight of 117,000 and is a hetero-oligomer composed of the products of bacteriophage genes *Nu1* and *A*.[17,19] Terminase binds to DNA at sites flanking *cos* and can recognize specific phage proheads to form ternary complexes. Circular monomers of λ DNA containing only one *cos* can be efficiently cut and packaged *in vitro*.[20] The *cos*-cleavage reaction requires ATP, polyamines, Mg ions, and, under certain conditions, host factor(s).[12] Terminase is also a DNA-dependent ATPase.[17]

Host Factor

Highly purified preparations of terminase require a factor present in extracts of uninfected *Escherichia coli* for activity.[11,12] This requirement is evident only in the *cos*-cleavage reaction, and the nature of the factor is

[19] M. Summer-Smith, A. Becker, and M. Gold, *Virology* **111,** 642 (1981).

[20] M. Gold, D. Hawkins, H. Murialdo, W. L. Fife, and B. Bradley, *Virology* **119,** 35 (1982).

unclear. Preliminary experiments indicate that one factor involved is a heat-stable very basic protein of molecular weight approximately 22,000. Its preparation is outlined below.[15] It shares some properties of the IHF protein,[21] which is required in the assay of λ integrase enzyme.[22]

Purification

The factor is prepared from *E. coli* K12 strain 1100. Frozen cells are disrupted, and the crude extract is treated with Polymin P exactly as outlined above for the purification of terminase except that the Polymin pellets prepared from 2 kg of frozen cells were eluted with 4 liters of 0.2 *M* ammonium succinate, pH 6.0, and the eluate was concentrated by the addition of solid ammonium sulfate to 80% saturation. After centrifugation, the pellets were dissolved in 0.02 *M* Tris-HCl, pH 8.0, 1 m*M* EDTA, 7 m*M* 2-mercaptoethanol, and 0.005 m*M* PMSF (all buffers used contained the latter three ingredients) and heated in 200-ml aliquots to 60° for 2 min. The precipitate was removed by centrifugation, and the supernatant was dialyzed extensively to yield the ammonium sulfate I fraction (ASI). The latter is very stable and can be prepared in good yield.

The ASI fraction was loaded on a 8 cm × 36 cm column of DEAE-Sephadex A-25 equilibrated with 0.02 *M* Tris-HCl, pH 8.0. The pass-through and wash were collected and precipitated with 80% ammonium sulfate. The pellets were collected and dissolved in 0.02 *M* Tris-HCl, pH 8.0, and dialyzed extensively to give the ammonium sulfate II fraction (ASII). A column of Cellex-P (Bio-Rad) 6 × 45 cm was equilibrated in 0.02 *M* Tris-HCl, pH 9.0; ASII (385 ml) was loaded, and the column was washed with 2 liters of 0.05 *M* potassium phosphate buffer, pH 6.5, containing 0.1 *M* KCl. The column was then developed with a 3-liter linear gradient of KCl from 0.1 to 2 *M* in the same buffer. Thirty fractions of 100 ml were collected and dialyzed briefly against 0.02 *M* Tris-HCl, pH 8.0. Subsequently each fraction was precipitated by the addition of solid ammonium sulfate to 80% concentration, the precipitates were collected in 0.02 *M* Tris-HCl, pH 8.0, and dialyzed against the same. Factor activity elutes late in the gradient. Active fractions were pooled and brought to 80% saturation in ammonium sulfate, spun and dialyzed; the resulting fraction was termed ASIII.

ASIII was loaded on a column of Sephadex G-100 (8 × 75 cm) prepared in 0.02 *M* Tris-HCl, pH 8.0, containing 0.1 *M* KCl. A peak of activity was obtained eluting at a position of approximately 20,000 daltons

[21] H. A. Nash and C. A. Robertson, *J. Biol. Chem.* **256,** 9246 (1981).
[22] H. A. Nash, *Annu. Rev. Genet.* **15,** 143 (1981).

molecular mass. A column (0.9 × 12 cm) of Pharmacia PBE118 was poured and equilibrated in 0.025 *M* triethylamine pH 11.0. ASIV was loaded on the column followed by a 15-ml wash with the above buffer. The column was developed with 115 ml of Pharmalyte 8–10.5 (Pharmacia) diluted 1 : 35. Subsequently fractions were brought to 80% saturation in ammonium sulfate, spun and resuspended, and dialyzed. Factor activity eluted at a pH of about 9.3. Suitable dilutions of this preparation can be added to the *cos*-cleavage assay to obtain full terminase activity.

[13] The Resolvase Protein of the Transposon $\gamma\delta$

By RANDALL R. REED

The transposable elements $\gamma\delta$ and Tn*3* encode an efficient site-specific recombination system essential in the normal transposition process.[1] Transposition of these elements appears to occur in a two-step process. An element-encoded protein, the transposase, is involved in the initial step that leads to a fused replicon or cointegrate. In this structure, two copies of the element are present in the same orientation at the junctions between the donor and target replicons.[1–3] The resolvase protein of $\gamma\delta$ or Tn*3* catalyzes a site-specific recombination reaction at a particular site, *res,* located within the element.[4] This recombination leads to conversion of the cointegrate to the two end products of transposition, the donor replicon and the target replicon, now containing a copy of the element. This second step in the transposition of Tn*3*-like elements, termed resolution, has been demonstrated in a purified, *in vitro* system.[5]

The site at which recombination occurs has been identified through *in vivo* and *in vitro* techniques.[4,6] It is centrally located within a 170 base-pair (bp) intercistronic region between the divergently transcribed transposase (*tnpA*) and resolvase (*tnpR*) genes. Early genetic studies with $\gamma\delta$ and Tn*3* identified a role for the *tnpR* gene product in regulation of transposition

[1] A. Arthur and D. Sherratt, *Mol. Gen. Genet.* **175,** 267 (1979).
[2] F. Heffron, P. Bedinger, J. J. Champoux, and S. Falkow, *Proc. Natl. Acad. Sci. U.S.A.* **74,** 702.
[3] R. Gill, F. Heffron, G. Dougan, and S. Falkow, *J. Bacteriol.* **136,** 742 (1978).
[4] R. R. Reed, *Proc. Natl. Acad. Sci. U.S.A.* **78,** 3428 (1981).
[5] R. R. Reed, *Cell* **25,** 713 (1981).
[6] R. Kostriken, C. Morita, and F. Heffron, *Proc. Natl. Acad. Sci. U.S.A.* **78,** 4041 (1981).

ISBN 0-12-182000-9

frequency.[7-9] It appears that the ability of resolvase to act as both a recombination enzyme and a repressor is a consequence of its binding at the *res* site.[10]

I have undertaken a detailed biochemical characterization of the mechanism of recombination and the role of resolvase in regulation of the element-encoded genes involved in transposition. In this chapter I shall outline the methods currently used to purify and assay the resolvase protein of $\gamma\delta$. The closely related protein from Tn*3* is being investigated elsewhere, and comparisons of the properties of these two enzymes should be enlightening.

Cloning of the $\gamma\delta$ Resolvase Gene

The resolvase protein of $\gamma\delta$ regulates its own gene expression and is normally present in relatively low quantities in a $\gamma\delta$-containing cell. As an initial step in developing an *in vitro* recombination system, I developed a method to increase resolvase gene expression and thus purify large quantities of the protein.

The major leftward promoter of bacteriophage $\lambda(P_L)$ has been used to increase expression of cloned genes.[11] In the particular construction that I used to amplify resolvase, an 1100 bp, *Bam*HI/*Bgl*II fragment encoding λ P_L and the N protein of λ was cloned into the *Bam*HI site of the plasmid pBR322. This vector, pλ8, is maintained in a strain (N4830) with a chromosomally encoded λ*cI* gene (*cI857*) as well as the *N* gene.[5] In this strain cI represses transcription from P_L at low temperature (30°) while allowing high-level expression when the cells are grown at 43°.

The resolvase protein of $\gamma\delta$ was cloned into pλ8 as a *Cla*I/*Eco*RI fragment. A *Cla*I site located approximately 400 bp downstream from P_L occurs four amino acids before the end of the λ*N* gene on pλ8[12] (see Fig. 1). Insertion of the fragment containing the $\gamma\delta$ *tnpR* gene at this site in *Cla*I/*Eco*RI cleaved pλ8 resulted in a novel COOH terminus for the λN protein and termination of N translation just preceding the Shine–Dalgarno region and the initiation codon of the resolvase protein (see Fig. 1). This close juxtaposition of the termination codon for the *N* gene and the initiation codon of resolvase may explain the efficiency of resolvase translation after shifting temperature of the culture from 30° to 43°. Under

[7] P. Kitts, L. Symington, M. Burke, R. Reed, and D. Sherratt, *Proc. Natl. Acad. Sci. U.S.A.* **79,** 46 (1982).
[8] R. Gill, F. Heffron, and S. Falkow, *Nature (London)* **282,** 797 (1979).
[9] J. Chou, P. Lemaux, M. Casadaban, and S. N. Cohen, *Nature (London)* **282,** 801 (1979).
[10] R. Reed, G. Shibuya, and J. Steitz, *Nature (London)* **300,** 5890 (1982).
[11] H.-U. Bernard and D. R. Helinski, this series, Vol. 68, p. 482. (1979).
[12] N. C. Franklin and G. N. Bennett, *Gene* **8,** 107 (1979).

```
                    Cla I
λN        AAGCGAAAATCGATTCCTCTTATCTAG
          LysGlyLysSerIleProLeuIle***

                    Cla I                                    S. D.
λN/tnpR hybrid AAGCGAAAATCGATTTTTTGTTATAACAGACACTGCTTGTCCGATATTTGATTAGGATACATTTTTATGCGA
          LysGlyLysSerIlePheCysTyrAsnArgHisCysLeuSerAspIle***                     MetArg
```

FIG. 1. Fusion of the *N* gene and the $\gamma\delta$ *tnpR* gene in the resolvase-producing plasmid. The DNA sequence at the COOH terminus of the *N* gene is from Franklin and Bennett.[12] The new COOH terminus predicted for the *N* gene after fusion to $\gamma\delta$ DNA at the *Cla*I site is indicated, as are the Shine-Dalgarno (S. D.) sequence and the initiation codon of resolvase.

the condition used here, $\gamma\delta$ resolvase represents 2–3% of total cell protein after 30 min at 43.

Purification of Resolvase

The procedure outlined below is a modification of a previously published method.[5] It has several advantages over earlier procedures; most notably, more than 50 mg of purified resolvase can be isolated in a single run.

Materials

TY broth: 16 g of tryptone, 10 g of yeast extract (Difco) per liter

TED buffer: 20 m*M* Tris-HCl, pH 7.5, 1 m*M* EDTA, 0.1 m*M* dithiothreitol (DTT)

Urea + TED buffer: 420 g of urea (Schwarz-Mann) adjusted to a final volume of 1 liter with TED buffer

Polymin P (polyethyleneimine, Pfaltz and Bauer) was titrated to pH 8.0 with concentrated HCl, and then the volume was adjusted to a 10% v/v solution with TED buffer.

Procedure. The strain RR1071 containing resolvase cloned under the control of P_L was grown in 8.5 liters of TY broth plus 50 μg of ampicillin per milliliter at 30° in a 141 fermentor with vigorous aeration. When the cells reached a density of 8×10^8 cells/ml the temperature was rapidly shifted to 43°, aided by the addition of 1.5 liters TY broth at 90°. Within 20 min of the temperature shift the density of the culture, judged spectrophotometrically, leveled off. After 45 min at 43° the cells were harvested (yield approximately 30 g wet weight) washed once in 100 m*M* NaCl, 10 m*M* Tris-HCl (pH 7.5), and resuspended in 100 ml of 10% sucrose in TED buffer. The following steps were performed at 0°.

The cells were sonicated with a regular tip at a setting of 7 for 2×1 min and then centrifuged at 12,000 *g* for 10 min. The polymin P solution was added dropwise to the resolvase containing supernatant to 0.5% final concentration. After 10 min on ice, the precipitate was collected by centrifugation, resuspended in 50 ml of 1 *M* NaCl + TED buffer, and allowed to sit 5 min on ice. The solution was clarified by centrifugation at 12,000 *g* for 10 min and dialyzed against 20 volumes of 100 m*M* NaCl + TED buffer for 2 hr. The precipitate collected at this stage was dissolved in 10 ml of 1 *M* NaCl + TED buffer and applied to a 100 × 7.5 cm BioGel P-60 column equilibrated with the same buffer. Five-milliliter fractions were collected at a flow rate of 50 ml/hr and monitored by absorbance at 280 nm. Two distinct peaks were observed, the first corresponding to the void volume and the second containing relatively pure resolvase. The fractions containing resolvase, as judged by SDS–polyacrylamide gel electrophoresis were pooled and precipitated by addition of $(NH_4)_3SO_4$ to 80% saturation at 0°. The precipitate was collected by centrifugation and dissolved in 25 ml of 7 *M* urea + TED buffer. This was applied to a 20 cm × 1 cm column of CM-Sepharose equilibrated with the same buffer.

Resolvase was eluted with a 50-ml gradient of 0 → 200 m*M* NaCl in urea + TED buffer at a 5 ml/hr flow rate. The peak of resolvase absorbance occurred at approximately 80 m*M*. Resolvase-containing fractions were pooled and dialyzed against four changes of 0.7 *M* NaCl in TED buffer before storage at 4°.

The purity of the resolvase was confirmed by several criteria. SDS–polyacrylamide gel electrophoresis yielded, on a single band of M_r 21,000, the size expected of γδ resolvase. Amino acid composition analysis of resolvase corresponds closely to that predicted from DNA sequence analysis of the *tnpR* gene (see Fig. 2). Specifically, analysis of the purified resolvase indicates less than one proline residue per 500 amino acids.

Activity of Purified γδ Resolvase

The activity of purified γδ resolvase is determined using a substrate containing two directly repeated *res* sites cloned into the plasmid pBR322. The conversion of substrate to product molecules can be easily monitored by the formation of unique *Eco*RI restriction fragments diagnostic of resolution.[5]

Resolvase activity was determined by the following procedure. The reaction mixture (20 μl) contained 20 m*M* Tris-HCl pH 7.5, 50 m*M* NaCl, 10 m*M* $MgCl_2$, 1 m*M* DTT, 0.25 g of plasmid DNA. To each reaction, 1 μl of resolvase serially diluted in 0.7 *M* NaCl was added, and the tube was incubated for 1 hr at 37°.[5] The amount of resolvase required to resolve 90% of the starting material was determined. The resolvase protein appears to be required in stoichiometric amounts and does not turn over under the conditions used here. Careful estimates indicate that 6–8 monomers of resolvase per resolution site are required to achieve recombination.

The amounts of resolvase present in stock solutions were determined by amino acid analysis or by absorption at 280 nm. The absorbance (A_{280}) of a 1 mg/ml solution of resolvase is 0.27.

```
                                  ILE                                                                 ILE
Tn3   C      A G                  A             T       G   C                   G                     A C
γδ   TTGATTTAGGATACATTTTT ATG CGA CTT TTT GGT TAC GCA CGG GTA TCA ACC AGC CAG CAA TCT CTC GAT ATT CAG GTT
                          MET ARG LEU PHE GLY TYR ALA ARG VAL SER THR SER GLN GLN SER LEU ASP ILE GLN VAL

                                                                                             THR
Tn3  A A   G           T       G   A       T   C               C               C              CA
γδ   CGG GCA CTC AAA GAC GCA GGA GTG AAA GCA AAT CGC ATC TTT ACT GAC AAG GCA TCG GGC AGT TCA AGC GAT CGG
     ARG ALA GLU LYS ASP ALA GLY VAL LYS ALA ASN ARG ILE PHE THR ASP LYS ALA SER GLY SER SER SER ASP ARG

     GLU
Tn3  G             T                                       T       T C             G   C       T C     C
γδ   AAA GGG CTG GAC TTG CTG AGG ATG AAG GTG GAG GAA GGT GAC GTC ATC TTG GTG AAG AAA CTT GAC CGC GTT GGG
     LYS GLY LEU ASP LEU LEU ARG MET LYS VAL GLU GLU GLY ASP VAL ILE LEU VAL LYS LYS LEU ASP ARG VAL GLY

                                                                             ALA VAL
Tn3        C   C   C               A C G                   T   T   G         G G G                 C
γδ   CGC GAT ACT GCT GAC ATG ATC CAG TTA ATA AAA GAG TTT GAC GCC CAA GGT GTA TCC ATT CGG TTT ATT GAT GAC
     ARG ASP THR ALA ASP MET ILE GLN LEU ILE LYS GLU PHE ASP ALA GLN GLY VAL SER ILE ARG PHE ILE ASP ASP

                             ASP         GLN                                                         ARG
Tn3    G               C   T   T       G C         G       C   C   G   G   T       A       T       C  G
γδ   GGA ATC AGT ACC GAT GGG GAG ATG GGT AAA ATG GTT GTC ACT ATT CTA TCT GCA GTG GCC CAG GCA GAA CGA CAG
     GLY ILE SER THR ASP GLY GLU MET GLY LYS MET VAL VAL THR ILE LEU SER ALA VAL ALA GLN ALA GLU ARG GLN

                                                         LYS LEU         ILE LYS             ARG     THR
Tn3    G   C           C   G       G   C   A   G   A      A  CTG         A C AAA         C C C  GG C T  CC
γδ   AGA ATA CTA GAG CGT ACC AAT GAA GGT CGC CAA GAG GCA ATG GCA AAA GGA GTT GTT TTT GGT AGA AAA AGA AAA
     ARG ILE LEU GLU ARG THR ASN GLU GLY ARG GLN GLU ALA MET ALA LYS GLY VAL VAL PHE GLY ARG LYS ARG LYS

     VAL         ASN VAL         THR LEU HIS     LYS     THR         THR GLU     ALA HIS GLN LEU SER
Tn3  G G   C   G A    T    G C G  CG C T CAT   G A     C ACT       A A G G A   T G T C T C G C C  G
γδ   ATA GAT AGA GAT GCA GTA TTA AAT ATG TGG CAA CAG GGG TTA GGT GCC TCA CAT ATA TCA AAA ACA ATG AAT ATT
     ILE ASP ARG ASP ALA VAL LEU ASN MET TRP GLN GLN GLY LEU GLY ALA SER HIS ILE SER LYS THR MET ASN ILE

                                 ILE LEU GLU ASP GLU ARG ALA SER***
Tn3    C   C   C   G   T         A T C T G A   C GAA  GG GCC TCGTG TACGCCTA T TT TAG TTAA GT A G T A AAT
γδ   GCT CGT TCA ACA GTA TAT AAA GTA ATA AAT GAA AGC AAC TAA CATAAAAGGTTGAGTGTGGAATTGAGTGTTACTTCAAAGTCTA
     ALA ARG SER THR VAL TYR LYS VAL ILE ASN GLU SER ASN ***
```

FIG. 2. The DNA sequence of the γδ *tnpR* gene and the predicted amino acid sequence. The DNA and protein sequence of the closely related transposon (Tn*3*) protein is also shown. Only those nucleotides or amino acids that differ between Tn*3* and γδ are indicated.

Properties of Purified $\gamma\delta$ Resolvase

The resolvase protein of $\gamma\delta$ has several properties that deserve further discussion. First, concentrated resolvase precipitates from solution when the NaCl concentration falls below 0.5 *M*. This may partially explain the observation that resolvase does not appear to catalyze multiple recombination events under the assay conditions (50 m*M* NaCl).[5] Second, resolvase appears to be a dimer when analyzed by gel filtration in 1 *M* NaCl. This is consistent with the results obtained from cross-linking studies using dimethyl suberimidate (J. J. Salvo and N. D. F. Grindley, unpublished observation).

Resolvase binds specifically to *res* DNA in a relatively complex manner.[13] There appear to be three separate but adjacent sites for resolvase binding, each site displaying twofold symmetry and suggesting binding of at least a dimer. This would lead to at least 6 monomers bound per *res* site or 12 monomers per substrate molecule, corresponding closely to the results obtained by titrating protein concentration in the resolution assay.

The resolvase-*res* recombination system described here is formally analogous to the lambda-encoded site-specific recombination system studies extensively by Nash and co-workers (see this volume [15]). Their analysis has revealed a nonspecific topoisomerase activity associated with the lambda *int* protein. Additionally, detailed analysis of the topological changes accompanying lambda *att* recombination has provided some insight into the mechanism of recombination. Characterization of the mechanisms of resolvase-mediated recombination are presently underway.[14] Preliminary studies have failed to reveal any topoisomerase activity associated with nonproductive recombination events.

The ability to purify large quantities of resolvase will allow structural and physicochemical analysis of protein DNA interactions. Presently, T. A. Steitz and co-workers have prepared resolvase crystals suitable for X-ray analysis. It is hoped that these studies in conjunction with analysis of intermediates in the recombination[14] will provide information on the active site of the protein.

The combination of biochemical, structural, and physicochemical analysis of the resolvase system now under way in several laboratories may reveal details of protein DNA interactions in general and, more specifically, the detailed mechanism of site-specific resolution of transposition intermediates.

[13] N. D. F. Grindley, M. R. Lanth, R. G. Wells, R. J. Wityk, J. J. Salvo, and R. R. Reed, *Cell* **30,** 19 (1982).

[14] R. R. Reed, and N. D. F. Grindley, *Cell* **25,** 721 (1981).

[14] Purification of *recA* Protein from *Escherichia coli*

By TAKEHIKO SHIBATA, LYNN OSBER, and CHARLES M. RADDING

recA protein of molecular weight 38,000[1–8] is important in various cellular functions, such as genetic recombination, DNA repair, mutagenesis, induction of prophages, cell division, and DNA replication in *Escherichia coli*.[9] Correspondingly, *recA* protein has various activities *in vitro*. Roberts *et al.* discovered that *recA* protein is a protease active on phage λ repressor.[6] Ogawa *et al.*[10] and Roberts *et al.*[11] found that *recA* protein has DNA-dependent ATPase activity. McEntee *et al.*[12] and Shibata *et al.*[13] found that *recA* protein promotes ATP-dependent pairing of double-stranded DNA and homologous single-stranded fragments to form D loops. Further work revealed that *recA* protein promotes (*a*) ATP-dependent pairing of homologous DNA molecules, if at least one of them has a single-stranded region[12–21]; (*b*) ATP-dependent unidirectional growth of

[1] K. McEntee, J. E. Hesse, and W. Epstein, *Proc. Natl. Acad. Sci. U.S.A.* **73,** 3979 (1976).
[2] P. T. Emmerson and S. C. West, *Mol. Gen. Genet.* **155,** 77 (1977).
[3] J. W. Little and D. G. Kleid, *J. Biol. Chem.* **252,** 6251 (1977).
[4] K. McEntee *Proc. Natl. Acad. Sci. U.S.A.* **74,** 5275 (1977).
[5] L. J. Gudas and D. W. Mount, *Proc. Natl. Acad. Sci. U.S.A.* **74,** 5280 (1977).
[6] J. W. Roberts, C. W. Roberts, and N. L. Craig, *Proc. Natl. Acad. Sci. U.S.A.* **75,** 4714 (1978).
[7] T. Horii, T. Ogawa, and H. Ogawa, *Proc. Natl. Acad. Sci. U.S.A.* **77,** 313 (1980).
[8] A. Sancar, C. Stachelek, W. Konigsberg, and W. D. Rupp, *Proc. Natl. Acad. Sci. U.S.A.* **77,** 2611 (1980).
[9] For review, see E. M. Witkin, *Bacteriol. Rev.* **40,** 869 (1976).
[10] T. Ogawa, H. Wabiko, T. Tsurimoto, T. Horii, H. Masukata, and H. Ogawa, *Cold Spring Harbor Symp. Quant. Biol.* **43,** 909 (1979).
[11] J. W. Roberts, C. W. Roberts, N. L. Craig, and E. M. Phizicky, *Cold Spring Harbor Symp. Quant. Biol.* **43,** 917 (1979).
[12] K. McEntee, G. M. Weinstock, and I. R. Lehman, *Proc. Natl. Acad. Sci. U.S.A.* **76,** 2615 (1979).
[13] T. Shibata, C. DasGupta, R. P. Cunningham, and C. M. Radding, *Proc. Natl. Acad. Sci. U.S.A.* **76,** 1638 (1979).
[14] R. P. Cunningham, T. Shibata, C. DasGupta, and C. M. Radding, *Nature (London)* **281,** 191; **282,** 426 (1979).
[15] T. Shibata, R. P. Cunningham, C. DasGupta, and C. M. Radding, *Proc. Natl. Acad. Sci. U.S.A.* **76,** 5100 (1979).
[16] C. DasGupta, T. Shibata, R. P. Cunningham, and C. M. Radding, *Cell* **22,** 437 (1980).
[17] R. P. Cunningham, C. DasGupta, T. Shibata, and C. M. Radding, *Cell* **20,** 223 (1980).
[18] E. Cassuto, S. C. West, J. Mursalim, S. Conlon, and P. Howard-Flanders, *Proc. Natl. Acad. Sci. U.S.A.* **77,** 3962 (1980).
[19] T. Shibata, C. DasGupta, R. P. Cunningham, J. G. K. Williams, L. Osber, and C. M. Radding, *J. Biol. Chem.* **256,** 7565 (1981).

ISBN 0-12-182000-9

heteroduplex joints[22-25]; (*c*) reciprocal strand exchange[26,27]; (*d*) ATPγS- and single-stranded DNA-dependent unwinding of duplex DNA[14]; (*e*) ATP-dependent extensive unwinding of duplex DNA[28,29]; (*f*) ATP- and single-stranded DNA-dependent proteolysis of *lexA* protein and some prophage repressors[6,11,30-33]; (*g*) single-stranded DNA-dependent ATP hydrolysis[10,11,34-36]; (*h*) double-stranded DNA-dependent ATP hydrolysis[34-36]; and (*i*) ATP hydrolysis dependent on the presence of both homologous double-stranded DNA and single-stranded DNA.[37] *recA* protein is also a potentially useful reagent in techniques for gene manipulation, such as site-specific mutagenesis.[38,39]

Assay of Activities of *recA* Protein

Reagents

DNA

Negative-superhelical closed-circular double-stranded DNA (form I DNA)[40] of coliphage fd or ϕX174 is prepared as described by Cun-

[20] R. P. Cunningham, A. M. Wu, T. Shibata, C. DasGupta, and C. M. Radding, *Cell* **24,** 213 (1981).
[21] C. DasGupta and C. M. Radding, *Proc. Natl. Acad. Sci. U.S.A.* **79,** 762 (1982).
[22] R. Kahn, R. P. Cunningham, C. DasGupta, and C. M. Radding, *Proc. Natl. Acad. Sci. U.S.A.* **78,** 4786 (1981).
[23] M. M. Cox and I. R. Lehman, *Proc. Natl. Acad. Sci. U.S.A.* **78,** 3433 (1981).
[24] M. M. Cox and I. R. Lehman, *Proc. Natl. Acad. Sci. U.S.A.* **78,** 6018 (1981).
[25] C. DasGupta and C. M. Radding, *Proc. Natl. Acad. Sci. U.S.A.* **79,** 762 (1982).
[26] C. DasGupta, A. M. Wu, R. Kahn, R. P. Cunningham, and C. M. Radding, *Cell* **25,** 507 (1981).
[27] S. C. West, E. Cassuto, and P. Howard-Flanders, *Proc. Natl. Acad. Sci. U.S.A.* **78,** 2100 (1981).
[28] T. Ohtani, T. Shibata, M. Iwabuchi, H. Watabe, T. Iino, and T. Ando, *Nature (London)* **299,** 86 (1982).
[29] A. M. Wu, M. Biauchi, C. DasGupta, and C. M. Radding, *Proc. Natl. Acad. Sci. U.S.A.*, in press.
[30] N. L. Craig and J. W. Roberts, *Nature (London)* **283,** 26 (1980).
[31] E. M. Phizicky and J. W. Roberts, *Cell* **25,** 259 (1981).
[32] N. L. Craig and J. W. Roberts, *J. Biol. Chem.* **256,** 8039 (1981).
[33] J. W. Little, D. W. Mount, and C. R. Yanische-Perron, *Proc. Natl. Acad. Sci. U.S.A.* **78,** 4199 (1981).
[34] G. M. Weinstock, K. McEntee, and I. R. Lehman, *J. Biol. Chem.* **256,** 8829 (1981).
[35] G. M. Weinstock, K. McEntee, and I. R. Lehman, *J. Biol. Chem.* **256** 8845 (1981).
[36] G. M. Weinstock, K. McEntee, and I. R. Lehman, *J. Biol. Chem.* **256,** 8856 (1981).
[37] T. Ohtani, T. Shibata, M. Iwabuchi, K. Nakagawa, and T. Ando, *J. Biochem. (Tokyo)* **91,** 1767 (1982).
[38] D. Shortle, D. Koshland, G. M. Weinstock, and D. Botstein, *Proc. Natl. Acad. Sci. U.S.A.* **77,** 5375 (1980).
[39] C. Green, and C. Tibbetts, *Proc. Natl. Acad. Sci. U.S.A.* **77,** 2455 (1980).
[40] Abbreviations: Form I DNA, form II DNA, and form III DNA denote negative-superheli-

ningham *et al.*[17] and heated at 68° for 50–60 sec in 10 m*M* Tris-HCl buffer (pH 7.5) containing 0.1 m*M* EDTA to reduce the background of the D-loop assay. A high background may be caused by D loops or R loops formed *in vivo*.[41]

Circular single-stranded phage DNA of fd or ϕX174 is prepared as described by Cunningham *et al.*[17]

Single-stranded fragments of phage DNA of fd or ϕX174 (840 μM[42]) dissolved in 10 m*M* Tris-HCl buffer (pH 7.5) containing 0.1 m*M* EDTA (140 μl) is put in an Eppendorf centrifuge tube (200 μl) and heated in a boiling water bath for 7 min.[14,43] The average chain length of the fragments is estimated by agarose gel electrophoresis. A restriction endonuclease *Hae*III digest of single-stranded DNA of fd or ϕX174 is used as a standard of molecular weight.[44,45]

Form III (linear duplex) DNA of ϕX174 is generated by cleavage of the form I DNA by *Pst*I. ϕX174 *am3* [^{3}H]DNA (113 nmol) is incubated with 0.6 unit of *Pst*I (New England Bio-Labs) for 2 hr at 37° in the presence of 50 m*M* NaCl, 13 m*M* Tris-HCl (pH 7.5), 0.7 m*M* EDTA, 6.7 m*M* $MgCl_2$, and 6 m*M* 2-mercaptoethanol. The total reaction volume is 280 μl. Aliquots of 10 μl are removed after 60, 80, and 120 min and assayed by the method of Kuhnlein *et al.*[46] to determine the extent of the reaction. The *Pst*I-DNA digest is extracted, at room temperature, with 300 μl of phenol, equilibrated with 10 m*M* Tris-HCl (pH 7.5) containing 1 m*M* EDTA. Centrifugation at 12,800 *g* for 10 min is required to separate the layers. The upper aqueous phase is removed and extracted three times with ether; each extraction is followed by centrifugation for 1 min. The form III DNA, about 200 μl, is dialyzed against 10 m*M* Tris-HCl (pH 7.5) containing 1 m*M* or 0.1 m*M* EDTA.

All DNA is dissolved in 10 m*M* Tris-HCl buffer (pH 7.5) containing 0.1 m*M* EDTA.

ATP and dithiothreitol are dissolved in distilled water at 26 m*M* and

cal closed-circular double-stranded DNA, nicked-circular double-stranded DNA, and linear double-stranded DNA, respectively.

[41] K. L. Beattie, R. C. Wiegand, and C. M. Radding, *J. Mol. Biol.* **116,** 783 (1977).

[42] Amount of DNA is expressed in moles of nucleotide residues in the DNA.

[43] T. Shibata, R. P. Cunningham, and C. M. Radding, *J. Biol. Chem.* **256,** 7557 (1981).

[44] E. Beck, R. Sommer, E. A. Auerswald, C. Kurz, B. Zink, G. Osterburg, H. Schaller, K. Sugimoto, H. Sugisaki, T. Okamoto, and M. Takanami, *Nucleic Acids Res.* **5,** 4495 (1978).

[45] F. Sanger, A. R. Coulson, T. Friedmann, G. M. Air, B. G. Barrell, N. L. Brown, J. C. Fiddes, C. A. Hutchinson III, P. M. Slocombe, and M. Smith, *J. Mol. Biol.* **125,** 225 (1978).

[46] U. Kuhnlein, E. E. Penhoet, and S. Linn, *Proc. Natl. Acad. Sci. U.S.A.* **73,** 1169 (1976).

1 *M*, respectively, and the pH of the solutions is adjusted by addition of 1 *N* NaOH, so that the pH becomes 7.5 after dilution to the same concentration as that in the reaction mixture. Since *recA* protein significantly changes its properties with a small shift in pH, the pH of the reaction mixture should be adjusted precisely.

Bovine serum albumin (BSA). Since commercial bovine serum albumin often contains DNase activity, the DNase activity in the preparation is tested before use. For this purpose, form I DNA or circular single-stranded phage DNA of fd or ϕX174 is incubated in the absence of *recA* protein under the standard condition and examined by gel electrophoresis through 2% agarose gel by using E-buffer.[13,47] Under these conditions, nicked-circular double-stranded DNA (form II DNA), linear double-stranded DNA (form III DNA), and form I DNA (from top to bottom in this order), or circular single-stranded DNA and linear single-stranded DNA (in this order from top to bottom) are separated.

Buffers

Buffer D (for dilution of *recA* protein): 50 m*M* Tris-HCl (pH 7.5), 0.3 m*M* EDTA, 5 m*M* dithiothreitol, 10% (v/v) glycerol

E-buffer[47] (for electrophoresis): 40 m*M* Tris-acetate (pH 7.9), 5 m*M* sodium acetate, 1 m*M* EDTA

DNA-Dependent ATPase Activity

The ATPase activity of *recA* protein can be observed in several different ways as described above. The single-stranded DNA-dependent ATPase activity is the least sensitive to change in the conditions for the reaction, such as pH and concentrations of mono- and divalent cations.[34,43] We assay single-stranded DNA-dependent ATPase activity to estimate the amount of active *recA* protein in the fractions throughout purification of the protein.

[^{3}H]ATP (26 nmol, 7 Ci/mol) is incubated at 37° with *recA* protein in the presence or the absence of 50 μM[42] single-stranded DNA of phage ϕX174 or fd, in 18 μl of reaction buffer that contains 35 m*M* Tris-HCl (pH 7.5), 6.7 m*M* $MgCl_2$, 100 μg of bovine serum albumin per milliliter and 2 m*M* dithiothreitol. *recA* protein is diluted with buffer D. A reaction mixture is built up in a well of a Microtiter plate (Linbro Titertek, 96 wells per

[47] P. A. Sharp, B. Sugden, and J. Sambrook, *Biochemistry* **12,** 3055 (1973).

plate) floating on an ice-water bath. The reaction is started by transferring the plate onto a water bath at 37°. After 30 min of incubation at 37°, the reaction is terminated by transferring the plate onto an ice-water bath, followed immediately by addition of 12 μl of 25 m*M* EDTA (pH 9) containing 3 m*M* each of unlabeled ATP, ADP, and AMP as carrier. A set of strips (1 × 10 cm) is prepared by scoring off the material in lines from a sheet (10 × 10 cm) of polyethyleneimine on a plastic film (Polygram Cel 300 PEI, Macherey-Nagel Co.). The strips are washed successively with 0.5 *M* LiCl and 1 *M* formic acid, rinsed extensively with distilled water, and dried before use. Then we spot a 10-μl aliquot of the sample in a line about 2 cm distant from the bottom, dry, and develop by ascending chromatography at room temperature for about 40 min in a solvent of 0.5 *M* LiCl, 1 *M* formic acid. Spots of ATP, ADP, and AMP (from the bottom to top) on the strip are located by illuminating under an UV lamp (at 254 nm) and cut out of the strip. The radioactivity in each spot is measured by scintillation counting in Econofluor (New England Nuclear). During the measurement of radioactivity, the layer of polyethyleneimine should face upward in the vial. If the layer faces downward, the efficiency of counting ^{3}H decreases by 30%.

We define 1 unit of ATPase activity as the amount that hydrolyzes 1 nmol of ATP at 37° for 30 min. The amount of hydrolysis of ATP is linearly dependent on the amount of *recA* protein up to about 10 units under the standard condition.

Homologous Pairing by recA Protein

recA protein promotes pairing of a couple of homologous DNA molecules to form stable joint molecules, if at least one of the molecules has a single-stranded region and at least one of them has a free end.[16,19] Free ends are not essential to homologous pairing but are required to stabilize the paired complex.[16,17,19] Suitable substrates for assaying homologous pairing are single-stranded fragments plus superhelical DNA or circular single strands plus linear duplex DNA. The products of both pairs of substrates can be assayed readily by the nitrocellulose filter method (D-loop assay).[19,21,43] The conditions for handling these two pairs of substrates, which are slightly different, are described below.

Pairing of Single-Stranded Fragments with Superhelical DNA. Form I [^{3}H]DNA (3.7 μM[42]) of phage fd or ϕX174 and 0.4 μM[42] single-stranded fragments (unlabeled or labeled with ^{32}P; chain length about 500 nucleotides) prepared from phage DNA of fd or ϕX174 is incubated at 37° with *recA* protein in 21 μl of a reaction buffer containing 31 m*M* Tris-HCl (pH 7.5), 13 m*M* $MgCl_2$, 1.3 m*M* ATP, 1.8 m*M* dithiothreitol, and 88 μg of bovine serum albumin per milliliter. *recA* protein is diluted with buffer D. After 30 min of incubation in a well of a Microtiter plate floating on a water bath at 37°, we transfer the plate onto an ice-water bath to terminate the reaction.

Duplex DNA passes through a nitrocellulose filter (pore size 0.45 μm) in the presence of 1.5 *M* NaCl, 0.15 *M* sodium citrate. Under this condition, single-stranded DNA is efficiently trapped by the filter. Duplex DNA which has a D loop is also retained very efficiently (almost 100%) by the filter, since this DNA has a single-stranded region.

Immediately after the transfer of the Microtiter plate onto an ice-water bath, we add 40 μl of 25 m*M* EDTA (pH 9.4) and 3 μl of 10% Sarkosyl (NL97, Ciba-Geigy Co.) and keep the plate on an ice-water bath for 30–60 min. Then the reaction mixture is diluted into 0.5 ml of 25 m*M* EDTA. An aliquot of 50 μl is spotted directly onto a nitrocellulose filter to measure total radioactivity. An aliquot of 200 μl is diluted into 1 ml of cold 1.5 *M* NaCl, 0.15 *M* sodium citrate, incubated at 41° for 4 min and then immediately diluted with 7 ml of cold 1.5 *M* NaCl, 0.15 *M* sodium citrate. The sample is filtered at 4 ml for 10 sec through a nitrocellulose filter (Sartorius membrane filter SM11306; pore size, 0.45 μm) that has been washed with 2 ml of 1.5 *M* NaCl, 0.15 *M* sodium citrate. The filter is washed successively with 1.5, 1.5, and 5 ml of cold 1.5 *M* NaCl, 0.15 *M* sodium citrate. Radioactivity retained on the filter is counted by scintillation counting in Econofluor. Another aliquot of 200 μl is used to estimate the fraction of nicked molecules by the method of Kuhnlein *et al.*[43,46]

Pairing of Circular Single Strands with Linear Duplex DNA. Complexes of form III DNA and single-stranded circles will also be trapped on a nitrocellulose filter with a 0.45-μm pore. These complexes are formed by incubating 4 μ*M* *recA* protein, 2 μ*M* linear duplex DNA, and 4 μ*M* homologous circular single strands at 37° for 15 min in the presence of 31 m*M* Tris-HCl (pH 7.5), 12 m*M* $MgCl_2$, 1.3 m*M* ATP, 1.8 m*M* dithiothreitol, and 88 mg of BSA per milliliter. From a reaction volume of 20 μl, a 9-μl aliquot is removed and added to 300 μl of 25 m*M* EDTA (pH 9.4) at 0°. Spotting 50 μl of the EDTA solution directly onto a dry filter yields the total radioactive counts. A dilution of 200 μl into 5 ml of cold 1.5 *M* NaCl, 0.15 *M* sodium citrate is filtered (without incubation at 41°) as described above to give the percentage of joint molecules formed. Up to 80–90% of duplex DNA can be converted to joint molecules under the above conditions.

The amount of D loops formed by *recA* protein is not a linear function of the amount of *recA* protein (see below).

Purification of *recA* Protein

Media and Buffers

Minimal 56 srl agar plate

Minimal 56 medium[48]: 0.55% KH_2PO_4, 0.85% Na_2HPO_4, 0.20% $(NH_4)_2SO_4$, 29 μg of $Ca(NO_3)_2 \cdot 4\ H_2O$ per milliliter, 0.5 μg of $FeSO_4 \cdot 7\ H_2O$ per milliliter, 0.02% $MgSO_4 \cdot 7\ H_2O$ at pH 7.4

D-Sorbitol, 0.2%

D-Biotin, 1 $\mu g/ml$

L-Arginine, 30 $\mu g/ml$

L-Lysine, 30 $\mu g/ml$

Agar, 2%

Sorbitol, biotin, arginine, lysine, and agar are sterilized separately. Biotin is sterilized by filtration through a nitrocellulose filter (pore size 0.20 μm).

K medium[49]

M9 buffer: 0.3% KH_2PO_4, 0.6% Na_2HPO_4, 0.1% NH_4Cl, 0.05% NaCl, at pH 7.4, 0.1 m*M* $CaCl_2$, 1 m*M* $MgSO_4$, 0.4% D-sorbitol, 1 $\mu g/ml$ D-biotin, 1% casamino acids (Difco), 10 $\mu g/ml$ tetracycline. $CaCl_2$, $MgSO_4$, sorbitol, biotin, casamino acids and tetracycline (in 95% ethanol) are added after sterilization.

SLBH medium: 0.29% KH_2PO_4, 1.21% K_2HPO_4, 1.0% Bacto-tryptone, 2.0% Bacto yeast extract, 0.45% NaCl, 0.38% (v/v) glycerol. The pH of the medium is roughly adjusted to 7 by addition of 6 ml of 1 *N* NaOH per liter before sterilization. Glycerol and phosphate buffer are added after sterilization.

Media are sterilized by autoclaving at 1 atm for 15–20 min unless otherwise stated.

Buffers

Buffer A: 50 m*M* Tris-HCl buffer (pH 7.5), 1 m*M* EDTA, 10 m*M* 2-mercaptoethanol, 10% (v/v) glycerol

Buffer B: 20 m*M* potassium phosphate buffer (pH 6.8), 10 m*M* 2-mercaptoethanol, 10% (v/v) glycerol

Buffer C: 20 m*M* potassium phosphate buffer (pH 6.8), 1 m*M* EDTA, 10 m*M* 2-mercaptoethanol, 10% (v/v) glycerol

[48] J. Monod, G. Cohen-Bazire, and M. Cohn, *Biochim. Biophys. Acta* **7,** 585 (1951).

[49] W. D. Rupp, C. E. Wilde III, D. L. Reno, and P. Howard-Flanders, *J. Mol. Biol.* **61,** 25 (1971).

Buffer D: See preceding section.
Buffer F: 20 mM Tris-HCl (pH 7.5), 50 mM NaCl, 0.1 mM EDTA, 5 mM $MgCl_2$, 1 mM dithiothreitol, 10% (v/v) glycerol

Reagents

Polymin P (50% polyethyleneimine, BDH Chemicals Ltd.) is diluted to about 15% with distilled water, and the pH of the solution is adjusted to 7.9 by adding concentrated HCl solution. The concentration of polyethyleneimine is adjusted to 10% by adding distilled water.
Hydroxyapatite: Bio-Rad BioGel HTP is suspended in buffer B and packed into a column as described in a manual published by Bio-Rad.[50] The column is equilibrated with buffer B.
Sephacryl S-200 Superfine (Pharmacia Fine Chemicals) is packed into a column and equilibrated with buffer A containing 0.3 M ammonium sulfate.
DEAE-cellulose: DE-52 (Whatman) is equilibrated with buffer C and packed into a column as described in Whatman's manual.[51]
ATP-agarose (type IV, P-L Biochemicals, Inc.) is packed into a column and then equilibrated with 10 bed volumes of buffer F.

Strain

Escherichia coli DR1453, which is a derivative of KM4104 (F^- *mtlA strA lysA argA* Δ(*lac*) × 74 (deletion of entire *lac* operon) Δ7(*srl-recA*) Δ2134(*gal-bio*),[52] contains plasmid pDR1453. Plasmid pDR1453 is a derivative of pBR322 in which the *srl-recA* region (8.6 kilobase pairs) of the *E. coli* chromosome is inserted at the *Pst*I site.[53]

Growth of the Cells

Since the srl^+-$recA^+$ region on pDR1453 is not maintained very stably, one needs to select cells whose phenotype is Srl^+ (or $RecA^+$) before use. Cells from stock culture are seeded on a minimal 56 srl agar plate, and the plate is incubated at 37° for 2–3 days. Srl^+-cells can grow on minimal 56 srl agar plates and are usually $RecA^+$. Cells from this plate are inoculated in 500 ml of K medium in a 2-liter flask, and incubated at 37° overnight with aeration. The culture is diluted into fresh 2.2 liters of SLBH medium so that $A_{590\ nm}$ is 0.2 to 0.4. The density of bacterial cells in the culture is

[50] Materials, equipment, and system for chromatography, electrophoresis, immunochemistry, and HPLC. (Catalogue G, 1981), Bio-Rad Laboratories, Richmond, California.
[51] Advanced ion exchange celluloses laboratory manual, Whatman, Kent, U.K.
[52] See K. McEntee, *J. Bacteriol.* **132,** 904 (1977).
[53] A. Sancar and W. D. Rupp, *Proc. Natl. Acad. Sci. U. S. A.* **76,** 3144 (1979).

measured by using a cell with a 1-cm optical path. The cell suspension is put into a culture vessel of a Hi-density fermentor (Lab-Line Instruments, Model No. 29500) and incubated at 30° with aeration (6 liters/min). The culture vessel is rotated at 250 rpm. When $A_{590\ nm}$ reaches 2.0, air is replaced by oxygen (4 liters/min). When $A_{590\ nm}$ reaches 5.0 to 8.0 (3 to 4 hr after the start of this incubation), 140 mg of nalidixic acid (Sigma) dissolved in 50 ml of 0.03 *N* NaOH is added to the culture. An aliquot (1 ml) of culture is withdrawn before the addition of nalidixic acid for test of UV resistance and Srl^+ phenotype. The incubation is continued for 1 hr after the addition, and cells are collected by centrifugation at 9000 rpm for 10 min in a Sorvall GS3 rotor at room temperature. About 35 g of cells (wet weight) are obtained. Cells are then resuspended in 50 m*M* Tris-HCl buffer (pH 7.5) containing 10% sucrose (2.9 ml of the buffer is added to 1 g of cells) at room temperature, and the cell suspension is frozen in a plastic beaker immersed in an ethanol–Dry Ice bath. The frozen cells are stored at −20° or −80°.

Procedures

All procedures for purification are carried out at 4° unless otherwise stated.

Cell-Free Extracts. The cell suspension, which contains 65 g of cells, is thawed in an ice-water bath for 12 hr. To this suspension, we added 1 *M* dithiothreitol, 0.1 *M* EDTA, and 1% lysozyme, so that the final concentrations are 10 m*M*, 1m*M*, and 0.05%, respectively. This suspension is kept in an ice-water bath for 30 min. The cells in the suspension are converted to spheroplasts, the formation of which may be monitored by light microscopy. When almost all the cells are converted to spheroplasts (within 30 min after addition of lysozyme), 3.5 *M* KCl and 8% Brij 58 (Sigma) are added, so that the final concentrations are 0.2 *M* and 0.42%, respectively. The suspension is incubated at 0° for 30 min more, and a viscous lysate is obtained. The cell lysate is centrifuged at 35,000 rpm for 60 min at 5° in a type 45Ti rotor (Beckman), and the supernatant is saved (fraction I, 276 ml containing 5.9 g of protein).

Polymin P Fractionation. A 10% solution of polymin P (pH 7.9) is added with stirring to fraction I (276 ml) over a 15-min period to a final concentration of 0.3%, and the solution is stirred for 20 min more. The precipitate is collected by centrifugation at 15,000 rpm for 10 min at 5° in a Sorvall SS34 rotor, and resuspended in 65 ml of 0.5 *M* NaCl dissolved in buffer A. The suspension is stirred intermittently with a glass rod over a 30-min period. The precipitate is again collected by centrifugation and resuspended in 100 ml of 1 *M* NaCl dissolved in buffer A. The suspension is homogenized with a Teflon pestle over a 60-min period and then centrifuged as above. To the supernatant (102 ml), ammonium sulfate is added to 50% saturation (0.31 g of ammonium sulfate to 1 ml of supernatant) over a 45-min period, and the suspension is stirred for 30 min more. The precipitate is collected by centrifugation, and dissolved in 100 ml of 1 *M* NaCl dissolved in buffer A. Proteins are again precipitated by addition of ammonium sulfate to 50% saturation. Resolution and precipitation are repeated one more time. The precipitate is collected by centrifugation and can be stored at −20°. The precipitate is dissolved in 22 ml of buffer B containing 50 m*M* potassium phosphate (pH 6.8), and dialyzed against 2 liters of the same buffer twice. If a precipitate appears during dialysis it is removed by centrifugation. The solution is fraction II, about 24 ml containing 600 mg of protein.

The dialysis against buffer B (20 m*M* potassium phosphate) often resulted in loss of *recA* protein by precipitation. In some cases, when we used an *E. coli* strain other than DR1453, a large fraction of *recA* protein was precipitated during dialysis even if we used 50 m*M* potassium phosphate in buffer B. In these cases, it is necessary to use the higher concentration of phosphate buffer to prevent precipitation.

Hydroxyapatite Chromatography. Fraction II is diluted 2.5-fold with 10% (v/v) glycerol solution containing 10 m*M* 2-mercaptoethanol and immediately loaded onto a column (3.2 × 26.5 cm) of hydroxyapatite. We wash the column with 250 ml of buffer B, and apply a 3.6-liter linear gradient (0.02 to 0.5 *M*) of potassium phosphate in buffer B to the column at 60 ml/hr. Fractions (20 ml each) are collected. *recA* protein (DNA-dependent ATPase activity) is eluted at 0.06 *M* potassium phosphate (fraction III, 111 ml containing 113 mg of protein) (Fig. 1).

Proteins in the fraction III are precipitated by ammonium sulfate at 75% saturation (0.52 g of ammonium sulfate to 1 ml of solution), collected by centrifugation, and dissolved in 7.4 ml of buffer A containing 0.3 *M* ammonium sulfate (fraction IIIa, 10 ml).

Sephacryl S-200 Gel Filtration. Fraction IIIa is loaded onto a column (3.2 × 42 cm) of Sephacryl S-200. Proteins are eluted with buffer A containing 0.3 *M* ammonium sulfate at 20 ml/hr and fractions (9 ml each) are collected. Fractions that contain DNA-dependent ATPase activity but contain little DNA-independent ATPase activity are pooled and dialyzed against buffer C (fraction IV, 28 ml containing 64 mg of protein).

DEAE-Cellulose Column Chromatography. An aliquot (10 ml) of fraction IV is loaded onto a column (1 × 13.5 cm) of DEAE-cellulose. We wash the column with 15 ml of buffer C and apply a 110-ml linear gradient (0 to 0.5 *M*) of KCl in buffer C at 12 ml/hr to the column. Fractions (2.5 ml each) are collected. *recA* protein is eluted between 0.23 and 0.26 *M* KCl

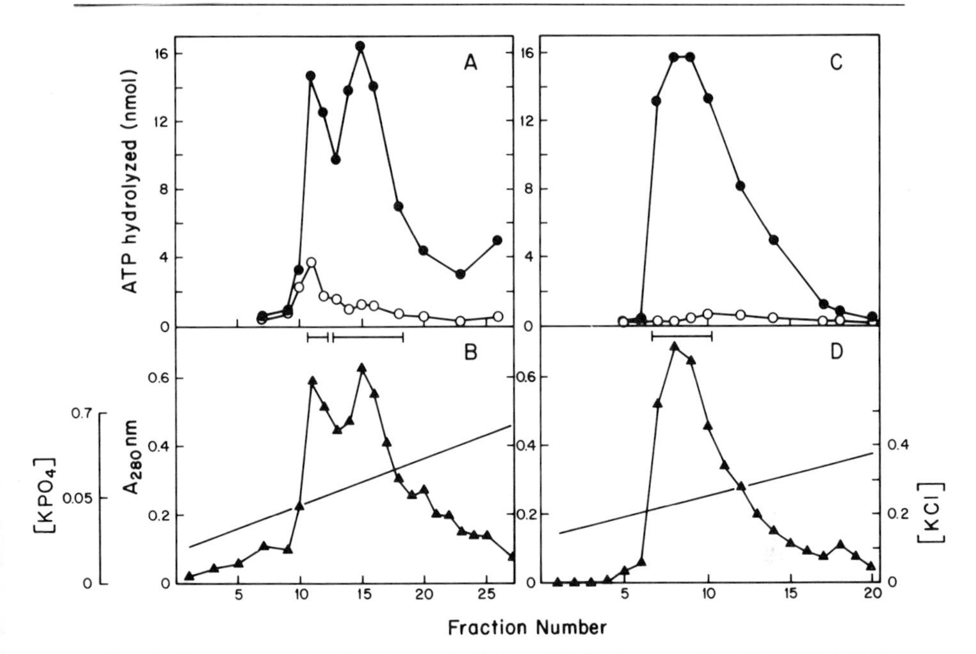

FIG. 1. Chromatography of *recA* protein.[43] (A and B) Hydroxyapatite: (C and D): DEAE-cellulose. One microliter of each fraction was assayed for ATPase activity under standard conditions. ●, ATPase activity in the presence of ϕX174 single-stranded phage DNA; ○, ATPase activity in the absence of DNA; ▲ $A_{280\ nm}$. The straight lines in (B) and (D) indicate the concentration gradients of potassium phosphate and KCl, respectively.

(Fig. 1), and the pooled fractions are dialyzed against buffer D (fraction V, 10 ml containing 12 mg of protein).

A typical purification of recA protein is summarized in the table.

ATP-Agarose Affinity Chromatography. In fraction V prepared as described above, only *recA* protein is detected by SDS-acrylamide gel elec-

PURIFICATION OF *recA* PROTEIN

	DNA-dependent ATPase				
Fraction	Units/mg $\times 10^{-3}$	Units/g cells	Recovery (%)	Fraction of total ATPase	Protein (mg/ml cells)
I. Extract					90
II. Polymin P–$(NH_4)_2SO_4$	4.3	40	100	0.80	9.2
III. Hydroxyapatite	7.3	14	34	0.89	1.9
IV. Sephacryl S-200	10.0	12	30	0.99	1.2
V. DEAE-cellulose	8.8	5.3	13	0.99	0.6

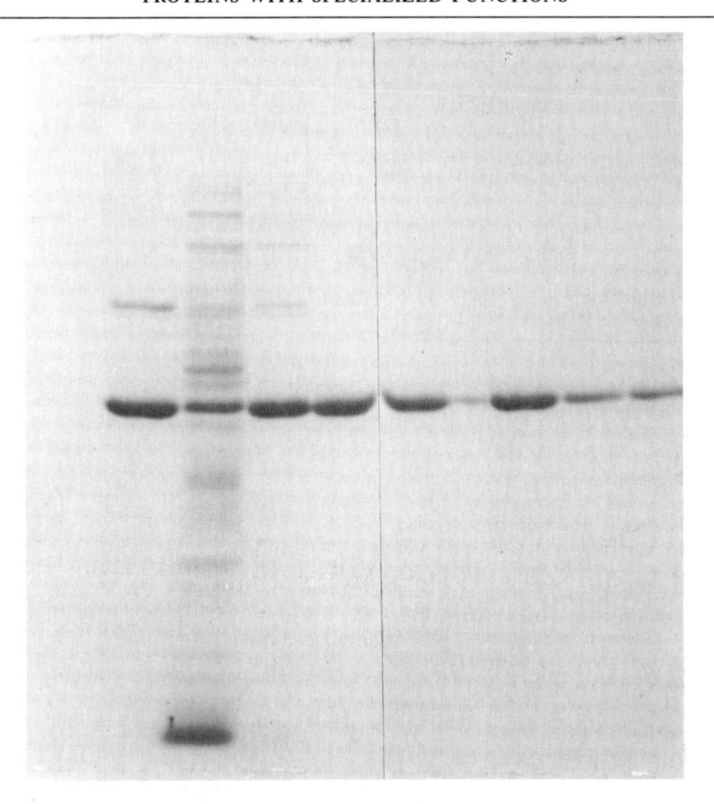

FIG. 2. Purification of *recA* protein.[43] Preparations from each step of purification were examined by electrophoresis through a slab (14 cm × 15 cm × 1.6 mm) of a 10% acrylamide gel containing 0.1% sodium dodecyl sulfate. Electrophoresis was at 20 mA for 5 hr. Protein in the gel was stained by Coomassie Brilliant Blue. About 100 units of DNA-dependent ATPase activity of each fraction were loaded on channels c, d, e, and g. a: Fourteen micrograms of the fraction at the first peak from hydroxyapatite column (Fig. 2); b: 60 μg of cell-free extract, fraction I; c: 23 μg of Polymin P-$(NH_4)_2SO_4$ fraction II; d: pooled fractions (14 μg) from the second peak from the hydroxyapatite column (Fig. 2), fraction III; e: pooled fractions (9.6 μg) from the Sephacryl S-200 column, fraction IV; f and g: 1 and 11 μg, respectively, of pooled fractions from the DEAE–cellulose column (Fig. 2), fraction V; h: 10 μg of the fraction that contained the greatest activity of DNA-independent ATPase activity from the Sephacryl S-200 column; i: 7 μg of fractions 13 and 14 from the DEAE-cellulose column (Fig. 2).

trophoresis (Fig. 2). However, when the purification is started from cells in which *recA* protein is not amplified as much, a minor band of a protein may be detected in fraction V by SDS–acrylamide gel electrophoresis.[13] ATP-agarose affinity chromatography is useful to remove such contaminating proteins.

Fraction V (0.5 mg protein) is dialyzed against buffer F and loaded onto a column [0.5 (diameter) × 1.1 (height) cm] of ATP-agarose. The column is washed with 2 ml of buffer F, and *recA* protein is eluted with 2 ml of buffer F containing 5 m*M* ATP at 1.4 ml/hr. Fractions (0.2 ml each) are collected. Active fractions (DNA-dependent ATPase activity) are pooled (fraction VI, 0.6 ml containing 0.15 mg of protein). ATP is removed by gel filtration through Sephadex G-50 (Pharmacia Fine Chemicals AB).

Properties of *recA* Protein

Stability. Fraction V can be stored in ice for more than a year without detectable loss of both ATPase activity and activity to promote homologous pairing.

Stoichiometry between recA Protein and DNA Substrates. The formation of D loops requires stoichiometric amounts of *recA* protein, and the amount of D loops formed by *recA* protein is not a linear function of the amount of *recA* protein[13,15,19]; with less than a certain amount of *recA* protein, no formation of D loops occurs. As more *recA* protein is added, the initial velocity of the formation increases and reaches a plateau.

When the double-stranded DNA is negatively superhelical, excess *recA* protein causes subsequent dissociation of D loops, and the kinetics of formation of D loops are anomalous.[19,54,55] When the double-stranded DNA is not superhelical, no distinct phase of dissociation of paired molecules is observed even though excess *recA* protein is present.[19,54]

Effect of pH and Cation. Single-stranded DNA-dependent ATPase activity of *recA* protein has a wide range of optimal pH between 6 and 9[34] and requires a concentration of Mg^{2+} greater than 1 m*M*.[19] Double-stranded DNA-dependent ATPase activity has a narrow optimum at pH 6.[34]

The formation of D loops does not require negative superhelicity but is accelerated by superhelicity more than 50-fold.[54] When the double-stranded DNA substrate is negatively superhelical, the formation of D loops requires Mg^{2+} between 7 m*M* and 25 m*M*, but when the double-stranded DNA substrate is not superhelical, the formation requires Mg^{2+} between 11 m*M* and 25 m*M*.[19]

Single-stranded DNA-dependent ATPase activity is resistant to high salt (a half inhibition is obtained by 300 m*M* NaCl[34]). However, formation of D loops is very sensitive to salt. About 30 m*M* NaCl gives a half inhibition of D-loop formation.[19]

[54] T. Shibata, T. Ohtani, P. K. Chang, and T. Ando, *J. Biol. Chem.* **257,** 370 (1982).

[55] T. Shibata, T. Ohtani, and M. Iwabuchi, unpublished observation.

[15] Purification and Properties of the Bacteriophage Lambda Int Protein

By HOWARD A. NASH

In living systems, recombinant DNA is often formed by a crossover between specific sequences on DNA. The integration of bacteriophage lambda DNA into the chromosome of *Escherichia coli* is a particularly well-studied example of such site-specific recombination. Cell-free extracts carry out integrative recombination between the specialized recombining site of the phage, *att*P, and that of the bacterium, *att*B.[1] *In vitro* recombination provides a functional assay for the components involved in the integration of phage lambda DNA. This approach has led to the discovery of two proteins that are encoded by the *E. coli* host: DNA gyrase[2] and integration host factor (IHF).[3] The role of DNA gyrase in integrative recombination is simply to provide a suitably supercoiled substrate DNA.[4] Recombination of supercoiled DNA requires only IHF and a viral protein, Int, the product of the phage *int* gene. Recent studies indicate that Int carries the active site responsible for the breakage and reunion of DNA at the crossover; IHF plays an important but secondary role. This chapter describes the purification of Int and provides a brief summary of the properties of the purified protein.

Assay Method

Principle. Integrative recombination between a supertwisted circle containing *att*P and a linear fragment containing *att*B results in the union of the two substrates. The product is a linear DNA whose molecular length is the sum of that of the two substrate DNAs.[5] Recombinant DNA is separated from the substrates by electrophoresis in agarose and is detected by ethidium bromide fluorescence.

Procedure. The assay mixture (0.02 ml) contains 50 mM Tris-HCl (pH 7.8), 1 mM Na_2EDTA, 70 mM KCl, 5 mM spermidine, 1 mg of bovine serum albumin (Pentex) per milliliter, 10 μg of plasmid pPAl DNA per

[1] H. A. Nash, *Proc. Natl. Acad. Sci. U. S. A.* **72,** 1072 (1975).
[2] M. Gellert, K. Mizuuchi, M. H. O'Dea, and H. A. Nash, *Proc. Natl. Acad. Sci. U.S.A.* **73,** 3872 (1976).
[3] Y. Kikuchi and H. A. Nash, *J. Biol. Chem.* **253,** 7149 (1978).
[4] K. Mizuuchi, M. Gellert, and H. A. Nash, *J. Mol. Biol.* **121,** 375 (1978).
[5] K. Mizuuchi, and M. Mizuuchi, *Cold Spring Harbor Symp. Quant. Biol.* **43,** 1111 (1979).

METHODS IN ENZYMOLOGY, VOL. 100 ISBN 0-12-182000-9

milliliter (3.3 n*M att* sites), 10 μg of plasmid pBB105 DNA per milliliter (2.9 n*M att* sites), 1 μl of crude IHF, and 1 μl of an appropriate dilution of Int. Int is diluted just prior to use in 50 m*M* Tris-HCl, pH 7.4, containing 600 m*M* KCl, 1 m*M* Na_2EDTA, 25 m*M* potassium phosphate buffer, pH 7.0, 10% glycerol, and 2 mg of bovine serum albumin per milliliter. IHF is the penultimate addition, and Int is the final addition to the assay mixture. Incubation is at 25° for 45 min. The reaction is terminated by addition of 5 μl of 25% (w/v) Ficoll 400 containing 5% SDS and 0.03% bromophenol blue. The samples are layered onto a 1% agarose gel containing 90 m*M* Tris-boric acid, pH 8.2, and 2.5 m*M* EDTA. After electrophoresis at 40 V for 16–18 hr, the gel is stained in 1 μg of ethidium bromide per milliliter for 1 hr and photographed in ultraviolet light.

Preparation of DNA. Supertwisted circles containing *att*B and *att*P are prepared after chloramphenicol amplification of plasmid-containing strains. Plasmid pPAl (constructed by K. Mizuuchi and M. Mizuuchi) consists of pBR322 with a 600 base pair (bp) *att*P insert in the *Hin*cII/*Sal*I site.[5] Plasmid pBB105 (constructed by W. Ross and A. Landy) consists of pBR322 with a 1600 bp *att*B insert replacing the *Eco*RI-*Bam*Hl segment of the vector. Each plasmid is contained in Meselson strain 204, a *recA thy* derivative of *E. coli* K12. The cells are grown overnight at 38° in broth containing 25 μg of ampicillin per milliliter; they are diluted in 100-fold in K medium (7 g of Na_2HPO_4, 3 g of KH_2PO_4, 1 g of NH_4Cl, 1 g of NaCl, 0.3 g of $MgSO_4$, 15 g of decolorized casamino acids, 2 g of glucose, and 0.1 mg of thiamin hydrochloride per liter) supplemented with 5 μg of thymidine per milliliter and grown by shaking at 38° to mid-log phase ($A_{650} = 0.6$). After centrifugation and resuspension at the same density in fresh K medium supplemented with 5 μg of thymidine per milliliter, chloramphenicol is added to 150 μg/ml. The cells are shaken overnight at 38° and collected by centrifugation. The cell pellet from 200 ml of cells is resuspended in 2 ml of ice-cold 10 m*M* Tris-acetic acid, pH 8.2, containing 20% sucrose and 100 m*M* sodium chloride. The mixture is transferred to a Corex tube, and to it is added 0.5 ml of a 4 mg/ml solution of egg white lysozyme in 100 m*M* Tris-acetic acid, pH 8.2, containing 100 m*M* Na_2EDTA. After 1 min on ice, 2.5 ml of 2 *M* NaCl containing 10 m*M* Na_2EDTA, 1% (w/v) Brij 58, and 0.4% sodium deoxycholate is added. The mixture is incubated at room temperature for 35 min and centrifuged for 30 min at 15,000 rpm. The supernatant is decanted to a 50-ml beaker; 6 g of cesium chloride and 1.0 ml of 10% (w/v) sodium lauryl sarcosinate are added. After swirling at room temperature for 1 hr, 1 ml of ethidium bromide (10 mg/ml) is added and the refractive index is adjusted to 1.396 with 50 m*M* Tris-acetic acid, pH 8.2. The mixture is stored at 0° in the dark for 1–2 hr and centrifuged in a polypropylene tube for 20 min at

10,000 rpm. Clear solution is removed with a Pasteur pipette from the region between aggregated material at the top and bottom of the tube. The plasmid DNA in this material is purified by two cycles of density centrifugation, each for 40 hr at 15°, 34,000 rpm. The portion of the final gradient containing supercoiled DNA is extracted 4–6 times with *n*-butanol and dialyzed against two changes of 50 m*M* Tris-HCl, pH 8.0, containing 1 m*M* Na_2EDTA. Two hundred milliliters of culture usually yields 200–400 μg of supercoiled DNA.

The *att*B-containing plasmid is linearized by digestion with *Eco*RI restriction endonuclease. After digestion, the mixture is adjusted to 1 *M* NaCl and 20 m*M* Na_2EDTA, extracted with phenol, and dialyzed exhaustively against 50 m*M* Tris-HCl, pH 8.0, containing 1 m*M* Na_2EDTA.

Preparation of Crude IHF. Strain HN356, a *recB21* derivative of *E. coli* K12 strain W3102 (constructed by R. Weisberg) is grown to mid-log phase at 31° in 10 g of Tryptone, 5 g of yeast extract, and 5 g of NaCl per liter. After centrifugation, the cell pellet is resuspended in 1/500th volume of 20 m*M* Tris-HCl, pH 7.4, containing 1 m*M* Na_2EDTA, 20 m*M* NaCl, and 10% glycerol. The cells are disrupted sonically and centrifuged for 20 min at 15,000 rpm. The supernatant is stored at 20° and diluted 5-fold in 50 m*M* Tris-HCl, pH 7.4, containing 10% glycerol prior to incorporation in assay mixtures. Several *E. coli* K12 strains, including rec^+ and *recA* isolates, are equivalent to HN356 as a source of crude IHF.

Unit. When increasing amounts of Int are assayed, recombination rises from an undetectable level to some maximal level. A unit of recombination activity is defined as the smallest amount of Int required to produce maximal recombination under the assay conditions described. From day to day there can be as much as a two-fold variation in the amount of recombination at the maximum. However, for a given batch of enzyme, the amount of protein required to reach the maximal level remains constant. Recombination is inhibited when more than 5 units of Int are included in assay mixtures.

Purification

Overproducing Strains. High-level expression of the *int* gene normally requires activation of its promotor by the bacteriophage *cII* gene product.[6,7] A variant of the *int* promoter that no longer requires activation has been isolated.[8,9] Honigman *et al.*[10] cloned a segment of the lambda

[6] N. Katzir, A. Oppenheim, M. Belfort, and A. B. Oppenheim, *Virology* **74,** 324 (1976).

[7] H. Shimatake and M. Rosenberg, *Nature (London)* **292,** 128 (1981).

[8] K. Shimada and A. Campbell, *Proc. Natl. Acad. Sci. U.S.A.* **71,** 237 (1974).

[9] R. H. Hoess, C. Foeller, K. Bidwell, and A. Landy, *Proc. Natl. Acad. Sci. U.S.A.* **77,** 2482 (1980).

[10] A. Honigman, S.-L. Hu, and W. Szybalski, *Virology* **92,** 542 (1979).

genome containing the *int* gene and this constitutive promoter in the plasmid RSF2124, an ampicillin-resistant derivative of ColE1.[11] Strain WCi22642,[10] a derivative of *E. coli* K12 strain W3350 carrying this plasmid, is routinely used as the source Int protein; similar levels of Int protein are also produced in several other K12 strains that have received this plasmid by transformation. The plasmid-containing cells grow vigorously and only occasionally yield subclones that no longer produce Int. In contrast, cells grow poorly and frequently are overgrown by nonproducing variants when they contain pBR322 plasmid derivatives that bear an *int* gene and a constitutive promoter.[12]

Growth of Cultures. The growth medium contains per liter: 10 g of Tryptone, 5 g of yeast extract, and 5 g of sodium chloride. Strain WCi22642 is grown to saturation at 31° in six flasks each containing 500 ml of growth medium plus 25 μg/ml of ampicillin. A fermentor containing 300 liters of growth medium without ampicillin is inoculated with this culture, and the cells are grown at 31° with aeration (7 cubic feet/min) and agitation (200 rpm). After exponential growth (doubling time = 45 min) to mid-log phase (A_{650} = 0.65) the cells are collected with two chilled Sharples AS-16P centrifuges, each operating at a flow rate of 150 liters/hr. The cell paste (350–400 g) is brought to 600 ml by addition of ice-cold 50 m*M* Tris-HCl, pH 7.4, containing 10% sucrose; the paste is homogenized slowly in a Waring blender. The mixture is distributed in 35-ml aliquots to polypropylene centrifuge tubes, frozen in liquid nitrogen and stored at −70°.

Preparation of Extract. Unless otherwise stated, the following procedures are carried out at 4° with solutions prepared at 25°. In a typical preparation, 4 tubes of frozen cells, corresponding to the yield from 75 liters of cells, are thawed at 31° and immediately placed on ice. To each tube is added 2 ml of a 10 mg/ml solution of lysozyme in 250 m*M* Tris-HCl, pH 7.4, and the tube is thoroughly mixed. After 35 min on ice, the mixture is centrifuged at 15,000 rpm for 40 min; the supernatant (15–20 ml/tube) is pooled (fraction I).

Differential Salt Precipitation. Fraction I is diluted with 50 m*M* Tris-HCl, pH 7.4, to 145 ml and distributed to four polyallomer centrifuge tubes (1 inch × 3½ inches). After centrifugation at 49,000 rpm for 200 min in a Beckman Ti rotor, the supernatant is decanted and discarded. The centrifuge tubes with adherent pellets are frozen in liquid nitrogen and stored overnight at −70°. After thawing, the pellets are resuspended with the aid of a Teflon pestle in buffer X (50 m*M* Tris-HCl, pH 7.4, containing 600 m*M* KCl, 1 m*M* Na_2EDTA, 1 m*M* 2-mercaptoethanol, and 10%, w/v, glycerol). After adjusting to a volume of 75 ml with buffer X, the mixture is stirred for 2 hr and centrifuged in a Beckman 60 Ti rotor as before. The

[11] M. So, R. Gill, and S. Falkow, *Mol. Gen. Genet.* **142,** 239 (1975).

[12] Y. Kikuchi and H. A. Nash, *Proc. Natl. Acad. Sci. U.S.A.* **76,** 3760 (1979).

PURIFICATION OF Int

Procedure	Fraction	Volume (ml)	Units $\times 10^{-3}$	Protein (mg)	Specific activity (U/mg)
Crude extract	I	86	864	2160	400
Differential salt precipitation	II	61	306	367	830
Phosphocellulose	III	7	236	6	39,330
Calcium phosphate-cellulose	IV	5	47	2	23,500

clear, straw-colored supernatant is carefully removed by pipetting from the underlying pellet and associated viscous material. The supernatant is pooled (60 ml), frozen in liquid nitrogen, and stored at −70° (fraction II).

Phosphocellulose Chromatography. A phosphocellulose (Whatman P-11) column (0.9 × 9.4 cm) is equilibrated with buffer X. Fraction II is then loaded at 15–25 ml/hr, and the column is washed with 18 ml of buffer X. The column is then developed with a 60-ml gradient (buffer X to buffer X supplemented to 1.7 *M* KCl). Int activity (fraction III) elutes between 0.7 and 0.9 *M* KCl. The fractions are either stored on ice for 2–3 days or frozen in liquid nitrogen and stored at −70°.

Calcium Phosphate-Cellulose Chromatography. Calcium phosphate gel is prepared by mixing 5800 ml of water and 600 ml of a 0.6 *M* solution of $CaCl_2 \cdot 2\ H_2O$ and then slowly adding 600 ml of a 0.4 *M* solution of $Na_3PO_4 \cdot 12\ H_2O$. The pH is adjusted to 7.4 with 1 *M* acetic acid; the gel is repeatedly rinsed with water, centrifuged briefly, and resuspended in 1 liter of water (30 mg of solid per milliliter). Washed cellulose is prepared by mixing 200 g of cellulose (Whatman CF-1) with 4 liters of water. After settling, the cellulose is washed twice with 4 liters of buffer (125 m*M* potassium phosphate, pH 6.8, containing 750 m*M* KCl). The cellulose is then incubated overnight at room temperature in 4 liters of double strength buffer, washed 5 times with water, and resuspended at 10 g/ml in fivefold diluted buffer. Equal volumes of calcium phosphate gel and washed cellulose are combined, and the mixture is stored at 4°. A column (1.5 × 3.4 cm) of the calcium phosphate-cellulose mixture is equilibrated with buffer X. Fraction III is thawed, diluted with sufficient 50 m*M* Tris-HCl, pH 7.4, containing 1 m*M* Na_2EDTA and 10% glycerol so as to match the conductivity of buffer X, and loaded onto the column at a rate of 0.2 ml/min. The column is washed with 3 volumes of buffer X and eluted with buffer X supplemented with 25 m*M* potassium phosphate buffer. Active fractions are frozen in liquid nitrogen and stored at −70° (fraction IV). The results of a typical purification are shown in the table. Although the final column chromatography results in substantial losses of activity and thus a decrease in specific activity, this step is essential to remove several contaminating proteins of low molecular weight.

Properties of the Purified Protein

Purity and Molecular Weight. Slab gel electrophoresis in the presence of SDS shows that the purified enzyme is more than 95% homogeneous.[13] By comparison with the electrophoretic mobility of suitable standards, the molecular weight of the denatured material is approximately 40,000. Hydrodynamic properties of the native enzyme, a sedimentation coefficient of 3.0 S, and a Stokes' radius of 29 Å,[14] indicate that Int is purified as a monomeric protein.

Requirements. To carry out integrative recombination under the assay conditions described above, Int requires that at least *att*P be present on a supercoiled circle.[4,5] However, under conditions of reduced ionic strength (20 m*M* instead of 70 m*M* KCl), nonsupertwisted DNA can recombine, at a rate 5 to 10-fold slower than supercoiled DNA under the same conditions.[15] IHF is a second requirement for recombination.[3] This *E. coli* protein has been extensively purified and shown to be a complex of two different polypeptides.[16] With purified IHF and purified Int, the spermidine requirement for recombination can be replaced by Mg^{2+} or partially replaced by monovalent cations at high ionic strength. With purified components, recombination is optimal between pH 7.7 and 8.7, at temperatures between 25 and 31° and at KCl concentrations between 50 and 150 m*M*. At least 20 Int molecules are required in the assay mixture to produce one recombinant molecule.[16]

Stability. Purified Int retains its activity for over 1 year at −70°. Small losses in activity during very long storage or repeated cycles of thawing and refreezing may be partially alleviated by addition of bovine serum albumin (2 mg/ml) to the purified preparation. The recombination activity of Int is inactivated by heating to 45° for 10 min or by incubation for 15 min at 25° in the presence of 7 m*M* *N*-ethylmaleimide.

Other Activities of Int. Int binds to double-stranded and single-stranded DNA. Specific binding of Int to double-stranded *att*P DNA has been known for several years.[17] This property has been used to assay the purification of Int protein; the purified material proved to be active in

[13] Y. Kikuchi and H. Nash, *Cold Spring Harbor Symp. Quant. Biol.* **43,** 1099 (1979).
[14] H. A. Nash, *in* "The Enzymes" (P. D. Boyer, ed.), 3rd ed., Vol. 14, p. 471. Academic Press, New York, 1981.
[15] T. J. Pollock, and K. Abremski, *J. Mol. Biol.* **131,** 651 (1979).
[16] H. A. Nash and C. A. Robertson, *J. Biol. Chem.* **256,** 9246 (1981).
[17] D. Kamp, Ph.D. Thesis, University of Cologne, 1973.

recombination.[18] Investigations with the DNase footprinting technique have revealed multiple interactions between Int and sequences within *att*P.[19,20] Some of these interactions are with parts of *att*P that are located 50–150 bp away from the site of the crossover; the role of these complexes in recombination is unknown. Int also binds to sequences in *att*P and *att*B that are located in the region of the crossover.[19] These interactions imply a direct role for Int in strand exchange. This hypothesis is strengthened by the finding that Int is a topoisomerase, i.e., has the capacity to break and reseal DNA.[12] Int is a type I topoisomerase,[21] similar in biochemical properties to enzymes described in rat liver[22] and HeLa cells.[23] The action of a topoisomerase in recombination was inferred early from the demonstration that the breakage and reunion step of recombination occur in the absence of a high-energy cofactor, such as ATP.[4,24] Recent studies have shown that cleavage of DNA by Int topoisomerase is highly specific, being restricted to sequences found at the crossover region of *att*P and *att*B and to related sequences found elsewhere in DNA (Craig and Nash, unpublished observations). Taken together, these facts strongly indicate that exchange of DNA strands is carried out by the topoisomerase function of Int. A model has been prepared that suggests how a crossover can be carried out by a topoisomerase.[12,21]

Acknowledgments

I acknowledge the indispensable aid of Dr. Joseph Shiloach, who carried out the fermentor-scale growth and collection of cells. Dr. Brenda Lange-Gustafson provided useful suggestions and information during the course of this work. I am grateful to Carol Robertson for dedicated and masterful technical assistance. Mindy Kaufman is thanked for the efficient preparation of this manuscript.

[18] M. Kotewicz, E. Grzesiuk, W. Courschene, R. Fischer, and H. Echols, *J. Biol. Chem.* **255,** 2433 (1980).

[19] W. Ross, A. Landy, Y. Kikuchi, and H. Nash, *Cell* **18,** 297 (1979).

[20] P.-L Hsu, W. Ross, and A. Landy, *Nature* (*London*) **285,** 85 (1980).

[21] H. A. Nash, K. Mizuuchi, L. W. Enquist, and R. A. Weisberg, *Cold Spring Harbor Symp. Quant. Biol.* **45,** 417 (1981).

[22] S. R. Martin, W. K. McCoubrey, Jr., B. L. McConaughy, L. S. Young, M. B. Been, B. J. Brewer, and J. J. Champoux, this volume [8].

[23] L. F. Liu, this volume [7].

[24] K. Mizuuchi, M. Gellert, R. A. Weisberg, and H. A. Nash, *J. Mol. Biol.* **141,** 485 (1980).

[16] Analysis of the ϕX174 Gene *A* Protein Using *in Vitro* DNA Replication Systems

By DAVID R. BROWN, DANNY REINBERG, THOMAS SCHMIDT-GLENEWINKEL, STEPHEN L. ZIPURSKY, and JERARD HURWITZ

The *A* gene of icosahedral single-stranded DNA phages encodes a protein that is required *in vivo* for duplex replicative form[1] (RF) and single-stranded circular [SS(c)] viral[2] DNA synthesis. The gene *A* protein was shown to be a *cis*-acting endonuclease *in vivo*[1,3] which converted supercoiled RF DNA (RFI) to the relaxed (RFII) form by introducing a single nick specifically into the viral (+) DNA strand.[3] The gene *2* protein of the filamentous single-strand DNA phages has an analogous function, both *in vivo*[4] and *in vitro*,[5] during phage RF[4,6] and viral[4,7] DNA synthesis. Indeed, the gene *A* protein of icosahedral phage such as ϕX174 may be prototypic of a variety of endonucleases in nature, associated with several aspects of DNA synthesis and metabolism including recombination and transposition.[8,9] The advent of recombinant DNA technology and the development of ϕX174-specific *in vitro* DNA replication systems have given rise to a powerful methodology for the study of proteins such as the gene *A* protein. Methods developed for the analysis of multiple activities associated with the ϕX *A* protein during DNA synthesis are described in this chapter.

Properties of the ϕX *A* Protein

Purified ϕX174 gene *A* protein has been obtained independently in several laboratories[10–13] from *Escherichia coli* infected with ϕX174 am3 (*lysis*-, gene *E*), and shown to be essential for specific ϕX DNA synthesis

[1] E. S. Tessman, *J. Mol. Biol.* **17,** 218 (1966).
[2] H. Fujisawa and M. Hayashi, *J. Virol.* **19,** 416 (1976).
[3] B. Franke and D. S. Ray, *Proc. Natl. Acad. Sci. U.S.A.* **69,** 475 (1972).
[4] N. S. C. Lin and D. Pratt, *J. Mol. Biol.* **72,** 37 (1972).
[5] T. F. Meyer and K. Geider, *J. Biol. Chem.* **254,** 12642 (1979).
[6] H. M. Fidanian and D. S. Ray, *J. Mol. Biol.* **72,** 51 (1972).
[7] B. Mazur and P. Model, *J. Mol. Biol.* **78,** 285 (1973).
[8] R. M. Harshey and A. I. Bukhari, *Proc. Natl. Acad. Sci. U.S.A.* **78,** 1090 (1981).
[9] D. J. Galas and M. Chandler, *Proc. Natl. Acad. Sci. U.S.A.* **78,** 4858 (1981).
[10] T. J. Henry and R. Knippers, *Proc. Natl. Acad. Sci. U.S.A.* **71,** 1549 (1974).
[11] S. Eisenberg and A. Kornberg, *J. Biol. Chem.* **254,** 5328 (1979).

ISBN 0-12-182000-9

in vitro.[14,15] The exact endonucleolytic cleavage site of the purified *A* protein has been mapped to the origin of viral strand replication,[16] within the *A* cistron, between nucleotide residues 4305 and 4306 of the viral strand sequence.[17] In addition to ϕX RFI DNA, the ϕX *A* protein specifically cleaved ϕX viral SS(c) DNA as well as RFI and SS(c) DNAs of the related icosahedral phages G4, St-1, α3, U3, G14, and ϕK.[18–21] In all cases, the cleavage occurred between the seventh and eight residues of an AT-rich 30-nucleotide conserved sequence.[16,18,20,21] Slight sequence divergence was found only in phages St-1 and α3, each of which shared two identical base substitutions within the conserved region.[20,21] This 70% AT, 30-nucleotide conserved sequence, flanked by GC-rich regions, when present in SS or RFI DNA, was clearly implicated as a target for the site-specific endonucleolytic activity of the ϕX *A* protein.

The cleavage reaction catalyzed by the ϕX *A* protein has been characterized *in vitro*. The reaction required Mg^{2+} and a specific recognition sequence, in single-stranded or negatively supercoiled form, and produced a free 3′-hydroxyl terminus at the nick, and a covalent (presumably phosphodiester) linkage between the *A* protein and the newly generated 5′ terminus (RFII · *A* complex).[11,12,22] A role for a noncovalent multimeric *A* protein complex in the cleavage reaction has been proposed,[22] but equimolar amounts of RFII DNA and *A* protein have been reported in RFII · *A* preparations purified by sucrose gradient velocity sedimentation.[11]

---

[12] J.-E. Ikeda, A. Yudelevich, and J. Hurwitz, *Proc. Natl. Acad. Sci. U.S.A.* **73,** 2669 (1976).

[13] S. A. Langeveld, G. A. van Arkel, and P. J. Weisbeek, *FEBS Lett.* **114,** 269 (1980).

[14] C. Sumida-Yasumoto, A. Yudelevich, and J. Hurwitz, *Proc. Natl. Acad. Sci. U.S.A.* **73,** 1887 (1976).

[15] S. Eisenberg, J. F. Scott, and A. Kornberg, *Proc. Natl. Acad. Sci. U.S.A.* **73,** 1594 (1976). (1976).

[16] S. A. Langeveld, A. D. M. van Mansfeld, P. D. Baas, H. S. Jansz, G. A. van Arkel, and P. J. Weisbeek, *Nature (London)* **271,** 417 (1978).

[17] F. Sanger, G. M. Air, B. G. Barrell, N. L. Brown, A. R. Coulson, J. C. Fiddes, C. A. Hutchison III, P. M. Slocombe, and M. Smith, *Nature (London)* **265,** 687 (1977).

[18] A. D. M. van Mansfeld, S. A. Langeveld, P. J. Weisbeek, P. D. Baas, G. A. van Arkel, and H. S. Jansz, *Cold Spring Harbor Symp. Quant. Biol.* **43,** 331 (1979).

[19] M. Duguet, G. Yarranton, and M. Gefter, *Cold Spring Harbor Symp. Quant. Biol.* **43,** 335 (1979).

[20] F. Heidekamp, S. A. Langeveld, P. D. Baas, and H. S. Jansz, *Nucleic Acids Res.* **8,** 2009 (1980).

[21] F. Heidekamp, P. D. Baas, and H. S. Jansz, *J. Virol.* **42,** 91 (1982).

[22] J.-E. Ikeda, A. Yudelevich, N. Shimamoto, and J. Hurwitz, *J. Biol. Chem.* **254,** 9416 (1979).

Results of several workers have suggested that the ϕX *A* protein is a multifunctional enzyme, as first proposed by Eisenberg *et al.*[23] Incubation of the purified RFII · *A* complex with the purified *rep* and single-strand binding (SSb) proteins of *E. coli* resulted in ATP-dependent unwinding of the DNA.[24] Further addition of the DNA polymerase III elongation system allowed net synthesis of ϕX SS(c) DNA in the absence of additional free *A* protein,[12] with yields in excess of 5 SS(c) product molecules per input RFII · A complex.[11] Collectively, these results suggested that the multiple functions of the ϕX *A* protein included: (*a*) initiation of viral strand DNA synthesis by placing an endonucleolytic incision specifically within a 30-nucleotide target sequence of RFI DNA; (*b*) interaction with *rep* and SSb proteins to effect ATP-dependent helix unwinding ahead of the replication fork; (*c*) termination of DNA synthesis by directing the sequence-specific circularization of the displaced viral strand; (*d*) regeneration of the RFII · *A* complex by means of a putative protein transfer event, associated with termination, thereby allowing catalytic utilization of the *A* protein.

Figure 1 summarizes the reaction pathways in which the ϕX *A* protein is thought to participate.

Materials

Resins and Reagents

Sepharose, DEAE-Sepharose, and DEAE-Sephadex A-50 were purchased from Pharmacia; cellulose (CF-11), DEAE-cellulose (DE-52), and phosphocellulose (P-1), from Whatman; DNA (calf thymus) for making DNA-cellulose, from Worthington; Bio-Rex 70 (100–200 mesh) from Bio-Rad; and hydroxyapatite from Bio-Rad or Serva. *Eco*RI restriction endonuclease was from Bethesda Research Laboratories, and all other restriction endonucleases were from New England BioLabs. Lysozyme was purchased from Calbiochem, proteinase K from E.M. Biochemicals, DNA polymerase I [both complete protein and large proteolytic (Klenow) fragment] from Boehringer Mannheim, and endonuclease S1 from Boehringer Mannheim or P-L Biochemicals. Exonuclease VII was a gift from Dr. J. Chase of this institution. *Eco*RI linkers were purchased from

[23] S. Eisenberg, J. Griffith, and A. Kornberg, *Proc. Natl. Acad. Sci. U.S.A.* **74,** 3198 (1977).
[24] J. F. Scott, S. Eisenberg, L. L. Bertsch, and A. Kornberg, *Proc. Natl. Acad. Sci. U.S.A.* **74,** 193 (1977).

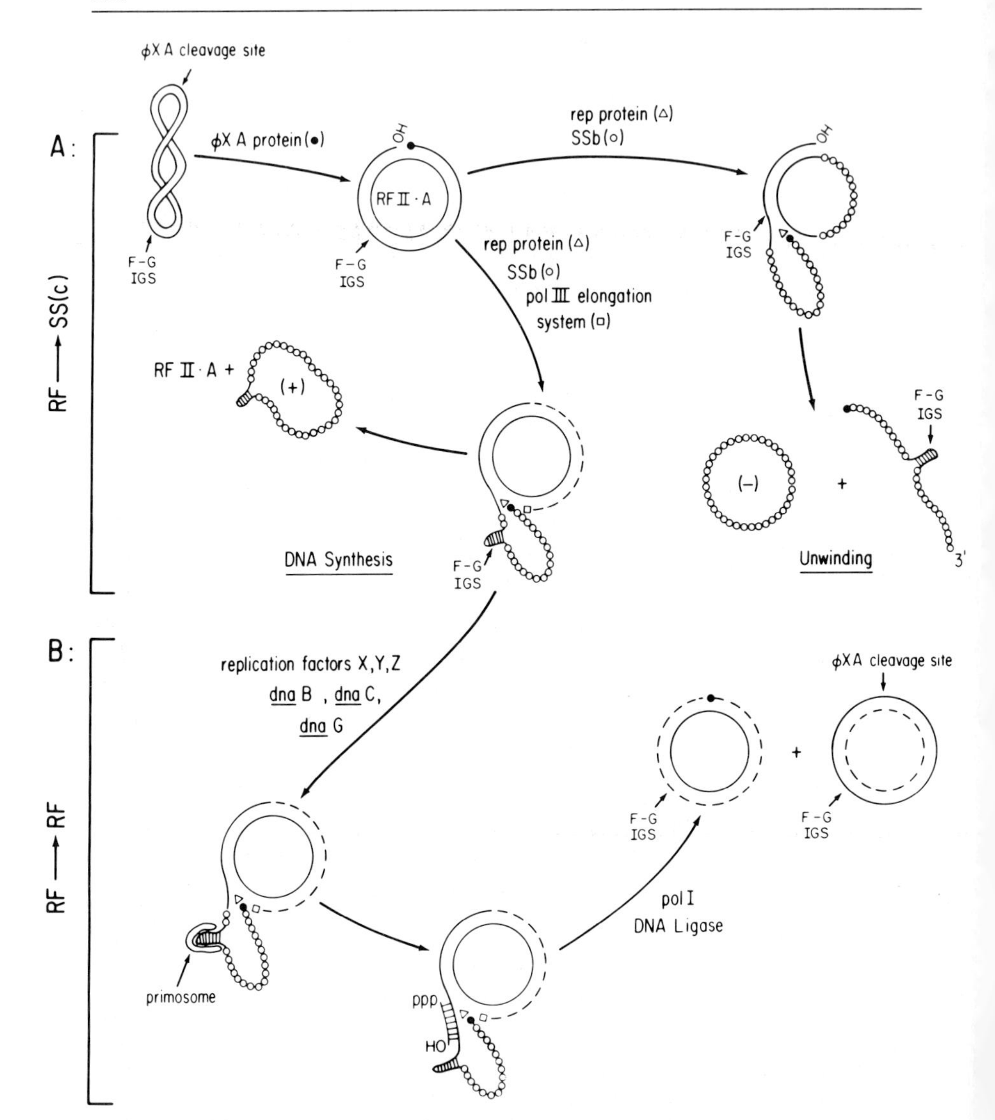

FIG. 1. Formation and fate of RFII · *A* complex. The formation and fate of the RFII · *A* complex is shown. Purified ϕX *A* protein (●) specifically cleaves ϕX174 RFI DNA and forms a covalent linkage with the 5′-phosphate terminus of nucleotide 4306. The addition of *Escherichia coli rep* (△) and SSb (○) proteins results in unwinding of the RFII · *A* complex, forming linear viral (+) and circular complementary strand (−) SS DNA products, completely coated with SSb protein. The linear viral strand remains linked to the ϕX *A* protein at the 5′ terminus and contains the lagging strand replication origin, the F-G intergenic space (F-G IGS). Further addition of the DNA polymerase III elongation system (□) allows viral strand synthesis to proceed behind helix unwinding. As the viral strand replication fork

Collaborative Research, *Bam*HI linkers were a gift from Dr. K. Marians of this institution, and T4 DNA ligase was purchased from Bethesda Research Laboratories. Deoxyribonucleoside triphosphates were from Boehringer Mannheim or Sigma, ribonucleoside triphosphates from Schwarz/Mann or Sigma, [^{3}H]- or [α-^{32}P]deoxyribonucleoside triphosphates from Amersham Searle, sodium dodecyl sulfate from Pierce Chemical Co., agarose (HGT) from Marine Colloids Division of FMC Corp., and acrylamide from Baker. All other chemicals were from Sigma, Fisher, Schwarz–Mann, or Mallinckrodt.

Bacterial and Bacteriophage Strains

Escherichia coli strains used were: H514(ϕX174^s, *endA, su, hcr, thy, arg*); CR34(*thr, leu, thi, thyA, lacY, dra, tonA, supE, rpsL*); SK2267(F-, *thi, gal, endA, sbcB, recA, hsdR, ton*); HMS83(*polA, polB, thyA, lys, lacZ, rpsL, rha*); BT1029(F-, *thy, endA, polA, dnaB, su*); PC22(F-, *thy, his, malA, xyl, arg, mth, thi, polA, dnaC, rpsL, su*); MV12(F-, *trpE, thr, leu, thi, recA, supE, lacY,* ColE1/*pLC44-7*); JA200(F$^+$, *trpE, thr, leu, thi, recA, supE, lacY,* ColE1/*pLC44-7*); CSR603(F$^-$, *uvrA, recA, phrl, rpsL, nalA, pDR2000*); D92(*thy, rep*); AX727(F$^-$, *thi, lac, dnaZ, rpsL*); and RLM782. The bacteriophages used were ϕX174 am3 (*lysis*$^-$, gene *E*), ϕX174 amH90 (gene *A*), and G4 am305. Cells were grown in L broth (10 g of Difco Bacto-tryptone, 5 g of Difco yeast extract, 5 g of NaCl, 10 g of glucose per liter) or Hershey broth (10 g of Difco Nutrient Broth, 5 g of Difco Bacteo-peptone, 5 g of NaCl, 10 g of glucose per liter).

---

migrates past the F-G IGS, the lagging strand replication origin is activated. Termination of viral strand replication results in the formation of SSb-coated single-strand circles [SS(c)] and regeneration of the RFII · *A* complex. The synthesis of SS(c) products from RF templates is known as the RF → SS(c) DNA replication reaction.

If the proteins required for the initiation of lagging strand DNA synthesis are present in addition to *rep* and SSb proteins and the DNA polymerase III elongation system, both viral (+) and complementary (−) strands are synthesized in a process known as RF → RF DNA replication. Replication factors X, Y, and Z (probably corresponding to proteins i, n′, and n + n″, respectively, in the nomenclature of Arai and Kornberg, together with the *dnaB, dnaC,* and *dnaG* proteins, are required for the assembly of a multienzyme priming complex (primosome) [K. Arai and A. Kornberg, *Proc. Natl. Acad. Sci. U.S.A.* **78,** 69 (1981)] on a specific region of the displaced viral strand in the F-G IGS. The primosome catalyzes the synthesis of oligonucleotide primers complementary to the viral strand, which are subsequently elongated by the DNA polymerase III system. Following the termination of leading strand synthesis, segregation of the regenerated RFII · *A* complex and a gapped circular duplex progeny molecule occurs. After gaps are filled in (probably by DNA polymerase I) and nicks are sealed by DNA ligase, a covalently closed circular duplex (RFI′) DNA product is formed. The RFI′ structure can be converted to the final RFI product by DNA gyrase.

General Methods

We are developing a system in which the single activities of the multifunctional ϕX *A* protein can be separated and studied individually. Our basic strategy involves the construction of recombinant plasmids containing inserts homologous to the conserved 30-nucleotide *A* protein target sequence, or portions of this sequence, and an analysis of their behaviors in ϕX-specific *in vitro* DNA replication systems. Several plasmids have been constructed that contain a single insert in attempts to define the minimum DNA sequence required for expression of the endonuclease activity (replication initiation function) of the ϕX *A* protein. Recombinant plasmids containing two inserts, widely separated and asymmetrically placed within the vector (at least one of which supports the expression of the initiation function), have been constructed in order to study the termination and putative protein transfer events.

Four *in vitro* reactions are used to study the behavior of recombinant plasmids in the presence of the ϕX *A* protein. First, the cutting reaction simply tests whether or not the ϕX *A* protein can recognize, bind, and specifically nick RFI plasmids containing target sequence inserts. The second reaction, the unwinding rection, is used to determine the effect of the combined actions of purified ϕX *A* protein, and the *E. coli* proteins *rep* and SSb, on recombinant plasmids. Questions that may be addressed include: (*a*) once cleaved by the A protein, can *rep*, SSb and A protein unwind the recombinant plasmid DNA? (*b*) in the case of double-insert recombinant plasmids, when a target sequence is encountered during unwinding, does circularization ensue? (*c*) again with double-insert recombinants, if circularization occurs at a target sequence, is this accompanied by protein transfer. (The last question has not yet been studied experimentally but will be addressed in the near future.) The third reaction contains all the components of the unwinding reaction with the additional inclusion of the *dnaN* gene product (previously known as elongation factor I, or the β subunit of DNA polymerase III holoenzyme in the nomenclature of Burgers *et al.*[25]) and the DNA polymerase III elongation system (including the *dnaE* gene product, the *dnaZ* gene product, and elongation factor III, EF-III). This reaction catalyzes the formation of SS(c) plasmid DNA from RFI templates [RF $\rightarrow$ SS(c) reaction] and is used to determine whether plasmids nicked by the ϕX *A* protein are in fact utilized as template-primers for DNA synthesis. Product analysis provides further information regarding the efficiency of the termination and circularization events. In the fourth reaction, crude extracts (referred to as receptor

[25] P. M. J. Burgers, A. Kornberg, and Y. Sakakibara, *Proc. Natl. Acad. Sci. U.S.A.* **78,** 5391 (1981).

fractions) derived from an *E. coli* strain (BT1029) carrying a thermosensitive allele for the *dnaB* gene, supplemented with the purified ϕX gene *A* and *E. coli dnaB* proteins, are used to catalyze duplex replication of recombinant plasmid DNA (RF $\rightarrow$ RF reaction). The *dnaB* protein (as well as *dnaC* and *dnaG* proteins) is required for lagging, but not leading, strand DNA synthesis from ϕX174 or recombinant plasmid templates in crude extracts.[26] Thus, RF $\rightarrow$ RF or RF $\rightarrow$ SS(c) DNA synthesis is observed depending on whether purified *dnaB* protein is added to or omitted from, respectively, the heat-inactivated BT1029 receptor. Using this reaction, the behavior of recombinant plasmids within a complex protein milieu may be characterized, perhaps lending more credibility to the notion that events observed *in vitro* are a reflection of *in vivo* phenomena. In addition, this reaction serves as a probe for mechanisms of initiation of lagging strand DNA replication, as the ϕX *A* protein-specific component only supports leading strand synthesis. Zipursky and Marians[27] have used this system to characterize replication factor Y-dependent initiation of DNA synthesis on pBR322 templates.

Purification of *in Vitro* System Components

In all cases, one unit of enzyme activity is defined as the amount that catalyzes the incorporation of 1 nmol of [^{3}H]dTMP into an acid-insoluble form under standard conditions. A detailed purification procedure for the ϕX *A* protein is described below. Procedures for other components required in the *in vitro* system are outlined.

ϕX A Protein

The ϕX *A* protein was prepared using the procedure of Ikeda *et al.*[22] with modifications. Complementation assays included ϕX *A* protein fractions, 1 nmol of ϕX RFI DNA, and receptor fractions from *E. coli* H514 (supplemented with fraction II[14] prepared from uninfected, or ϕX amH90-infected, *E. coli* H514) or *E. coli* PC22 (supplemented with crude *dnaC* protein fractions from *E. coli* HMS83). RF $\rightarrow$ RF reaction mixtures (50 μl) containing [^{3}H]-dTTP (200–1000 cpm/pmol) were incubated for 15–30 min, stopped by the addition of 0.1 ml of sodium pyrophosphate (0.1 *M*), 0.1 ml of heat-denatured calf thymus DNA (1 mg/ml), 4 ml of cold trichloroacetic acid (5%), and acid-insoluble ^{3}H radioactivity was determined. A typical preparation is summarized in the table.

Escherichia coli H514 were infected with ϕX am3 as described,[24] cen-

[26] D. Reinberg, S. L. Zipursky, and J. Hurwitz, *J. Biol. Chem.* **256,** 13143 (1981).
[27] S. L. Zipursky and K. J. Marians, *Proc. Natl. Acad. Sci. U.S.A.* **78,** 6111 (1981).

Purification of ϕX *A* Protein[a]

Fraction	Total units	Specific activity
I. Lysate	92,550	70
II. Ammonium sulfate	76,075	208
III. DEAE-Sephadex	96,200	1,749
IV. DEAE-cellulose	67,310	6,625
V. Phosphocellulose	6,000	10,000

[a] ϕX *A* protein was assayed in RF → RF DNA replication reaction mixtures containing ϕX RFI DNA and *Escherichia coli* PC22 receptor fractions supplemented with *dnaC* protein.

trifuged (yielding 15 g wet weight), washed in 50 m*M* Tris-HCl, pH 7.5, recentrifuged and resuspended (1 : 1, w/v) in 50 m*M* Tris-HCl, 1 m*M* EDTA, and frozen at −80°. Cells were thawed (15 g, original wet weight), adjusted to 80 ml with 50 m*M* Tris-HCl, pH 7.5, and distributed in 20-ml portions to Beckman type 30 tubes. Mixtures were adjusted to 10 m*M* EDTA, 2 m*M* dithiothreitol, 0.4 *M* NaCl, 0.1 mg of lysozyme per milliliter, and incubated on ice for 20 min. Brij 58 was added to 0.15%, and incubations were continued for 20 min. Lysates were brought to 1.0 *M* NaCl, incubated on ice an additional 20 min, and centrifuged at 100,000 *g* for 60 min. The supernatant (fraction I) was diluted with 10 m*M* Tris-HCl, pH 7.5, 2.5 m*M* EDTA, to a conductivity equivalent to 0.4 *M* NaCl. Neutralized ammonium sulfate (saturated, 4°) was added to 45% saturation during a 15-min period while stirring gently at 0°. The precipitate was stirred an additional 15 min, collected by centrifugation at 18,000 *g* for 30 min, and resuspended in buffer A (50 m*M* Tris-HCl, pH 7.5, 1 m*M* EDTA, 2.5 m*M* dithiothreitol) to a conductivity equivalent to 0.1 *M* NaCl.

The preparation (fraction II) was loaded on a 200-ml DEAE-Sephadex A-50 column (previously equilibrated with buffer A containing 0.1 *M* NaCl), washed with 200 ml of buffer A containing 0.1 *M* NaCl, and eluted with a 1.6-liter linear gradient of 0.1 to 1.0 *M* NaCl in buffer A. The ϕX *A* protein activity eluted with 0.45 *M* NaCl, relatively free of nucleic acids (the A_{260} peak eluted with 0.65 *M* NaCl). Active fractions were pooled and adjusted to 75% saturation with neutralized ammonium sulfate (saturated, 4°) during a 15-min period while stirring gently at 0°. The precipitate was stirred an additional 15 min, collected by centrifugation at 18,000 *g* for 30 min, resuspended in 20 ml of buffer A containing 10% glycerol, and dialyzed for 2 hr against four changes (500 ml each) of the same buffer.

The preparation (fraction III) was loaded on a 100-ml DEAE-cellulose column (previously equilibrated with buffer A containing 10% glycerol), washed with 100 ml of buffer A containing 10% glycerol, and eluted with a

0.5 liter linear gradient of 0 to 0.5 *M* NaCl in buffer A plus 10% glycerol. We have found that *A* protein activity could be recovered from the DEAE-cellulose column with a 0.15 *M* NaCl batch elution following a 0.1 *M* NaCl wash (150 ml each, in buffer A plus 10% glycerol). In this case, the dialysis buffer used prior to DEAE-cellulose chromatography included 0.1 *M* NaCl. Active DEAE-cellulose fractions were pooled and dialyzed for 3 hr against three changes (1 liter each) of 0.4 *M* NaCl in buffer B (40 m*M* sodium phosphate, pH 6.0, 1 m*M* EDTA, 5 m*M* dithiothreitol).

The preparation (fraction IV) was loaded on a 50-ml phosphocellulose column, washed with 100 ml of buffer B containing 0.4 *M* NaCl, and eluted with a 0.5 liter linear gradient of 0.4 to 1.2 *M* NaCl in buffer B.

Smaller phosphocellulose columns (10–40 ml, eluted with 100 to 400-ml gradients) have been employed in efforts to maximize protein concentrations in active fractions. The active phosphocellulose fractions (0.55–0.7 *M* NaCl) were pooled and concentrated by dialysis against 30% polyethylene glycol (20,000) in buffer C (20 m*M* Tris-HCl pH 7.5, 1 m*M* EDTA, 5 m*M* dithiothreitol, 1.0 *M* NaCl, and 10% glycerol). After dialysis in buffer C without polyethylene glycol, the preparation (fraction V) was divided in aliquots and frozen at −80°. Final fractions had site-specific endonuclease activities directed against ϕX174 SS(c) and RFI, but not against bacteriophage f1 SS(c) or pBR322 RFI DNAs. These fractions were also active in supporting ϕX174 RF → SS(c) and RF → RF DNA replication *in vitro*.

The Rep Gene Protein

The *rep* protein was purified as described by Scott and Kornberg[28] from *E. coli* MV12 or JA200, transformed with the ColE1-derived plasmid pLC44-7 (Clarke and Carbon collection[29]) which carries the *rep* gene. Presence of the plasmid allowed approximately 10-fold overproduction of the protein. Complementation assays contained receptor fractions of *E. coli* D92 (*rep* amber mutant, rep_{38}), the ϕX A protein, 1 nmol of ϕXRFI and *rep* protein fractions. ATPase activity in the presence and in the absence of 1 nmol of ϕX viral DNA was measured by the method of Conway and Lipmann.[30] Protein from crude lysates was precipitated with ammonium sulfate and chromatographed on DEAE-cellulose, Bio-Rex 70, and SS DNA-cellulose. SS DNA-cellulose fractions were adequate for routine use and were stable to repeated freezing and thawing. Further chromatography on a second DEAE-cellulose column resulted in a more pure fraction which lost activity with repeated freezing and thawing.

[28] J. F. Scott and A. Kornberg, *J. Biol. Chem.* **253,** 3292 (1978).
[29] L. Clarke and J. Carbon, *Cell* **9,** 91 (1976).
[30] T. W. Conway and F. Lipmann, *Proc. Natl. Acad. Sci. U.S.A.* **52,** 1462 (1964).

SSb Protein

The *E. coli* SSb protein was a gift of Dr. J. Chase of this institution, purified as previously described from a 12-fold SSb protein overproducer using a filter-binding assay.[31] Lysates were prepared from *E. coli* CSR603 carrying the plasmid pDR2000 (containing the $ssbA^+$ gene). DNA was precipitated in 10% polyethylene glycol, and the supernatant was dialyzed, resulting in precipitation of the SSb protein. The SSb protein precipitate was collected, chromatographed on SS DNA cellulose and DEAE-Sephadex A-50, and concentrated by 65% ammonium sulfate precipitation. Final fractions were resuspended to 0.5–0.7 mg of SSb protein per milliliter and were stable to repeated freezing and thawing.

dnaN Protein

The *dnaN* protein (elongation factor I) and the DNA polymerase III elongation system (see below) have been purified according to unpublished procedures, details of which are available upon request. Activity was measured with an elongation assay containing ϕX DNA · RNA hybrids as template-primers. Crude lysates were prepared from *E. coli* nucleic acids were HMS83, and nucleic acids were precipitated with streptomycin sulfate. Protein was precipitated in 40% ammonium sulfate and chromatographed on DEAE-Sepharose, ω-amino-C_8-alkyl-Sepharose, phosphocellulose, and Bio-Rex 70. The final preparation was nearly homogeneous and stable to multiple freezing and thawing.

DNA Polymerase III Elongation System (Containing dnaE *and* dnaZ *Proteins and Elongation Factor III)*

Polymerase activity was measured with an elongation assay (see *dnaN* protein) and with a replication assay containing nicked salmon sperm DNA templates. Receptor fractions from *E. coli* AX727, thermosensitive in the *dnaZ* allele, were used in complementation assays for the *dnaZ* protein. Crude lysates were prepared from *E. coli* HMS83, and DNA was precipitated with streptomycin sulfate. Protein was precipitated in 40% ammonium sulfate and chromatographed on DEAE-Sepharose and phosphocellulose. Phosphocellulose fractions remained somewhat heterogeneous, but were highly active in the replication and elongation assays. The preparation was stable to multiple freezing and thawing.

dnaB Protein

The *dnaB* protein used in these studies has been purified by one of two procedures. In the procedure of Reha-Krantz *et al.*,[32] lysates were pre-

[31] J. W. Chase, R. F. Whittier, F. Auerbach, A. Sancar, and W. D. Rupp, *Nucleic Acids Res.* **8,** 3215 (1980).

[32] L. J. Reha-Krantz and J. Hurwitz, *J. Biol. Chem.* **253,** 4043 (1978).

pared from *E. coli* MV12/28, a strain carrying a ColE1-derived plasmid (Clarke and Carbon collection[29]) containing the *dnaB* gene. Complementation assays included receptor fractions from *E. coli* BT1029 (*dnaB*^{ts}), 1 nmol of ϕX viral DNA, and *dnaB* fractions. After precipitation of nucleic acids in 4% streptomycin sulfate, protein was precipitated in 40% ammonium sulfate and chromatographed on DEAE-cellulose and phosphocellulose. Phosphocellulose fractions were used routinely and were stable to approximately five freeze-thaw cycles.

The *dnaB* protein was also purified by Dr. K. Marians (this department) using a modification of the procedure of Arai *et al.*[33] The protein source was *E. coli* RLM782 (from Dr. R. McMacken, Johns Hopkins University), a 10-fold *dnaB* protein overproducer that carries a pBR322-related plasmid containing the *dnaB* gene. Protein was precipitated from lysates with ammonium sulfate, chromatographed on DEAE-cellulose, reprecipitated with ammonium sulfate, and chromatographed on ATP-Sepharose using the method of Lanka *et al.*[34] The procedure was found to be relatively rapid and gave a high yield of very pure *dnaB* protein.

BT1029 Receptor

Extracts were made according to the procedure described by Reinberg *et al.*[26] from *E. coli* BT1029 (*dnaB*^{ts}). Lysis was found to be a critical step; variations in procedure were therefore avoided. A vigorous lysis was required, but extensive disruption of cell membranes and DNA was avoided. Cells were harvested, resuspended 1 : 1 (w : v) in 10% sucrose, 10 m*M* Tris · HCl, pH 7.5, and stored at −80°. Thawed cells were adjusted to 1 m*M* dithiothreitol, 20 m*M* EDTA, 0.15 *M* KCl, 0.2 mg of lysozyme per milliliter, and brought to pH 7.5–8.0 with Tris base. Components were added rapidly and mixed by gentle inversion. After incubation on ice for 10–20 min, the mixture was adjusted to 0.1% Brij 58. The incubation was continued on ice for approximately 20 min, interrupted every 5 min by 10 sec of vigorous shaking by hand. When the mixture acquired a viscous and somewhat marbled appearance, lysis was considered complete, and cell debris was pelleted by centrifugation at 100,000 *g* for 60 min. Pellets were large and flocculent, but little or no visible cell debris remained in the supernatant. After precipitation of nucleic acids in 4% streptomycin sulfate, protein was collected by precipitation in 40% ammonium sulfate, resuspended, and reprecipitated in 45% ammonium sulfate, resuspended again, and frozen at −80°. Fresh aliquots were thawed each day and the *dnaB*^{ts} gene product was heat inactivated prior to use by incubating at 39° for 5 min.

[33] K. Arai, S. Yasuda, and A. Kornberg, *J. Biol. Chem.* **256,** 5247 (1981).

[34] E. Lanka, C. Edelbluth, M. Schlicht, and H. Schuster, *J. Biol. Chem.* **253,** 5847 (1978).

The *in Vitro* Reactions

General Comments

Proteins were removed from storage at −80° or −20°, thawed on ice for 30–60 min, and gently mixed just prior to assembling reactions. Receptors were incubated on ice for at least 20 min after heat treatment at 39°. Addition of thawed proteins immediately after the buffer melted gave poor results; presumably several minutes were required for protein renaturation after thawing. Reactions were assembled at room temperature as quickly as possible and were mixed gently (shaking by hand) after the addition of each component. The order of additions for the unwinding reaction was: water, reaction buffer, *rep* protein, SSb protein, RFII · *A* complex. The order for the RF → SS(c) reaction was: water, reaction buffer, and nucleotides, DNA, *A* protein, SSb protein, *dnaN* protein, DNA polymerase III elongation system, *rep* protein. For the RF → RF reaction, the order was: water, reaction buffer and nucleotides, DNA, *A* protein, BT1029 receptor, *dnaB* protein.

φX A Protein Endonuclease Assay

Reaction mixtures (50 μl) containing 20 m*M* Tris-HCl, pH 7.5, 10 m*M* $MgCl_2$, 4 m*M* dithiothreitol, 0.1 mg of heat-denatured bovine serum albumin per milliliter, 0.06–0.12 pmol (as molecules) of φX174, recombinant plasmid, or pBR322 RFI DNA, and 0.02–0.20 unit of the φX *A* protein, were incubated at 30° for 15–30 min. Reactions were terminated by adjusting mixtures to 15 m*M* EDTA, 0.1% sodium dodecyl sulfate and 0.1 mg of proteinase K per milliliter, and were incubated at 37° an additional 1–2 hr. After phenol (hydrated in 50 m*M* Tris-HCl, pH 7.5)–chloroform (1 : 1 v/v) extraction, ether extraction, and ethanol precipitation, products were electrophoresed through 1% agarose slab gels (3 mm thick, 10 cm long) at 9 V/cm for 2 hr in TAE buffer (50 m*M* Tris base, 40 m*M* sodium acetate, 1 m*M* EDTA, adjusted to pH 7.9 with HCl). Gels were stained in 0.5 μg of ethidium bromide per milliliter, and photographed. DNA in gel slices was quantified by heating in 1 ml of water at 100°, adding Aquasol and measuring ^3H radioactivity.

Unwinding Reaction

Reaction mixtures (50 μl) containing 20 m*M* Tris-HCl, pH 7.5, 10 m*M* $MgCl_2$, 4 m*M* dithiothreitol, 2 m*M* ATP, 10 μg of rifampicin per milliliter, 0.026 unit of *E. coli rep* protein, 0.4–0.8 μg of SSb protein, and 0.25 μg of

^{3}H-labeled ϕX174 or recombinant plasmid RFII · *A* complex (purified by sucrose gradient velocity sedimentation, specific activity, 10–20 cpm per picomole of nucleotide) were incubated at 30° for 20 min. Reactions were terminated by adjusting mixtures to 20 m*M* EDTA, 0.1% sodium dodecyl sulfate, and 0.1 mg of proteinase K per milliliter, and were incubated at 37° for an additional 60 min. Products were electrophoresed through 1% agarose gels (3 mm thick, 30 cm long) in TAE, pH 7.9, or TBE, pH 8.4 (50 m*M* Tris base, 40 m*M* boric acid, 1 m*M* EDTA) buffer at 6 V/cm. Gels were treated with a scintillation fluid (Enhance, New England Nuclear) and autoradiographic exposures for 2–4 weeks at −80° were made in the presence of Cronex (DuPont) intensifying screens.

RF → SS(c) DNA Synthesis

Reaction mixtures (50 μl) containing 20 m*M* Tris-HCl, pH 7.5, 10 m*M* $MgCl_2$, 4 m*M* dithiothreitol, 40 μM each of dATP, dCTP, dGTP, and [^{3}H]- or [α-^{32}P]dTTP (400–4000 cpm/pmol), 2 m*M* ATP, 10 μg of rifampicin per milliliter, 0.04–0.17 pmol (as molecules) of ϕX174, recombinant plasmid, or pBR322 RFI DNA, 0.02–0.20 unit of ϕX *A* protein, 0.4–2.5 μg of SSb protein, 0.1–0.3 unit of EFI, and 6.7 units of the DNA polymerase III elongation system, and 0.026 unit of *E. coli rep* protein, were incubated at 30° for 5–90 min. Reactions were terminated by adjusting mixtures to 15 m*M* EDTA, 0.1% sodium dodecyl sulfate, and 0.1 mg of proteinase K per milliliter, and incubated at 37° an additional 1–2 hr. Aliquots were precipitated in 5% trichloroacetic acid and total [^{3}H]- or [α-^{32}P]dTMP incorporation (pmol) was determined. Remaining DNA was extracted in phenol–chloroform, ether extracted, ethanol precipitated, electrophoresed through 1.5% agarose gels (3 mm thick, 10 cm long) in TBE buffer, pH 8.4, at 7 V/cm for 12 hr, and visualized by autoradiography.

RF → RF DNA Synthesis

Reaction mixtures (50 μl) containing 20 m*M* Tris-HCl, pH 7.5, 10 m*M* $MgCl_2$, 4 m*M* dithiothreitol, 40 μM each of dATP, dCTP, dGTP, and [^{3}H]- or [α-^{32}P]dTTP (100–4000 cpm/pmol), 2 m*M* ATP, 0.1 m*M* each of UTP, CTP, and GTP, 10 μg of rifampicin per milliliter, 0.1 m*M* NAD^+, 0.5 μg of ϕX174, recombinant plasmid, or pBR322 RFI DNA, 0.10–0.17 unit of ϕX *A* protein, 4 or 5 μl of heat-inactivated (39°, 5 min) BT1029 receptor fraction (0.10–0.14 mg of protein), and 0.02 unit of *dnaB* protein, were incubated at 30° for 30 min. In some experiments, an additional 0.01 unit of *dnaB* protein was added after 15 min. Product analysis was performed as described for the RF → SS(c) reaction.

Characterization of Replication Products as SS(c)

HaeIII Restriction Endonuclease Digestion. Individual [α-^{32}P]-labeled RF → SS(c) replication products were electroeluted from agarose gel slices into Spectrapor No. 3 dialysis tubing in TBE buffer, pH 8.4, at 4°, 8 V/cm, for 12 hr. DNA was mixed with 2 μg of unlabeled ϕX174 SS(c) viral carrier DNA, concentrated by butanol extraction, dialyzed against 10 m*M* Tris-HCl, pH 7.5, 1 m*M* EDTA, 10 m*M* NaCl (TEN), precipitated in ethanol, and restricted with *Hae*III endonuclease. Reactions (50 μl) containing 20 units of endonuclease, and the buffer mixture specified by the vendor, were incubated at 37° for 23 hr. Restriction products were electrophoresed in 5% polyacrylamide (20 : 1 acrylamide : bisacrylamide, w/w) and 50% urea, at 30 V/cm in TBE buffer, pH 8.3 (50 m*M* Tris base, 50 m*M* boric acid, 1 m*M* EDTA), and visualized by autoradiography.

Endonuclease S1 Digestion. Individual [α-^{32}P]-labeled RF → SS(c) reaction products were purified as described for *Hae*III restriction studies and incubated in 0.1-ml reaction mixtures containing 50 m*M* sodium acetate, pH 4.5, 0.1 m*M* $ZnSO_4$, 20 μg of unlabeled or ^{3}H-labeled (0.3 cpm/pmol nucleotide) heat-denatured *E. coli* DNA, and 100–200 units of endonuclease S1, at 37° for 1 hr. In some reactions, NaCl (0.2 *M*) was also included. Reactions were terminated by the addition of carrier DNA and trichloroacetic acid to 5%; acid-insoluble and/or acid-soluble ^{3}H and α-^{32}P radioactivity was measured, and the percentage of solubilization by enzyme treatment of RF → SS(c) reaction products was determined.

Exonuclease VII Digestion. Individual [α-^{32}P]-labeled RF → SS(c) reaction products were purified as described for *Hae*III restriction studies and incubated in 50 μl reaction mixtures containing 50 m*M* potassium phosphate, pH 7.9, 50 m*M* Tris-HCl, pH 7.9, 2 m*M* dithiothreitol, 8 m*M* EDTA, 0.1 mg of heat-denatured bovine serum albumin per milliliter, 1 μg of unlabeled or 3 μg of [^{3}H]-labeled (0.3 cpm/pmol nucleotide) heat-denatured *E. coli* DNA, and 1.4–2.8 units of exonuclease VII, at 37° for 60 min. Reactions were terminated and percentage of solubilization by enzyme treatment of RF → SS(c) reaction products was determined as described for endonuclease S1 digestion studies.

Construction of Recombinant Plasmids

Zipursky *et al.*[35] first demonstrated that recombinant plasmids containing the conserved 30 nucleotide ϕX *A* protein target sequence supported RF → RF and RF → SS(c) DNA synthesis in ϕX-specific *in vitro*

[35] S. L. Zipursky, D. Reinberg, and J. Hurwitz, *Proc. Natl. Acad. Sci. U.S.A.* **77,** 5182 (1980).

DNA replication systems. These plasmids, referred to as G5 and G39, were constructed by isolating the 281 base-pair *Hae*III-6b restriction fragment from ϕX RF DNA, modifying the termini with *Bam*HI linkers, and inserting the fragment into the *Bam*HI site of pBR322. *Escherichia coli* strains CR34 and SK2267 were transformed with recombinant molecules in these and subsequent experiments.

Further studies have been concerned with defining the minimum sequences required for expression of the various activities associated with the ϕX *A* protein. Toward this end, a computer analysis of the two known icosahedral phage DNA sequences (ϕX174 and G4) was performed in order to determine a scheme (Fig. 2), using commercially available restriction endonucleases, for the isolation of the smallest possible DNA fragment that contained the 30-nucleotide ϕX *A* protein target sequence. Phage G4 RFI DNA (4 μg) was digested with restriction endonuclease *Hgi*AI, and 3'-extended termini of digestion products were modified with the 3' → 5' exonuclease activity of the large (Klenow) fragment of *E. coli* DNA polymerase I to form blunt-ended fragments. The 175 base-pair fragment containing the G4 viral strand replication origin was isolated, redigested with restriction endonuclease *Mnl*I, and the resulting 5'-extended termini were made blunt by DNA polymerase I treatment. Products were modified by the terminal addition of *Bam*HI linkers, and one fragment, containing 49 G4-derived base pairs (including the conserved 30-nucleotide *A* protein target sequence), was isolated and inserted into the *Bam*HI site of pBR322. The resulting plasmids, A40 and B57, were shown to be fully active templates in the ϕX-specific *in vitro* DNA replication systems. B57 DNA has been subsequently used to construct various plasmids containing yet smaller inserts. The single-stranded *Hae*III-6b restriction fragment of ϕX viral DNA was hybridized to the complementary strand of the B57 *Bam*HI insert. Only the conserved 30-nucleotide sequence could potentially form a stable duplex segment in this heteroduplex structure. The heteroduplex was treated with exonuclease VII and endonuclease S1; products were modified with DNA polymerase I, blunt-end ligated to *Bam*HI linkers, and inserted into the *Bam*HI site of pBR322. The resulting plasmids are currently being analyzed in the various *in vitro* systems.

A complementary approach to the problem of defining minimum sequence requirements for the various ϕX *A* protein activities has been taken by Heidekamp *et al.*[36] and van Mansfeld *et al.*[37] These workers have

[36] F. Heidekamp, P. D. Baas, J. H. van Boom, G. H. Veeneman, S. L. Zipursky, and H. S. Jansz, *Nucleic Acids Res.* **9,** 3335 (1981).

[37] A. D. M. van Mansfeld, S. A. Langeveld, P. D. Baas, H. S. Jansz, G. A. van der Marel, G. H. Veeneman, and J. H. van Boom, *Nature (London)* **288,** 561 (1980).

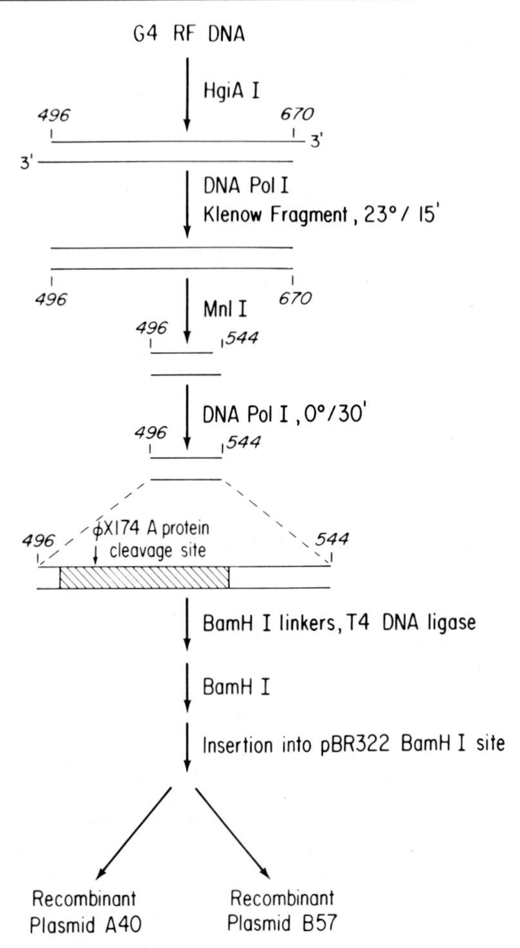

FIG. 2. Construction of recombinant plasmids A40 and B57. The scheme used in construction of recombinant plasmids A40 and B57 is shown. The 30 nucleotide conserved sequence is indicated by the shaded box. See the text for further details.

constructed synthetic oligonucleotides homologous to portions of the 30-nucleotide conserved sequence,[37] and cloned them into the plasmid pACYC177.[36] Resulting plasmids all contained the *A* protein cleavage site, with insert sequences homologous to up to 20 residues of the conserved sequence, but none were nicked by the *A* protein.

Plasmids containing two widely separated and asymmetrically placed copies of the 30 nucleotide conserved sequence have been constructed in order to develop a system in which the initiation, termination, and putative protein transfer activities of the ϕX *A* protein can be studied individually.[38] In preliminary studies, the ϕX174 *Hae*III-6b fragment has been modified with *Eco*RI linkers and inserted into the *Eco*RI site of the previously constructed recombinant plasmid G39. The resulting plasmids, G39-27 and G27-4, have the ϕX-derived fragments inserted in opposite, and identical orientations, respectively. While the G39-27 template gave rise to complex replication products, both inserts of G27-4 templates supported apparently normal initiation and termination functions of the ϕX *A* protein *in vitro*. A second double-insert plasmid (pPR903.1, constructed by P. Weisbeek from pFH903[36]), carrying the ϕX174 *Taq*I-T2 fragment (includes the conserved 30-nucleotide target sequence) and a second sequence homologous to only the first 16 residues, has been tested in the *in vitro* system. No initiation or termination at the 16 nucleotide sequence has been detected.[38]

Results

Results of selected experiments are presented as examples of applications of the described methods.

RF → SS(c) DNA Replication Directed by Single-Insert Recombinant Plasmid Templates

An autoradiograph of products synthesized from recombinant plasmid A40 and B57 templates by RF → SS(c) reaction mixtures in the presence and in the absence of the ϕX *A* protein is shown in Fig. 3. DNA synthesis required the ϕX *A* protein, and gave rise to several discrete products when analyzed by native agarose gel electrophoresis. However, when gel samples were heated at 65° for 5 min immediately prior to electrophoresis, only one product species was observed. Furthermore, DNAs from bands labeled SS(c)1, SS(c)2, SS(c)3, and SS(c)4 in Fig. 3 were individually purified and analyzed by digestion with restriction endonuclease *Hae*III, exonuclease VII (single-strand specific, requires termini), and endonuclease S1 (single-strand specific, no requirement for termini). Results showed that all four bands represented unit length SS(c) DNA. We conclude that the 49 base pair G4-derived inserts are recognized as origins of replication by the *in vitro* RF → SS(c) system. Furthermore, the SS(c) replication products, composed primarily of single-stranded pBR322 DNA sequences (which may potentially form a variety of stem-loop structures) apparently migrate in agarose gels as a series of noncovalently constrained topoisomers under nondenaturing conditions.

[38] D. Reinberg, S. L. Zipursky, P. Weisbeek, D. R. Brown, and J. Hurwitz, *J. Biol. Chem.* **258,** 529 (1983).

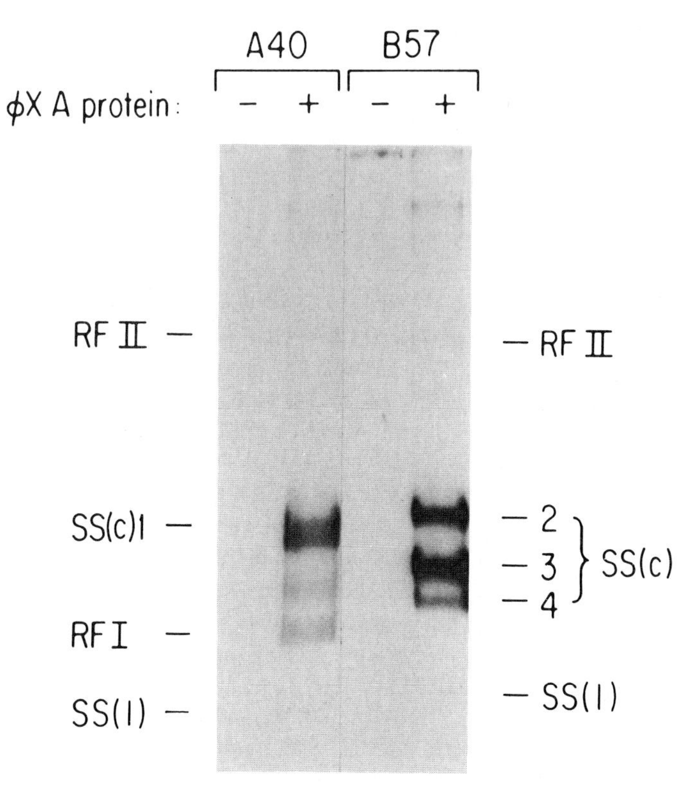

FIG. 3. Products of A40 and B57 RF → SS(c) DNA replication. Reaction conditions and product purification procedures are described in the text. The indicated RFI DNA templates (0.2 μg) were incubated in 50-μl reaction mixtures containing *Escherichia coli* SSb (2.5 μg) and *rep* (0.026 unit) proteins, EFI (0.1 unit) and the DNA polymerase III elongation system (6.7 units), with or without the ϕX *A* protein (0.08 unit), at 30° for 20 min. Aliquots of reaction products were precipitated in 5% trichloroacetic acid, and total [α-^{32}P]dTMP (4100 cpm/pmol) incorporation (pmol) was determined. Products were purified and subjected to electrophoresis in 1.5% agarose gels (TBE buffer, pH 8.4) at 7 V/cm for 12 hr.

FIG. 4. Products of G5, G39, G39-27, and G27-4 RF → SS(c) DNA replication. Reaction conditions and product purification procedures are described in the text. Lanes 1 and 2 show the products of reactions directed by recombinant plasmid G5 in the absence of ϕX *A* protein and with 0.115 unit of ϕX *A* protein, respectively. Lanes 3–6 show the products derived from the G39-27 DNA template in the absence of ϕX *A* protein and with 0.023, 0.069, and 0.115 unit of ϕX *A* protein, respectively. Lanes 7–10 show the products derived from the G27-4 DNA template in the absence of ϕX *A* protein and with 0.023, 0.069, and 0.115 unit of ϕX *A* protein, respectively. Lanes 11 and 12 show the products of reactions directed by recombinant plasmid G39 in the absence of ϕX *A* protein and with 0.115 unit of ϕX *A* protein, respectively. Aliquots of reaction products were precipitated in 5% trichloroacetic acid and total [α-^{32}P]-dTMP (3200 cpm/pmol) incorporation (pmol) was determined. Products were purified and subjected to electrophoresis in 1.5% agarose gels (TBE buffer, pH 8.4).

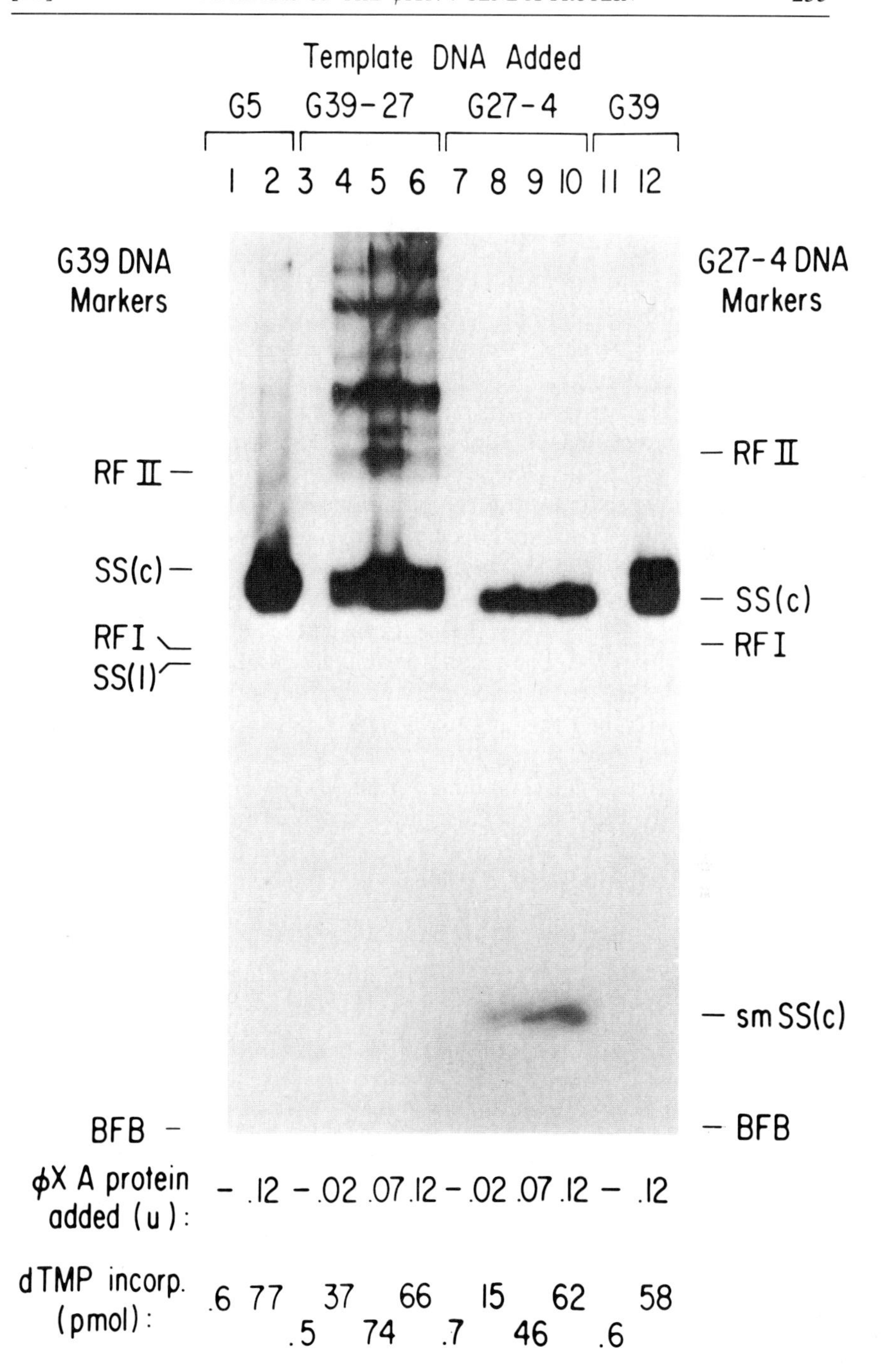

Template DNA Added
G5
G39-27
G27-4
G39
1 2 3 4 5 6 7 8 9 10 11 12
G39 DNA Markers
G27-4 DNA Markers
RF II
SS(c)
RF I
SS(I)
RF II
SS(c)
RF I
sm SS(c)
BFB
BFB
φX A protein added (u):
– .12 – .02 .07 .12 – .02 .07 .12 – .12
dTMP incorp. (pmol):
.6 77 .5 37 74 66 .7 15 46 62 .6 58

RF → SS(c) DNA Replication Directed by Double-Insert Recombinant Plasmid Templates

An autoradiograph of products synthesized from recombinant plasmid G5, G39, G39-27, and G27-4 templates by RF → SS(c) reaction mixtures in the presence and in the absence of the ϕX *A* protein is shown in Fig. 4.[38] DNA synthesis required the ϕX *A* protein, and in the presence of G27-4 templates, gave rise to two discrete DNA products (samples were heated prior to electrophoresis). The larger product, lgSS(c), approximately comigrating with unit length SS(c) markers, and the smaller product, smSS(c), were both shown to be circular and single stranded by their sensitivity to endonuclease S1 digestion and resistance to exonuclease VII digestion. Experiments using the blotting technique of Southern[39] showed that fewer than 5% of lgSS(c) product molecules contained smSS(c) DNA sequences. Therefore, both inserts of recombinant plasmid G27-4 were active in supporting the initiation and termination functions of the ϕX *A* protein. (We do not expect that both inserts of the same plasmid molecule were active in the initiation reaction; rather, one of the two inserts was probably utilized at random. Following cleavage at one insert, loss of superhelicity should preclude cleavage at the other insert.) Preliminary experiments with recombinant plasmid G27-4 in the unwinding reaction suggested that termination (circularization of the unwound (+) strand) occurred in presence of only the ϕX *A* protein, and the *E. coli rep* and SSb proteins.[38]

In contrast to G27-4, recombinant plasmid G39-27 gave rise to a complex series of products which comigrated with, or migrated slower than, unit length SS(c) marker DNA in agarose gels.[38] The mechanism by which these products were formed is unclear, however they arose as a consequence of the ϕX *A* protein target sequences being present in opposite orientations in the template DNA. One possibility may involve a switching of the strand utilized as template by the replication apparatus at these "inverted" recognition sequences.

[39] E. M. Southern, *J. Mol. Biol.* **98,** 503 (1975).

FIG. 5. Products of ϕX174, A40, and B57 RF → RF DNA replication. Reaction conditions and product purification procedures are described in the text. Reaction mixtures (50 μl) were incubated at 30° for 30 min in the presence of the indicated RFI DNA templates (0.5 μg) and 4 μl of *Escherichia coli* BT1029 (*dnaB*^{ts}) receptor fractions. Reactions were started with the addition of the ϕX gene *A* (0.10 unit) and the *dnaB* (0.02 unit) proteins. The *dnaB* protein (0.01 unit) was added again after 15 min. Aliquots of reaction products were precipitated in 5% trichloroacetic acid, and total [^{32}P]dCMP (650 cpm/pmol) incorporation (pmol) was determined. Products were purified and subjected to electrophoresis in 1.5% agarose gels (TBE buffer, pH 8.4) at 8 V/cm for 9 hr.

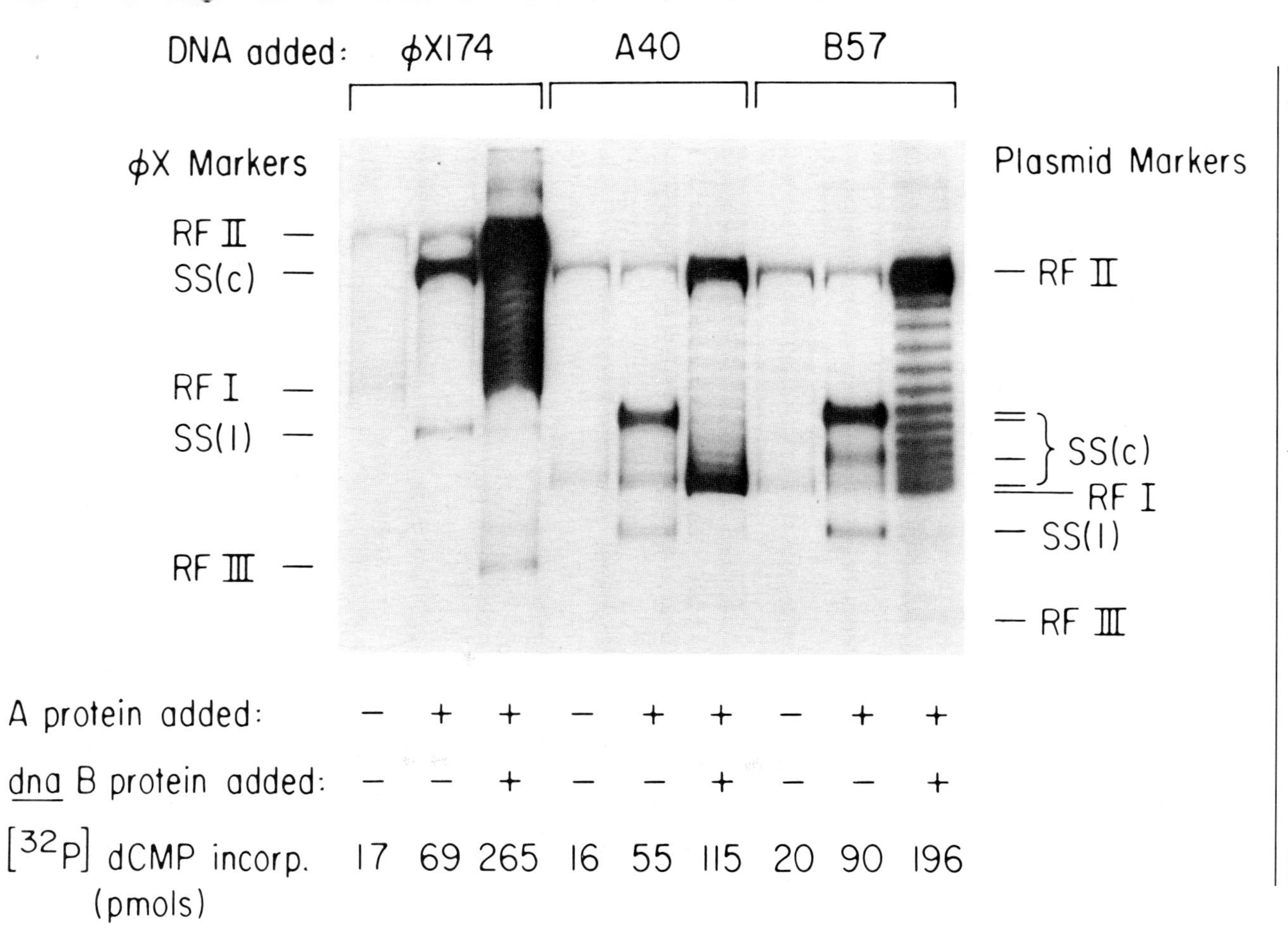
DNA added:
φX174
A40
B57
φX Markers
RF II
SS(c)
RF I
SS(l)
RF III
Plasmid Markers
RF II
SS(c)
RF I
SS(l)
RF III
A protein added: − + + − + + − + +
dna B protein added: − − + − − + − − +
[32P] dCMP incorp. (pmols) 17 69 265 16 55 115 20 90 196

RF → RF DNA Replication Directed by Single-Insert Recombinant Plasmid Templates

An autoradiograph of products synthesized from ϕX174, A40, and B57 DNA templates by RF → RF reaction mixtures containing BT1029 receptor fractions, in the presence and in the absence of the ϕX gene *A* and *dnaB* proteins, is shown in Fig. 5. A low level of nonspecific synthesis was observed in the absence of the ϕX *A* protein, probably due to the combined actions of endonucleases, exonucleases, and DNA polymerases present in the extracts. When extracts were supplemented only with the ϕX *A* protein, all templates supported SS(c) DNA synthesis (gel samples were not heat-treated in this case). Duplex molecules (RFII, RFI, and molecules of intermediate superhelical density) were formed in all reactions supplemented with both the ϕX gene *A* and *dnaB* proteins.

Conclusion

The ϕX *A* protein is involved in several important aspects of DNA replication, including the initiation of rolling-circle DNA replication, helix unwinding ahead of the replication fork, termination of leading strand synthesis resulting in segregation of covalently closed unit length circular progeny molecules, and a putative protein transfer event, following termination, that leads to catalytic regeneration of the RFII · *A* complex. *In vitro* systems, requiring purified proteins, crude cell extracts, and recombinant plasmids, have been described that allow detailed analyses of the various activities of the ϕX *A* protein. These are currently the only *in vitro* systems in which both initiation and termination of DNA replication are so readily amenable to analysis.

The initiation, termination, and putative transfer activities of the ϕX *A* protein are all directed against the same primary DNA structure, the 30-nucleotide target sequence. But secondary and tertiary target structures, as well as the configurations of other proteins (e.g., *rep* and SSb proteins) and associated DNA–protein interactions, are distinctly different during initiation, termination, and putative transfer events. It is hoped that the system described will allow a detailed dissection of these events and a better understanding of the contributions, to the replication process, of primary and higher-order DNA structures, and of the various replication proteins.

The systems described here have additional applications. ϕX *A* protein-directed leading strand DNA synthesis *in vitro,* from recombinant plasmid templates, has been used as a probe for studying mechanisms of initiation of lagging strand DNA replication.[27] In addition, *in vitro* systems similar to those described may have immediate applications in the analysis of molecular mechanisms of transposition.

Acknowledgments

Computer analysis was performed using the MOLGEN project on the SUMEX-AIM system at Stanford University. This research was supported by NIH Grants 5R01-GM-13344-17 and 5T32GM7288 (from NIGMS) and by NSF Grant PCM 78-16550.

Section IV

New Methods for DNA Isolation, Hybridization, and Cloning

[17] A Rapid Alkaline Extraction Method for the Isolation of Plasmid DNA

By H. C. BIRNBOIM

Plasmids are double-stranded circular DNA molecules that have the property of self-replication, independent of chromosomal DNA. In bacteria, they carry genes that may specify a variety of host properties. In recent years naturally occurring plasmids have been modified to produce new plasmids, which are used as cloning vehicles in recombinant DNA research. Although the presence of a plasmid in a bacterial cell may be detected genetically as a change in phenotype (e.g., resistance to a particular antibiotic), often it is necessary to isolate plasmid DNA for molecular studies, such as size determination, restriction enzyme mapping, and nucleotide sequencing, or for the construction of new hybrid plasmids. The degree of purification required will depend upon the intended use. Highly purified material can be prepared by the "cleared lysate" method, which involves a long period of centrifugation in a dye–CsCl gradient.[1] Less purified plasmid DNA is often satisfactory for recombinant DNA studies, and a large number of shorter and simpler methods have been developed (see Birnboim and Doly[2] and references therein). This chapter describes one such method that uses an alkaline extraction step. It is rapid enough to be used as a screening method, permitting 50–100 or more samples to be extracted in a few hours. The DNA is sufficiently pure to be digestible by restriction enzymes, an important advantage for screening. A preparative version that allows isolation of larger quantitites of more highly purified material is also described.

Principle

Isolation of plasmid DNA requires that it be separated from host-cell chromosomal DNA as well as other macromolecular components. Alkaline extraction exploits the covalently closed circular (CCC) nature of plasmid DNA and the very high molecular weight of chromosomal DNA. When a cell extract is exposed to conditions of alkaline pH in the range 12.0–12.6, linear (chromosomal) DNA will denature but CCC DNA will not.[3] pH adjustment is simplified by using glucose as a buffer. On neutral-

[1] D. B. Clewell and D. R. Helinski, *Proc. Natl. Acad. Sci. U.S.A.* **62,** 1159 (1969).

[2] H. C. Birnboim and J. Doly, *Nucleic Acids Res.* **7,** 1513 (1979).

[3] P. H. Pouwels, C. M. Knijnenburg, J. van Rotterdam, and J. A. Cohen, *J. Mol. Biol.* **32,** 169 (1968).

ISBN 0-12-182000-9

izing the extract in the presence of a high concentration of salt, precipitation of chromosomal DNA occurs. We presume this is because interstrand reassociations occur at multiple sites owing to the very high molecular weight of the DNA, which then leads to the formation of an insoluble DNA network.[4] CCC DNA remains in the soluble fraction. The bulk of cellular RNA and protein will also precipitate under these conditions if protein is first complexed with an anionic detergent, sodium dodecyl sulfate (SDS). By combining reagents appropriately, precipitation of most of the chromosomal DNA, RNA, and protein can be accomplished in a single step.

Materials and Reagents

Bacterial Strains and Plasmids. For the experiments to be described here as an illustration of the method, *Escherichia coli* K12 Strain RR1 containing a dimeric form of plasmid pBR322 was used.[5] Plasmids containing *Eco*RI fragments of coliphage T4 in *E. coli* K12 strain K802 were from E. Young, University of Washington.

Equipment for the Screening Method

Bench-top centrifuge capable of generating r.c.f. of 8000–13,000 *g*, such as the Eppendorf Model 5412 microcentrifuge. This accommodates 12 polypropylene tubes (1.5 ml capacity).

Racks to support 60 centrifuge tubes; these can be constructed by drilling holes 11 mm in diameter in sheets of 2 mm-thick aluminum.

Pasteur pipette, drawn out to a fine tip and flame-polished is useful in aspirating supernatants after centrifugation.

Repetitive pipettor capable of delivering 0.1–0.2 ml volumes is also helpful when large numbers of samples are to be processed.

Equipment for the Preparative Method

Refrigerated preparative centrifuge such as a Sorval RC-5

Rotors to accommodate 50-ml and 250-ml bottles

Solutions

Lysozyme solution: glucose (50 m*M*), CDTA (10 m*M*), and Tris-HCl (25 m*M*) (pH 8.0); the solution can be kept for many weeks at room temperature. Lysozyme is added at a concentration of 1 mg/ml shortly before use and dissolved by gentle mixing. The chelating agent, cyclohexane diaminetetraacetic acid (CDTA), is available from Sigma or Aldrich Chemical Company and can be rendered

[4] R. J. Britten, D. E. Graham, and B. R. Neufeld, this series, Vol. 29, p. 363.

[5] F. Bolivar, R. L. Rodriguez, P. J. Greene, M. C. Betlach, H. L. Heyneker, and H. W. Boyer, *Gene* **2,** 95 (1977).

colorless if necessary by treatment with charcoal. EDTA can be substituted, but CDTA has the following advantages over EDTA: at a concentration of 1 m*M* or higher, it appears to inhibit growth of microorganisms, allowing solutions to be stored at room temperature; it chelates metal ions much more effectively; it is more soluble in alcoholic and acidic solutions.

Alkaline SDS. This solution contains 0.2 *N* NaOH, 1% SDS. It is used to lyse cells and denature chromosomal DNA. Because of the buffering capacity of glucose, which is added in the first solution, proper pH for denaturation can be obtained without use of a pH meter. The shelf life at room temperature is about a week or longer, depending upon the source of SDS; reagent grade sodium dodecyl sulfate from BDH Chemicals has been satisfactory.

High-salt solution: 3 *M* potassium acetate, 1.8 *M* formic acid. For 100 ml of solution, use 29.4 g of potassium acetate and 5 ml of 90% formic acid; store at room temperature. If turbid, the solution should be clarified by filtration. It is used to neutralize the alkali used in the previous step and provide conditions under which chromosomal DNA, RNA, and protein will precipitate. Formic acid was substituted for acetic acid, used in the earlier procedure,[2] to make this solution easier to prepare and easier to adjust to pH 8 in later steps. Potassium acetate was substituted for sodium acetate because it is slightly more effective in precipitating denatured chromosomal DNA and SDS–protein complexes. Potassium acetate has also been used by other workers.[6]

Acetate–MOPS: 0.1 *M* sodium acetate, 0.05 *M* MOPS, adjusted to pH 8.0 with NaOH. It is stable at room temperature if stored over a drop of chloroform. MOPS (morpholinopropanesulfonic acid) is from Sigma.

CDTA–Tris: 1 m*M* CDTA, 10 m*M* Tris-HCl (pH 7.5); stable indefinitely at room temperature

Additional Solutions Required for the Preparative Method

LiCl solution: 5 *M* LiCl, 0.05 *M* MOPS, adjusted to pH 8.0 with NaOH. Filter through 0.45-μm membrane filter if turbid.

Ribonucleases: Ribonuclease A stock solution, 1 mg/ml in 5 m*M* Tris-HCl (pH 8.0); ribonuclease T1 stock solution, 500 units/ml in 5 m*M* Tris-HCl (pH 8.0). Both are available from Sigma, and solutions are heated at 80° for 10 min after preparation to inactivate contaminating deoxyribonuclease, if present. Store at −20°.

[6] D. Ish-Horowicz and J. F. Burke, *Nucleic Acids Res.* **9,** 2989 (1981).

CG-50 Ion-exchange resin: Amberlite CG-50 (200–400 mesh), a carboxylic acid-type cation-exchange resin. It is prepared by washing with 1 *N* HCl, water, 1 *N* NaOH, water and is finally equilibrated with 10 m*M* MOPS, 1 m*M* CDTA at pH 8. Fines are removed during washing. It is stored as a 50% (v/v) slurry.

Screening Method

Clones for plasmid extraction are grown to saturation in small volumes (2–3 ml) of medium such as L broth in the presence of the appropriate antibotic. Alternatively, single colonies (about 4 mm in diameter) from an agar plate can be used. The number that can be screened is usually limited by the number of slots available in an agarose gel electrophoresis apparatus; extracting 60 samples at a time is convenient. All manipulations and centrifugations (in a microcentrifuge) are at room temperature unless otherwise indicated.

1. One-half milliliter of each culture is transferred to a 1.5-ml Eppendorf tube for extraction; the remainder is stored at −20° for future use after addition of an equal volume of 80% (v/v) glycerol. If single colonies are to be used, they can be scraped with a sterile toothpick and suspended in 0.5 ml of water. Tubes are centrifuged for 15 sec. Longer times may make the pellet difficult to suspend. The supernatant is carefully removed using a fine-tip aspirator, and the cell pellet is loosened from the wall by vortexing two tubes together on a mixer while allowing their tips to clatter.
2. Each pellet is suspended thoroughly in 0.1 ml of lysozyme solution by vortexing immediately after each addition. The suspension is held at 0° for 5 min.
3. Alkaline SDS (0.2 ml) (at room temperature) is added. The sample is mixed gently by inversion several times and held at 0° for 5 min. The lysate should become almost clear initially, but will become cloudy on standing as SDS precipitates.
4. High-salt solution (0.15 ml) (at room temperature) is added. It is again mixed gently and held at 0° for 15 min. A curdlike precipitate will form. Centrifuge for 2 min.
5. Part of the supernatant (0.35 ml) is transferred into another tube, care being taken to avoid disturbing the bulky pellet. Cold ethanol (0.9 ml) is added to the sample, which is held at −20° for 15 min, then centrifuged for 1 min. Each tube should be oriented in the centrifuge so that the position of the pellet (which may not be obvious) will be known in order that it not be disturbed as the supernatant is removed. The pellet is dis-

solved in 0.1 ml of acetate–MOPS and reprecipitated with 0.2 ml of ethanol. This washing step should be repeated once more if the DNA is to be digested with enzymes or used for transformation. The final pellet is suspended in 0.04 ml of water or CDTA–Tris. It is suitable for analysis by gel electrophoresis directly or after digestion by restriction enzymes, or can be used for transformation of other cells. One-fifth of each sample is usually sufficient to give a discernible band on a gel.

Comments on the Screening Procedure. The method has been used successfully with other *E. coli* strains, with other bacteria, and with very large plasmids. Some cells lyse adequately in alkaline SDS without prior lysozyme treatment. The times indicated at each step are approximate minimum times and appear not to be very critical. Gentle handling at steps 3 and 4 helps to eliminate chromosomal DNA (presumably by preserving its high molecular weight), but some loss of plasmid may occur. RNA present in samples prepared for screening can obscure short fragments of DNA, which may be released by restriction nuclease treatment. If this is anticipated, samples dissolved in CDTA–Tris can be incubated with ribonuclease A (50 μg/ml) for 10 min prior to addition of concentrated buffer for restriction-enzyme digestion. Note that ribonuclease can activate trace amounts of endonuclease I (if present) by eliminating tRNA, a powerful inhibitor of the enzyme.[7]

Preparative Method

The first steps of the preparative method are similar to those of the screening method, and the reagents are identical. Either a nonamplified or an amplified culture can be used. Advantages of the latter are that smaller volumes of reagents at the early steps are needed and the yield (per liter of culture) appears to be higher. pBR322 in *Escherichia coli* can be amplified as follows. Twenty milliliters of an overnight culture is used to inoculate 1 liter of L broth containing 100 μg of ampicillin per milliliter. When the optical density at 600 nm reaches 0.8–1.0, chloramphenicol (170 μg/ml) is added and incubation continued for 18–20 hr. Alternatively, a nonamplified, overnight culture can be used. The volumes of extraction solutions given below are for 1 liter of culture of amplified cells.

1. Cells are harvested by centrifugation in 250- or 500-ml bottles at 6000 g for 10 min at 0°. Higher speeds may make the pellets difficult to suspend. The pellets are brought up in about 50 ml of water and recentrifuged. The cell pellet is first suspended as well as possible in 1 ml of

[7] I. R. Lehman, G. G. Roussos, and E. A. Pratt, *J. Biol. Chem.* **237**, 819 (1962).

glucose–CDTA–Tris at 0°, then 9 ml of cold glucose–CDTA–Tris (containing 10 mg of lysozyme) is added. For nonamplified cells, four times the amount of glucose–CDTA–Tris and lysozyme are used. The cell suspension is kept at 0° for 30 min.

2. Twenty milliliters of alkaline SDS (at room temperature) is added. The mixture is stirred fairly gently with a glass rod until nearly homogeneous and clear. It is kept at 0° for 10 min, then 15 ml of high-salt solution is added. The mixture is stirred a little more vigorously than before for several minutes until a coarse white precipitate forms. After standing at 0° for 30 min, the precipitate is removed by centrifugation at 12,000 *g* for 10 min at 0°. For nonamplified cells, the volumes used are four times those indicated.

3. The supernatant is transferred into another tube, and 2 volumes of cold ethanol is added. The precipitate of nucleic acids that forms on standing at −20° for 20 min is collected by centrifugation and dissolved in 5 ml of acetate–MOPS (12 ml for nonamplified cells). Nucleic acids are again precipitated with 2 volumes of ethanol and dissolved in 2 ml of water for both amplified and nonamplified cells. Up to this stage, the procedure is very similar to that used for screening.

4. The volume of solution is measured, and an equal volume of LiCl solution is added. The sample is held at 0° for 15 min, and the heavy precipitate that forms is removed by centrifugation at 12,000 *g* for 10 min at 0°. The clear supernatant is heated at 60° for 10 min, and a small amount of additional precipitate that may form is removed by centrifugation as before. Plasmid DNA is precipitated by the addition of 2 volumes of cold ethanol to the supernatant solution. After holding for 15 min at −20°, the precipitate is collected by centrifugation and redissolved in 2.5 ml of acetate–MOPS. After another ethanol precipitation, plasmid DNA (with some contaminating low molecular weight RNA at this stage) is dissolved in 2 ml of CDTA–Tris.

5a. Contaminating RNA is removed by treatment with ribonuclease and precipitation with isopropanol as follows. Ribonucleases A and T1 are added to 10 μg and 5 units per milliliter, respectively. After incubation at 37° for 15 min, 0.04 ml of 10% SDS and 2 ml of acetate–MOPS are added. Four milliliters of isopropanol is added dropwise with mixing, and the precipitate of plasmid DNA that forms after 15 min at room temperature is collected by centrifugation at 20°. Acetate–MOPS (2 ml) is added and plasmid DNA (which may be difficult to dissolve after isopropanol precipitation) is reprecipitated with 2 volumes of ethanol. Plasmid DNA is dissolved in 2 ml of CDTA–Tris and can be stored either frozen or at 4° over a drop of chloroform.

5b. An alternative choice of procedures at this point, which avoids steps 5a and 6, involves binding and elution of plasmid DNA from glass powder.[8]

6. In *E. coli* strains containing endonuclease I (*endA*$^{+}$), traces of this enzyme may survive previous treatments. The enzyme is readily eliminated by binding to an ion-exchange resin as follows.[7] Plasmid DNA (2 ml) from the previous step is combined with 2 ml of CG-50 slurry and 0.08 ml of LiCl solution. The mixture is gently shaken at room temperature for 30 min, then centrifuged to recover the supernatant. The pellet is washed with 1 ml of acetate–MOPS. The supernatants are combined, and purified plasmid DNA is precipitated with ethanol and dissolved in 2 ml of CDTA–Tris.

Additional Purification Where Required. Some preparations of plasmid DNA contain what appears to be cell wall carbohydrate material (that is, it gives a positive reaction in a phenol–sulfuric acid test[9]). If present, it can be detected after step 5 by its faintly colloidal appearance or by the present of insoluble material. Much of it can be removed by vortexing the sample for 2 min with a few drops of chloroform and then centrifuging at 12,000 *g* for 15 min at 20°. Plasmid DNA remains in the aqueous phase. Traces of chromosomal DNA can be removed by repeating the alkali denaturing step,[8] by extraction with phenol under acidic conditions,[10] by centrifugation to equilibrium in ethidium bromide/CsCl gradients,[11] or by chromatography on acridine yellow.[12]

Characterization of Plasmid DNA by Agarose Gel Electrophoresis

Hybrid Plasmids Extracted by the Screening Method. Twenty-eight individual clones of hybrid plasmids containing fragments of coliphage T4 DNA inserted into pBR322 were prepared by alkaline extraction as an illustration of the method (Fig. 1). The results are typical of those that may be expected with the screening procedure. Bands of plasmid DNA are detectable in nearly every slot, although the intensity of the individual band varies. Thirty samples can be handled readily on a 20 cm-wide gel. Overall, the mobility of the CCC form of the plasmid should provide a good indication of the size of the inserted fragment.

[8] M. A. Marko, R. Chipperfield, and H. C. Birnboim, *Anal. Biochem.* **121,** 382 (1982).
[9] G. Ashwell, this series, Vol. 8, p. 85.
[10] M. Zasloff, G. D. Ginder, and G. Felsenfeld, *Nucleic Acids Res.* **5,** 1139 (1978).
[11] R. Radloff, W. Bauer, and J. Vinograd, *Proc. Natl. Acad. Sci. U.S.A.* **57,** 1514 (1967).
[12] W. S. Vincent, III and E. S. Goldstein, *Anal. Biochem.* **110,** 123 (1981).

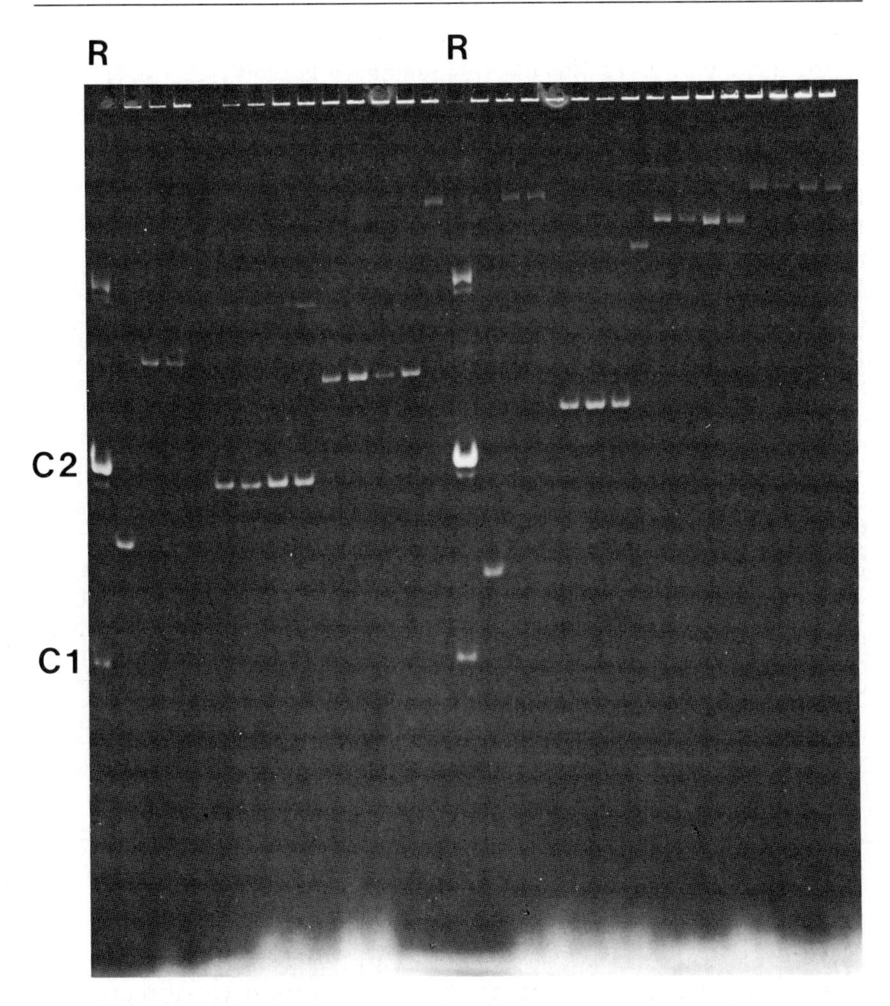

FIG. 1. Screening of coliphage T4–pBR322 recombinant plasmids by gel electrophoresis of alkali-extracted plasmid DNA. A total of 28 clones, representing 7 different hybrid plasmids, was extracted as described in the text, and 10 μl of each extract was applied to a vertical 0.8% agarose gel slab (0.3 × 20 × 20 cm); the electrophoresis buffer contained 40 m*M* Tris base, 20 m*M* sodium acetate, 2 m*M* EDTA, adjusted to pH 7.8 with acetic acid. Electrophoresis was carried out at room temperature for 15 hr with an applied voltage of 55 V. Gels were stained for 30 min at room temperature with ethidium bromide (1 μg/ml in water) and photographed through a Kodak No. 24 filter with 300 nm UV illumination [C. F. Brunk and L. Simpson, *Anal. Biochem.* **82,** 455 (1977)] using Polaroid type 665 pos/neg film. UV lamps were obtained from Fotodyne, Inc., New Berlin, Wisconsin. R is a reference mixture of the monomer and dimer form of pBR322 DNA. Other designations are described in the legend to Fig. 2.

Preparative Method. This version of the method is illustrated by preparing dimeric pBR322 plasmid DNA from a 1-liter culture of amplified cells. Samples were taken after steps 3, 4, and 5 of the purification procedure and are shown in slots A, B, and C, respectively, of Fig. 2. The principal differences between samples that can be seen on the gel are the elimination of intensely staining material (RNA) near the bottom of the gel and a small increase in the amount of open-circular form after ribonuclease treatment in slot C. After step 5, the preparation is virtually free of RNA fragments, as determined by Sephadex G-50 chromatography.[8] Very little contaminating chromosomal DNA can be seen on this agarose gel, but a little more can be expected if nonamplified cells were used instead as a source of plasmid DNA. Contaminating chromosomal DNA is seen more readily in some slots of Fig. 1 as material that remains at the origin or is distributed diffusely near the top of the gel.

Identification of Bands. Plasmid DNA in extracts can exist in a number of different forms that can make bands on a gel sometimes difficult to identify unambiguously. It is necessary to be aware that a plasmid may exist as a dimer or even higher multimer and that DNA can be present in the CCC form (with varying number of superhelical turns), the open-circular (nicked) form, or the linear form. Fortunately, the complexity is not usually a problem, since all these molecular species should give rise to identical fragments after restriction nuclease digestion. The principal band seen in Fig. 2 is the CCC dimer of pBR322. This can be deduced because a small amount of the CCC monomer can be seen, and because both a linear monomer form generated by complete *Hin*dIII digestion and a linear dimer form resulting from partial digestion appear (see Fig. 4). Other slowly migrating bands are presumed to be higher multimers. In addition to the molecular species discussed, a small amount of another form can be generated as a result of the alkaline extraction step. This is the "irreversibly denatured" form, which can be seen as a faint band running ahead of the CCC band in Fig. 2, slot A. It is removed at the next step of purification or on digestion with small amounts of nuclease S1, a single-strand specific nuclease (Fig. 3). The nicked form and a small amount of the linear form are also produced by S1, so this enzyme can be used to help in the identification of minor bands.

Digestion with Restriction Endonuclease. An important next step in the characterization of plasmids is often an examination of the fragments produced by digestion with different restriction enzymes. If a large number of samples is to be digested, such as in screening, susceptibility of plasmid DNA in the crude extract to restriction enzymes is an essential feature of a rapid extraction method. An example of the digestibility of pBR322 at different stages of purification is shown in Fig. 4. Slots A–C

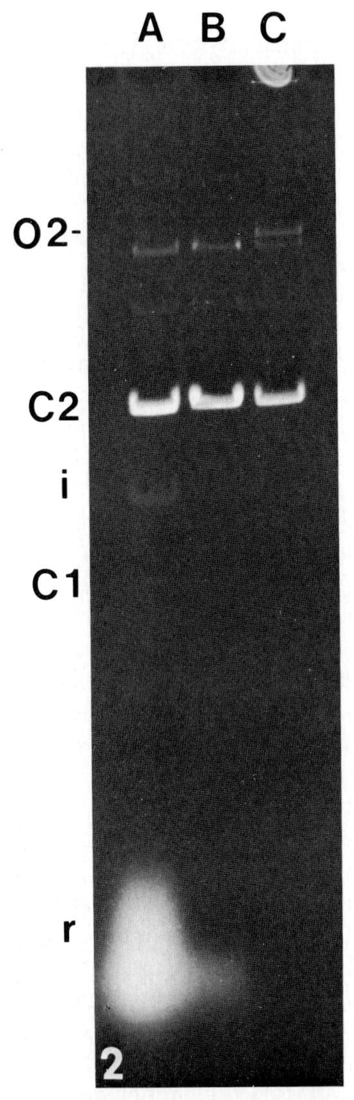

FIG. 2. Purification of dimeric pBR322 DNA by the preparative method. Samples (corresponding to 0.15 ml of culture) were taken after steps 3, 4, and 5 and applied to an agarose gel (slots A–C, respectively). Electrophoresis was carried out as in Fig. 1. The designation of bands in this and other figures is as follows: r, ribosomal and transfer RNA; i, irreversibly denatured form of CCC DNA; C, covalently closed circular form of plasmid DNA; O, open circular form; L, linear form; 1 and 2 refer to the monomeric and dimeric forms.

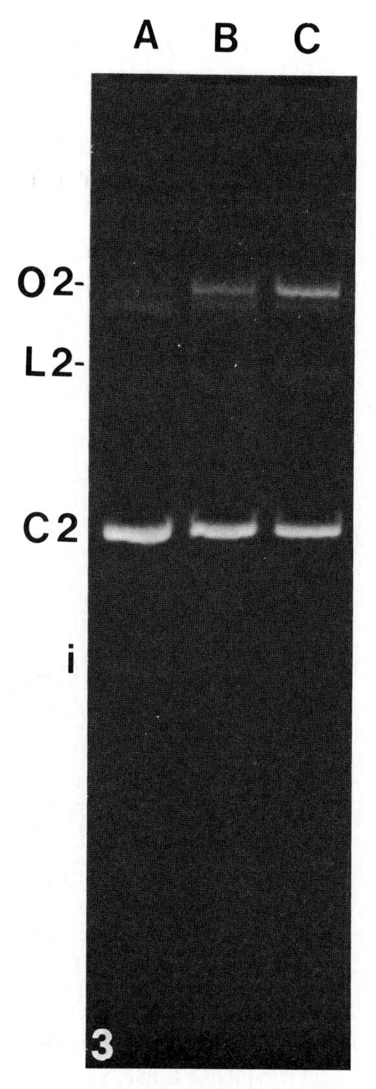

FIG. 3. Treatment of plasmid DNA with nuclease S1. A sample of pBR322 DNA taken after step 3 (see Fig. 2) was treated lightly with nuclease S1 to assist in identifying minor bands. Undigested DNA is in slot A; the effects of treatment for 10 min at 37° with 30 units/ml (slot B) or 60 units/ml (slot C) of nuclease S1 are shown. The enzyme was from Miles Laboratories, and the buffer contained sodium acetate (50 m*M*), $ZnCl_2$ (1 m*M*), pH 4.5; the DNA concentration was approximately 20 μg/ml. On treatment with nuclease S1, the faint band of irreversibly denatured CCC DNA disappeared and the open-circular and linear forms were generated.

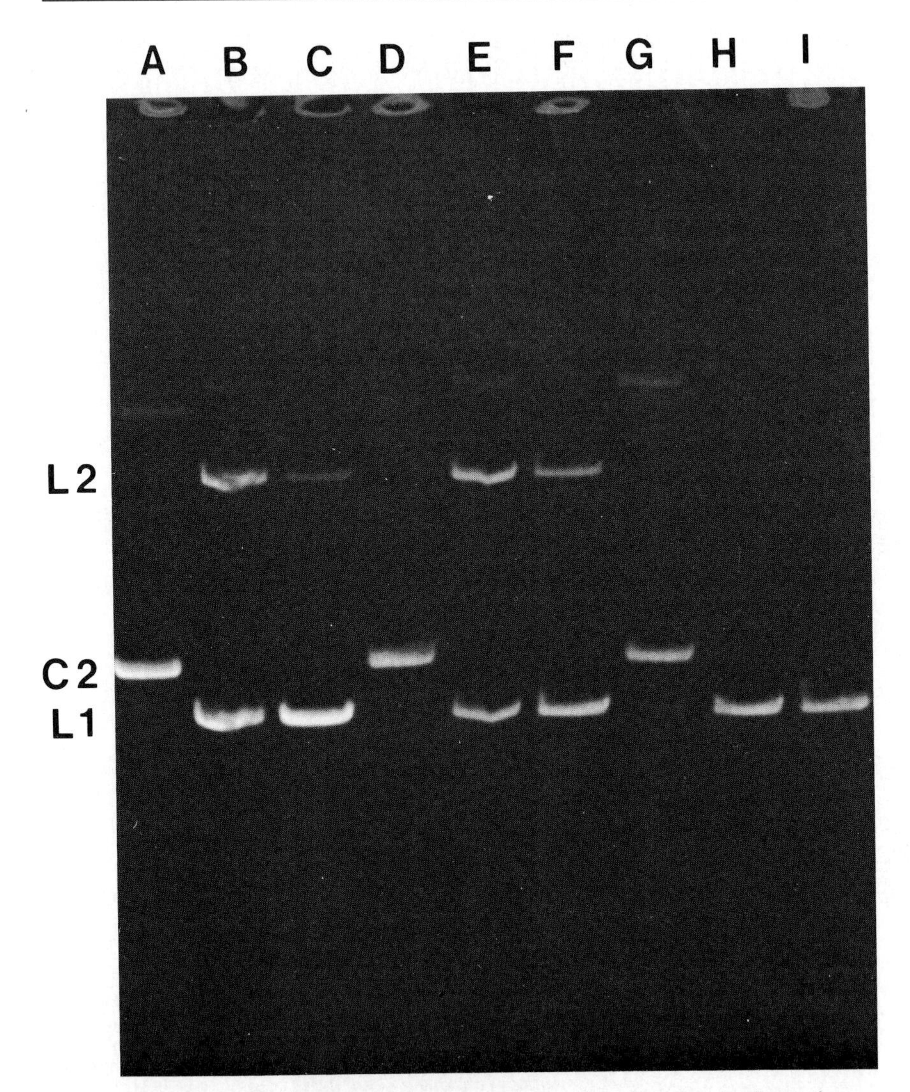

FIG. 4. Treatment of plasmid DNA with restriction nuclease *Hin*dIII. Undigested samples in slots A, D, and G correspond to samples in slots A, B and C, respectively, of Fig. 2. Each sample, taken at successively later stages in the purification procedure, was treated with either 0.35 unit (B, E, H), or 0.70 unit (C, F, I) of *Hin*dIII (from Boehringer) at 37° for 30 min, then analyzed by gel electrophoresis as in Fig. 1. Designation of bands is as in Fig. 2.

contain plasmid DNA that corresponds to the purity obtained in the screening method. The CCC-dimer form is quite readily cleaved to the linear monomer form. Not unexpectedly, highly purified plasmid DNA in slot G is more readily digested than the less pure samples in slots A and D.

Conclusions

Alkaline extraction has proved to be a useful method for the isolation of covalently closed circular DNA from bacterial cells. It is sufficiently simple and reliable to be useful for screening clones containing hybrid plasmids and also for the preparation of more highly purified plasmid DNA.

Acknowledgments

I thank Dr. J. D. Childs for helpful suggestions and discussion and for permitting the use of the experiment shown in Fig. 1, and Dr. E. Young for supplying plasmids of coliphage T4 DNA in pBR322.

[18] Hybridization of Denatured RNA Transferred or Dotted to Nitrocellulose Paper

By PATRICIA S. THOMAS

The technique of Southern[1] for transferring electrophoretically separated DNA fragments to nitrocellulose paper for subsequent hybridization with radioactive RNA or DNA probes has proved to be a powerful tool for the analysis of DNA. Until recently, the only comparable procedure for RNA has been to transfer and covalently couple it to activated cellulose paper (diazobenzyloxymethyl paper, DBM paper) according to the method of Alwine *et al.*[2,3] Although DBM paper does offer the advantage of covalent attachment, in my own hands it also has had several drawbacks. First, I estimate that the amount of RNA in a single band that is

[1] E. M. Southern, *J. Mol. Biol.* **98,** 503 (1975).

[2] J. C. Alwine, D. J. Kemp, and G. R. Stark, *Proc. Natl. Acad. Sci. U.S.A.* **74,** 5350 (1977).

[3] J. C. Alwine, D. J. Kemp, B. A. Parker, J. Reiser, J. Renart, G. R. Stark, and G. M. Wahl, this series, Vol. 68, p. 220.

ISBN 0-12-182000-9

just detectable (after several days of hybridization and autoradiography using a DNA probe with a specific activity of 10^8 cpm/μg), is about 500 pg using DBM paper. Using the same probe under similar conditions I can easily detect 1 pg of DNA per band on nitrocellulose paper. I suspect the increased sensitivity of nitrocellulose is due to its higher binding capacity for DNA (about 80 $\mu g/cm^2$) compared to the capacity of most of my preparations of DBM paper (1–2 $\mu g/cm^2$). Second, I found the preparation and activation of DBM paper to be expensive, time-consuming, often variable, and somewhat hazardous to health.

For these reasons I decided to reinvestigate whether RNA could be bound to nitrocellulose and retained during hybridization and stringent washing procedures. Although early reports claimed that RNA would not bind to nitrocellulose, later work indicated that at least some RNAs [poly(A)$^+$mRNA] could be bound.[4,5] I found that by ensuring proper denaturation any RNA could be bound to nitrocellulose with the same ease and reproducibility as was true for DNA. In the original method[6] I was able to detect about 50 pg of a specific RNA per band. The method has now been extended to detect 1 pg or less of a specific RNA. I also describe a related method for dotting RNA onto nitrocellulose pretreated with high salt. Thus, all the advantages that nitrocellulose blotting previously afforded for DNA analysis are now available for RNA analysis as well.

Principle of the Method

Both RNA and DNA can be bound to nitrocellulose with equal tenacity. However, reproducible binding of RNA requires more careful attention to the manner in which it is denatured. I have tested several denaturants (glyoxal, dimethyl sulfoxide, methylmercuric hydroxide, alkali, and heat) and have found glyoxal and methylmercuric hydroxide to be the most effective in promoting binding. Glyoxal was selected for general use because it is safer to use and because it covalently modifies nucleic acids, forming a stable adduct with guanine at neutral or acidic pH.[7,8] Thus, glyoxylated RNA is kept totally denatured throughout the electrophoresis and transfer steps. Once the denatured RNA is bound to nitrocellulose,

[4] S. V. Lee, J. Mendecki, and G. Brawerman, *Proc. Natl. Acad. Sci. U.S.A.* **68,** 1331 (1971).

[5] G. Brawerman, J. Mendecki, and S. V. Lee, *Biochemistry* **11,** 637 (1972).

[6] P. S. Thomas, *Proc. Natl. Acad. Sci. U.S.A.* **77,** 5201 (1980).

[7] N. E. Broude and E. I. Budowsky, *Biochim. Biophys. Acta* **254,** 380 (1971).

[8] G. K. McMaster and G. G. Carmichael, *Proc. Natl. Acad. Sci. U.S.A.* **74,** 4835 (1977).

the glyoxal groups can be quantitatively removed by treating the blot with pH 8.0 buffer at 100°.

Reagents

Crude glyoxal (40% ethanedial in aqueous solution) was obtained from Matheson, Coleman and Bell (Norwood, Ohio), diluted to 30% (about 4 *M*), and deionized as described by McMaster and Carmichael.[8,9] The crude solution was passed over a mixed-bed ion-exchange resin (Bio-Rad AG 501-X8) until the pH of the purified glyoxal solution (less than 3.0 before purification) reached 5.5–6.0, at which point the glyoxylic acid contaminant in the crude glyoxal solution has been removed. The purified glyoxal is stable for several months stored at −20° in tightly capped tubes; however, it was not reused after exposure to air.

Dimethyl sulfoxide was analytical grade obtained from Mallinckrodt. Formamide (ACS grade, Eastman) was deionized with a mixed-bed resin (AG-501-X8) and stored at −20°. Agarose, medium, was obtained from SeaKem; nitrocellulose paper, 0.45 μm and 0.20 μm, was obtained from Schleicher & Schuell.

[^{32}P]RNA (specific activity 5×10^6 cpm/μg) was prepared as described[10] from a line of *Xenopus laevis* kidney cells labeled for 2.5 days with $H_3{}^{32}PO_4$. Unlabeled RNA was prepared from total nucleic acid isolated using sodium dodecyl sulfate and proteinase K[11] followed by removal of the DNA with DNase I pretreated with proteinase K.[12] Erythrocytes were prepared from adult anemic chickens or from 5-day chicken embryos and fractioned into nuclei and cytoplasm.[13] RNA was isolated from the cytoplasmic fraction using sodium dodecyl sulfate and proteinase K.[11] Glyoxal reacts with protein as well as nucleic acid, and care must be taken thoroughly to deproteinize all samples. Residual protein may cause nucleic acid to stick at the top of the gel during electrophoresis. Since glyoxal treatment causes DNA as well as RNA to bind to nitrocellulose, care must be taken to remove the DNA if only the RNA component is of interest.

The preparation of nick-translated DNA probes (specific activity, 2 to 4×10^6 cpm/μg) was as described by Weinstock *et al.*[14]

[9] G. G. Carmichael and G. K. McMaster, this series, Vol. 65, p. 380.

[10] M. S. N. Khan and B. E. H. Maden, *J. Mol. Biol.* **101,** 235 (1976).

[11] G. S. McKnight, *Cell* **14,** 403 (1978).

[12] H. R. Tullis and H. Rubin, *Anal. Biochem.* **107,** 260 (1980).

[13] S. Weisbrod and H. Weintraub, *Proc. Natl. Acad. Sci. U.S.A.* **76,** 631 (1979).

[14] R. Weinstock, R. Sweet, M. Weise, H. Cedar, and R. Axel, *Proc. Natl. Acad. Sci. U.S.A.* **75,** 1299 (1978).

RNA Transfer Method

Denaturation. RNA is denatured essentially as described by McMaster and Carmichael.[8,9] RNA (up to 10 μg per 8 μl of reaction mixture) is incubated in 1 *M* glyoxal–50% (v/v) dimethyl sulfoxide to 10 m*M* sodium phosphate buffer, pH 6.5–7.0, at 50° for 1 hr (dimethyl sulfoxide is not essential for denaturation or transfer but has been included for increasing the density of samples for easy loading). The reaction mixture is cooled on ice, and 2 μl of sample buffer containing 50% (v/v) glycerol, 10 m*M* sodium phosphate buffer at pH 6.5–7.0, and bromophenol blue is added. The samples are electrophoresed on horizontal 1.1% agarose gels (3 mm thick and 20 cm long) in 10 m*M* phosphate buffer, pH 6.5–7.0. The buffer covers the gel to a depth of about 3–5 mm. RNA (9 S) migrates about 9 cm after electrophoresis at 90 V for 6 hr. Rapid recirculation of the buffer is required throughout electrophoresis to maintain a constant pH: this is critical, since the glyoxal adduct readily dissociates from RNA at a pH of 8.0 or above.[7]

Transfer. Glyoxalated RNA is transferred from agarose gels to nitrocellulose paper using 3 *M* NaCl–0.3 *M* trisodium citrate (20 × NaCl–citrate), essentially as described for transfer of DNA by Southern[1] with some modifications. After electrophoresis the gel (without prior treatment of any kind) is placed over two sheets of Whatman 3 MM paper saturated with 20 × NaCl–citrate. The nitrocellulose paper is wetted with H_2O, laid over the gel, covered with two sheets of 3 MM paper and a layer (5–7 cm thick) of paper towels, a glass plate, and a weight (it is unnecessary to equilibrate the nitrocellulose paper with 20 × NaCl–citrate). Transfer of RNA is essentially complete in 12 hr. I do not treat the gel with alkali to reduce the size of the RNA, because treatment of the gel with alkali and neutralization dramatically reduces the efficiency of transfer of RNAs from the gel to the nitrocellulose paper, particularly for large RNA species (18 S or greater). I also find that presoaking the gel with 20 × NaCl–citrate or staining the gel with ethidium bromide reduces the amount of transfer. I obtain essentially quantitative transfer of RNA if the gel is in very low salt (10 m*M* phosphate), using a transfer buffer that is high salt (20 × NaCl–citrate).

Baking. The blots are dried under a lamp and baked in a vacuum oven for 2 hr at 80°. Baking is required for retention of the RNA on the nitrocellulose. Prewashing the nitrocellulose blot with lower salt (6 × NaCl–citrate) before baking removes most of the bound RNA.

Removal of the Glyoxal Adduct. The baking step, although partially effective in removing the glyoxal adduct from RNA,[6] is not completely effective. Treating the RNA blots after baking with 20 m*M* Tris, pH 8.0 at

100° for 5 min, greatly increased the efficiency of hybridization of the RNA on the nitrocellulose blots. The blots may be added to about 200 ml of 20 m*M* Tris buffer, pH 8.0, at 100° and allowed to cool to room temperature. There is no loss of bound RNA due to this procedure.

Hybridization. I have used the following buffers as described by Wahl *et al.*[15] The prehybridization buffer contains 50% (v/v) formamide, 5 × NaCl–citrate, 50 m*M* sodium phosphate at pH 6.5, sonicated denatured salmon sperm DNA at 250 μg/ml, and 0.02% each of bovine serum albumin, Ficoll, and poly(vinylpyrrolidone). The RNA blots are prehybridized for 8–20 hr at 42°. The hybridization buffer contains 4 parts of the same buffer and 1 part 50% (w/v) dextran sulfate. The nick-translated probes are denatured in H_2O at 100° for 5–10 min, cooled, and added to the hybridization buffer, and the blots are hybridized for about 20–48 hr at 42° in "seal and save" plastic bags. The RNA blots are washed with four changes of 2 × NaCl–citrate–0.1% sodium dodecyl sulfate for 5 min each at room temperature and then washed with two changes of 0.1 × NaCl–citrate–0.1% sodium dodecyl sulfate for 15 min each at 50° (these wash conditions are stringent and largely exclude cross-hybridization of embryonic and adult β-globin chicken mRNAs; thus, lower wash temperatures may be required if cross-hybridization of related RNAs is desired). The blots are wrapped while still damp in Saran Wrap (Dow) and exposed to X-ray film at −70°, using a Cronex Hi-plus intensifying screen (Kodak).

Probe Removal and Rehybridization. Removal of the hybridized probe is accomplished by placing the RNA blot in a glass tray of 100° sterile distilled H_2O for 5–10 min, or the blot may be left in the H_2O as it cools to room temperature. The blot is then prehybridized and hybridized with the desired probe.

Visualization of RNA. I have preferred to run duplicate lanes of RNA samples for staining. RNA may be successfully stained by ethidium bromide after treatment with NaOH and neutralization as described by Alwine *et al.*[3] or may be stained with acridine orange as described by McMaster and Carmichael.[8,9]

RNA Dot Blot Method

RNA (up to 10 μg per 4-μl reaction) is incubated in 1 *M* glyoxal–10 m*M* phosphate buffer, pH 6.5–7.0, at 50° for 1 hr. The reaction mixture is cooled on ice and dilutions are made with sterile distilled H_2O or 0.1% sodium dodecyl sulfate (dimethyl sulfoxide is omitted from the reaction mixture, since it dissolves nitrocellulose). The pretreated RNA samples

[15] G. M. Wahl, M. Stern, and G. R. Stark, *Proc. Natl. Acad. Sci. U.S.A.* **76,** 3683 (1979).

and dilutions are spotted directly onto dry nitrocellulose paper that has been pretreated with H_2O, equilibrated with 20 × NaCl–citrate and dried under a lamp. A uniform small volume (4 μl) of each RNA sample is spotted in one application onto the nitrocellulose. After all samples have been spotted, the blot is dried under a lamp, baked for 2 hr at 80°, treated with 20 m*M* Tris at pH 8.0 for 5–10 min at 100°, and prehybridized and hybridized as described for RNA transfers.

Results and Discussion

Transfer of RNA to Nitrocellulose Paper; Capacity and Sensitivity

The capacity of nitrocellulose to bind RNA was investigated using total RNA from *X. laevis* tissue culture cells labeled with $H_3{}^{32}PO_4$. Figure 1 shows the pattern of [^{32}P]RNA (10^5 cpm, 0.02 μg) in samples containing 0, 0.1, 0.5, 1, 5, or 10 μg of unlabeled RNA (lanes 1–6, respectively), (a) transferred and retained on nitrocellulose paper, or (b) the pattern of [^{32}P]RNA remaining in the gel at the end of transfer. It is apparent from Fig. 1 that the transfer of [^{32}P]RNA in all samples is very efficient (the efficiency of transfer was estimated visually from the autoradiographs to be greater than 95% for all samples). Transfer of RNA is essentially quantitative in 12 hr. Trace amounts of 28 S rRNA are left in the gel only in samples containing the largest amounts of carrier total RNA. Attempts to increase transfer of large RNAs by pretreating the gel with alkali and neutralizing with salt before transfer, resulted in reduction of the amount of transfer of RNA by more than 50%. Not only are the *X. laevis* RNAs transferred efficiently to nitrocellulose paper, but after glyoxal denaturation they are retained on the nitrocellulose paper throughout prehybridization, 2 days of hybridization, and the stringent washes employed during a typical blot hybridization experiment. Visual comparison of autoradiographs from several experiments of [^{32}P]RNA on nitrocellulose paper before and after the typical hybridization and washing procedures showed that greater than 90% of the [^{32}P]RNA remains bound. [An example of retention of [^{32}P]RNA by nitrocellulose is shown in Fig. 3 by dot blots. Compare filter (a) after washing with filter (c) before washing.]

Although large RNAs (9 S or larger) were very efficiently retained by nitrocellulose paper (0.45 μm), smaller RNAs (4 S), were less efficiently retained (50%). The use of smaller pore size (0.1 or 0.2 μm) papers increased retention of 4 S RNA. Because I have assayed mainly rRNA, it is clear that poly(A) is not needed for retention of RNA on nitrocellulose.

In the procedure originally described, I indicated that the glyoxal adduct was probably effectively removed by baking the RNA blots at 80°,

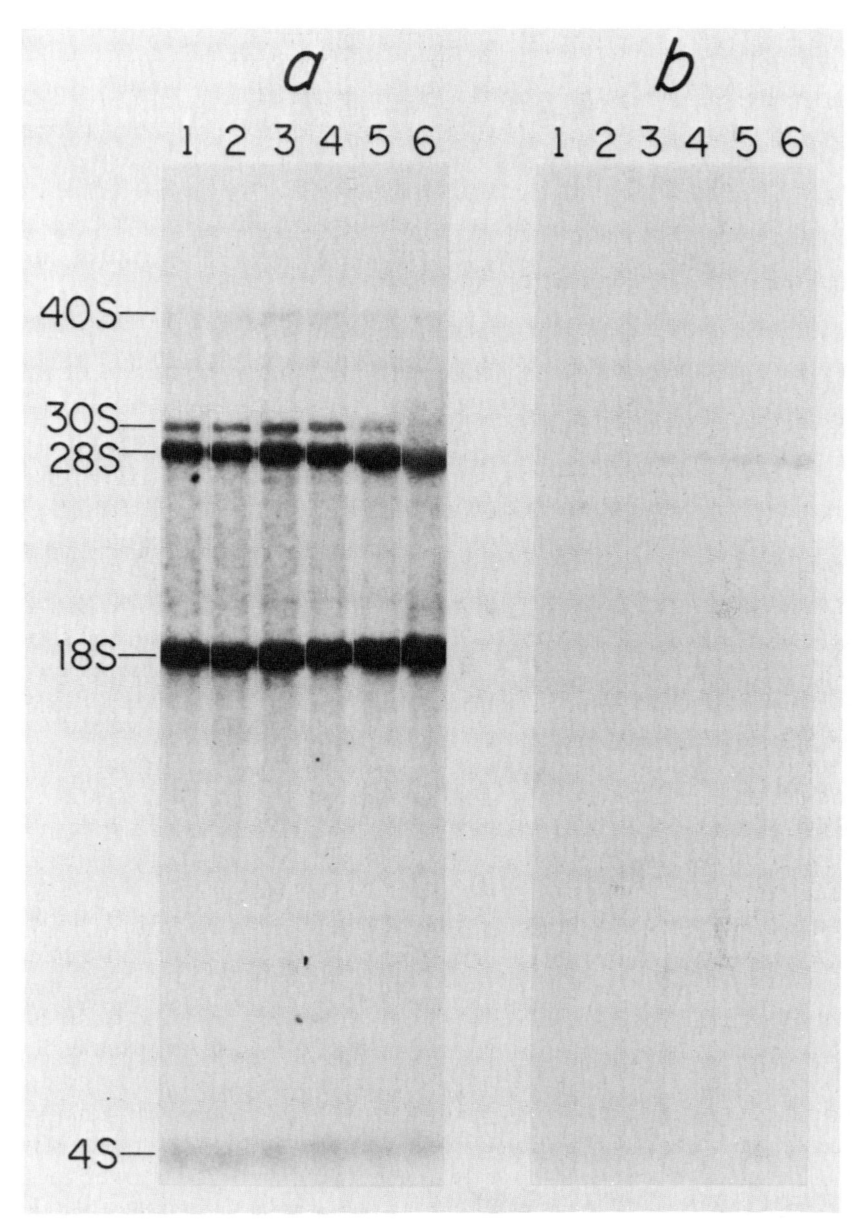

FIG. 1. Capacity of nitrocellulose paper to bind RNA transferred from gels. ^{32}P-labeled total *Xenopus laevis* RNA (10^5 cpm, 0.02 μg) was added to RNA samples (1–6) containing 0, 0.1, 0.5, 1, 5, or 10 μg of unlabeled total *X. laevis* RNA, respectively, denatured with glyoxal, fractionated on a 1.1% agarose gel, and transferred to nitrocellulose paper according to the procedure described in the text. (a) A 2-hr autoradiograph of the nitrocellulose paper after transfer for 12 hr, baking, treating for 10 min in 100° buffer, pH 8.0, prehybridization (12 hr), hybridization (48 hr), and washing. (b) A 2-hr autoradiograph of the gel after 12 hr transfer to nitrocellulose paper.

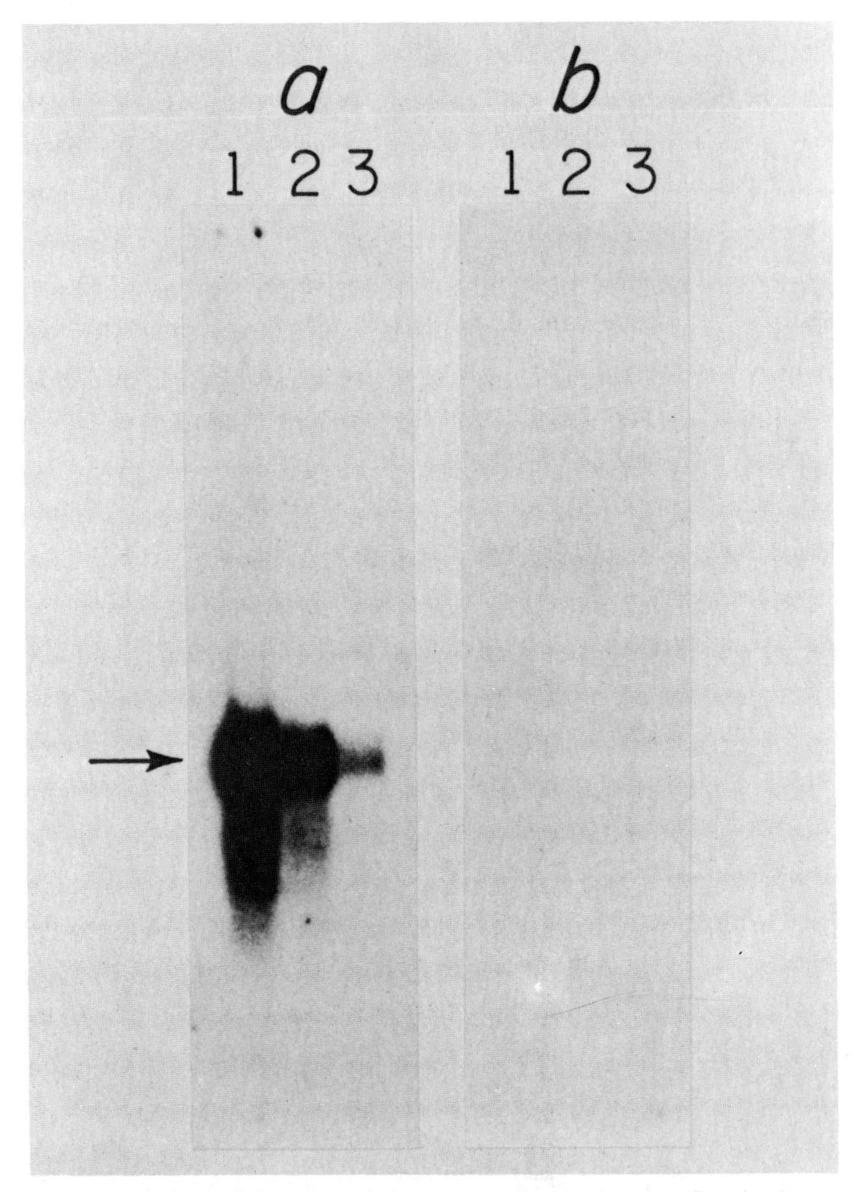

FIG. 2. Sensitivity of RNA sequence detection on nitrocellulose paper. Samples (lanes 1–3) containing 0.025, 0.005, or 0.001 μg, respectively, of total RNA from 5-day chick embryonic red blood cells together with 10 μg of carrier RNA (total RNA from chick embryonic fibroblast cells), were denatured with glyoxal, fractionated on a 1.1% agarose gel, and transferred to nitrocellulose paper. (a) A 20-hr autoradiograph of the nitrocellulose paper hybridized for 48 hr with a nick-translated 0.8 kb DNA fragment containing a portion of the embryonic β-globin gene of clone λCβG-2 [M. Groudine, M. Peretz, and H. Weintraub, *Mol. Cell. Biol.* **1,** 281 (1981)] and washed as described. (b) A 20-hr autoradiograph of the same nitrocellulose paper after removal of the hybridized probe by washing with 100° H_2O for 10 min. The arrow shows the position of 9 S RNA.

since I could detect as little as 50 pg of β-globin mRNA by this procedure.[6] Additionally, prehybridization of the blots for 20 hr at pH 8.0, conditions that should completely remove the glyoxal adduct from RNA,[7] did not increase the hybridization efficiency of the glyoxalated RNA on the nitrocellulose blots. I have since found that the glyoxal adduct is apparently more efficiently removed by treating the RNA blots with 20 m*M* Tris buffer, pH 8.0 (pH 8.0 at room temperature) for 5–10 min at 100° before hybridization. Using these conditions, the hybridization of an embryonic β-globin probe to 0.025, 0.005, and 0.001 μg of total RNA from 5-day embryonic chicken red blood cells is shown in Fig. 2a (lanes 1–3, respectively). Embryonic β-globin mRNA is clearly detectable in as little as 0.001 μg of total erythrocyte RNA; this represents the detection of about 1 pg of β-globin mRNA in the 20-hr autoradiogram. This is a 50-fold increase over the sensitivity originally described[6] and indicates that the same high sensitivity as has been previously possible for detection of specific DNA sequences has been achieved for RNA. The blot shown in Fig. 2a was treated at 100° H_2O for 10 min, to remove the hybridized probe and autoradiographed for 20 hr and shown in Fig. 2b. The probe is efficiently removed by this procedure. I have routinely rehybridized RNA blots with no measurable loss of hybridization signal. The RNA blots may be stored dry at room temperature.

RNA Dot Blots; Capacity and Sensitivity

A nitrocellulose dot blot assay for measuring the relative concentration of a specific mRNA would provide a rapid screening method of much usefulness. The same genomic clones used for the analysis of DNA sequences on Southern blots could then be used for detecting mRNA sequences. I described a dot blue assay[6] that could be used for RNA and DNA preparations. I have further tested the requirements for assaying RNA on nitrocellulose, and report these results here. The capacity of nitrocellulose paper for binding RNA in the dot blot assay[6] was tested using ^{32}P-labeled total *X. laevis* RNA. The [^{32}P]RNA (10^5 cpm, 0.02 μg) with 0, 0.1, 0.5, 1, 5, or 10 μg of unlabeled total *X. laevis* RNA (lanes 1–6, respectively), was denatured (a) with glyoxal or (b, c) denatured by heating to 65° and cooling on ice, and spotted in triplicate to dry nitrocellulose paper that was prepared as described in Methods. The nitrocellulose blots (a, b) were dried, baked, treated at 100° for 10 min, and then compared by autoradiography to the same RNA samples spotted to nitrocellulose (c) and given no further treatment. It is evident from Fig. 3 that nitrocellulose paper has a high capacity for binding denatured RNA (10 μg of glyoxalated RNA or 5 μg of heat-denatured RNA per 4-μl sample). The glyoxa-

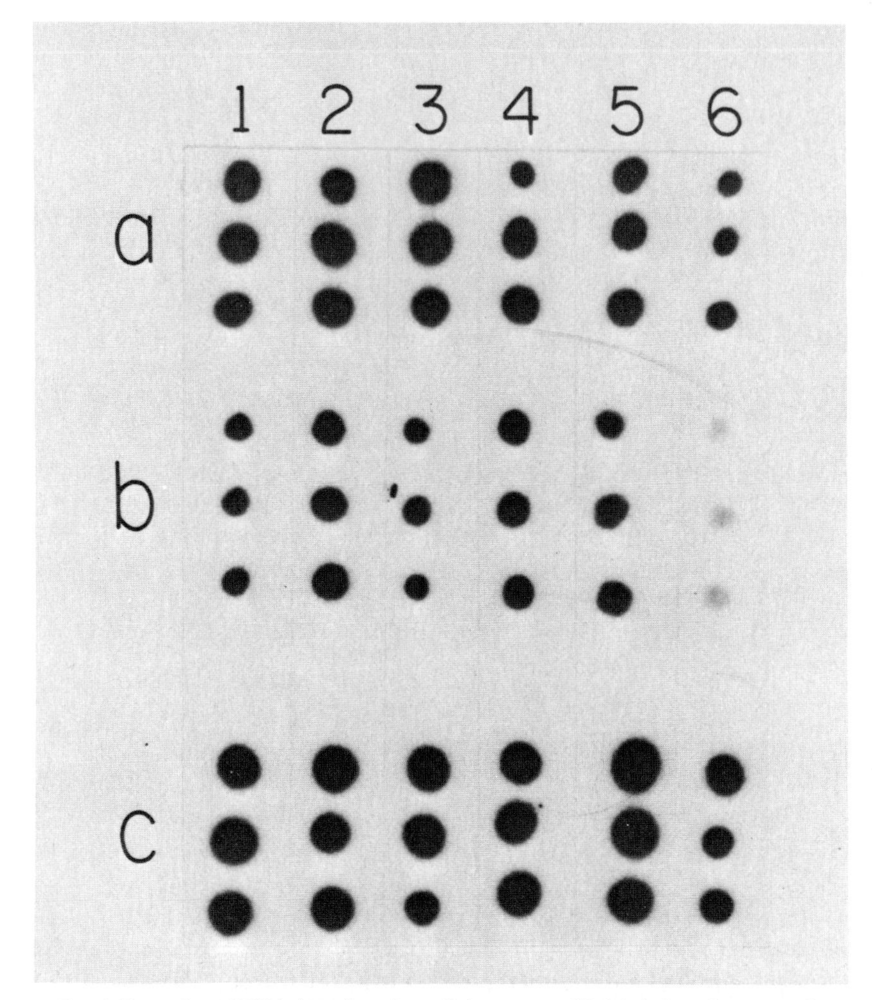

FIG. 3. Retention of RNA dotted to nitrocellulose paper. ^{32}P-labeled total *Xenopus laevis* RNA (10^5 cpm, 0.02 μg) was added to unlabeled RNA samples (1–6) containing 0, 0.1, 0.5, 1, 5, or 10 μg of total *X. laevis* RNA, respectively. The RNA samples were (a) treated with glyoxal, (b) heat denatured and applied in 4 μl, in triplicate, directly to the nitrocellulose paper pretreated as indicated in Methods. (c) A 2-hr autoradiograph of RNA after dotting to nitrocellulose. The nitrocellulose blots (a, b) were further treated for 10 min with 100° Tris buffer (pH 8.0) after baking and autoradiographed for 2 hr.

lated RNA appears to be somewhat more efficiently and uniformly bound than the heat-denatured RNA.

To test the specificity of the dot blot assay for detecting a specific mRNA sequence, I compared the hybridization of *X. laevis* total RNA

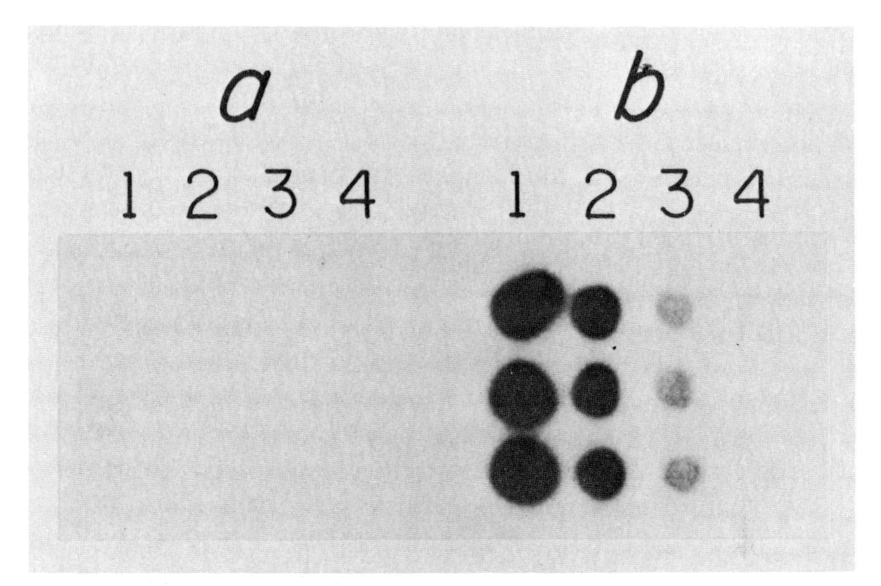

FIG. 4. Sensitivity and specificity of RNA sequence detection using dot blots. (a) Samples of *X. laevis* total RNA (lanes 1–4) containing 10, 1, 0.1, or 0.01 μg of RNA, respectively, in 4 μl, were denatured with glyoxal and dotted to nitrocellulose paper, in triplicate. (b) Samples of 5-day chick embryonic red blood cell total RNA (lanes 1–4) containing 0.1, 0.01, 0.001, or 0.0001 μg, respectively, each with 10 μg of carrier RNA (total chick embryo fibroblast cell RNA), in 4 μl, were denatured with glyoxal and dotted to nitrocellulose paper, in triplicate. The nitrocellulose dot blots (a, b) were hybridized with a nick-translated 0.8 kb DNA fragment containing a portion of the embryonic β-globin gene of clone λCβG-2. [M. Groudine, M. Peretz, and H. Weintraub, *Mol. Cell. Biol.* **1,** 281 (1981)]. A 24-hr autoradiograph of the blot after hybridization (48 hr) and washing is shown.

with that of embryonic total RNA from 5-day chicken red blood cells. The RNA dot blots were hybridized with an embryonic β-globin DNA probe. Figure 4a shows *X. laevis* total RNA (10, 1, 0.1, 0.01 μg, lanes 1–4, respectively) spotted to nitrocellulose paper in triplicate; Fig. 4b shows total red blood cell RNA from 5-day chicken embryos (0.1, 0.01, 0.001, and 0.0001 μg, lanes 1–4, respectively). Although no hybridization of the β-globin DNA probe with as much as 10 μg of *X. laevis* total RNA is detectable, hybridization of the probe to as little as 0.001 μg of total 5-day red blood cell RNA is readily apparent after 24-hr autoradiography. Since the embryonic red blood cell RNA was spotted in 10 μg of carrier RNA, and we estimate that 0.001 μg of total red blood cell RNA represents about 1 pg of β-globin mRNA, this indicates that it is possible to detect a specific mRNA sequence present at only 1×10^{-7} in an RNA preparation using the dot blot assay described here. This assay has been successfully used in several laboratories.[16,17]

Acknowledgment

The author thanks R. Reeder for helpful suggestions on this manuscript. This work was supported by a grant from the National Institutes of Health to Harold Weintraub. P. S. T. is a Postdoctoral Fellow of The National Institutes of Health.

[16] M. Groudine, R. Eisenman, and H. Weintraub, *Nature* (*London*) **292,** 311 (1981).
[17] S. F. Wolf and B. R. Migeon, *Nature* (*London*) **295,** 667 (1982).

[19] Isolation of Multigene Families and Determination of Homologies by Filter Hybridization Methods

By GERALD A. BELTZ, KENNETH A. JACOBS, THOMAS H. EICKBUSH, PETER T. CHERBAS, and FOTIS C. KAFATOS

A high proportion of genes in eukaryotes are now known to be members of multigene families. Although initial indications were provided by protein sequencing, the ubiquity of multigene families has been revealed most convincingly by recombinant DNA methods. A list of some of the best known multigene families, by no means inclusive, would include the genes for histones, globins, immunoglobulins, histocompatibility antigens, actins, tubulins, seed storage proteins, chorion proteins, keratins, collagens, cuticle proteins, yolk proteins, heat shock proteins, and salivary glue proteins. Even genes for proteins that appear homogeneous may belong to multigene families; e.g., the chicken ovalbumin gene is now known to be expressed coordinately with two homologous genes, *X* and *Y*. In the field of hormone research, analysis of gene families related to well known effectors such as insulin or growth hormone is one of the most exciting areas of current investigation. Many members of multigene families show stage- or tissue-specific expression: a familiar example is the family of globin genes. Thus, the study of multigene families is important for understanding the mechanisms of physiological regulation, differential gene expression, and molecular evolution in eukaryotes.

In the study of multigene families by recombinant DNA methods, usually the first step is to isolate clones (cDNA or genomic) that represent all members of the family. The next step is to discriminate among clones that correspond to different genes. The third step is to relate each gene to

ISBN 0-12-182000-9

a particular type of protein product or to a specific RNA transcript. Ultimately, structural characterization of the genes, including sequence analysis, is undertaken. Because of their sequence homologies, members of a multigene family often cross-hybridize. In the first three steps of such studies, careful control of the hybridization conditions is necessary—both to isolate the family members as cross-hybridizing species in the first step, and to permit their discrimination and characterization in the second and third steps. A number of filter hybridization methods are used in these studies: phage plaque[1] and bacterial colony hybridization,[2] dot blot hybridization,[3] Southern hybridization,[4] Northern hybridization,[5] and hybrid-selected translation.[6] Their specific features are discussed elsewhere in this volume. Here, we wish to emphasize their relatedness and to point out how deliberate control of the hybridization stringency maximizes their utility. We shall give examples from our studies of the chorion gene families in silkmoths.

The silkmoth chorion system has been reviewed.[7] More than a hundred genes belonging to several multigene families encode the structural proteins of the chorion. These families may themselves be related, constituting a superfamily,[8] but the interfamily homologies are low and can be neglected in this discussion. Within each family, both closely and distantly related genes are found: observed degrees of mismatching range from less than 1% to as much as 50%.[9] Based on this variation, members of each family have been classified into distinct *types* (distantly related genes, observed mismatching 10 to 50%) and into *copies* of the same gene type (more closely related genes, <1 to 5% mismatching). In our experience, even gene copies differ by one or more substitutions that lead to amino acid replacements, and thus have already taken the first step toward evolving into distinct genes, by the criterion of differences in the encoded proteins. The sequence variations are not uniformly distributed within the chorion genes: in each family the sequence corresponding to a central protein domain is substantially conserved, whereas flanking se-

[1] W. D. Benton and R. W. Davis, *Science* **196,** 180 (1977).

[2] M. Grunstein and D. S. Hogness, *Proc. Natl. Acad. Sci. U.S.A.* **72,** 3961 (1975).

[3] F. C. Kafatos, C. W. Jones, and A. Efstratiadis *Nucleic Acids Res.* **7,** 1541 (1979).

[4] E. M. Southern, *J. Mol. Biol.* **98,** 503 (1975).

[5] J. C. Alwine, D. J. Kemp, and G. R. Stark, *Proc. Natl. Acad. Sci. U.S.A.* **74,** 5350 (1977).

[6] R. P. Ricciardi, J. S. Miller, and B. E. Roberts, *Proc. Natl. Acad. Sci. U.S.A.* **76,** 4927 (1979).

[7] F. C. Kafatos, *Am. Zool.* **21,** 707 (1981).

[8] J. C. Regier, F. C. Kafatos, and S. J. Hamodrakas, *Proc. Natl. Acad. Sci. U.S.A.* (in press).

[9] C. W. Jones and F. C. Kafatos, *J. Mol. Evol.* **19,** 87 (1982).

quences that encode protein "arms" (NH_2-terminal and COOH-terminal domains) are much more variable.[10]

Parameters Affecting Nucleic Acid Hybridizations

A Qualitative Summary

By definition, the course of any hybridization reaction is determined by the concentrations of the reacting species and the second-order rate constant k. The stability of the resulting duplex is measured by the melting temperature, T_m. For reassociation reactions in solution involving perfectly matched complementary strands, the effects of various reaction conditions have been investigated in detail, and thus T_m and k values can be calculated with some precision. Neither calculation can be extended with confidence to cases involving imperfect hybrids or to solid-phase reactions. Nevertheless, the effects of reaction conditions can be expected to be qualitatively similar, and thus they bear review here.

In general, T_m depends on the composition of the duplex (T_m increases with G + C content), on the duplex length (T_m increases with length), on the ionic strength (T_m increases with salt concentration up to a plateau), and on the concentration of any organic denaturants present (e.g., formamide, urea). T_m is higher for RNA : DNA hybrids than for the corresponding DNA : DNA duplexes, and the differential effect of increasing (G + C) is greater for the hybrids.

Parameters such as (G + C), duplex length, ionic strength, and concentration of any denaturants generally affect k in the same direction as they affect T_m. Therefore, an additional parameter, which is of paramount importance for k, is the temperature of the reaction, T_i. In general, k increases with T_i, reaches a broad maximum at 25 to 20° below T_m, and decreases thereafter, becoming severely depressed at $T_i \geq T_m - 5°$. An example of such temperature dependence is shown in Fig. 1, for the reassociation of T4 DNA in solution.[11]

The qualitatively similar effects of most reaction conditions on k and T_m have led to the practice of summarizing the reaction conditions in terms of the *criterion,* which is numerically equal to T_m minus T_i. When the criterion is large, the reaction is described as permissive; when it is small, the reaction is stringent. While the *absolute* values of k vary drastically as a function of individual parameters (e.g., salt), the *relative* values as a function of criterion are thought to be described in general by the

[10] S. J. Hamodrakas, C. W. Jones, and F. C. Kafatos, *Biochim. Biophys. Acta* **700,** 42 (1982).

[11] T. I. Bonner, D. J. Brenner, B. R. Neufeld, and R. J. Britten, *J. Mol. Biol.* **81,** 123 (1973).

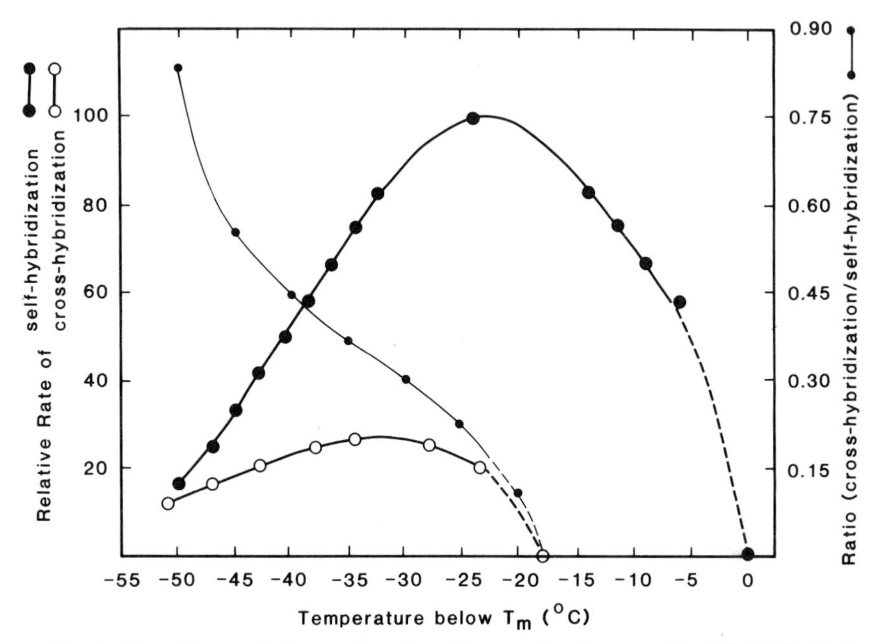

FIG. 1. Rate of reassociation as a function of temperature for normal (self-hybridization) and mismatched DNA (cross-hybridization). Data are replotted from Fig. 3 of Bonner *et al.*[11]; they are derived from normal T4 DNA and T4 DNA partially deaminated with nitrous acid. The large open and filled circles are from the best-fit curve plotted in the original figure, and dotted lines are extrapolations assuming that at T_m the rate of reassociation is zero. Under the conditions of the experiment (0.12 *M* phosphate buffer), the T_m of normal T4 DNA is 81°. The discrimination ratio, i.e., k of cross-hybridization divided by k of self-hybridization, is also presented.

curve of Fig. 1, for the reassociation of perfectly matched strands in solution.[11]

The problem of discrimination among related nucleic acid sequences is one of devising hybridization conditions that are stringent for some sequences and permissive for others. Although it is known that mismatching depresses both T_m and k, its effect on the relationship between k and criterion has not been studied in detail. Figure 1 presents one of the few sets of systematic data available for hybridization between normal T4 DNA and T4 DNA partially deaminated with nitrous acid (resulting in purine–pyrimidine mismatches that lower the T_m value by 18°). The temperature dependence of k for this "cross-hybridization" (T4 DNA vs deaminated T4 DNA) shows some similarity to that for "self-hybridization" (T4 DNA vs T4 DNA), but the rate constant is lower and reaches its optimal value at a lower temperature.

From data such as those of Fig. 1, one may calculate what we will call the "distribution ratio" for each temperature, i.e., the ratio between the rate constants for cross-hybridization and for self-hybridization reactions. The higher the value of this ratio, the easier it is to detect even distant homologs; the lower the ratio, the easier it is to discriminate between self-hybrids and cross-hybrids. It can be seen that the ratio varies with criterion in a sigmoidal manner. Although the exact shape of the sigmoid is not based on extensive empirical data, its general shape is not affected by minor variations in the individual k vs temperature curves: it depends on the bell shape of the curves and the shift of the cross-hybridization curve to lower temperatures.

The sigmoidal curve for discrimination ratio vs temperature has important implications for the study of multigene families. If we imagine a set of such curves, corresponding to different degrees of mismatching, we can understand why it is not possible to find a compromise criterion that would permit both detection of distant homologs and discrimination between close homologs. If both goals are to be attained, the following two-step procedure is necessary.

Step 1. To recover all members of a multigene family, an initial screen should be performed at a very permissive criterion. From Fig. 1, it can be seen that a criterion 40° to 50° below T_m is desirable if very distant homologs are to be recovered: at these temperatures the rate of self-hybridization is reduced two- to six-fold, but even distant homologs give a relatively strong signal. A practical limit is imposed by background problems. For example, if the probes have been cloned by G : C tailing, nonspecific hybridization due to the tails becomes significant at very permissive criteria.

Step 2. To discriminate between homologs, a very stringent criterion should be used in a second hybridization step. Since it appears that the discrimination ratio drops steeply near the T_m of the mismatched hybrids (Fig. 1), we recommend the use of criteria near the T_m of the relevant cross-hybrids. Temperatures as high as 5° below the T_m of the self-hybrids can be used without difficulty.

In theory, these two steps might be compressed into a single two-step experiment: After an initial hybridization under permissive conditions, poorly matched hybrids might be recognized by their release in a subsequent wash under stringent conditions. In practice, although this protocol is useful for discriminating between distant homologs,[3] we have found it not to be useful when the homologs are closely related. In part, the problem is technical. Although probes are for the most part reversibly hybridized, in filter hybridization a variable portion of the probe may become irreversibly bound. This problem can be minimized by scrupulously avoiding drying the filter until after the melting reaction, but the precaution is not always sufficient.

A more important consideration affecting the design of the experiment is that a stringent hybridization and a stringent wash are not entirely equivalent. *A priori* the stability of a hybrid during washing depends on the T_m of the best-matched region of duplex. By contrast k reflects the nucleation frequency, which is an unknown function of the distribution of matched and unmatched stretches. In practice we find that a stringent hybridization is more discriminatory than an equally stringent wash.

Quantitative Considerations

It is clear from the preceding discussion that in the study of multigene families, selection of appropriate conditions requires attention to the parameters that govern T_m and k. We emphasize that systematic empirical data for matrix-supported reactions, and for mismatched sequences, are extremely limited. Thus, the appropriate conditions must be determined empirically in each case. However a starting point can be selected by calculations, and we summarize here the pertinent information from the literature for convenience in doing so.

The rate constant for renaturation of randomly sheared DNA in solution is given by the Wetmur–Davidson equation[12]

$$k = (k_N L^{0.5})/N$$

where L is the mean length of the reassociated duplex per nucleation; N is the complexity of the DNA; and k_N is a length-independent nucleation rate constant. The length factor is quite uncertain, because it attempts to incorporate effects of length on both nucleation frequency and yield of duplex per nucleation. Therefore, the equation can be used only for a rough estimate if the hybridizing strands differ in length. The constant k_N is influenced by such environmental factors as the incubation temperature, the salt concentration, organic solvents, and various polymers.[13] In an aqueous solution at an incubation temperature of $T_m - 25°$ and 1 M Na^+, $k_N = 3 \times 10^5$ liters/mole-second. Attachment of one of the reacting species to a matrix decreases k_N, probably by an order of magnitude.[13a] In addition, if a large amount of DNA is filter-bound, the hybridization may be limited by diffusion of the probe to the filter, unless the reaction vessel is shaken.[13a]

The effect of individual parameters on k is as follows. The change in

[12] J. G. Wetmur and N. Davidson, *J. Mol. Biol.* **31,** 349 (1968).

[13] R. Wieder and J. G. Wetmur, *Biopolymers* **20,** 1537 (1981).

[13a] R. A. Flavell, E. J. Birfelder, J. P. Sanders, and P. Borst, *Eur. J. Biochem.* **47,** 535 (1974).

optimal k at $T_m - 25°$ as a function of salt has been tabulated,[12,14] and is most dramatic below 0.4 M Na^+. An 1% increase in G + C increases[12] optimal k by a factor of approximately 0.018. Mismatching is reported[11,14] to depress the optimal k by a factor of 2 per 10° reduction in T_m. Concentrations of 80% formamide are thought to depress k by a factor of 3 for DNA–DNA hybrids and by a factor of twelve for RNA–DNA hybrids, although the data are limited.[15] Volume-excluding inert polymers can be used to increase k.[13,15a] In an aqueous solution of 1 M NaCl at 70°, the renaturation rate increases 100-fold when dextran sulfate is added to 40%.[13]

Concerning T_m, a large number of quantitative studies can be summarized as follows[11,15,16]: For a perfectly matched DNA duplex,

$$T_m = 81.5 + 0.41\,(G + C) + 16.6 \log(Na^+) - 0.63(\%\ \text{formamide}) - \frac{300 + 2000(Na^+)}{d}$$

where G + C = percentage of guanine + cytosine; Na^+ = molarity of (Na^+) or equivalent monovalent cation; and d = the length of the hybridized duplex in nucleotides.

It is prudent to bear in mind the limits of each term in this formula. The dependence on G + C is accurate over the range 30–75% G + C.[17] The salt dependence[16] is valid for 0.01 to 0.40 M (Na^+), only approximately so at higher (Na^+); T_m is maximal at about 1.0–2.0 M (Na^+). The depression of T_m by formamide[15] is greater for poly(dA : dT) (0.75°/1% formamide) than for poly(dG : dC) (0.50°/1% formamide), and the value used in the formula was derived from human rDNA (58–67% G + C). The effect of polynucleotide length on T_m is somewhat controversial, but the correction listed[14] seems valid over the range 0.05–0.5 M salt. Finally, the formula applies to the "reversible" T_m, e.g., as assayed by optical measurements. The "irreversible" T_m, which is of relevance for autoradiographic detection, is usually higher by 7–10° in aqueous solutions.[18]

For estimating the T_m of RNA–DNA hybrids, the paper of Casey and Davidson[15] should be consulted. The effect of formamide on such hybrids is nonlinear (unlike the effect on DNA–DNA hybrids), leading to the well-known preferential formation of RNA–DNA hybrids at high formamide concentrations. At 80% formamide, 0.3 M Na^+, RNA–DNA hybrids are

[14] R. J. Britten, D. E. Graham, and B. R. Neufeld, this series, Vol. 29, p. 363.

[15] J. Casey and N. Davidson, *Nucleic Acids Res.* **4,** 1539 (1977).

[15a] G. M. Wall, M. Stern, and G. R. Stark, *Proc. Natl. Acad. Sci. U.S.A.* **76,** 3683 (1979).

[16] C. Schildkraut and S. Lifson, *Biopolymers* **3,** 195 (1965).

[17] J. Marmur and P. Doty, *J. Mol. Biol.* **5,** 109 (1962).

[18] K. Hamaguchi and E. P. Geiduschek, *J. Am. Chem. Soc.* **84,** 1329 (1962).

reported to be 20–30° more stable than DNA–DNA hybrids. At 50% formamide, 0.3 *M* Na^+, we find that the difference is approximately 11° in the case of β-globin sequences.[3]

The effect of mismatching on T_m is obviously very important for studies on multigene families. The usual simplification is that T_m decreases by 1° for every 1 ± 0.3% mismatch,[14] reckoned for DNA–DNA duplexes of 40% G + C with randomly placed mismatches.[14] In another study, percentage homologies for BK and SV40 DNA fragments determined by heteroduplex analysis gave the best fit to actual sequence data when 0.5°/1% mismatch was used.[18a] Obviously the distribution of mismatched bases will have a critical determining effect. Consider an extreme example: sequence A differs from sequence B by 50%. If the mismatch is clustered so that extended regions of A and B are identical is sequence, a high T_m will be observed; by contrast, if the mismatch is dispersed so that every second nucleotide differs, no hybridization will occur. The effect of sequence on the hybridization of short oligonucleotides with mismatches has been studied.[19,20] The extrapolation of these very interesting results to real-life problems will require some judgment and additional experimentation. Conditions that permit detection of homologies between various unequally diverged regions of sequenced viral genomes have been investigated.[21] In the chorion system, we have noted that sequences that are highly divergent overall often show a smaller T_m depression than expected.[22] As an example, the B family clones pc401 and pc408 differ by 37.2% overall, but only by 13.6% in the conserved central domain[9]; the change in T_m of the cross hybrids, as determined from dot blots,[22] is approximately 10°.

It must be stressed again that the formulas given above were derived for solution hybridization, and thus are only rough approximations for filter hybridization. For example, it would not be surprising if the effective length of filter-bound hybrids is reduced by steric constraints. In agreement with this possibility, we have the impression that for filter self-hybridizations *k* declines more rapidly than expected from Fig. 1, when the temperature is raised to within 15° to 5° of the experimentally determined T_m. The T_m itself may be slightly depressed for filter-bound hy-

[18a] R. C. Yang, A. Young, and R. Wu, *J. Virol.* **34,** 416 (1980).

[19] R. B. Wallace, M. Schold, M. J. Johnson, P. Bembek, and K. Itakura, *Nucleic Acids Res.* **9,** 3647 (1981).

[20] S. Gillam, K. Waterman, and M. Smith, *Nucleic Acids Res.* **2,** 625 (1975).

[21] P. M. Howley, M. A. Israel, M.-F. Law, and M. A. Martin, *J. Biol. Chem.* **254,** 4876 (1979).

[22] G. K. Sim, F. C. Kafatos, C. W. Jones, M. D. Koehler, A. Efstratiadis, and T. Maniatis, *Cell* **18,** 1303 (1979).

brids. For example, for β-globin DNA–DNA dot hybrids of 51% G + C, 580 bp at 0.3 M Na^+ and 50% formamide, we have observed[3] a T_m of 56°, whereas the formula presented above would predict 60.6° for reversible T_m, and therefore at least 67° for irreversible T_m.

Two commonly overlooked variables in differential hybridization, which affect discrimination among nucleic acid sequences, are the extent of the reaction and the mass ratio of the radioactive probe to the unlabeled, filter-bound nucleic acid. In general, the hybridization of the probe will show the following kinetics when two (or more) filter-bound sequences react in the same bag with a probe in solution

$$dC/dt = -kC^2 - k_f IC - k_i H_i C \tag{1}$$

where C is the concentration of single-stranded probe at time t; k is the reassociation rate constant for the probe sequence in solution; k_f is the reassociation rate constant between the probe and the identical sequence, I, bound to the filter; and k_i is the reassociation rate constant between the probe and the cross-hybridizing sequence H_i, bound to the filter.

The kinetics will differ significantly, depending on whether the filter-bound nucleic acid or the probe is in excess, and in the latter case whether or not the probe can itself reassociate. For simplicity, let us assume that all filter-bound sequences are equal in mass ($I = H_i$), and that $k_f = k$. (In fact $k_f < k$, but the exact magnitude of the difference is not known, and it does not change the results qualitatively.) Then, if the filter-bound nucleic acid is in vast excess, for example, in typical dot blots,[3] the hybridization reactions follow pseudo-first-order kinetics; when they go to completion, the fraction of the probe hybridized to sequence i becomes

$$k_i \Big/ \sum_{i=1}^{m} k_i \tag{2}$$

where m is the number of filter-bound sequences.

At any time during the reaction, the ratio of the amount of probe hybridized to sequence i relative to sequence j is given by k_i/k_j (Fig. 2); i.e., the ratio is invariant with respect to time. In this case, discrimination between self- and cross-hybrids is not affected by the extent of the reaction (Fig. 2), because all the filter-bound sequences continuously compete for the same, limiting probe. (Discrimination will be affected if different samples are hybridized separately, but that situation does not normally apply.)

If the radioactive probe is in excess over the filter-bound sequences (e.g., typical Northern blots,[5] genomic Southern blots,[4] and initial screens of genomic or cDNA libraries), discrimination will generally decline as

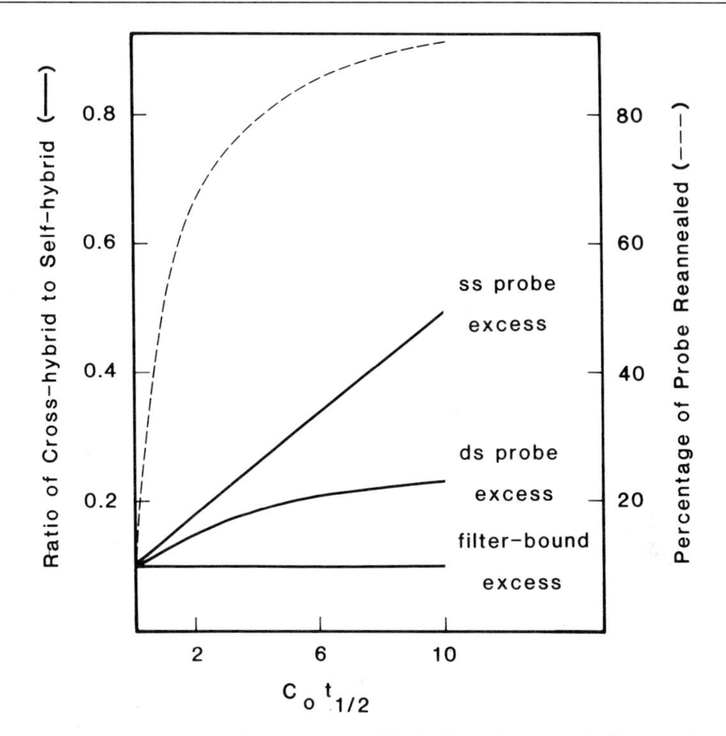

FIG. 2. Effective discrimination between self-hybrids and cross-hybrids, as a function of the extent of the reaction. The solid line shows the ratio of the amounts of probe hybridized to filter-bound heterologous and homologous sequences, $E_i(t)/E(t)$, for each of the three cases discussed in the text: filter-bound sequence in excess, double-stranded (ds), denatured probe in excess, and single-stranded (ss) (e.g., M13) probe in excess. The discrimination ratio, k_i/k, is assumed to be the same in all cases, 0.1. The dashed line shows normal C_0t kinetics of denatured, double-stranded DNA reannealing in solution.

the reaction proceeds. If the probe in solution can self-anneal, the relevant equation is

$$E_i(t)/E(t) = [1 - (1 + kC_0t)^{-n}]\,[1 - (1 + kC_0t)^{-1}]^{-1} \tag{3}$$

where $E(t)$, $E_i(t)$ = extent of the reaction, i.e. the fraction of filter-bound DNA that has hybridized to the probe, for the self-hybrid and cross-hybrid i, respectively; C_0 = concentration of single-stranded probe at time 0; $n = k_i/k$, i.e., the discrimination ratio.

It can be seen that the effective discrimination, i.e., extent of cross-hybridization over self-hybridization, equals the discrimination ratio k_i/k very early in the reaction and deteriorates with the kinetics described above and as shown in Fig. 2.

If the probe in solution cannot self-anneal (e.g., M13 probes), the effective discrimination is given by

$$\frac{E_i(t)}{E(t)} = [1 - e^{-k_i C_0 t}]\,[1 - e^{-k C_0 t}]^{-1} \quad (4)$$

The parameters are defined above. Effective discrimination again equals k_i/k very early in the reaction, but deteriorates very rapidly (Fig. 2).

In summary, hybridizations should ideally be done for very short times: discrimination among various clones will then be maximal, and the same under conditions of either probe excess or filter-bound DNA excess. In practice, however, short hybridization times may not generate a sufficient autoradiographic signal. Thus, to distinguish among related sequences, it is best to use conditions of filter-bound DNA excess, as in typical dot blots[3] (see below).

A Strategy for Systematic Analysis of Multigene Families

Recovery of the Family

To begin the study of a multigene family for which a probe already exists, we recommend that a sublibrary be prepared by screening the appropriate library (cDNA or genomic) at a very permissive criterion [$k_i \simeq k$, $n = 1$ in Eq. (3)], using the Grunstein–Hogness[2,23] or Benton–Davis[1] procedures. At this stage we do not attempt to achieve discrimination, but only to recover as complete a collection of family members as possible; as discussed above, these two goals cannot be pursued simultaneously. A criterion of 30–35° below T_m and high salt (e.g., 0.6 *M* NaCl) for speeding up the hybridization reaction are recommended. To avoid unnecessary increase in the background, which is inherently high in the initial screening, the criterion should not be relaxed more than is necessary for the particular family. For completeness of the sublibrary, even very weakly hybridizing clones should be collected—especially since the plaque or colony size is frequently not uniform.

Such an initial screen will usually produce a mixture of true and false positives, which are distinguished by rescreening. Typical sublibraries that we have recovered in our study of chorion genes include from 200 to 600 clones, too many to make isolation of DNA from each clone practical. For sublibraries using bacterial plasmids as vectors, bacterial cultures are arranged in Microtiter plates, replica plated onto a nitrocellulose filter, placed on nutrient agar, and rescreened by the method of Grunstein and Hogness.[2] We have developed a similar replica plating method to rescreen

[23] M. Grunstein and J. Wallis, this series, Vol. 68, p. 379.

bacteriophage λ genomic sublibraries (see below). An example of the power and reproducibility of this method is shown in Fig. 3. Part a of Fig. 3 shows a rescreening of a genomic sublibrary under the same hybridization conditions as for the initial screen of the total library. This sublibrary had been selected to include genes closely homologous to the cDNA clone pc401, a B-family chorion sequence from *Antheraea polyphemus*.[9,22] APc110, a genomic clone containing two copies of chorion gene 401, and APc173, another genomic clone containing two copies of gene 10, a distantly related B-family sequence, were included as positive and negative controls.[9] A few blank spots were seen, corresponding to false positives of the initial screen, and a few clones grew poorly. Among the rest, some intensity differences were evident. However, even the negative control (APc173), showed significant hybridization under these conditions.

The next step is plaque or colony hybridization under stringent conditions. Figure 3b shows typical results, for the same sublibrary as in Fig. 3a. In this case, stringency was enhanced by a 15° increase in temperature and by use of a more specific probe. Instead of the complete pc401 DNA (565 bp), the probe contained a 257 bp fragment from the region corresponding to the 3′ end of pc401 mRNA. This probe encodes only a small portion of the conserved central protein domain, and consists largely of the more variable COOH-terminal arm sequence and nonconserved 3′ untranslated region.[9] Only 122 bp of this 3′-specific probe is well conserved within the B family (9% mismatch relative to pc10), whereas the entire pc401 probe includes a total of 255 bp of well conserved sequence. Clearly, the experiment of Fig. 3b achieved high discrimination. Hybridization with APc173 and with many of the clones in the sublibrary was essentially undetectable; only a few clones hybridized as intensely as APc110 itself, and others could be assigned to 3 or 4 classes on the basis of their degree of hybridization.

We perform plaque hybridizations to bacteriophage λ genomic sublibraries by the following procedure.

Preparation of Filters. Bacterial lawns are prepared by mixing 0.25 ml of 10 m*M* $MgCl_2$–10 m*M* $CaCl_2$, 0.5 ml of fresh bacterial overnight culture, and 6.5 ml of medium in 0.8% agarose. The mixture is poured over bottom agar (medium with 1.2% agar) in 150-mm petri dishes. Care should be taken to avoid air bubbles. We find that the bottom agar should be poured at least 2 days in advance to avoid condensation during phage growth. Allow the top agar mixture to gel for 10 min at room temperature. Using a replica plater,[24] transfer a small aliquot of each phage stock from

[24] M. Brenner, D. Tisdale, and W. F. Loomis, *Exp. Cell Res.* **90,** 249 (1975).

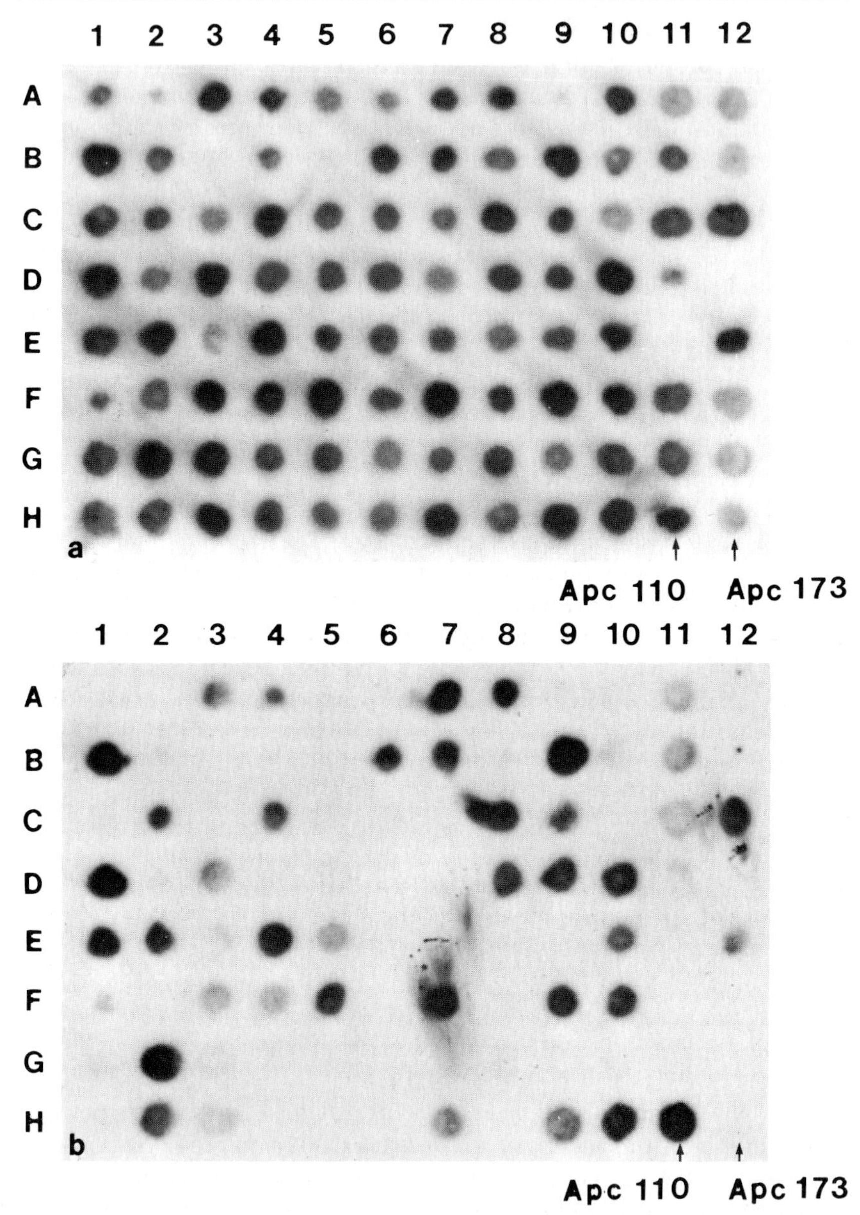

FIG. 3. An *Antheraea polyphemus* genomic DNA library was screened as described by Benton and Davis[1] with a combination of two chorion cDNA probes, pc18 and pc401. (The genes encoding chorion protein 18 are members of the A family and are each paired with a B-family gene encoding chorion protein 401.) Plaque-purified phage stocks from all positive clones were prepared. Aliquots of each stock were delivered into Microtiter wells and plated

Microtiter plates to the bacterial lawn. Allow the liquid to be absorbed into the agar (about 30 min), invert the plates, and incubate overnight at 37°. Uniform plaques of about 6 mm should develop. Chill the plates at 4° for 10–30 min before placing a sheet of dry nitrocellulose directly on the plaque-containing lawn. Precut nitrocellulose circles are commercially available, but we find it more economical and convenient to use rectangular sheets cut to 8.5 × 12.5 cm. Allow the DNA to transfer for 60 to 90 sec, remove the filter from the plate, and denature and fix the DNA to the nitrocellulose by soaking the filter in 0.1 *M* NaOH, 1.5 *M* NaCl for about 30 sec. Neutralize the filter by soaking in 0.5 *M* Tris-HCl, pH 7.5, 0.5 *M* NaCl for about 30 sec. Blot the excess buffer on 3 MM paper, air-dry, and bake under vacuum for 2 hr at 80°. We find that the same plate can be used to prepare at least four filters with no apparent loss in signal.

Prehybridization. Filters are prehybridized in 2–4× SET buffer (1× SET buffer is 0.15 *M* NaCl, 0.03 *M* Tris-HCl, pH 8, 1 m*M* EDTA), 10× Denhardt's solution, 0.2% SDS, and 50 μg/ml herring sperm DNA for at least 3 hr at the hybridization temperature.

Hybridization and Washing. Filters are hybridized in prehybridization mixture supplemented with the radioactive probe and dextran sulfate to a final concentration of 10%. Although we normally hybridize overnight (12–16 hr), we find that the signals are sufficiently strong after hybridizing for only 5 hr. Filters are washed for 20 min each in two changes of 2× SET, 0.2% SDS and two changes of 1× SET, 0.1% SDS at the hybridization temperature.

Discrimination by Dot Hybridization

Once experiments such as those of Fig. 3b have focused attention on a limited number of clones, more refined analysis can be undertaken by dot-blot hybridization. The original paper describing the method[3] should be consulted, as well as a more recent review.[25] In this procedure, DNAs are purified from the clones of interest and immobilized on filters in precisely

[25] F. C. Kafatos, G. T. Thireos, C. W. Jones, S. G. Tsitilou, and K. Iatrou, *in* "Gene Amplification and Analysis," Vol. 2: "Structural Analysis of Nucleic Acids" (J. G. Chirikjian and T. P. Papas, eds.), p. 537. Elsevier/North-Holland, Amsterdam, 1981.

onto a lawn of bacteria. DNAs from the resulting plaques were transferred to nitrocellulose and hybridized to: (*a*) nick-translated pc401, at 65°, 0.6 *M* NaCl; (*b*) a nick-translated *Kpn*I fragment of pc401 representing the 3′ half of the mRNA, at 80°, 0.3 *M* NaCl. In each case hybridized filters were washed at the hybridization temperature in 0.3 *M* NaCl and 0.15 *M* NaCl. APc110 (a genomic clone containing two copies of 401) and APc173 (a genomic clone containing two copies of the distantly related, B-family chorion gene 10) were used as positive and negative controls.

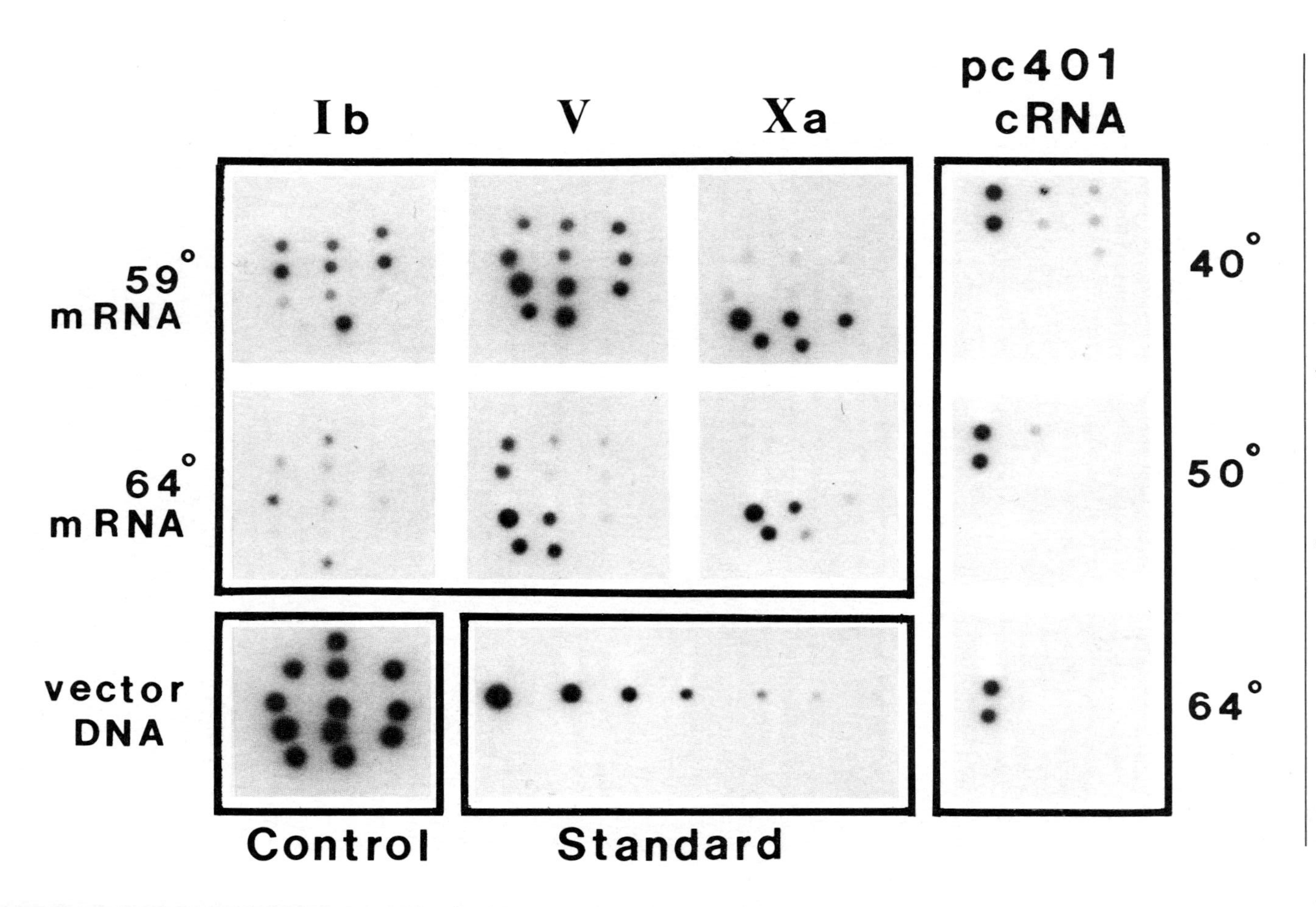
Ib
V
Xa
pc401
cRNA
59°
mRNA
64°
mRNA
vector
DNA
40°
50°
64°
Control
Standard

equal amounts and in dots of uniform diameter. As a result, after hybridization the intensity of the various dots can be estimated semiquantitatively, by visual comparison to standards consisting of a dilution series of radioactive DNA directly spotted on a similar filter.

Figure 4 shows examples of dot-blot hybridizations.[22] The panel labeled "Standard" is a 2-fold dilution series that spans a 64-fold range of radioactivity. The panel on the right exemplifies the discrimination that can be achieved between homologs by judicious choice of conditions. The experiment was performed in 50% formamide, and pc401 cRNA was used as the probe. At 40°, all seven members of the B family included in this filter hybridized, albeit to different extents; five clones of the A family were also present in the filter, but did not detectably hybridize. At 50° and 64° two dots hybridized intensely and almost equally; one was pc401 itself (top) and the other pc602, which differs from pc401 by less than 1% mismatching. Weak hybridization with pc10 was observed at 50° (about 10-fold less than the self-hybrid), but none at 64°.

The remainder of Fig. 4 exemplifies a different, widely applicable use of dot-blot hybridization: assaying the concentrations of various sequences in a series of samples. Such assays are enormously more convenient and more sensitive than assays by liquid hybridization and, by careful attention to the conditions, are only slightly less accurate. The example shown permitted classification of 12 chorion cDNA clones into "early," "middle," and "late" developmental classes, depending on their preferential hybridization to mRNA preparations from the corresponding developmental stages (Ib, V, and Xa, respectively). In this case, the cloned DNAs were fixed to the filter and the labeled probes were mRNAs containing these sequences in unknown amounts. It is also possible to attach the unknown samples (either DNA or RNA) to the filter and probe them with a specific cloned sequence; examples of this approach are given by

FIG. 4. Discrimination of homologous sequences by dot hybridization.[22] *Top left:* Changing concentrations of specific chorion RNA sequences during development. Poly(A)-containing cytoplasmic RNA from *Antheraea polyphemus* follicles at stages Ib, V, or Xa was isolated, end-labeled with ^{32}P after alkali treatment, and hybridized to replicate filters containing equal amounts of spotted DNA from 12 different chorion cDNA clones. Hybridizations were performed in 50% formamide, 0.6 *M* NaCl for 28 hr (59°) or 40 hr (64°). As a control (*lower left*) one filter was hybridized with ^{32}P labeled pML-21 DNA, the vector used in cloning. *Right:* Cross-hybridization of chorion cDNAs. DNAs from 12 chorion cDNA clones (7 from the B family and 5 from the A family) were again spotted on nitrocellulose. ^{32}P-labeled 401 cRNA was prepared using *E. coli* RNA polymerase and the nuclease S1 excised insert of pc401. Hybridizations were performed in 50% formamide, 0.6 *M* NaCl at the temperatures indicated. *Standard:* A two-fold dilution series of [^{32}P]DNA spotted directly on nitrocellulose.

Weisbrod and Weintraub.[26] In the case of multigene families, it is important to distinguish the concentrations of identical vs related sequences. In Fig. 4, this was accomplished by performing the experiment at both 59° and 64°.

Exactly the same approach can be used to characterize members of the multigene families by hybrid-selected translation[6]: one hybridizes DNA dots under discriminating conditions with "probe" consisting of unlabeled mRNA and then melts the hybrids and translates the RNA in a cell-free system by standard procedures. If hybridization is performed under conditions that permit formation of some cross-hybrids, this technique can be used to detect the degree of homology between protein species for which sequence information is not available.[27,28]

An apparatus is now available from Bethesda Research Laboratories and Schleicher & Schuell for performing dot blots with multiple samples stored in Microtiter plates. Dot blots can also be performed manually, and we recommend the following procedure.[25]

Preparation of Filters. All filters are washed in water for 1 hr. A plain nitrocellulose filter (22 mm in diameter) is mounted on top of two nitrocellulose filters with grids, on a sintered-glass platform connected to a water aspirator. The filters are washed 3–4 times with 1 *M* ammonium acetate before spotting with DNA.

Plasmid DNA is linearized by restriction endonuclease treatment and digested with proteinase K (200 μg/ml in 50 m*M* Tris-HCl, pH 7.5; at 37°, 30 min each with 0.2% and 2% SDS). After phenol extraction the DNA is denatured in 0.3 *N* NaOH for 10 min and chilled; when needed, it is diluted with an equal volume of cold 2 *M* ammonium acetate to a concentration of 1.4 μg/ml. It is then taken up in a capillary pipette attached to a micropipette filler (Clay Adams suction apparatus No. 4555), and is spotted on the filter under light vacuum; once the pipette touches the filter, contact is maintained continuously, while the DNA solution is delivered slowly with a combination of vacuum suction and positive pressure from the filler. We routinely spot 0.7 μg of DNA in 50 μl per dot, using a 100-μl micropipette. The dot is washed with a drop of 1 *M* ammonium acetate, as is the area to be spotted next. After all the dots are made, the filter is washed again with 1 *M* ammonium acetate under suction, air dried, treated with 2× Denhardt's solution for 1 hr, drained, air-dried, and baked at 80° for 2 hr.

Prehybridization. The dried filters are wetted evenly by slow immersion in 10× Denhardt's–4× SET buffer and are shaken in that solution for

[26] S. Weisbrod and H. Weintraub, *Cell* **23,** 391 (1981).

[27] N. K. Moschonas, Ph.D. Thesis, Univ. of Athens, Athens, Greece, 1980.

[28] G. Thireos and F. C. Kafatos, *Dev. Biol.* **78,** 36 (1980).

at least 1 hr. They are transferred to sterile, siliconized scintillation vials containing blank hybridization mixture, e.g., 50% deionized formamide, 2× Denhardt's solution, 4× SET, 0.1% SDS, 100 μg of yeast tRNA and 125 μg of poly(A) per milliliter, and are incubated for at least 1 hr at the hybridization temperature.

Hybridization and Washing. The filters are hybridized for 16–48 hr in hybridization mixture (prehybridization mixture supplemented with radioactive probe). The temperature is selected to give the desired criterion of stringency. Washes are performed at the same temperature, twice each with 4×, 2×, 1×, 0.4×, and 0.2× SET, all with 0.1% SDS. Two final washes are performed at room temperature, with 0.1× SET without SDS. The filters are covered with Saran wrap and autoradiographed, either dry or moist (if melts are to be undertaken).

Discrimination by Southern and Northern Analysis

Discrimination between members of a multigene family, by careful control of hybridization conditions and selection of appropriately specific probes, can also be accomplished in Southern and Northern transfer experiments. While dot hybridization is preferable for quantitation and is somewhat more convenient, Northern analyses permit detection of the sizes of the hybridizing transcripts (which may vary during development), and Southern experiments using genomic DNA clones can identify restriction fragments that contain the cross-hybridizing sequences. Thus, these three filter hybridization procedures are complementary, and all are apt to be used in the study of a multigene family.

Figure 5 shows an example of discrimination by low- and high-stringency Southerns. The probe in this case was the high-cysteine cDNA clone m2574 of *Bombyx mori,*[29] and the filter-bound DNAs were derived from 18 overlapping genomic clones (B1 to B18) totaling 270 kb of continuous DNA spanning the high-cysteine region of the chorion locus.[30] Fifteen m2574-like genes were detected by low-stringency hybridization (75°, 0.6 *M* NaCl). High-stringency hybridization permitted identification of gene 12 as the one from which m2574 was derived, genes 14 and 15 as very close homologs, genes 8 and 10 as moderately close homologs, and the remaining 10 genes as more distant homologs. In this case, high stringency was accomplished by high temperature (85°, 0.3 *M* NaCl) and by the use of a relatively high concentration of DNase during nick-translation of the probe. The high DNase approach is recommended for maximizing

[29] K. Iatrou, S. G. Tsitilou, and F. C. Kafatos, *J. Mol. Biol.* **157,** 417 (1982).
[30] T. H. Eickbush and F. C. Kafatos, *Cell* **29,** 633 (1982).

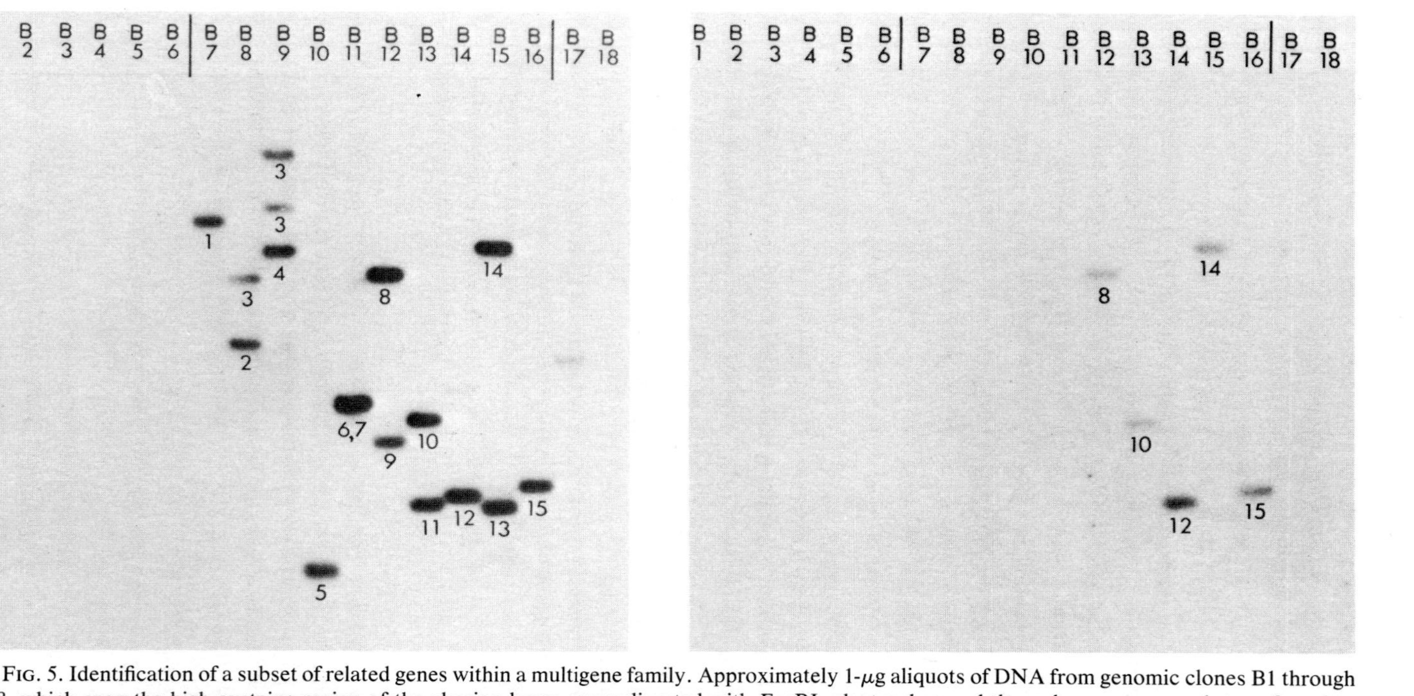

FIG. 5. Identification of a subset of related genes within a multigene family. Approximately 1-μg aliquots of DNA from genomic clones B1 through B18, which span the high-cysteine region of the chorion locus, were digested with *Eco*RI, electrophoresed through an agarose gel, transferred to nitrocellulose paper by the Southern method,[4] and hybridized with nick-translated cDNA probe m2574. The low-stringency hybridization (*left*) was at 75°, 0.6 *M* NaCl with the probe nick-translated in the presence of 5 ng of DNase I per milliliter. The high-stringency hybridization (*right*) was at 85°, 0.3 *M* NaCl, with the probe nick-translated in the presence of 125 ng of DNase I per milliliter. The numbers under each hybridized fragment refer to the m2574-like genes as numbered in Eickbush and Kafatos.[30]

discrimination when sequence information is not available, and therefore specific probes cannot be tailored as in Fig. 3b. High DNase results in short probe lengths, and, as a result, if mismatching is nonrandomly distributed, only the short fragments that might contain a conserved sequence will cross-hybridize at stringent conditions. By contrast, with long probes, signals from hybrids that are stable because of a short conserved sequence will be enhanced by the presence of covalently attached unpaired tails, resulting in decreased discrimination.

In conclusion, deliberate control of the stringency of hybridization is possible in all filter hybridization procedures and is a powerful tool in the isolation and characterization of multigene families.

Acknowledgments

Original work was supported by grants from NIH, NSF, and the American Cancer Society.

[20] Synthesis of ds-cDNA Involving Addition of dCMP Tails to Allow Cloning of 5′-Terminal mRNA Sequences

By HARTMUT LAND, MANUEL GREZ, HANSJÖRG HAUSER, WERNER LINDENMAIER, and GÜNTHER SCHÜTZ

Cloning of mRNA sequences after reverse transcription into cDNA copies[1–3] plays an important role in the analysis of gene structure and function. A variety of methods for cloning cDNAs in bacterial plasmids has been described.[4–10] In the method most commonly used,[7–9] the ability

[1] I. Verma, G. F. Temple, H. Fan, and D. Baltimore, *Nature (London) New Biol.* **235,** 163 (1972).

[2] D. L. Kacian, S. Spiegelman, A. Bank, M. Terada, S. Metafora, L. Dow, and P. A. Marks, *Nature (London) New Biol.* **235,** 167 (1972).

[3] J. Ross, H. Aviv, E. Scolnick, and P. Leder, *Proc. Natl. Acad. Sci. U.S.A.* **69,** 264 (1972).

[4] F. Rougeon, P. Kourilsky, and B. Mach, *Nucleic Acids Res.* **2,** 2365 (1975).

[5] T. H. Rabbits, *Nature (London)* **260,** 221 (1976).

[6] K. O. Wood and J. C. Lee, *Nucleic Acids Res.* **3,** 1961 (1976).

[7] T. Maniatis, S. G. Kee, A. Efstratiadis, and F. C. Kafatos, *Cell* **8,** 163 (1976).

[8] R. Higuchi, G. V. Paddock, R. Wall, and W. Salser, *Proc. Natl. Acad. Sci. U.S.A.* **73,** 3146 (1976).

[9] F. Rougeon and B. Mach, *Proc. Natl. Acad. Sci. U.S.A.* **73,** 3418 (1976).

[10] S. Zain, J. Sambrook, R. J. Roberts, W. Keller, M. Fried, and A. R. Dunn, *Cell* **16,** 851 (1979).

ISBN 0-12-182000-9

of single-stranded cDNA to form a hairpin-like structure at its 3′ end[9,11,12] is exploited to prime the synthesis of the second DNA strand. Subsequently S1 nuclease is used to open the single-stranded hairpin loop before inserting the double-stranded cDNA into a plasmid via homopolymeric tailing[13–15] or via molecular linker molecules.[16–18] S1 nuclease digestion invariably results in the loss of sequences corresponding to the extreme 5′-terminal region of the mRNA.[7,16–19] Cloning of the entire 5′-untranslated region of a mRNA is, however, highly desirable. The presence of the entire 5′ end would facilitate mapping of the mRNA onto the corresponding genomic DNA, particularly with respect to the start site of transcription,[20] and it would also increase the probability of expressing cloned structural genes in bacteria.[21] Moreover, in several cases construction artifacts like inversions and inverted duplications have been reported.[19,22–25] The method described in this chapter provides the possibility to clone the entire sequence of an mRNA efficiently and to avoid the introduction of sequence artifacts into cloned cDNA sequences.[26]

Principle of the Method

By inserting cDNA–mRNA hybrids into a plasmid vector, Zain *et al.*[10] were able to clone a complete copy of the leader segments of adenovirus 2

[11] W. A. Salser, *Annu. Rev. Biochem.* **43,** 923 (1974).
[12] A. Efstratiadis, F. Kafatos, A. Maxam, and T. Maniatis, *Cell* **7,** 279 (1976).
[13] D. A. Jackson, R. H. Symons, and P. Berg, *Proc. Natl. Acad. Sci. U.S.A.* **69,** 2904 (1972).
[14] P. E. Lobban and A.-D. Kaiser, *J. Mol. Biol.* **78,** 453 (1973).
[15] P. C. Wensinck, D. J. Finnegan, J. E. Donelson, and D. S. Hogness, *Cell* **3,** 315 (1974).
[16] A. Ullrich, J. Shine, J. Chirgwin, R. Pictet, E. Tischer, W. J. Rutter, and H. M. Goodman, *Science* **196,** 1313 (1977).
[17] P. H. Seeburg, J. Shine, J. A. Martial, J. D. Baxter, and H. M. Goodman, *Nature (London)* **270,** 486 (1977).
[18] J. Shine, P. H. Seeburg, J. A. Martial, J. D. Baxter, and H. M. Goodman, *Nature (London)* **270,** 494 (1977).
[19] A. E. Sippel, H. Land, W. Lindenmaier, M. C. Nguyen-Huu, T. Wurtz, K. N. Timmis, K. Giesecke, and G. Schütz, *Nucleic Acids Res.* **5,** 3275 (1978).
[20] M. Grez, H. Land, K. Giesecke, G. Schütz, A. Jung, and A. E. Sippel, *Cell* **25,** 743 (1981).
[21] A. Y. Chang, H. A. Erlich, R. P. Gunsalus, J. H. Nunberg, R. J. Kaufman, R. T. Schimke, and S. N. Cohen, *Proc. Natl. Acad. Sci. U.S.A.* **77,** 1442 (1980).
[22] R. J. Richards, J. Shine, A. Ullrich, J. R. E. Wells, and H. M. Goodman, *Nucleic Acids Res.* **5,** 1137 (1979).
[23] J. B. Fagan, I. Pastan, and B. de Crombrugghe, *Nucleic Acids Res.* **8,** 3055 (1980).
[24] S. Fields and G. Winter, *Gene* **15,** 207 (1981).
[25] G. Volckaert, J. Tavernier, R. Derynck, R. Devos, and W. Fiers, *Gene* **15,** 215 (1981).
[26] H. Land, M. Grez, H. Hauser, W. Lindenmaier, and G. Schütz, *Nucleic Acids Res.* **9,** 2251 (1981).

fiber mRNA. Because this technique gives only low yields of recombinants, we have chosen a procedure similar to the one used by Rougeon *et al.*[4] and Cooke *et al.*,[27] in which S1 nuclease digestion of double-stranded cDNA is avoided for construction of recombinant plasmids.

Poly(A)-containing RNA is reverse-transcribed by AMV reverse transcriptase under conditions that result in a high yield of full-length cDNA. After alkaline hydrolysis of the mRNA the 3′ end of the cDNA is dCMP-tailed with terminal deoxynucleotidyltransferase. The synthesis of the second cDNA strand is primed by oligo$(dG)_{12-18}$ hybridized to the 3′-homopolymer tail. Full length ds-cDNA copies are purified on a preparative agarose gel. After a second tailing step with dCTP, the cDNA is annealed to plasmid DNA tailed with dGTP. The hybrid DNA is introduced and amplified in bacteria. The protocol is summarized in Fig. 1.

Cloning of chicken lysozyme cDNA into pBR322 by this procedure without a sizing step is described here as an example.[26] Sequence analysis revealed that at least 9 out of 19 randomly isolated plasmids contained the entire 5′-untranslated mRNA sequence.[26]

Materials and Reagents

Avian myeloblastosis virus reverse transcriptase was kindly provided by Dr. J. W. Beard through the Office of Program Resources and Logistics, National Cancer Institute of the United States. Terminal transferase was obtained from Bethesda Research Laboratories, endonuclease S1 from *Aspergillus oryzae* was type III enzyme from Sigma, and *Pst*I was supplied by Renner, Darmstadt. Oligo$(dT)_{12-18}$ and oligo$(dG)_{12-18}$ were obtained from Collaborative Research.

Method

Synthesis of cDNA

Reverse transcription of 10 μg of poly$(A)^+$ RNA (maximum 100 μg/ml) is carried out in a siliconized Eppendorf tube in 50 m*M* Tris-HCl, pH 8.3, 50 m*M* KCl, 8 m*M* $MgCl_2$, 0.4 m*M* DTT, 30 μg of oligo(dT) and 0.1 mg of actinomycin D per milliliter, 1 m*M* each dGTP, dCTP, and dTTP, and 0.2 m*M* $[^{32}P]$dATP (0.75 Ci/mmol). The reaction is started with the addition of 50–100 units of AMV reverse transcriptase. The mixture is kept for 2 min at room temperature followed by incubation at 42° for 60 min. The reac-

[27] N. E. Cooke, D. Coit, R. J. Weiner, J. D. Baxter, and J. A. Martial, *J. Biol. Chem.* **255,** 6502 (1980).

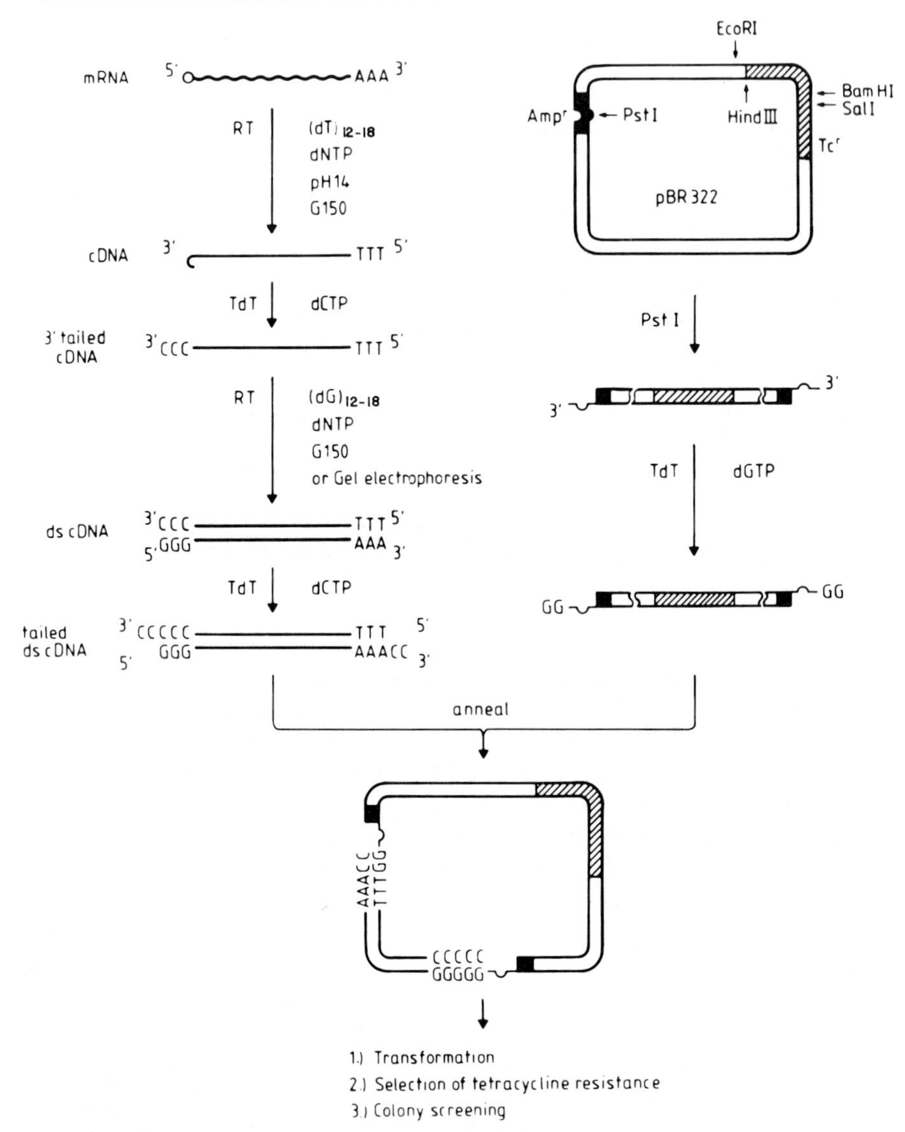

FIG. 1. Schematic diagram of the construction of recombinant plasmids containing eukaryotic mRNA sequences.

tion is stopped by adjusting the mixture to 0.5 *M* NaCl, 20 m*M* EDTA, and 0.2% SDS.

cDNA synthesis is followed by precipitation of 1-μl aliquots at 0 time and 60 min with trichloroacetic acid and liquid scintillation counting after collection of precipitable nucleic acids on nitrocellulose filters. The reaction should yield 1–2 μg of cDNA with a specific activity of about 1×10^6 cpm/μg.

The reaction mixture is extracted once with one volume of chloroform–isoamyl alcohol (24 : 1, v/v). The high salt concentration prevents notable losses of cDNA during extraction. The aqueous phase is combined with 2.5 volumes of ethanol and chilled for 10 min in a Dry Ice–ethanol bath. The nucleic acid precipitate is collected by centrifugation for 15 min in an Eppendorf microfuge. After briefly drying the pellet under vacuum, it is dissolved in 40 μl of freshly prepared 0.4 *N* NaOH to hydrolyze the RNA during an overnight incubation at 25°. Nucleotides and oligodeoxynucleotides are separated from the cDNA by gel filtration through Sephadex G-150 packed into a siliconized Pasteur pipette. The chromatography is carried out in 10 m*M* NaOH, 0.3 *M* NaCl. The use of a high pH reduces the losses of cDNA during this purification step.

The fractions containing the excluded material are pooled and neutralized by addition of an equimolar amount of HCl and of 0.1 volume of 1 *M* Tris-HCl, pH 7.0 (check with pH paper). The cDNA is precipitated without carrier by adding 2.5 volumes of ethanol, chilling for 10 min in Dry Ice-ethanol, and centrifuging for 15 min in an Eppendorf microfuge. The addition of any RNA carrier is not recommended because trace amounts of DNA may disturb the determination of the extent of the following tailing reaction with terminal transferase. If the cDNA concentration is lower than 2 μg/ml after the G-150 column a partial lyophilization step should be introduced prior to ethanol precipitation. Alternatively 5 μg of deproteinized, DNA- and RNA-free glycogen[28] can be added as a carrier; glycogen does not interfere with any of the following enzymic reactions. The pelleted cDNA is dissolved in 20 μl of TE (10 m*M* Tris-HCl, pH 7.5, 0.1 m*M* EDTA), and its amount is determined by Cerenkov counting. The molar amount of cDNA is calculated after estimation of the average length of the cDNA molecules. This can be done by electrophoresis of the cDNA on a 2% alkaline agarose gel at 80 mA in 20 m*M* NaOH, 2 m*M* EDTA and autoradiography of the dried gel. cDNA copies with a length as expected from the size of the corresponding mRNA have been observed for lysozyme, ovomucoid, and ovalbumin. The proportion of full-length cDNA molecules decreased with increasing size of the mRNA templates: lysozyme mRNA (630 bases long) yielding 75%, ovomucoid mRNA (860 bases long) yielding 50%, and ovalbumin mRNA (1900 bases long) yielding 5% of full-length cDNA copies.[26]

[28] W. Roewekamp and W. Schmid, personal communication.

Tailing of the cDNA with Terminal Transferase

cDNA (1 μg; about 5 pmol) is tailed at the 3′ end with dCMP residues using terminal transferase (TdT).[29,30] The reaction is performed in 100 μl of 140 m*M* cacodylic acid (free acid), 30 m*M* Tris-base adjusted with KOH to a pH of 7.6. $CoCl_2$, DTT, and [^{3}H]dCTP (500 cpm/pmol) are added separately to final concentrations of 1 m*M*, 0.1 m*M*, and 0.1 m*M*, respectively, to the diluted buffer system. In order to avoid precipitates occurring and subsequent high backgrounds in liquid scintillation counting of trichloroacetic acid-precipitated aliquots, the mixture is preincubated for 5–10 min at 37° without cDNA and enzyme. After chilling in ice, the cDNA is added and the reaction is started with 50 units of TdT and kept for 2–5 min at 15°. The relatively high concentration of dCTP was chosen to allow reduction of the temperature and the time of incubation, which should minimize possible degradation of the cDNA by impurities in the enzyme preparation. Since the tails should not be longer than 25 nucleotides, it is recommended to test the tailing reaction in a 10-μl pilot experiment. Take 2-μl aliquots at 0, 2, 5, and 10 min for trichloroacetic acid precipitation and liquid scintillation counting. The conditions chosen appear to be optimal to tail nearly all the input cDNA molecules,[31] so that a simple calculation of the incorporated radioactivity versus the molar amount of cDNA is a good estimate for the length of the synthesized tails; 15–25 nucleotides are added per cDNA molecule under these conditions. Nucleotide sequence analysis of the corresponding tails of chicken lysozyme cDNA clones[26] are in good agreement with this calculation.

The incubation of the preparative sample is stopped by the addition of 10 m*M* EDTA, the mixture is adjusted to 0.3 *M* sodium acetate, heated for 5 min to 70°, and extracted with one volume of chloroform–isoamyl alcohol. The organic phase is reextracted with 50 μl of 0.3 *M* sodium acetate, and the combined aqueous phases are precipitated with ethanol. The pellet is dissolved in TE and, after addition of sodium acetate to 0.3 *M*, the cDNA is reprecipitated with ethanol. A marked reduction of the size of the pellet should be observed. After one washing with 70% ethanol, the pellet is dried briefly and dissolved in 20 μl of TE.

Synthesis of Double-Stranded cDNA on 3′-dCMP-Tailed cDNA

In order to prime the synthesis of the second strand, oligo$(dG)_{12-18}$ is hybridized to the 3′-dCMP-tailed cDNA. Hybridization is carried out with 0.5–1 μg of tailed cDNA in 50 μl of 100 m*M* Tris-HCl, pH 8.3, 60 m*M* KCl, 20 m*M* $MgCl_2$, 60 μg of oligo(dG) per milliliter. The mixture is kept

[29] R. Roychoudhury, E. Jay, and R. Wu, *Nucleic Acids Res.* **3,** 863 (1976).

[30] R. Rouchoudhury and R. Wu, this series, Vol. 65, p. 43.

[31] G. Deng and R. Wu, *Nucleic Acids Res.* **9,** 4173 (1981).

at 70° for 5 min and at 43° for 30 min. After chilling on ice, the reaction mixture is diluted to 100 μl with H_2O and adjusted to 10 m*M* DTT and 1 m*M* each dATP, dGTP, dCTP, and dTTP. Second-strand synthesis is started by adding 30 units of reverse transcriptase followed by incubation for 10 min at 37° and for 60 min at 42°. The reaction is stopped, and the sample is extracted and ethanol precipitated as described above.

The synthesis of the second strand is followed by determining the amount of radioactive cDNA that has become resistant to S1 nuclease digestion; 1-μl aliquots are taken before adding reverse transcriptase and at the end of the incubation time. They are added to 1 ml of 30 m*M* sodium acetate, pH 4.5, 3 m*M* $ZnSO_4$, 0.3 *M* NaCl, and 10 μg of denatured salmon sperm DNA. The samples are divided into two 0.5-ml aliquots, to one of which 300 units of S1 nuclease are added.[32] After incubation for 60 min at 37°, the trichloroacetic acid-precipitable radioactivity is measured by liquid scintillation counting; 70–80% of the cDNA is protected when tailed cDNA is used in combination with the oligonucleotide primer. In the absence of primer about 30% of the cDNA becomes resistant to S1 nuclease. In comparison, second-strand synthesis with nontailed cDNA as template leads to 40–60% S1 nuclease-resistant cDNA.[26]

Sizing of the ds-cDNA

To remove the oligodeoxynucleotides from the resulting ds-cDNA, the sample can be purified by gel filtration through Sephadex G-150 in 10 m*M* Tris-HCl, pH 7.5, 0.3 *M* NaCl, 1 m*M* EDTA. However, a more effective purification step is the separation of the ds-cDNA on a 2% agarose gel in 40 m*M* Tris-acetate, pH 8.0, 4 m*M* EDTA at 30 mA for 2 hr. According to markers run in parallel slots that are stained with ethidium bromide, the ds-cDNA (not stained) of a discrete size class can be cut out from the gel. It is recovered by the glass binding method described by Gillespie and Vogelstein.[33] The advantage of this method is the ability to elute the glass-bound DNA in a small volume of buffer (50% formamide, 10 m*M* Tris-HCl, pH 7.5, 0.1 *M* NaCl, 2 m*M* EDTA) so that even very small amounts of DNA can be ethanol precipitated. If about 0.8 μg of ds-cDNA are loaded onto the gel, about 0.2 μg is recovered after collecting the material larger than 500 bp.

Tailing of the ds-cDNA and Construction of Recombinant Plasmid DNA

To add about 10 dCMP residues to each 3′ end of the ds-cDNA, up to 5 pmol of 3′ ends of ds-cDNA are tailed in 100 μl under the same buffer

[32] H. M. Goodman and R. J. MacDonald, this series, Vol. 68, p. 75.

[33] D. Gillespie and B. Vogelstein, *Proc. Natl. Acad. Sci. U. S. A.* **76,** 615 (1979).

conditions as described above, except that the reaction is incubated at 37°. The reaction is stopped by adding 10 m*M* EDTA, 0.3 *M* NaCl. After addition of 10 μg of tRNA, the terminal transferase is heat-inactivated at 70° for 5 min. The sample is precipitated with ethanol twice and redissolved in TE to a concentration of 0.7 ng/μl. Since tails longer than 20 nucleotides cause a marked decrease in the cloning efficiency,[34] it is recommended to tail only small aliquots of the ds-cDNA to different extents. This pilot experiment allows to determine the optimal tail length for forming recombinant clones: 0.05 pmol of ds-cDNA 3′ ends (about 10 ng) are tailed in an assay volume of 10 μl.

DNA of plasmid pBR322 is digested for 1 hr with 10 units of *Pst*I per microgram to ensure complete digestion. The DNA is extracted once with one volume of phenol, extracted three times with ether, and ethanol precipitated. About 10 dGMP residues per 3′ terminus are added to the linearized DNA with terminal transferase; 5 pmol of 3′ ends per 100 μl are tailed under conditions described above. After incubation at 37°, the reaction is stopped by adjustment to 10 m*M* EDTA, 0.3 *M* NaCl, heated at 70° for 5 min and extracted with phenol and ether. The DNA is precipitated with ethanol and resuspended in TE (10 m*M* Tris-HCl, pH 7.5, 0.1 m*M* EDTA) at a concentration of 5 ng/μl. Tailed plasmid DNA and tailed ds-cDNA are mixed in a molar ratio of 1 : 1 at a final concentration of 250 ng of vector per milliliter in 10 m*M* Tris, pH 7.5, 100 m*M* NaCl, 1 m*M* EDTA. The mixture is incubated at 68° for 5 min, then at 43° for 2 hr; it is finally allowed to cool down to room temperature during a 2-hr period. Transformation of *E. coli* 5K[35] is carried out as described by Dagert and Ehrlich.[36] Starting with 1 μg of poly(A)$^+$ RNA, 10^4 to 10^5 transformants, of which 99% carry cDNA sequences, are usually obtained.

Desired recombinant clones can be isolated by colony screening,[37] hybrid-arrested and hybrid-promoted translation,[38,39] and immunological screening of colonies for expression of the cloned cDNA.[40,41]

[34] W. Roewekamp and R. A. Firtel, *Dev. Biol.* **79,** 409 (1980).

[35] J. Hubatek and S. W. Glover, *J. Mol. Biol.* **50,** 111 (1970).

[36] M. Dagert and S. D. Ehrlich, *Gene* **6,** 23 (1979).

[37] M. Grunstein and D. S. Hogness, *Proc. Natl. Acad. Sci. U.S.A.* **72,** 3961 (1975).

[38] M. Rosbash, D. Blank, G. Fahrner, L. Hereford, R. Ricciardi, B. Roberts, S. Ruby, and J. Woolford, this series, Vol. 68, p. 454.

[39] B. M. Paterson, B. E. Roberts, and E. L. Kuff, *Proc. Natl. Acad. Sci. U.S.A.* **74,** 4370 (1977).

[40] H. A. Erlich, S. N. Cohen, and H. O. McDevitt, *Cell* **13,** 681 (1978).

[41] S. Broome and W. Gilbert, *Proc. Natl. Acad. Sci. U.S.A.* **75,** 2746 (1978).

[21] β-Galactosidase Gene Fusions for Analyzing Gene Expression in *Escherichia coli* and Yeast

By MALCOLM J. CASADABAN, ALFONSO MARTINEZ-ARIAS, STUART K. SHAPIRA, and JOANY CHOU

The β-galactosidase structural gene *lacZ* can be fused to the promoter and controlling elements of other genes as a way to provide an enzymic marker for gene expression.[1] With such fusions, the biochemical and genetic techniques available for β-galactosidase can be used to study gene expression.[2] Gene fusions can be constructed either *in vivo,* using spontaneous nonhomologous recombination or semi-site specific transposon recombination,[3] or *in vitro,* with recombinant DNA technology.[4,5] Here we describe *in vitro* methods and list some recently developed β-galactosidase gene fusion vectors. With these *in vitro* methods, gene controlling elements from any source can be fused to the β-galactosidase structural gene and examined in the prokaryote bacterium *Escherichia coli* or the lower eukaryote yeast *Saccharomyces cerevisiae* (see this series, Vol. 100: Rose and Botstein [9], Guarente [10], and Ruby *et al.* [16] for additional descriptions of β-galactosidase gene fusion systems in yeast).

β-Galactosidase gene fusions can be constructed both with transcription initiation control signals and with transcription plus translation initiation control signals. This is done by removing either just the promoter region from the β-galactosidase gene or both the promoter and translation initiation regions. The β-galactosidase gene is convenient for making translational fusions because it is possible to remove its translation initiation region along with up to at least the first 27 amino acid codons without affecting β-galactosidase enzymic activity. In a gene fusion this initial part of the β-galactosidase gene can be replaced by the promoter, translation initiation site, and apparently any number of amino-terminal codons from another gene. Here we focus on these transcription–translation fusions because they provide all the gene initiation signals from the other gene.

[1] P. Bassford, J. Beckwith, M. Berman, E. Brickman, M. Casadaban, L. Guarante, I. Saint-Girons, A. Sarthy, M. Schwartz, and T. Silhavy, *in* "The Operon" (J. H. Miller and W. S. Reznikoff, eds.), p. 245. Cold Spring Harbor Laboratory, Cold Spring Harbor, New York, 1978.

[2] J. Beckwith, *in* "The Operon" (J. H. Miller and W. S. Reznikoff, eds.), p. 11. Cold Spring Harbor Laboratory, Cold Spring Harbor, New York, 1978.

[3] M. Casadaban and S. Cohen, *Proc. Natl. Acad. Sci. U.S.A.* **76,** 4530 (1979).

[4] M. Casadaban and S. Cohen, *J. Mol. Biol.* **138,** 179 (1980).

[5] M. Casadaban, J. Chou, and S. Cohen. *J. Bacteriol.* **143,** 971 (1980).

ISBN 0-12-182000-9

For a description of *in vitro* β-galactosidase transcriptional fusions, see Casadaban and Cohen.[4]

β-Galactosidase expression from a gene fusion can be used not only to measure gene expression and regulation, but also to isolate mutations and additional gene fusions. Mutants with altered gene regulation, for example, can be used to identify regulatory genes and to help decipher regulatory mechanisms. Mutants with increased gene expression, or constructions that join the fused gene to a stronger gene initiation region, can be used to isolate large amounts of the original gene product following removal of the β-galactosidase fused gene segment.[6]

An additional result of translational β-galactosidase gene fusions is the formation of hybrid proteins that can be used in protein studies. The hybrid protein can readily be identified by its β-galactosidase enzymatic activity and purified for studies of protein functions located in the amino terminus, such as DNA or membrane-binding functions.[7,8] These hybrid proteins can also be used to determine amino terminal sequences[9,10] and to elicit antibody formation.[11]

Principle of the Methods

A cloning vector, containing the β-galactosidase structural gene segment without its initiation region, is used to clone a DNA fragment that contains a gene's transcription–translation initiation region and amino terminal codons to form the gene fusion. The fragment either can be purified or can be selected from among the resulting clones. Alternatively, fusions can be made by inserting a DNA fragment (cartridge) containing the β-galactosidase structural gene into a gene already on a vector by cloning (inserting) the β-galactosidase gene into a site within the other gene. For a functional fusion, the gene segments must be aligned in their direction of transcription; and, for a translational fusion, their amino acid codons must be in phase. Gene control regions that cannot conveniently be fused to β-galactosidase by cloning in a single step can be cloned in a nearby position followed by an additional event to form the fusion, such as a deletion to remove transcription termination or to align the codons. These additional events can be obtained either *in vitro,* as by nuclease

[6] M. Casadaban, J. Chou, and S. Cohen, *Cell* **28,** 345 (1982).

[7] B. Müller-Hill and J. Kania, *Nature (London)* **249,** 561 (1974).

[8] T. Silhavy, M. Casadaban, H. Shuman, and J. Beckwith, *Proc. Natl. Acad. Sci. U.S.A.* **73,** 3423 (1976).

[9] A. Fowler and I. Zabin, *Proc. Natl. Acad. Sci. U.S.A.* **74,** 1507 (1977).

[10] M. Ditto, J. Chou, M. Hunkapiller, M. Fennewald, S. Gerrard, L. Hood, S. Cohen, and M. Casadaban, *J. Bacteriol.* **149,** 407 (1982).

[11] H. Schuman, T. Silhavy, and J. Beckwith, *J. Biol. Chem.* **255,** 168 (1980).

digestion, or *in vivo* using genetic screens or selections for lactose metabolism.

Once fusions have been made, they can be used to measure gene expression under different regulatory conditions, such as with or without an inducer or repressor, or in different growth media or physical conditions of growth (such as temperature[12]). Self-regulation can be investigated by providing (or removing) in *trans* an expressed wild-type copy of the gene (such as on another plasmid, episome, lysogenic prophage or in the chromosome[13]) and checking the effect on β-galactosidase expression.[14] Linked or unlinked, *cis* or *trans* acting mutations can be sought with appropriate mutagenesis, selection, and screenings using the lactose phenotype.[2,13]

Materials

Escherichia coli strain M182 and its derivatives (see the table) are currently used because they are deleted for the *lac* operon, can be efficiently transformed, and yield good plasmid DNA preparations. Their *gal* mutations may contribute to their transformation properties, since they are missing polysaccharides containing galactose. Strains M182 and MC1060 have no nutritional requirements and so can be used to select for growth on minimal media with a single carbon and energy source (such as lactose) without the addition of any nutritional factor that could be used as an alternative carbon or energy source. Several of these strains have the *hsdR*$^-$ host restriction mutation so that DNA unmodified by *E. coli* K12 is not degraded. All strains are *hsdS*$^+$ M$^+$ so that DNA in them becomes K12 modified and can be transferred efficiently to other *E. coli* K12 strains. Some MC strains have a deletion of the *ara* operon, so they can be used with plasmids containing the *ara* promoter.[4] Strain MC1050 has an amber mutation in the *trpB* gene (for tryptophan biosynthesis) so that amber mutation suppressors can be introduced, a procedure that is useful in studies of regulatory genes.[13] MC1116 is a *recA*$^-$ and streptomycin-sensitive derivative. Casadaban and Cohen[4] describe additional derivative strains with features such as the early *lacZ* M15 deletion, which can be intercistronically complemented by the α segment of β-galactosidase.

[12] M. Casadaban, T. Silhavy, M. Berman, H. Schuman, A. Sarthy, and J. Beckwith, *in* "DNA Insertion Elements, Plasmids and Episomes" (M. Bukhari, J. Shapiro, and S. Adhya, eds.), p. 531. Cold Springs Harbor Laboratory, Cold Spring Harbor, New York, 1977.

[13] J. Chou, M. Casadaban, P. Lemaux, and S. Cohen, *Proc. Natl. Acad. Sci. U.S.A.* **76,** 4020 (1979).

[14] F. Sherman, G. Fink, and C. Lawrence, "Methods in Yeast Genetics." Cold Spring Harbor Laboratory, Cold Spring Harbor, New York, 1979.

STRAINS USED FOR ANALYZING GENE EXPRESSION[a]

	Strain	Genotype	Reference
E. coli	M182	Δ(*lacIPOZYA*)*X74*,*galU*,*galK*,*strA*^{r}	4
	MC1000	M182 with Δ (*ara*,*leu*)	4
	MC1050	MC1000 with *trpB9604 amber*	4
	MC1060	M182 with *hsdR*$^-$	4
	MC1061	MC1060 with Δ (*ara*,*leu*)	4
	MC1116	MC1000 with *strA*^{s},*spcA*^{r} and *recA56*	
	MC1064	MC1061 with *trpC9830*	
	MC1065	MC1064 with *leuB6*, *ara*$^+$	
	MC1066	MC1065 with *pyrF74*::Tn*5*(Km^r)	
	AMA 1004	MC1065 with Δ (*laclPOZ*)*C29*,*lacY*$^+$	
Yeast	M1-2B	α, *trp1*, *ura3-52*	R. Davis
	M1-2BA2	M1-2B with an increased permeability to β-galactosidase substrates	
	JJ4	*a*, *trp1*, *lys1*, *ura3-52*, Δ (*gal7*) 102, *gal2*$^+$	J. Jaehning

[a] *ara*, *gal*, and *lac* refer to mutations that abolish the ability of the cell to use arabinose, galactose, and lactose as carbon and energy source for growth; *leu*, *trp*, and *pyr* (*ura*) refer to mutations creating a requirement for leucine, tryptophan, or pyrimidine (uracil) to be added to minimal media for growth. For further references, see text.

Strains MC1064, MC1065, and MC1066 are derivatives of M182 with *hsdR*$^-$ and Δ (lac) that have auxotropic markers that can be complemented by the yeast genes *leu2* (*leuB*), *ura3* (*pyrF*), and *trp1* (*trpC*). The *leuB6* and *trp*C9830 are point mutations that revert infrequently, and the *pyrF* is a Tn*5* kanamycin-resistance transposon insertion that reverts at a higher frequency. AMA1004 is an MC1065 derivative with a deletion of the *lacZ* gene on the chromosome, which leaves the *lacY* gene expressed from an unknown chromosomal promoter. This strain is useful for complementing β-galactosidase plasmids without the *lacY* gene. The yeast strains in the table can be efficiently transformed with plasmid DNA. The *ura3* mutation does not revert, and the *trp1* reverts at a low rate. Strain M1-2B has a heterologous pedigree, so that diploids derived from it do not sporulate efficiently.

Escherichia coli medium is described by Miller.[15] β-Galactosidase expression *in vivo* can be detected with three types of agar media: medium with colorimetric β-galactosidase substrates, lactose (or other β-galactoside) minimal medium, and fermentation indicator medium. The colorimetric substrate medium is the most sensitive and can be used to detect β-galactosidase levels below those necessary for growth on lactose minimal

[15] J. Miller, "Experiments in Molecular Biology." Cold Spring Harbor Laboratory. Cold Spring Harbor, New York, 1972.

medium. The XG (5-bromo-4-chloro-3-indolyl-β-D-galactoside) substrate is commonly used since its β-galactosidase hydrolysis product dimerizes and forms a blue precipitate that does not diffuse through the agar.[15] The blue color is most easily seen on a clear medium such as M63. XG dissolved in dimethylformamide solution at 20 mg/ml can be conveniently stored for months at $-20°$ and used by spreading on the agar surface just before use, or by adding to the liquid agar just before pouring. Acid-hydrolyzed casamino acids can also be added (to 250 μg/ml) to minimal *E. coli* or yeast media to speed colony growth without affecting the detection of low levels of blue color. Note that the acid-hydrolyzed casamino acids can serve as a source for leucine, but not for pyrimidine or tryptophan, for growth on minimal medium.

β-Galactosidase detection with XG in yeast colonies can be done with the standard yeast SD minimal medium[14] if it is buffered to pH 7 with phosphate buffer to allow β-galactosidase enzymic activity. The phosphate buffer can be prepared as described for M63 medium[15] but must be added separately (usually as a 10-fold concentrate) to the cooled agar medium just prior to pouring in order to avoid a phosphate precipitate. The carbon energy source sugar also is usually prepared and added separately to avoid caramelization. Less efficient carbon sources for yeast than glucose, such as glycerol plus ethanol,[14] result in bluer yeast colonies on SD-XG medium, presumably because less acidity is formed in the growing colony.

Yeast cells are not very permeable to β-galactosidase, and so lower levels of β-galactosidase can be better detected by first permeabilizing the yeast cells. This can conveniently be done by replica plating or growing yeast colonies on filters (such as Millipore No. 1) and permeabilizing the yeast by immersion in liquid nitrogen. The filters can then be placed in β-galactosidase assay Z buffer[15] with the XG substrate. This process also overcomes the problem of the inhibition of β-galactosidase activity by the acid formed within the yeast colony.

Higher levels of β-galactosidase are indicated if lactose (or another β-galactoside) can be utilized for growth on minimal media (such as M63 medium[15]). For detecting even higher levels of β-galactosidase, the fermentation indicator plates, such as lactose MacConkey (Difco), tetrazolium, or EMB[15] can be used. On fermentation indicator plates, both lactose utilizing (Lac^+) and nonutilizing (Lac^-) cells form colonies, but the colonies are colored differently. Lactose utilization also requires expression of the *lacY* permease gene unless very high levels of β-galactosidase are made. Yeast *S. cerevisiae* does not have a permease for lactose, and so media of these types cannot be used. There is a Lac^+ yeast species *Kluveromyces lactis,* with a lactose permease, but no transformation system has yet been devised for it.[16] The M1-2B A2 *S. cerevisiae* mutant we have isolated (see the table, unpublished results), however, does allow a patch of yeast cells expressing β-galactosidase, but not a single yeast cell, to grow on lactose minimal medium. This mutant seems to result in a lethal permeabilization of about 80% of the cells, which can then take up lactose and cross-feed their neighbors with lactose hydrolysis products (unpublished results).

β-Galactosidase assays are as described.[15] For yeast, cells are broken either by freeze-thawing three times with ethanol–Dry Ice, by vortexing with glass beads, or by sonicating. Procedures involving glusulase to form spheroplasts cannot be used, since glusulase contains a β-galactosidase activity. However, the more purified spheroplasting enzyme zymolase can be used.[14] It is important not to centrifuge away broken yeast particulate matter, since much of the β-galactosidase activity can be associated with it. After incubation with the colorimetric β-galactosidase substrate ONPG[15] and stopping the reaction by addition of 1 *M* $NaCO_3$, the particulate matter can then be removed by centrifugation before reading the optical density of the yellow product at 420 nm. Units are expressed as nonomoles of ONPG cleaved per minute per milligram of protein in the extract, where 1 nmol is equivalent to 0.0045 OD_{420} unit with a 10-mm light path.[15]

Methods

β-Galactosidase Vectors. The prototype vector pMC1403 (Fig. 1) has been described.[5] pMC1403 contains the entire *lac* operon but is missing the promoter, operator, and translation initiation sites as well as the first eight nonessential codons of the *lacZ* gene for β-galactosidase. It contains three unique restriction endonuclease cleavage sites, for *Eco*RI, *Sma*I, and *Bam*HI, into which DNA fragments containing promoter and translation initiation sites can be inserted to form gene fusions (Fig. 1). Figure 2 lists derivatives of pMC1403 with additional restriction sites. Figure 3 lists vectors with the *lac* promoter and translation initiation sites before a series of restriction sites aligned in various translation phases with the β-galactosidase codons. These vectors (unpublished results) are derived from the M13mp7, 8, and 9 cloning vectors.[17] Additional vectors with the p15a plasmid replicon are described by Casadaban *et al.*[5] p15a is a high copy number plasmid that is compatible with pBR322 and ColE1.

[16] R. Dickson, *Gene* **10,** 347 (1980).

[17] J. Messing, R. Crea, and P. Seeburg, *Nucleic Acids Res.* **9,** 309 (1981).

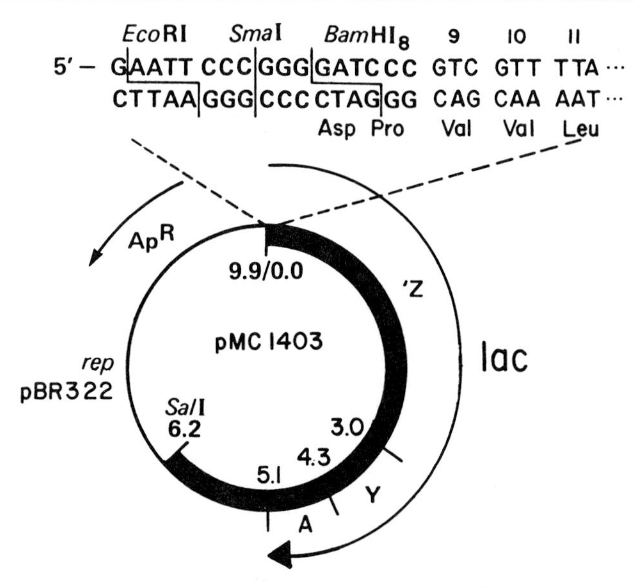

FIG. 1. The prototype β-galactosidase hybrid protein gene fusion vector pMC1403 with three unique cloning sites *Eco*RI, *Sma*I (*Xma*I), and *Bam*HI.[5] It contains the *lac* operon with the *lacZ* gene for β-galactosidase, the *lacY* gene for lactose permease in *Escherichia coli,* and the nonessential *lacA* gene for transacetylase. It is missing the *lac* promoter, operator and, translation initiation site, as well as the first seven and one-third nonessential amino acid codons; codons are labeled from eight onward. The non-*lac* segment is from the *Eco*RI to *Sal*I sites of pBR322, which has been entirely sequenced [G. Sutcliffe, *Cold Spring Harbor Symp. Quant. Biol.* **43,** 77 (1979)]. Fragments inserted into the unique restriction sites form gene fusions if they contain promoters and translation initiation sites with amino terminal codons that align with the β-galactosidase gene codons. Note that the *Sma*I site yields blunt ends into which any blunt-ended DNA fragment can be integrated, and that any such inserted fragment will become flanked by the adjacent *Eco*RI and *Bam*HI sites contained in the synthetic *Eco*RI-*Sma*I-*Bam*HI linker sequence on the plasmid. These adjacent sites can be used separately to open the plasmid on either side of the inserted fragment for making exonuclease-generated deletions into either side of the fragment. There is a unique *Sac*I (*Sst*I) site in *lacZ* and a unique *Bal*I site (blunt ended) in the pBR322 sequence just after the *lac* segment. This plasmid has no *Bgl*II, *Hin*dIII, *Kpn*I, *Xba*I, or *Xho*I sites. There are two *Bal*I and *Hpa*I sites and several *Acc*I, *Ava*I, *Ava*II, *Bgl*I, *Bst*EII, *Eco*RII, *Hin*cII, *Pvu*I, and *Pvu*II sites.

Figure 4 describes vectors containing yeast replicons *ars1* and *2* with the selectable yeast genes *trp1, ura3,* and *leu2*. All these vectors are derived from the pBR322 cloning vector. Unique sites are present for *Eco*RI, *Bam*HI, and *Sma*I. pMC1790 and 2010 have all these three unique cloning sites, are relatively small, and are deleted for the *lacY* and *lacA* genes.

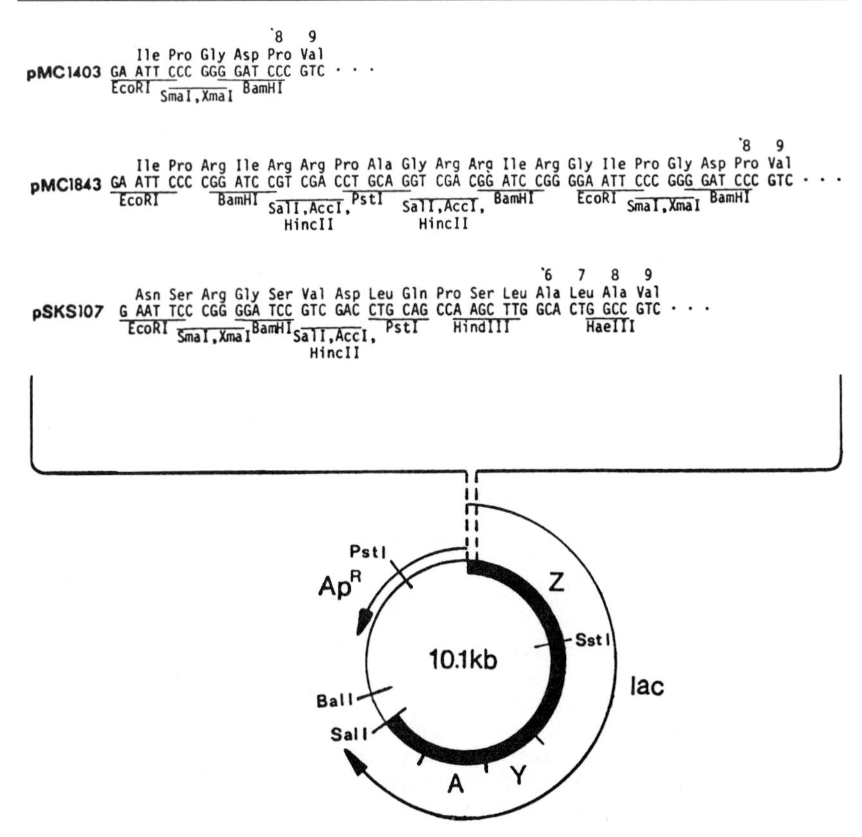

FIG. 2. Translation fusion vectors without a promoter. The numbers above refer to the original *lacZ* codons. pMC1403 is from Fig. 1. pMC1843 is pMC1403 with the polylinkers from M13mp7[17] inserted into the *Eco*RI site of pMC1403. pSKS107 is from pSKS105 (Fig. 3) with the removal of the *Eco*RI fragment containing the *lac* promoter. Additional *Acc*I, *Hin*cII, and *Hae*III sites are not shown.

Alternative β-Galactosidase Insertion Fragments. Instead of forming β-galactosidase fusions by inserting gene control fragments into β-galactosidase vectors, it is also possible to insert β-galactosidase fragments without replicons ("cartridges") into sites within genes already on replicons. Figure 5 describes plasmids from which β-galactosidase gene fragments can be excised with restriction sites on both sides of *lacZ*. pMC931[5] has in addition the *lacY* gene, whereas pMC1871 has only the *lacZ* gene with no part of *lacY*. The *lac* fragment from pMC931 can be obtained with *Bam*HI or *Bam*HI plus *Bgl*II, which yields identical sticky ends. The *lac* fragment from pMC1871 can be obtained with *Bam*HI, *Sal*I, or *Pst*I.

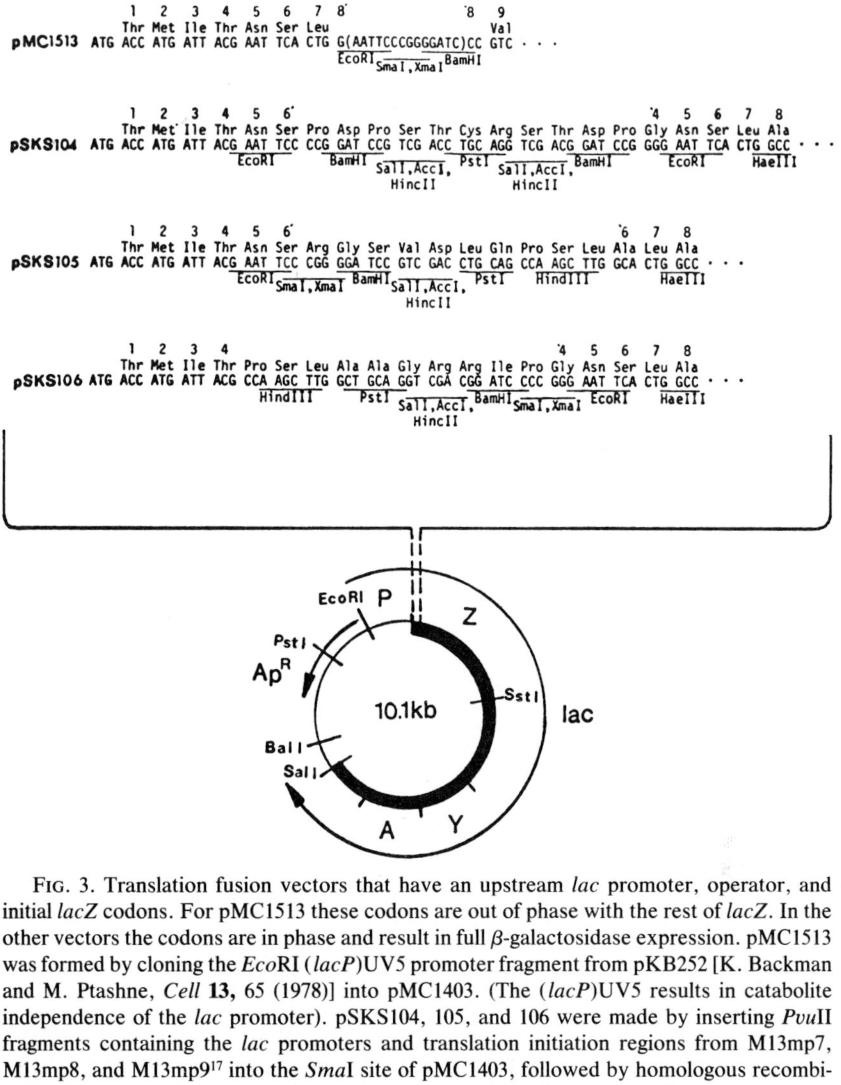

```
                 1   2   3   4   5   6   7   8'               '8  9
                 Thr Met Ile Thr Asn Ser Leu                      Val
pMC1513      ATG ACC ATG ATT ACG AAT TCA CTG G(AATTCCCGGGGATC)CC GTC . . .
                                             EcoRI SmaI,XmaI BamHI

                 1   2   3   4   5   6'                                                          '4  5   6   7   8
                 Thr Met Ile Thr Asn Ser Pro Asp Pro Ser Thr Cys Arg Ser Thr Asp Pro Gly Asn Ser Leu Ala
pSKS104      ATG ACC ATG ATT ACG AAT TCC CCG GAT CCG TCG ACC TGC AGG TCG ACG GAT CCG GGG AAT TCA CTG GCC . . .
                               EcoRI         BamHI  SalI,AccI,  PstI  SalI,AccI, BamHI       EcoRI         HaeIII
                                                    HincII            HincII

                 1   2   3   4   5   6'                                              '6  7   8
                 Thr Met Ile Thr Asn Ser Arg Gly Ser Val Asp Leu Gln Pro Ser Leu Ala Leu Ala
pSKS105      ATG ACC ATG ATT ACG AAT TCC CGG GGA TCC GTC GAC CTG CAG CCA AGC TTG GCA CTG GCC . . .
                               EcoRI SmaI,XmaI BamHI SalI,AccI,  PstI  HindIII          HaeIII
                                                     HincII

                 1   2   3   4                                                   '4  5   6   7   8
                 Thr Met Ile Thr Pro Ser Leu Ala Ala Gly Arg Arg Ile Pro Gly Asn Ser Leu Ala
pSKS106      ATG ACC ATG ATT ACG CCA AGC TTG GCT GCA GGT CGA CGG ATC CCC GGG AAT TCA CTG GCC . . .
                                   HindIII   PstI   SalI,AccI, BamHI SmaI,XmaI EcoRI    HaeIII
                                                    HincII
```

EcoRI P Z PstI Ap^R 10.1kb SstI lac BalI SalI A Y

FIG. 3. Translation fusion vectors that have an upstream *lac* promoter, operator, and initial *lacZ* codons. For pMC1513 these codons are out of phase with the rest of *lacZ*. In the other vectors the codons are in phase and result in full β-galactosidase expression. pMC1513 was formed by cloning the *Eco*RI (*lacP*)UV5 promoter fragment from pKB252 [K. Backman and M. Ptashne, *Cell* **13,** 65 (1978)] into pMC1403. (The (*lacP*)UV5 results in catabolite independence of the *lac* promoter). pSKS104, 105, and 106 were made by inserting *Pvu*II fragments containing the *lac* promoters and translation initiation regions from M13mp7, M13mp8, and M13mp9[17] into the *Sma*I site of pMC1403, followed by homologous recombination between the *lacZ* segments to result in full β-galactosidase expression. In all cases an *Eco*RI site remains to the left (upstream) of *lac*P.

Alternatively, the 1871 fragment can be obtained with a combination of restriction enzymes (see Fig. 5). Fragments can also be obtained with different restriction ends from the plasmids of Figs. 1–4. Note that a fragment with blunt ends on both sides can be obtained using *Sma*I plus *Bal*I from many of these plasmids. In addition, *lac* fragments can easily

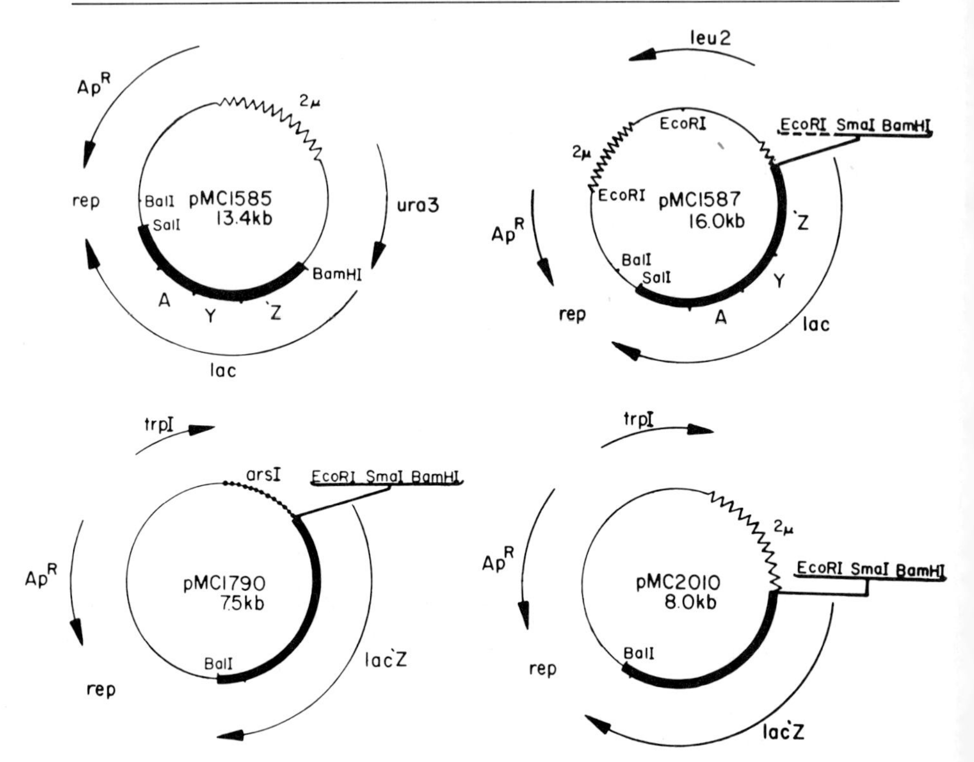

FIG. 4. Yeast replicon β-galactosidase vectors. All are derivatives of pMC1403 with the yeast 2μ or *ars*I replicons and the yeast *leu2, ura3,* or *trp1* gene.[24] These genes can be selected in either *Escherichia coli* or yeast as complementing appropriate auxotrophs (see the table and the Materials section). Unique cloning sites are underlined. pMC1585 and 1587 have the same *lac* segment as on pMC1403 of Fig. 1. pMC1790 and 2010 are similar but have a deletion of *lacY* and *A* gene formed by removing an *Ava*I fragment extending from *lacY* into pBR322, thereby also removing the *Sal*I site but keeping the *Bal*I site. This *Ava*I deletion was first isolated by M. Berman (Fredrick Cancer Research Center, Fredrick, Maryland).

be combined using a unique *Sst*I (*Sac*I) site in *lacZ* so that any sites before *lacZ* can be combined with any sites after the *lac* segment.

Choice of Fragment. To form a gene fusion with a β-galactosidase fusion vector it is necessary to obtain a fragment with the control elements of the gene of interest. This is most conveniently done if the sequence or restriction map is known so that appropriate sites can be identified. The fragment can have ends that match the ends of cleaved β-galactosidase fusion vectors, or the ends can be altered to match (as by filling in, chewing back, or adding linkers). One cleavage site should be well upstream to all the controlling sequences, and the other should be within the coding part of the gene such that the codons of the gene are in phase with the β-galactosidase codons. If the codons are not aligned, they can be made to align in a second step by obtaining a frameshift mutation (see below). Also, if the second site is beyond the gene, a second step deletion can be formed to fuse the gene.

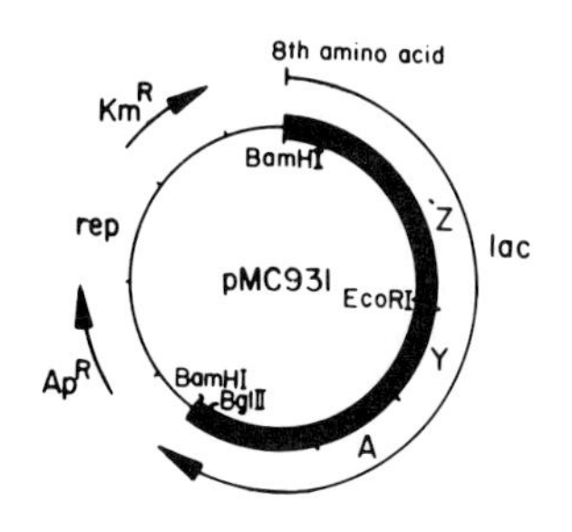

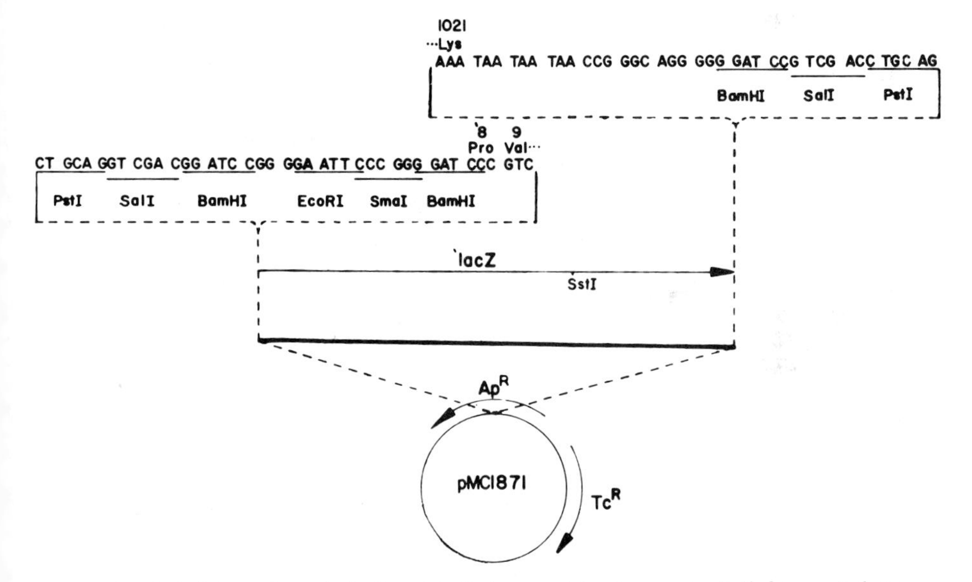

FIG. 5. β-Galactosidase hybrid protein fusion cartridge fragments. Both fragments have the *lacZ* gene without the control region of the promoter, operator, translation initiation region, and first eight nonessential *lacZ* codons, as in pMC1403 of Fig. 1. The *lac* fragment on pMC931[5] has the end of the *lac* operon (*lacY* and *A*) on the pACYC177 vector. The *lac* fragment on pMC1871 has only *lacZ*, without any part of *lacY* inserted in the *Pst*I site of pBR322. *lacY* was removed by cleaving at an *Hae*III site located between *lacZ* and *Y* [D. Buchel, B. Gronenborn, and B. Müller-Hill, *Nature* (*London*) **283,** 541 (1980)] and joining to a polylinker sequence on a plasmid derived from M13mp7.[17] The beginning of the pMC1871 *lac* fragment is from pMC1403 (Fig. 1) with additional sites from the M13mp7 polylinker.

The vectors of Fig. 1–5 have several different cloning sites. The *Sma*I site can be used with blunt-ended fragments. The *Bam*HI site can be used with *Bgl*II, *Bcl*I, or *Sau*3A (*Mbo*I) digested DNA since all these enzymes yield identical sticky ends. *Sau*3A recognizes only four base pairs and so should cleave within an average gene, whereas the other enzymes recognize six nucleotides and so should not cleave in an average gene. The *Eco*RI site can be used with fragments formed by *Eco*RI* digestion. The *Sal*I site can be used with *Xho*I-cleaved DNA. The *Eco*RI, *Bam*HI, *Xma*I, *Sal*I, and *Hin*dIII sites can also be used with blunt-ended DNA fragments either by making them blunt by filling in their ends with a polymerase or by adding synthetic linker DNA fragments onto the blunt fragment ends. Each of these manipulations can yield different fusion frames.

Selection of Fusion Clones. Fusions can be made with purified fragments or with mixtures of fragments followed by a screening of the resulting clones to identify those with the desired fragment containing the gene control region. Fusion clones can be identified by their resulting *lac* expression (Lac phenotype), or they can be selected by their physical structure using standard DNA cloning techniques without regard to their Lac phenotype. Fragments containing promoters which are expressed in *E. coli* and for which translation is aligned with the β-galactosidase codons can be detected as a resulting clone that forms red colonies on lactose MacConkey agar, dark blue colonies on XG agar, or colonies on lactose minimal medium. These clones can easily be distinguished from the original vector if a vector is used for which the *lac* genes were not expressed, as is the use for the vectors of Fig. 1, 2, and 4. The lactose MacConkey red colony phenotype is the most diagnostic since only efficient *lac* fusions have enough expression for this lactose fermentation indicator test.

For promoters that are expressed in yeast, only the XG indicator can be used, since there is no lactose permease in yeast. (Note that some but not all yeast promoters are expressed in *E. coli*[18]). Since yeast cells are also relatively impermeable to colorimetric β-galactosidase substrates, high levels of β-galactosidase expression are needed for yeast colonies to turn blue with XG. Lower levels of β-galactosidase formed by less efficient yeast fusions can be detected by permeablizing the yeast cells or by placing them into pH 7 buffered medium (see Materials).

In *E. coli,* inefficient fusions can be detected with XG agar. Fusions with good promoters but with codons out of phase or without any translation initiation site, or fusions without good promoters but with codons in phase, can usually be detected as blue colonies on XG but white colonies on lactose MacConkey.

[18] L. Guarante and M. Ptashne, *Proc. Natl. Acad. Sci. U.S.A.* **78,** 2199 (1981).

Some β-galactosidase vectors such as those of Fig. 3 already have promoters and translation start sites for expression of β-galactosidase. These can be used to detect clones of fragments that do not result in efficient β-galactosidase expression. Such a system has been used for transcriptional fusions with the pMC81 vector,[5] which has the arabinose *ara* operon promoter; fragments that do not express β-galactosidase well can be detected, as they result in less expression of β-galactosidase than is the case for the original vector when the *ara* promoter is induced by L-arabinose. This vector can also be used in the standard way in the absence of the *ara* inducer L-arabinose to detect fragments that promote β-galactosidase expression.

Each DNA fragment in a particular orientation yields a specific level of β-galactosidase expression. Thus, clones from a mixture of fragments can be sorted by their levels of *lac* expression, i.e., by their colors on *lac* indicator plates. This facilitates the identification of clones of different fragments or of a fragment in different orientations.[4]

Fusions can also be made by cloning DNA fragments without regard to their *lac* expression using standard cloning techniques: cloning can be done with an excess of fragments followed by screening of candidates for their plasmid DNA size and structure or by colony hybridization. Alternatively, the ends of a cleaved vector can be treated with phosphatase to prevent their ligation without an inserted fragment.

Frame-Shifting. For a DNA fragment with a promoter and translation initiation signals there are three possible phases with which it can be aligned with the *lacZ* codons. Fusions that are out of phase can be detected by their low levels of β-galactosidase but with high levels of the downstream *lacY* gene product. Expression of the *lacY* lactose permease can be detected using melibiose MacConkey agar at 37° or above. This classical *lac* procedure relies on the fact that the sugar melibiose can be transported by the *lacY* gene permease and that the original melibiose permease in *E. coli* K12 is temperature sensitive.[2]

An out of phase fusion can still be used to study regulation merely by assaying the low levels of β-galactosidase expression. The fusion can be made in phase by *in vitro* or *in vivo* manipulation. *In vitro* the fragment can be recloned into another β-galactosidase vector with the same cloning site in a different phase, or its ends can be altered and inserted into a different site to achieve a different phase. Alternatively, an out of phase fusion can be opened at or near its β-galactosidase joint and the phase changed by removing (as with the exonuclease *Bal*31) or by adding a few nucleotides (as with a linker).

Frames can be shifted *in vivo* by selecting for mutants with increased *lac* expression. This is most conveniently done by selecting for mutants that allow growth on lactose minimal media or yield red "papilla" growth from a colony incubated several days on lactose MacConkey agar. Normally we select spontaneous mutations formed without the use of a mutagen. These frame-shifts can be small changes or large deletions. Frameshift mutations can even be selected from fusions without a good promoter or translation initiation: the Lac^+ selection is powerful enough to select double genetic changes, one to frame shift and one to improve a promoter or translation initiation site.[6]

An Example. An example of a fusion construction that employs some of the points we have discussed is our fusion of the yeast *leu2* gene promoter to the β-galactosidase gene. At first we did not know the sequence of this gene. It had been cloned on a plasmid, and its major restriction sites were mapped.[19] We noted that there was an *Eco*RI site within the gene. From the location of sites on the restriction map we assumed that the *Eco*RI site was within the coding region of *leu2,* not in a control element. The *leu2* gene was known to be expressed in *E. coli,* but its direction of transcription was not known. We chose a clone of *leu2,* YEp13,[19] which had two *Eco*RI sites in addition to the one in *leu2,* such that when digested with *Eco*RI, one of the three fragments would contain the promoter of the *leu2* gene. We chose the vector pMC1403 (Fig. 1) because it had a unique *Eco*RI cloning site. (At this point the yeast replicon vectors pMC1790 and pMC2010 had not yet been constructed.) The mixture of the three *Eco*RI digested fragments of YEp13 was ligated with *Eco*RI-cleaved pMC1403 vector DNA, and the resulting ligation mixture was used to transform the efficiently transformable, *lac* deletion, *E. coli* strain M182 (see the table). Transformants were selected on both M63-glucose–XG-ampicillin and on lactose MacConkey ampicillin. About 10% of the transformants made blue colonies on the XG plates, and all of these were equally blue. On lactose MacConkey all the colonies were initially white, but after prolonged incubation for 2 days about 10^{-3} to 10^{-4} of the colonies had a small red papilla. Several of the blue colonies and the red papilla colonies were picked and purified by streaking to single colonies. Plasmid DNA from all of them was indistinguishable on 0.7% agarose gels. They each had two *Eco*RI bands, one the size of the pMC1403 vector and the other the size of one of the three *Eco*RI fragments from the *leu2* plasmid. Furthermore, digestion with other enzymes revealed that all clones had the same fragment orientation, and that this orientation placed the end of the *Eco*RI fragment that was within *leu2* nearest to β-galactosidase. From this we concluded that the *leu2* promoter was within the cloned *Eco*RI fragment and that the *Eco*RI site was probably out of phase

[19] B. Ratzkin and J. Carbon, *Proc. Natl. Acad. Sci. U.S.A.* **74,** 487 (1977).

with the *lacZ* codons from the *Eco*RI site in pMC1403. The plasmids from the papilla cells probably had a small frame-shift, since they did not have a DNA structure that was altered as judged on the agarose gel. The fusion was then joined to a yeast replicon and placed into yeast, where it was found that β-galactosidase expression was repressed by leucine and threonine as expected for the *leu2* gene. This showed that the *Eco*RI fragment cloned contained the gene control signals for regulation in yeast. These observations were verified when the sequence was obtained (P. Schimmel and A. Andreadis, personal communication).

The sequence predicted new restriction sites that would yield a smaller *leu2* gene control fragment of approximately 650 base pairs between an *Xho*I and a *Bst*EII site. These enzymes yielded a fragment with sticky ends that were filled in with *E. coli* DNA polymerase I. This fragment was ligated into the blunt *Sma*I site of pMC1790 (Fig. 4), which yielded an in-frame fusion that was detected as a red colony clone on lactose MacConkey ampicillin agar using strain MC1066. Strain MC1066 is deleted for *lac,* transforms well, is $hsdR^-$ restriction minus so as not to cleave unmodified DNA, and has the *E. coli trpC9830* mutation, which is complemented to Trp^+ by the *trp1* gene from yeast on pMC1790. Plasmid DNA of this clone was used to transform yeast strain M1-2B to Trp^+, (SD-glucose plates). These transformants were assayed for β-galactosidase after being grown with and without leucine and threonine (1 m*M* each). As expected β-galactosidase expression was repressible. Yeast cells containing this plasmid did not make enough β-galactosidase to form blue colonies on SD-glucose-XG plates, therefore they were assayed by replica plating onto filters and permeabilizing by freeze-thawing (see above).

The *leu2* promoter inserted into the *Sma*I site of pMC1790 did not have the *Sma*I site regenerated, but it could be excised at the adjacent *Eco*RI and *Bam*HI sites. These sites are the same as on pMC1403 (Fig. 1). Sequences upstream to the *leu2* promoter on this fragment could readily be deleted on this plasmid by using the unique *Eco*RI site on the vector. The plasmid clone was opened in the upstream site with *Eco*RI, partially degraded with the exonuclease *Bal*31, and the remaining fragment was excised by cutting on the other side with *Bam*HI. These deleted fragments, with a blunt end on one side formed by the *Bal*31 and a *Bam*HI end on the other side, were then ligated into the original pMC1790 vector, which had been cleaved with the blunt *Sma*I and with *Bam*HI. This resulted in a series of deletions of sequences on the upstream side of the *leu2* promoter, without any removal of vector sequences. These deletions are now being examined for their size and their *leu2* promoter regulation function in yeast.

Comments

We note that it may not be possible to fuse some gene control elements to the β-galactosidase gene with these plasmids. Some fragments may be lethal to the cell on a high copy plasmid. A promoter may be so strong that too much of a *lac* gene product is made, or transcription may proceed into the pBR322 replicon to repress replication (Renaut *et al.*[20] discuss this problem). Some hybrid protein β-galactosidases are lethal to the cell when synthesized at a high level, as is known for some membrane-bound or membrane-transported protein fusions.[21] In addition, the regulation of a gene may be altered when it is present in high copy number.

Some of these problems can be reduced by incorporating the fusion into a chromosome or a stable low copy number plasmid. In *E. coli,* a fusion can be incorporated onto a λ *lac* phage and integrated into the chromosome by lysogenization.[22] Alternatively, a fusion on a ColE1 replicon plasmid (as are the plasmids in this chapter) can be directly integrated by homologous recombination following introduction into a *pol*A^- cell that cannot replicate ColE1 replicons.[23] In yeast, plasmids without a yeast replicon can be integrated by homologous recombination after transformation into a strain with selection for a gene on the plasmid.[24]

Constructing a β-galactosidase gene fusion is more than a way to measure gene expression. Once it is made, it can be used as a source for further studies such as to isolate mutants with altered gene expression or regulation.[25] A translational fusion can also be used as a source of hybrid protein (see above).

β-Galactosidase fusions can be used in more species for which DNA transformation is possible, such as in gram-positive bacteria *Bacillus subtilis* and *Streptomyces* and in higher eukaryotes, such as plants and mammalian tissue culture cells.

Acknowledgments

We would like to acknowledge suggestions, help, and comments from our colleagues M. Berman, M. Ditto, M. Malamy, D. Nielsen, S. J. Suh, and H. Tu. This work was supported by NIH Grant GM 29067. S.K.S. is a Medical Scientist Trainee supported by NIH grant PHS 5T32 GM07281.

[20] E. Remaut, P. Stanssens, and W. Fiers, *Gene* **15,** 81 (1981).

[21] M. Hall and T. Silhavy, *Annu. Rev. Genet.* **15,** 91 (1981).

[22] M. Casadaban, *J. Mol. Biol.* **104,** 591 (1976).

[23] D. Kingsbury and D. Helinski, *Biochem. Biophys. Res. Commun.* **41,** 1538 (1970).

[24] D. Botstein, C. Falco, S. Stewart, M. Brennan, S. Scherer, D. Stinchcomb, K. Struhl, and R. Davis, *Gene* **8,** 17 (1979).

[25] J. Beckwith, *Cell* **23,** 307 (1981).

[22] Efficient Transfer of Small DNA Fragments from Polyacrylamide Gels to Diazo or Nitrocellulose Paper and Hybridization

By ERICH FREI, ABRAHAM LEVY, PETER GOWLAND, and MARKUS NOLL

A major breakthrough in DNA sequence analysis has been achieved by the gel to paper transfer technique developed by Southern.[1,2] It permits a rapid and precise analysis of the size distribution of a few specific DNA sequences among a mixture of many different sequences. The inherent simplicity of the method is captivating. A mixture of defined DNA fragments that have been separated electrophoretically in an agarose gel is denatured and transferred to a nitrocellulose filter by blotting. The fragments of interest are then selectively detected by annealing with a specific radioactive probe. The main limitation of the original method, as pointed out by Southern,[1,2] is that binding to nitrocellulose of DNA fragments shorter than a few hundred base pairs decreases rapidly with size. An improved method of transfer to nitrocellulose[3] overcomes this problem.

The replacement of nitrocellulose by diazobenzyloxymethyl (DBM) paper introduced by Alwine *et al.*[4,5] represents an important complementation of Southern's technique. In contrast to nitrocellulose, DBM paper binds single-stranded DNA covalently and hence also retains DNA fragments smaller than 150 base pairs very efficiently.[6-8] Therefore, for the analysis of small DNA fragments according to Southern's technique, DBM paper may be used as matrix. Since satisfactory resolution of small DNA fragments is achieved only in polyacrylamide gels, we need a procedure that transfers DNA from polyacrylamide gels to DBM paper with high efficiency. This has become feasible either by blotting[7] or by electrophoretic transfer[8,9] directly from polyacrylamide gels without the use of

[1] E. M. Southern, *J. Mol. Biol.* **98,** 503 (1975).
[2] E. Southern, this series, Vol. 68, p. 152.
[3] G. E. Smith and M. D. Summers, *Anal. Biochem.* **109,** 123 (1980).
[4] J. C. Alwine, D. J. Kemp, and G. R. Stark, *Proc. Natl. Acad. Sci. U.S.A.* **74,** 5350 (1977).
[5] J. C. Alwine, D. J. Kemp, B. A. Parker, J. Reiser, J. Renart, G. R. Stark, and G. M. Wahl, this series, Vol. 68, p. 220.
[6] J. Reiser, J. Renart, and G. R. Stark, *Biochem. Biophys. Res. Commun.* **85,** 1104 (1978).
[7] A. Levy, E. Frei, and M. Noll, *Gene* **11,** 283 (1980).
[8] M. Bittner, P. Kupferer, and C. F. Morris, *Anal. Biochem.* **102,** 459 (1980).
[9] E. J. Stellwag and A. E. Dahlberg, *Nucleic Acids Res.* **8,** 299 (1980).

METHODS IN ENZYMOLOGY, VOL. 100

ISBN 0-12-182000-9

composite gels that lengthen the manipulations of the gel before transfer and hence result in a low transfer efficiency.[6]

Principle of the Method

The probable mechanism of the covalent attachment of the single-stranded DNA to DBM paper has been described in detail.[4,5] It is thought that the negatively charged DNA interacts electrostatically with the positively charged diazonium groups on the paper first. At a much slower rate the diazonium groups form covalent bonds. These occur principally at the 2-position of deoxyguanosine and less frequently of thymidine residues, whereas no significant coupling to adenosine or cytidine is observed.[10,11] The finding that DBM paper, after incubation for 1 day at 4°, is still able to bind nucleic acids quantitatively yet has lost its ability to form covalent bonds[9] suggests, however, that the primary interaction with the paper is not mediated only by the positively charged diazonium groups. The diazotized form of the paper is stabilized by low pH whereas the coupling reaction is favored by alkaline conditions. The highest transfer efficiencies have been obtained by compromising on a pH between 4.0 and 6.5.[4,5,7-9]

After linkage of the DNA to the DBM paper, specific DNA sequences are detected by hybridization with a radioactive probe. The efficiency of hybridization drops dramatically with decreasing size of paper-bound small DNA fragments.[7] This reduction in hybridization efficiency, however, is attributable not only to the decrease in DNA size. An appreciable factor probably represents also the number of links by which a DNA molecule is coupled to the DBM paper. Thus, the efficiency of hybridization can be improved significantly by reducing the density of the diazonium groups on the paper. Since normally only the combined efficiency of DNA transfer and hybridization is of interest, optimal conditions do not coincide with those for highest transfer efficiency, as will be demonstrated in the Methods section.

Materials and Reagents

Synthesis of 1-[(*m*-nitrobenzyloxy)methyl]pyridinium chloride (NBPC) is carried out as described by Alwine *et al.*[4,5] It is important to store NBPC in small portions under N_2 in sealed tubes at −30°. After repeated warming up and exposure to air, NBPC deteriorates quickly;

[10] B. E. Noyes and G. R. Stark, *Cell* **5**, 301 (1975).

[11] It has been reported, however, that AMP and CMP bind covalently to DBM paper with relative affinities of 40–45% and 20% of that of GMP, respectively.[9]

this manifests itself in DBM paper of much reduced transfer and hybridization efficiency. For the reaction of NBPC with Whatman 540 paper and preparation of DBM paper, we have slightly modified the procedure of Alwine *et al.*[4,5] to reduce the background and to improve both the efficiency and the capacity of DNA binding.[7] This procedure has been worked out for large DNA fragments that are transferred by blotting.[12] The main modification consists in the repeated application of NBPC to the Whatman 540 paper.[7] As will be shown below, in order to generate optimal conditions for transfer as well as for subsequent hybridization of *small* DNA fragments, it is better to reduce the density of the diazonium groups on the paper by lowering the amount of NBPC applied to the paper.

Preparation of DBM Paper

A sheet of Whatman 540 paper (20.5 cm × 17 cm) is saturated in a freshly prepared solution of NBPC (0.8 g of NBPC in 8 ml of 70% ethanol) and rubbed with gloves to remove small bubbles. In order to achieve an even distribution of the NBPC and thus of the transfer, it is important to place the stretched paper in front of a strong fan for drying. For this purpose, we use a plastic frame into which the edges of the paper are inserted. After about 10 min, the paper is transferred to an oven at 60° for 10 min. This first treatment of the paper with NBPC is skipped for the preparation of DBM paper that produces best results with small DNA fragments (≤300 base pairs).

The paper is removed from the frame, saturated again in NBPC (a solution of 8 ml of 70% ethanol containing 0.8 g of NBPC[13] and 0.25 g of sodium acetate), and the previous steps are repeated except that drying at 60° continues for 20 min or slightly longer. When the paper is dried completely, it is transferred to 130–135° for 35–40 min, washed with water 3 to 5 times during 20 min at room temperature, dried completely at 60° (for at least 30 min), washed twice for 5 min with about 300 ml of acetone,[14] and air-dried. This NBM (nitrobenzyloxymethyl) paper is relatively stable and can be stored for many months in a desiccator at 4°.[5]

For conversion to DBM paper, the NBM paper is shaken for 30 min at 60° in about 350 ml of 20% (w/v) sodium dithionite, followed by washes in water, 30% acetic acid, and again water, for 5 min each. The following steps are carried out on crushed ice. The paper is shaken twice for 15 min in a solution made by mixing 6.4 ml of a freshly prepared solution of

[12] A. Levy and M. Noll, *Nature (London)* **289,** 198 (1981).

[13] The optimal amount of NBPC used in this step for transfer and hybridization of small DNA is about 0.16 g, as shown in Fig. 5.

[14] G. M. Wahl, M. Stern, and G. R. Stark, *Proc. Natl. Acad. Sci. U.S.A.* **76,** 3683 (1979).

$NaNO_2$ (10 mg/ml) with 240 ml of cold 1.2 *M* HCl. At this stage, the paper can be stored at −85° after freezing in liquid nitrogen.[5] Immediately before transfer, the paper is washed 5 times for 0.5 min each in about 300 ml of ice-cold water and finally for 5 min in ice-cold transfer buffer (21 × transfer buffer: 31 ml of 2 *M* citric acid, 80 ml of 1 *M* Na_2HPO_4, 889 ml of H_2O; this represents a threefold dilution of the buffer suggested to us by G. Stark and used in previous experiments (Figs. 1–4).[7]

Preparation of Labeled DNA

Specific DNA probes used for hybridization were labeled by nick translation according to Maniatis *et al.*[15]

*Hpa*II-restricted DNA of pBR322 and of M13mp8 was labeled with [α-^{32}P]dCTP of the Klenow fragment of DNA polymerase I (Boehringer, Mannheim).

DNA extracted from nuclei of *Drosophila* Kc cells that had been digested with DNase I[16] was treated with phosphate and end-labeled with [γ-^{32}P]ATP by polynucleotide kinase as described.[17]

Methods

Preparation of Polyacrylamide Step Gels

On examining the transfer of DNA fragments from polyacrylamide gels to DBM paper, we have noticed that whereas large DNA molecules are transferred by blotting rather inefficiently, it is possible to blot small DNA fragments if the polyacrylamide concentration is sufficiently low (4% or less). Unfortunately, such low concentrations are not suitable for achieving a high resolution of small DNA fragments. Therefore, we have combined the advantage of high resolution obtained in a gel of high polyacrylamide concentration with that of efficient DNA transfer from a gel of low polyacrylamide concentration by using a step gel.[7,18] Step gels are prepared as slab gels (25 cm × 18 cm × 0.15 cm) and consist of a short (4.2

[15] T. Maniatis, A. Jeffrey, and D. G. Kleid, *Proc. Natl. Acad. Sci. U.S.A.* **72,** 1184 (1975).

[16] M. Noll, *Nucleic Acids Res.* **1,** 1573 (1974).

[17] M. Noll, *J. Mol. Biol.* **116,** 49 (1977).

[18] More recently and in the experiment shown in Fig. 5, we switched to electrophoretic transfer[8,9,19] because it is faster than blotting and also DNA fragments larger than 300 base pairs are transferred with high efficiency from polyacrylamide gels.[9] For electrophoretic transfer, step gels are not necessary, although their range of high resolution is probably wider than that of a normal high percentage polyacrylamide gel (cf. Figs. 1 and 5).

[19] H. Towbin, T. Staehelin, and J. Gordon, *Proc. Natl. Acad. Sci. U.S.A.* **76,** 4350 (1979).

cm) polyacrylamide gel of high percentage (8%) above a long one (20.8 cm) of low percentage (4%). We have used step gels for nondenaturing as well as for denaturing urea gels.

Urea Gel. Denaturing gels are prepared according to Maniatis *et al.*[20] For the 4% bottom gel, 25.2 g of urea (ultrapure, Schwarz–Mann) are dissolved in 6 ml of 40% acrylamide, 4 ml of 2% bisacrylamide, 6 ml of 10 × TBE buffer (108 g Tris base, 55 g boric acid, 9.3 g Na_2EDTA per liter), 2.4 ml of 1.6% ammonium persulfate, and water is added to a volume of 60 ml. The solution is degassed, supplemented with 20 μl of TEMED, poured to a height of 21 cm, and overlayered carefully with TBE buffer. It is important to keep the interface perpendicular to the future direction of electrophoresis. After polymerization, the interface is washed several times with TBE buffer, and the 8% upper gel is poured. For preparation of this gel, 6.3 g of urea are dissolved in 3 ml of 40% acrylamide, 2 ml of 2% bisacrylamide, 1.5 ml of 10 × TBE buffer, and 0.6 ml of 1.6% ammonium persulfate; water is added to a volume of 15 ml. After degassing, polymerization is initiated by the addition of 5 μl of TEMED. During polymerization of the upper part, the gel is maintained in a slanted position to allow the addition of an excess of solution above the inserted comb. This ensures that polymerization occurs also above the level of the comb teeth.

The gel is prerun in TBE buffer for 90 min at 300 V. The lyophilized DNA samples are dissolved in sample buffer (0.5 g of Ficoll dissolved in 7.5 ml of 98% formamide, 20 m*M* sodium phosphate, pH 7.0, bromophenol blue), heated for 30 sec at 100°, chilled quickly, and loaded onto the gel. Electrophoresis is carried out at 300 V for about 4 hr at the end of which the bromophenol blue is about 6 cm from the bottom of the step gel.

Nondenaturing Gel. Nondenaturing step gels are prepared and prerun as urea step gels except that the urea is omitted. The DNA samples are dissolved in 0.25 × TBE buffer, 3% Ficoll, bromophenol blue and applied to the gel. Electrophoresis occurs at 200 V for about 6 hr (bromophenol blue is 6 cm from the bottom of the step gel).

Transfer of DNA to DBM Paper

The procedures of DNA transfer from polyacrylamide gels to DBM papers differ for urea and nondenaturing gels because only single-stranded DNA is bound covalently to DBM paper.[5]

Transfer of Small DNA Fragments Run under Nondenaturing Condition. After electrophoresis one of the two glass plates enclosing the gel is removed, and the 4% polyacrylamide portion of the gel supported by the other plate is rocked in transfer buffer (see Preparation of DBM Paper for

[20] T. Maniatis, A. Jeffrey, and H. van deSande, *Biochemistry* **14,** 3787 (1975).

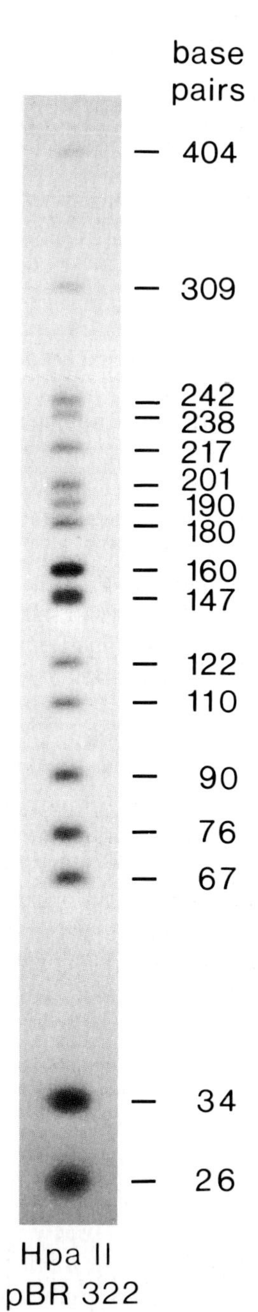

FIG. 1. Small DNA fragments were separated in a nondenaturing polyacrylamide step gel and transferred to diazobenzyloxymethyl (DBM) paper.[7] *Hpa*II restriction fragments of pBR322 DNA (1 μg) labeled at their ends with [α-^{32}P]dCTP were separated in a step gel and transferred to DBM paper. An autoradiogram of the DBM paper is shown. Fragment lengths are known from their sequence.[22]

composition) for 10 min at room temperature. The gel on the glass plate is slid into a plastic bag (Seal-N-Save, Sears), the bag is sealed, and the gel is heated for 10 min in boiling water to denature the DNA and chilled quickly in an ice-water bath.[21] The following steps are carried out in the cold room. The gel is removed from the bag on the glass plate and overlayered with 3–4 sheets of Whatman 3 MM paper saturated with transfer buffer. Trapping of air bubbles between the gel and the papers must be avoided. The gel, sandwiched between the glass plate and the paper, is turned over quickly. It is carefully detached from the glass plate and placed with the supporting Whatman papers on a 2 cm-thick rubber foam sponge (of high soaking capacity) immersed in cold transfer buffer in a tray.

To prevent the buffer from bypassing the gel, its edges are covered on all sides with Saran strips. The wet DBM paper is laid on the gel. Again no air bubbles should occur between the gel and the DBM paper. The DBM paper is covered with two layers of dry Whatman 3 MM paper and a thick layer of blotting papers pressed down by a small weight. The wet blotting papers are removed every hour, and the tray is refilled with transfer buffer 3 or 4 times. Transfer is continued for 1 day at 4° and for another half-day at room temperature.

The choice of a 8%/4% step gel results in an excellent resolution of DNA fragments smaller than a few hundred base pairs as evident from Fig. 1.[22] It shows *Hpa*II-restricted DNA of pBR322 after its transfer from the gel to DBM paper. The DNA fragments transferred from the 4% polyacrylamide gel are highly resolved up to at least 250 base pairs (238 and 242 bp are well separated) in a run where 400 base pair fragments just enter the lower gel. Transfer appears to occur with fairly constant efficiency for fragments shorter than 250 base pairs (the bands of 26, 34, 147, and 160 bp are doublets) although the efficiency clearly drops for longer DNA (Fig. 1).[18]

This conclusion is supported by quantitative analysis of the transfer efficiency as a function of DNA size shown in Fig. 2a. Transfer efficiency was measured for each restriction fragment by the ratio of the amount of labeled DNA bound to DBM paper to that of the DNA in the gel after electrophoresis.[23] To ensure that only single-stranded DNA that is bound

[21] This procedure of denaturing the DNA in the gel represents a slight modification of that described previously.[7] It is unlikely to affect the efficiency of DNA transfer.

[22] J. G. Sutcliffe, *Nucleic Acids Res.* **5,** 2721 (1978).

[23] Equal amounts of the same DNA samples were run in parallel in two lanes of a polyacrylamide step gel. The DNA of one lane was transferred to DBM paper. For *Hpa*II-restricted pBR322 DNA, individual bands were cut out of the gel and the paper, whereas for DNA obtained from DNase I digests of *Drosophila* nuclei, the entire lane of the gel and the paper were divided into adjacent sections. The amount of ^{32}P-labeled DNA was

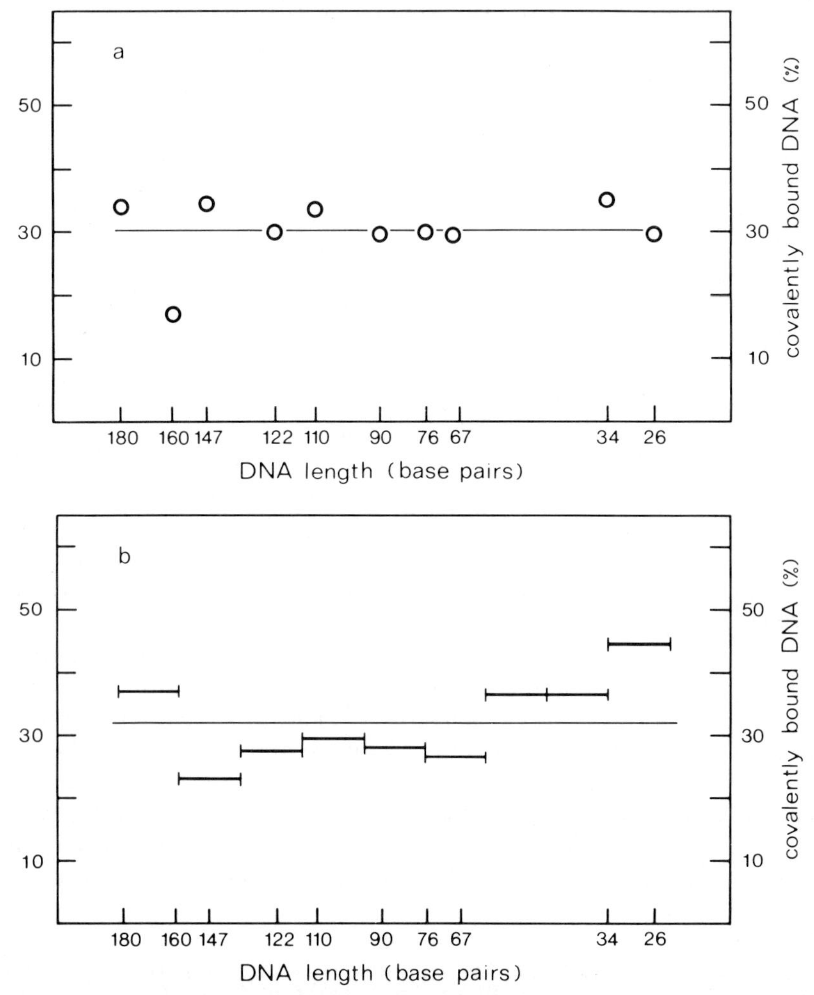

FIG. 2. Transfer efficiency to diazobenzyloxymethyl (DBM) paper of small DNA fragments separated in a nondenaturing polyacrylamide step gel.[7] (a) Labeled *Hpa*II-restricted pBR322 DNA (1 μg). (b) DNA (50 μg) extracted from DNase I-digested *Drosophila* nuclei and end-labeled with [γ-^{32}P]ATP. Efficiency of transfer was measured as described in footnote 23.

covalently to the paper was measured, the paper was treated, after complete mock hybridization, for 1 hr with 0.5 *M* NaOH. The average transfer efficiency for DNA smaller than 180 base pairs is about 30% (Fig. 2a).

determined by Cerenkov counting and corrected for differences in counting efficiencies (28% for DNA bound to DBM paper, 43% and 47% for DNA in the nondenaturing and in the 7 *M* urea 4% polyacrylamide gel, respectively).

Similar efficiencies were obtained when the alkaline treatment was omitted (not shown), indicating that the DNA does not reanneal significantly during the transfer. The high transfer efficiency demonstrates that the procedure of denaturing the DNA before transfer quickly by heating the gel at 100° is not only simple but also efficient. When larger DNA fragments are transferred from a 4% polyacrylamide gel, the efficiency drops considerably. It is reduced to 20% in the range of 300 to 400 base pairs, and it decreases to 10% between 500 and 600 base pairs.[18]

The same average transfer efficiency is observed when 30 times less of *Hpa*II-restricted pBR322 DNA (0.03 μg) is loaded onto the gel (our unpublished observation), which suggests that the efficiency of 30% is not limited by the binding capacity of the DBM paper. Even when 50 μg of DNA extracted from DNase I-digested *Drosophila* nuclei (of which about 80% is smaller than 200 bp) are separated on a step gel, the transfer efficiency remains about the same (Fig. 2b).[23]

Transfer of Small DNA Fragments from Urea Gel. Transfer from a urea gel is carried out like that from a nondenaturing gel except that no denaturation of the DNA by heating the gel at 100° is required. Thus, after rocking the gel supported by a glass plate in transfer buffer for 5 min at 4°, transfer is started immediately as described above.

A 8%/4% step gel containing 7 *M* urea is also suitable for separating small single-stranded DNA. Transfer from urea gels to DBM paper is even better (Fig. 3) than from corresponding nondenaturing gels (Fig. 2b). Again the binding capacity of the DMB paper does not limit the transfer efficiency to 50% when 50 μg of DNA are loaded because the efficiency is not increased at considerably lower DNA inputs (our unpublished observation). Furthermore, the transfer efficiency may be considered constant in the range examined, i.e., for DNA smaller than 200 nucleotides (Fig. 3).

Hybridization with a Specific Probe

Since the DNA strands bound to the DBM paper are rather short, it is important to test whether they are still able to hybridize with a specific probe. To this end, 50 μg of unlabeled DNA of DNase I-digested *Drosophila* nuclei are separated in a denaturing step gel, transferred to DBM paper, and hybridized with a ^{32}P-labeled cloned *Drosophila* DNA sequence.

After transfer, the DBM paper is prehybridized for 2–3 hr at 65° in about 100 ml of 6 × SSC containing 1 × Denhardt's solution[24] (0.02% each of Ficoll, poly(vinylpyrrolidone), and bovine serum albumin), 1% glycine,[4] and 2 mg of denatured herring sperm DNA per milliliter. Thereafter, SDS is added to a concentration of 0.1%, and prehybridization is

[24] D. T. Denhardt, *Biochem. Biophys. Res. Commun.* **23,** 641 (1966).

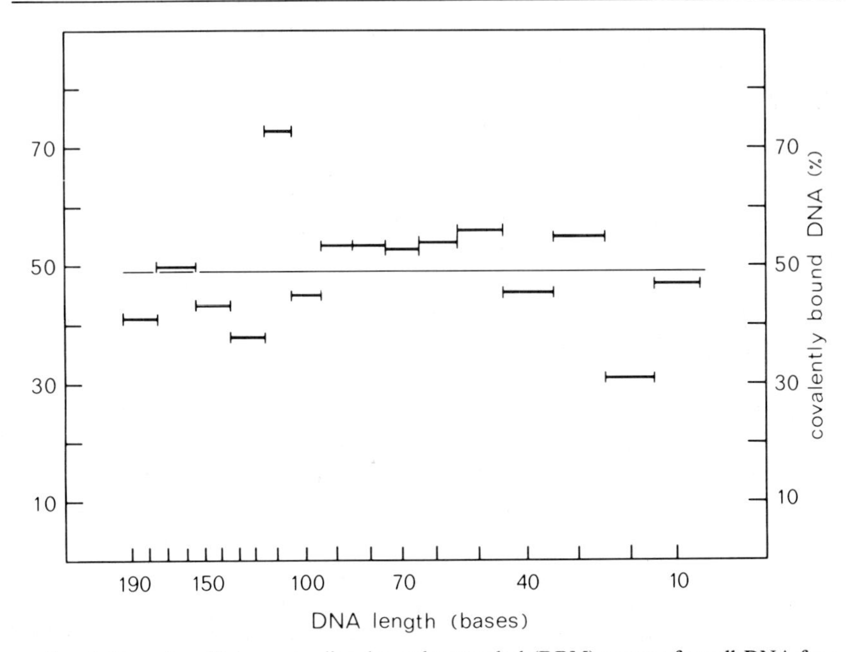

FIG. 3. Transfer efficiency to diazobenzyloxymethyl (DBM) paper of small DNA fragments separated in a denaturing polyacrylamide step gel containing 7 *M* urea.[7] End-labeled DNA fragments (50 μg) of *Drosophila* (see Fig. 2b) were separated in a step gel containing 7 *M* urea, and transferred to DBM paper. Transfer efficiency was determined as indicated in footnote 23.

continued overnight at 65°. Hybridization of the DBM paper with the radioactively labeled probe is carried out in a sealed plastic bag (Seal-N-Save, Sears) in 10 ml of 5 × SSC containing 1 × Denhardt's solution,[24] 1 m*M* EDTA, 0.1% SDS, and 50 μg of denatured herring sperm DNA per milliliter for 2 days at 37°, 50°, or 65°.[25] After hybridization, the paper is washed at room temperature 3 times for 30 min in 100 ml of 2 × SSC, 0.5% SDS, 1 m*M* EDTA, 16.6 m*M* sodium phosphate, pH 7.4, 0.05% $Na_4P_2O_7 \cdot 10\ H_2O$ and finally overnight in 100 ml of 2 × SSC, 0.5% SDS, 1 m*M* EDTA. For autoradiography, the paper is briefly washed with water, and, after removal of excess liquid, it is placed on a glass plate, covered with Saran wrap, and exposed.

The gel stained with ethidium bromide before transfer and the autoradiogram of the hybridizations carried out at the various temperatures are

[25] Hybridization can be improved and shortened to 8 hr by including 10% dextran sulfate in the hybridization mixture and by raising the concentrations of Ficoll, poly(vinylpyrrolidone), and bovine serum albumin to 0.1% both during prehybridization and hybridization.[14]

shown in Fig. 4. The bands revealed are multiples of about 10 nucleotides and represent the familiar pattern obtained after digestion of chromatin with DNase I.[16] They are well resolved up to a size of 200 nucleotides.

The optimal temperature for hybridization under the conditions chosen is about 50° (Fig. 4). At this temperature, the efficiency of hybridization is highest, and hybridization is detectable to fragments as short as 30 nucleotides (20 nucleotides fail to form stable hybrids; not shown). Since the efficiency of transfer is fairly constant below 200 nucleotides (Fig. 3), the ethidium bromide pattern (which reflects the DNA mass) may be compared with the patterns of hybridization (which measure the number of hybridized DNA molecules) at 50° or 65° (Fig. 4). It is evident from such a comparison that the efficiency of hybridization decreases drastically below 83 base pairs. This effect is most pronounced at 65°. Hybridization at 37° is poor over the entire size range examined. For consideration of hybridization efficiencies, we have assumed that the sequences hybridizing to the probe[26] exhibit the same sensitivity to DNase I as bulk chromatin, because above 100 nucleotides no differences in sensitivity are observed between sequences represented by the probe (50° and 65° in Fig. 4) and those of bulk chromatin (EthBr in Fig. 4).

Since no hybridization is detected when pBR322 DNA is used as a probe (not shown), hybridization conditions do not permit unspecific reannealing. Unspecific binding, however, is observed at 50° or lower temperatures when the carrier DNA is replaced by guanosine (our unpublished observation).

Efficiency of Hybridization as a Function of DBM Concentration

Because the transferred DNA molecules are coupled to the DBM paper through their bases, multiple links to the DBM paper of a DNA molecule might interfere with hybridization to the radioactive probe. In addition, hybridization to DNA molecules linked at multiple sites to the DBM paper could decrease the stringency of hybridization. Hence, it is desirable to optimize hybridization in relation to the DBM concentration on the paper.

Since the diazonium groups are relatively short lived,[9] their effective concentration depends on the amount of NBPC applied to the paper as well as on the velocity and conditions of DNA transfer. Rapid transfer favors multiple covalent coupling of DNA molecules. Therefore, transfer by blotting requires larger amounts of NBPC for the preparation of DBM paper to achieve optimal hybridization than the faster electrophoretic

[26] L. Moran, M.-E. Mirault, A. Tissières, J. Lis, P. Schedl, S. Artavanis-Tsakonas, and W. J. Gehring, *Cell* **17,** 1 (1979).

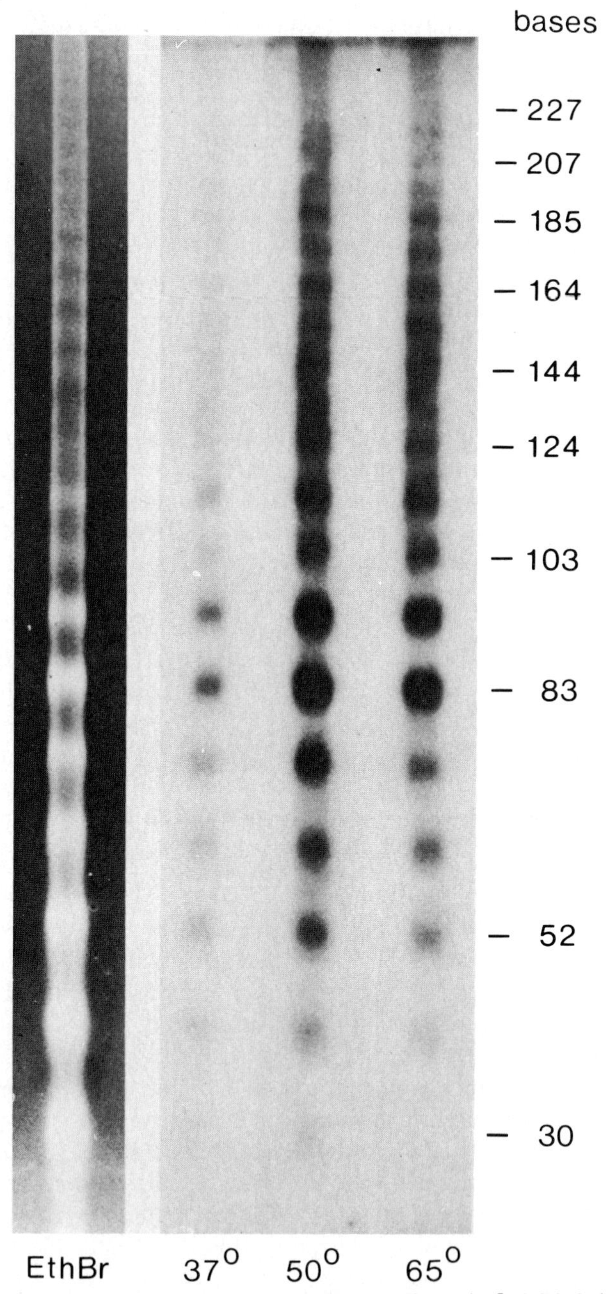

FIG. 4. Hybridization of transferred DNA with a specific probe.[7] Unlabeled DNA fragments, derived from the DNase I-digested *Drosophila* nuclei, were separated under denatur-

transfer (our unpublished observation). As electrophoretic transfer is the more economical method with respect to NBPC consumption, we examine the efficiency of hybridization as a function of the NBPC applied to the paper only for this method.

For electrophoretic transfer from polyacrylamide gels to DBM paper, we follow a procedure similar to that described by others.[8,9,19] We use a destaining apparatus whose electrodes consist of two parallel 18 cm × 20 cm metal grids 5.5 cm apart and mounted in a plastic vessel with inside dimensions of 19.5 cm × 19.5 cm × 7.5 cm. The DNA is denatured in the gel as for transfer by blotting. The DBM paper wetted in transfer buffer is carefully laid on top of the gel such that no air bubbles get trapped. The gel and the DBM paper are sandwiched first between two wet Whatman 3 MM papers on each side, then between two Scotch-Brite pads soaked in ice-cold transfer buffer, and finally between two plastic grids. The whole assembly is held together tightly by rubber bands and submerged in ice-cold transfer buffer in the destaining chamber which is cooled in an ice-water bath. Transfer occurs at 32 V resulting in a current of 0.8 A, which increases during the 2 hr of transfer to 1.2 A.[27]

The effect of the diazonium group concentration on transfer efficiency, binding capacity, and efficiency of hybridization is shown in Fig. 5. A set of eight DBM papers differing in diazonium group concentration was prepared. One paper was treated twice with NBPC as described above (2 ×) whereas for the preparation of the other papers the first treatment was omitted and decreasing amounts of NBPC were applied in the remaining treatment (1 ×, 0.5 ×, 0.25 ×, 0.1 ×, 0.05 ×, 0.025 ×, and 0 × the amount of NBPC indicated under Preparation of DBM Paper). Capacity and transfer efficiency are tested for each paper in four lanes containing in addition to end-labeled tracer DNA (3000 dpm per lane) 0, 1.5, 5, and 10 μg of *Hpa*II-restricted M13mp8 DNA. On each side of these lanes, the efficiency of hybridization is assessed from hybridization of a radioactive probe with a *Hae*III–*Rsa*I–*Taq*I digest of *Drosophila* DNA from Kc cells. As probe the hybrid plasmid 12D1 was used, which contains about 23 *Drosophila* 5 S gene repeats.[28] In addition to the final

[27] Transfer is nearly complete after 30 min; during prolonged transfer covalent links between the DNA and the DBM paper are formed.[9]

[28] S. Artavanis-Tsakonas, P. Schedl, C. Tschudi, V. Pirrotta, R. Steward, and W. J. Gehring, *Cell* **12,** 1057 (1977).

ing conditions. The DNA (50 μg) of one gel was stained with ethidium bromide (EthBr) while the DNA (25 μg) of another gel was transferred to diazobenzyloxymethyl (DBM) paper and hybridized with a ^{32}P-labeled specific probe of *Drosophila* DNA at the temperature indicated. The DNA probe used consists of the *Bgl*II–*Bgl*II fragment of the hybrid plasmid 132E3.[26]

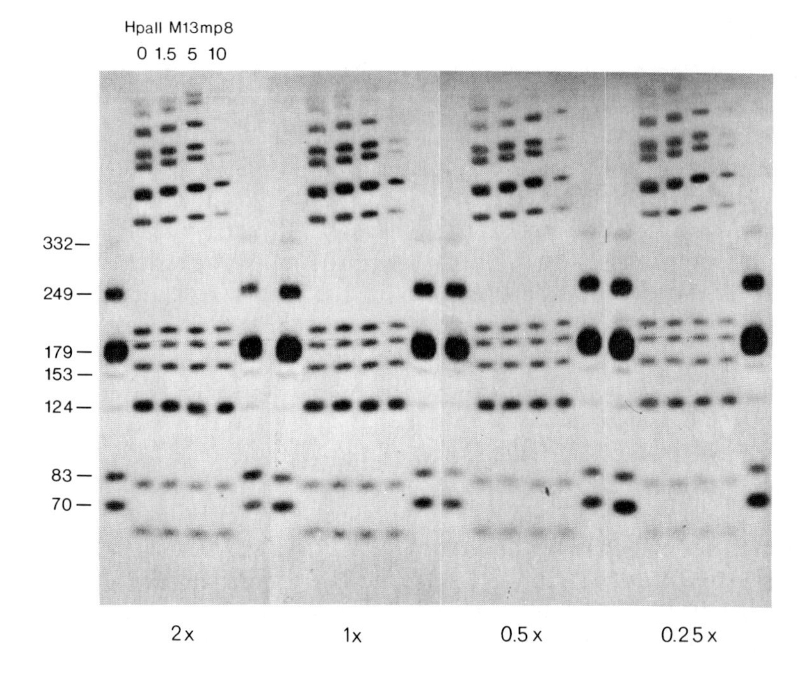

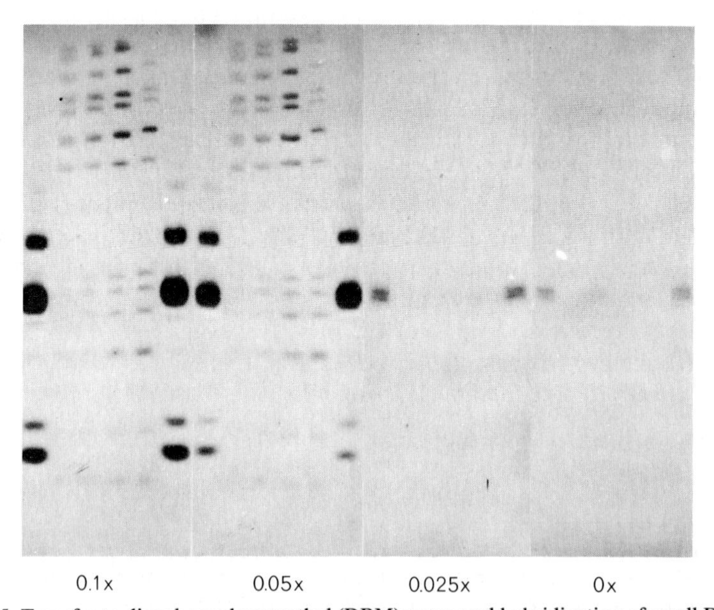

FIG. 5. Transfer to diazobenzyloxymethyl (DBM) paper and hybridization of small DNA as a function of DBM concentration. Various amounts of *Hpa*II-restricted M13mp8 DNA and a *Hae*III–*Rsa*I–*Taq*I genomic digest of *Drosophila* DNA (1 μg per lane) were separated

products [41 (not visible), 70, 83, and 179 ± 7 bp], fragments resulting from partial digestion with *Rsa*I and *Taq*I are obtained (124, 153, 249 ± 7, and 332 ± 7 pb) whose lengths agree well with those derived from the *Drosophila* 5 S gene sequence.[29]

It is evident from Fig. 5 that transfer efficiency decreases with the amount of NBPC applied to the cellulose paper (about 4-fold between 2 × and 0.05 × NBPC), yet is not abolished completely in the absence of DBM groups (0 ×) as signals of hybridization are still detectable. This reduction in transfer efficiency is not caused by a limiting binding capacity of the DBM paper, i.e., the probability for a small DNA molecule (less than 300 bp) to be bound to the DBM paper (≳0.05 × NBPC) during transfer remains constant over the input range examined (≲10 μg of *Hpa*II-restricted M13mp8 DNA). Hybridization efficiency, on the other hand, increases strongly with decreasing amounts of NBPC, such that the overall efficiency of transfer and hybridization reaches an optimum around 0.25 × to 0.1 × NBPC (Fig. 5). Considering that transfer efficiency is reduced about 2- to 4-fold at this DBM concentration (compared to 2 × NBPC), we estimate that hybridization efficiency is enhanced approximately 10-fold since the overall efficiency appears to be increased about 3-fold.

The enhancement of the efficiency of transfer and hybridization observed at lower diazonium group concentrations is not the same for all DNA fragments and probably depends on base composition. In addition, we find reproducibly that the background is elevated at 1 × NBPC (as evident from longer exposures) yet is reduced if the treatment is repeated (2 ×). We have compared DBM papers treated once with 0.1 × or 0.2 × NBPC or twice with 0.1 × NBPC and found that in this range of optimal diazonium group concentration repeated treatment also slightly reduces the background. However, DMB paper treated once with 0.2 × NBPC showed a better signal-to-noise ratio after hybridization than DBM paper treated twice with 0.1 × NBPC (not shown). Thus, optimal DBM paper to be used for the electrophoretic transfer of small DNA fragments is obtained after a single treatment with about 0.2 × NBPC.

[29] C. Tschudi and V. Pirrotta, *Nucleic Acids Res.* **8**, 441 (1980).

electrophoretically in nondenaturing 6% polyacrylamide gels in TBE buffer[20] and transferred electrophoretically to DBM papers of various diazonium group concentrations as described in the text. Hybridization with 10^7 dpm of nick-translated 12D1 DNA per paper occurred for 24 hr at 55° as described above except that 2 mg of carrier DNA per milliliter were present. Kodak XAR-5 films were preflashed [see R. A. Laskey and A. D. Mills, *FEBS Lett.* **82,** 314 (1977)] and exposed for 24 hr at −70° with a Kyokko intensifying screen.

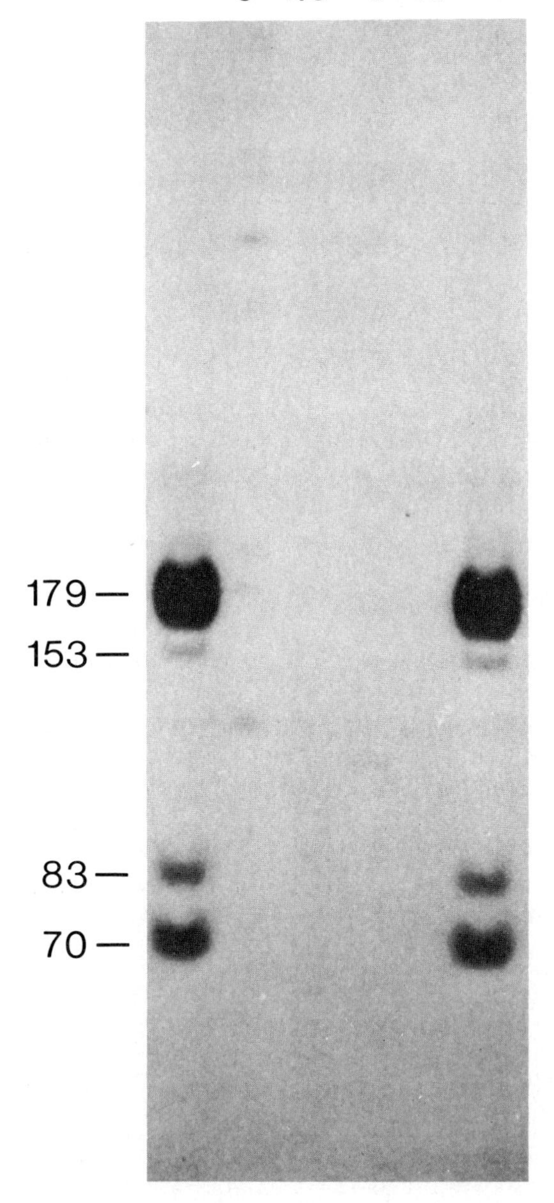
HpaII M13mp8
0 1.5 5 10
179—
153—
83—
70—

Brian Seed[30] has developed a simpler method to prepare diazo paper. We have prepared and tested this so-called DPT(diazophenyl thioether) paper for transfer and hybridization efficiency of small DNA fragments and have found it clearly less efficient than DBM paper.

Efficient Transfer of Small DNA to Nitrocellulose Paper and Hybridization

It has been shown that binding to nitrocellulose of small, denatured DNA fragments can be improved significantly if they are transferred in 1 *M* NH_4 acetate, 0.02 *N* NaOH[3,31] rather than in 20 × SSC. Essentially following the procedure of Smith and Summers,[3] we have transferred DNA fragments identical to those shown in Fig. 5 to nitrocellulose paper and hybridized with the same radioactive probe (Fig. 6).

Before transfer, the 6% polyacrylamide gel was washed for 10 min in 0.25 *N* HCl to reduce the DNA size[14] (this treatment is necessary only for DNA fragments $\gtrsim$600 bp), and the DNA was denatured by heating the gel for 10 min in a sealed bag containing 200 ml of 1 *M* NH_4 acetate, 0.02 *N* NaOH and immersed in boiling water. The gel was chilled in an ice-water bath and laid on two Whatman 3 MM papers of the same size as the gel and wetted in 1 *M* NH_4 acetate, 0.02 *N* NaOH. A sheet of wet nitrocellulose paper (Schleicher & Schuell), two wet sheets of Whatman 3 MM paper, and a stack of blotting paper were placed on top and pressed down by a small weight. Care was taken to avoid air bubbles between the Whatman papers, the gel, and the nitrocellulose filter. After 24 hr of transfer at room temperature, the filter was baked for 2 hr at 80°, washed in 100 ml of 4 × SSC, 1 × Denhardt's solution,[24] 0.1% SDS, 16.6 m*M* sodium phosphate, pH 7.4, 0.05% $Na_4P_2O_7 \cdot 10\ H_2O$ for 1 hr at 55°, and prehybridized in 100 ml of 4 × SSC, 5 × Denhardt's solution,[24] 0.1% SDS, 16.6 m*M* sodium phosphate, pH 7.4, 0.05% $Na_4P_2O_7 \cdot 10\ H_2O$, 20 μg of denatured herring sperm DNA per milliliter for 1 hr at 55°. The nitrocellulose paper was dried at room temperature, baked at 65° for 15 min, and

[30] B. Seed, Nucleic Acids Res. **10,** 1799 (1982).

[31] F. C. Kafatos, C. W. Jones, and A. Efstratiadis, *Nucleic Acids Res.* **7,** 1541 (1979).

FIG. 6. Transfer of small DNA fragments to nitrocellulose paper and hybridization. The same amounts of *Hpa*II-restricted M13mp8 DNA containing labeled tracer DNA and of a *Hae*III–*Rsa*I–*Taq*I genomic digest of *Drosophila* DNA were separated in a 6% polyacrylamide gel as in the experiment shown in Fig. 5. The DNA was transferred to nitrocellulose paper and hybridized with 2.6 × 10^7 dpm of nick-translated 12D1 DNA for 24 hr at 55° as described in the text. Preflashed [see R. A. Laskey and A. D. Mills, *FEBS Lett.* **82,** 314 (1977)] Kodak XAR-5 film was exposed for 24 hr at −70° with a Kyokko intensifying screen.

hybridized with labeled 12D1 DNA of the same specific activity as in the experiment shown in Fig. 5.

As evident from a comparison of Figs. 5 and 6, nitrocellulose retains small DNA fragments less efficiently than the best DBM paper. Particularly the binding capacity of nitrocellulose paper is much lower (no bands are detected when the labeled DNA is diluted with $\gtrsim 1.5$ μg of unlabeled *Hpa*II-restricted M13mp8 DNA). However, an enhanced efficiency of hybridization with DNA bound to nitrocellulose compensates for the reduced transfer efficiency so that the overall efficiencies of transfer and hybridization for nitrocellulose and DBM paper are comparable.

In conclusion, nitrocellulose and diazo papers are similar in sensitivity even for transfers of small DNA fragments. Transfer to nitrocellulose paper is probably preferable in most instances because it is simpler and cheaper. Transfer to diazo paper is recommended, however, when it is important that the *same* transfer be hybridized repeatedly with different labeled probes[12] or when conditions (e.g., urea gels) interfere with transfer to nitrocellulose.

Acknowledgment

This work has been supported by the Swiss National Science Foundation Grant 3.466.79.

[23] Electrophoretic Transfer of DNA, RNA, and Protein onto Diazobenzyloxymethyl Paper

By ALBERT E. DAHLBERG and EDMUND J. STELLWAG

The covalent coupling of discrete species of DNA and RNA, separated by gel electrophoresis, to diazobenzyloxymethyl (DBM) paper, and the subsequent hybridization with specific radioactive probes, has been a major advance in the purification and analysis of gene sequences.[1-4] DBM paper, unlike nitrocellulose, is capable of binding DNA fragments of all

[1] G. R. Stark and J. G. Williams, *Nucleic Acids Res.* **6,** 195 (1979).
[2] J. C. Alwine, D. J. Kemp, and G. R. Stark, *Proc. Natl. Acad. Sci. U.S.A.* **74,** 5350 (1977).
[3] J. Reiser, J. Renart, and G. R. Stark, *Biochem. Biophys. Res. Commun.* **85,** 1104 (1978).
[4] G. M. Wahl, M. Stern, and G. R. Stark, *Proc. Natl. Acad. Sci. U.S.A.* **76,** 3683 (1979).

ISBN 0-12-182000-9

sizes and does not have the problem of DNA loss during posthybridization washes. In addition, DBM paper is able to couple covalently with RNA and proteins.[2,5]

One of the difficulties associated with the use of DBM paper, however, has been the lability of reactive diazonium groups responsible for covalent coupling to macromolecules.[2,5] DNA, RNA, and protein transfers from agarose or acrylamide gels via blotting procedures may be slow and result in poor transfer and relatively low covalent coupling efficiencies.[3,6] Efficient transfer of large DNA molecules is achieved only by *in situ* DNA cleavage.[4] Other techniques intended to achieve faster blotting from gels to DBM paper have been, in our hands, slow and resulted in losses or chemical alteration of the macromolecules being transferred.

An electrophoretic transfer of macromolecules to DBM paper[7,8] avoids the limitations imposed by blotting techniques. This procedure allows efficient transfer of all sizes in intact DNA, RNA, protein, and ribonucleoprotein particles from a variety of gels to DBM paper without any special treatment of the gel or the macromolecules to be transferred. The process is rapid, thus achieving complete transfer within the time the paper is capable of forming covalent bonds. The transfer is direct, quantitative, and reproducible, preserving the sharpness of the original gel pattern on the DBM paper, and permitting the detection of very small amounts of DNA and RNA. The technique also applies to transfer to other papers, such as DPT paper, with appropriate adjustment in buffer and pH.[9]

Here we describe the electrophoretic transfer to DBM paper in a series of steps that include the preparation of the gel for transfer, the preparation of the DBM paper, the setting-up of the transfer apparatus, and the actual transfer itself. Gel methods for the initial separation of DNA, RNA, and protein are adequately described elsewhere and therefore are not discussed here. The transfer technique is essentially the same regardless of the component being transferred or the composition of the gel (agar, acrylamide, or a composite of agarose and acrylamide). Specific differences, however, are detailed below. In almost all cases a phosphate buffer is used for the transfers as it is inexpensive and has a good buffering capacity at the pH (5.5–6) where the DBM paper is stable. An

[5] J. Renalt, J. Reiser, and G. R. Stark, *Proc. Natl. Acad. Sci. U.S.A.* **76,** 3116 (1979).

[6] N. Rave, R. Crkvenjakov, and H. Boedtker, *Nucleic Acids Res.* **6,** 3559 (1979).

[7] E. J. Stellwag and A. E. Dahlberg, *Nucleic Acids Res.* **8,** 299 (1980).

[8] M. Bittner, P. Kupferer, and C. F. Morris, *Anal. Biochem.* **102,** 459 (1980).

[9] An excellent summary, "Methods for the Transfer of DNA, RNA and Protein to Nitrocellulose and Diazotized Paper Solid Supports" by M. Barinaga *et al.*, The Salk Institute, is available from Schleicher & Schuell, Keene, New Hampshire 03431.

exception, the transfer of polyribosomes and ribonucleoprotein particles, is described below.

Preparation of Gels for Electrophoretic Transfer

RNA Gels. Large RNAs in agarose or composite gels and small RNAs (tRNA) in high-percentage acrylamide gels are prepared for electrophoretic transfer by incubating the gel in 2 ml of 50 m*M* sodium phosphate, pH 5.5, per square centimeter of gel for 45–60 min at 4°. Gels containing rRNA species larger than 30 S are incubated in 3 ml of 25 m*M* sodium phosphate, pH 5.5, per square centimeter of gel for 90 min before transfer.

DNA Gels. DNA electrophoretically separated in agarose gels may be stained first with ethidium bromide for 30 min to identify bands, and then destained in distilled water for 1 hr at 20°. Prior to transfer, the DNA is denatured by incubation in 1 ml of 0.5 NaOH per square centimeter of gel for 20–30 min at 20°. The NaOH is decanted, and the gel is rinsed with distilled water. The gel is then incubated successively in 500, 50, and finally 25 m*M* sodium phosphate (pH 5.5) for 15, 15, and 30 min, respectively, at 4° in a volume of 1 ml per square centimeter of gel. If denaturation is not required (gels containing single-stranded DNA or double-stranded DNA with a poly(A) "tail"—see below), the NaOH incubation is omitted.

Protein Gels. Protein gels without SDS are incubated in 50 m*M* sodium phosphate, pH 5.5, as described above for RNA gels. A longer incubation (2–3 hrs) is necessary for gels containing urea to avoid swelling of the gel and consequent streaking of the pattern during the transfer. Protein gels containing a Tris–glycine–SDS buffer require the removal of the Tris and glycine prior to transfer to prevent their binding to the reactive diazonium groups on the paper. These gels are soaked in 50 m*M* sodium phosphate, pH 6.5, plus 0.1% SDS for 90 min at 20°.

Polyribosome and Ribonucleoprotein Gels. Protein–nucleic acid complexes separated in composite gels[10] require a buffer at higher pH plus magnesium ion for maintenance of structural integrity. The gels are incubated in 3 ml of 50 m*M* 2-(*N*-mopholino)ethanesulfonic acid (MES), 2 m*M* $MgCl_2$, pH 6.5 (or 7.0), per square centimeter of gel for 60 min at 4° prior to transfer.

[10] A. E. Dahlberg, *in* "Gel Electrophoresis of Nucleic Acids" (D. Rickwood and B. D. Hames, eds.), pp. 199–225. IRL Press Ltd., Oxford, 1982.

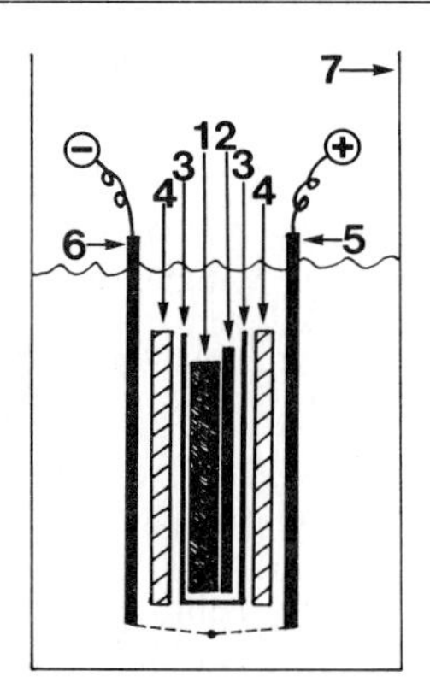

FIG. 1. Apparatus for electrophoretic transfer of macromolecules from slab gels to diazobenzyloxymethyl paper.

Preparation of DBM Paper

DBM paper may be prepared using nitrobenzyloxymethylpyridinium chloride (NBPC) (Pierce Chemical Co.) and Whatman 540 paper, and ABM paper may be purchased commercially and activated to DBM paper as described by Alwine *et al.*[2] (also see this volume [22]). The DBM paper may be prepared while the gel is being soaked in transfer buffer.

Electrophoretic Transfer Apparatus

A diagram of the apparatus for electrophoretic transfer of macromolecules from slab gels to DBM paper is shown in Fig. 1. The apparatus consists of a 23 × 15 cm electrophoretic destainer (E-C Apparatus Corp., St. Petersburg, Florida) with a porous stainless steel cathode plate (6) and platinum or palladium wire anode (5) woven in plastic screen, held by a plexiglas frame, and separated by a distance of 2 cm. The gel from which samples are to be transferred (1) is placed directly on freshly prepared DBM paper (2) and then wrapped in Whatman 3 MM paper (3) presoaked in electrophoresis buffer. Care is taken to avoid trapping air bubbles between the gel and paper, as they will disrupt the electric field and alter the pattern of migration onto the paper. The wrapped gel is then placed between two presoaked Scotch Brite pads (4) positioned between the anode and the cathode of the destaining apparatus so as to prevent slippage of the gel during electrophoresis. Up to three gels, separately wrapped in 3 MM paper, may actually be stacked together for transfer at one time. The anode and cathode plates of the destainer, containing the wrapped gel(s) and Scotch Brite pads, fit into a plexiglass destainer box

(7) (E-C Apparatus Corp.) (17 × 24 × 19 cm) containing sufficient buffer to submerge the gel completely. The buffer, previously degassed and cooled, is circulated by an external pump from cathode to anode (200 ml/min) to maintain constant temperature and pH. A stir bar may also be used at the base of the buffer chamber. Current is applied using an EC277 power supply (E.C. Apparatus Corp.) or the more versatile EC420 on the 25 V adjustable output range setting.

Electrophoretic Transfer to DBM Paper

DNA Gels. DNA is electrophoretically transferred from agarose gels to freshly prepared DBM paper in 2–4 liters of 25 m*M* sodium phosphate, pH 5.5, at 10 V/cm for 4–5 hr, 4°, using the apparatus of Fig. 1.

RNA Gels. RNA is similarly transferred from gels in 25 m*M* or 50 m*M* sodium phosphate buffer (depending on size of the RNA; see above) at 5–10 V/cm for 1–6 hr at 4°.

Protein Gels. Proteins are transferred from non-SDS-containing gels to DBM paper in 50 m*M* sodium phosphate, pH 5.5, at 10 V/cm for 2 hr at 4°. Higher temperature (20°) is required for transfer from SDS-containing gels. In addition, after the gel is placed on DBM paper, as in Fig. 1, it is surrounded by a thin plastic sheet with outer dimensions the size of the destainer electrodes and an inner rectangular area (6 × 12 cm) cut out to fit around the gel. The plastic sheet is sandwiched between the Whatman 3 MM papers or, if the papers are omitted, between the Scotch Brite pads. The plastic sheet thus directs the current entirely through the gel. The protein is electrophoretically transferred from the gel to DBM paper in 2 liters of 50 m*M* sodium phosphate, pH 6.5, (*no SDS*) at 170 mA, at about 1–2 V/cm for 6 hr at 20°.

Ribonucleoprotein Gels. Polyribosomes and ribonucleoprotein particles are transferred to DBM paper as described for transfer of large RNAs (10 V/cm for 1–6 hr at 4°), but at a higher pH in the MES–$MgCl_2$ buffer as described above.

Efficiency of Transfer to DBM Paper

Less than 2% of labeled DNA restriction fragments remain in the gel after 6 hr of electrophoresis using the conditions described above. The efficiency of binding to the DBM paper is 85%,[7] while DNA migration through the paper is negligible. RNAs ranging in size from 4 S to 23 S are completely transferred to DBM-paper in 4 hr at 5 V/cm, the rate of transfer depending on the size of the RNA. Larger RNAs, such as 40 S, require 6 hr at 10 V/cm. The covalent coupling efficiency of RNA varies from 65%

for 4 S and 5 S RNAs to 80% for higher molecular weight RNAs. Proteins ranging in size from 6000 to 65,000 daltons are transferred to DBM paper from 10% acrylamide urea gels in 2 hr at 10 V/cm. There is essentially complete transfer of the proteins. Similar results are obtained with two-dimensional urea gels. A less than quantitative transfer is achieved from SDS gels although the original gel pattern is preserved. This was also observed by Towbin *et al.*[11] with electrophoretic transfer of proteins to nitrocellulose. This transfer requires modifications as described above; i.e., it is necessary to keep SDS in the soak buffer while removing Tris and glycine to avoid precipitation of the protein in the gel. However, SDS is omitted from the reservoir buffer during electrophoretic transfer to DBM paper to prevent the SDS protein complexes from passing through the DBM paper.

Covalent binding of nucleic acids to DBM paper requires single-stranded regions, containing adenine or guanine bases to bind the diazonium groups. An alternative to denaturation of double-stranded DNA is the addition of a poly(A) "tail" to the 3′-OH terminal group by terminal deoxynucleotidyltransferase prior to electrophoresis.[12] The poly(A) tail is sufficient to bind the double-stranded DNA to the DBM paper.[7] This may be useful for studying interactions of proteins with discrete species of double-stranded DNAs.

The efficiency of covalent coupling is very dependent on speed of transfer. Diazonium groups are not stable, and, for example, covalent coupling of 4S RNA begins to decline rapidly 3–4 hr after diazotization, with approximately 50% of the initial coupling activity lost during the first 6–8 hr and less than 2% remaining after 24 hr.[7] The buffer pH also affects the stability of diazonium groups.[2,7] There is a rapid decline in covalent coupling activity above pH 7. The DBM paper is most stable at pH range 4–5.

Efficiency of Hybridization

The hybridization efficiency of DNA to DBM-paper carrying RNA or of RNA to DBM-paper carrying DNA is as good, if not better than, that seen by blot transfers. The electrophoretic transfer has no detrimental effect on hybridization. The background radioactivity resulting from non-specific binding of RNA or DNA to the DBM paper during hybridization is comparable to or lower than that obtained with nitrocellulose.[7]

Proteins electrophoretically transferred to DBM-paper from both urea and SDS gels bind antibody in the manner already described for blot-

[11] H. Towbin, T. Staehelin, and J. Gordon, *Proc. Natl. Acad. Sci. U.S.A.* **76,** 4350 (1979).
[12] R. Roychoudhury, E. Jay, and R. Wu, *Nucleic Acids. Res.* **3,** 863 (1976).

transferred proteins.[5] These "blots" may be used repeatedly for analysis of different antisera. DPT-paper electrophoretic blots have also been used as a means of isolating antibodies specific for a particular antigen.[13]

Noncovalent Binding to DBM Paper

DNA and RNA initially bind to DBM paper by an ionic interaction of negatively charged nucleic acids with positively charged diazonium groups. This ionic interaction is then superseded more slowly by covalent linkages within 2.5 hr.[2,7] DBM paper that has lost its covalent binding capacity still retains considerable noncovalent binding capacity for electrophoretically transferred samples. DNA and RNA bound in this manner may be quantitatively removed by overnight incubation of DBM paper in 2 × SSC, 0.1% SDS. This provides a preparative method for recovery of materials that are essentially unaltered in susceptibility to enzymic hydrolysis and electrophoretic mobility.[7]

Comments

It is important to keep the buffer at constant temperature and pH during electrophoresis. This is achieved by buffer recirculation. During long runs check the buffer pH at intervals of several hours. The buffer should be degassed prior to use to avoid excessive formation of air bubbles, particularly between the gel and the DBM paper, where they will disrupt the flow of the electric current and affect the transfer of the material to the paper.

The transfer of proteins from SDS-containing gels is probably the most inefficient of the transfers described here. It is essential to keep SDS in the gel to avod precipitation of the protein, and yet excessive SDS will cause the proteins to pass right through the DBM paper. Hence, SDS is omitted from the reservoir buffer.

The freshly activated DBM paper is bright yellow. This color will change to dark orange over a period of time owing to electrophoresis or increase in pH. As long as the transfer is carried out within the times described here, preferably at low temperatures and low pH, change in color does not appear to appreciably to affect the transfer and covalent coupling.

Careful selection of transfer buffer is important. Tris, glycine, and other amine buffers are to be avoided, as they compete for the diazonium groups on the DBM paper. Zwitterionic buffers such as MES or MOPS, at pH values equal to their isoelectric points, appear to be equally as effec-

[13] J. Olmsted, *J. Biol. Chem.* **256,** 11955 (1981).

tive as phosphate, but they are quite expensive. These buffers are particularly useful for transfers involving ribonucleoprotein particles. Denaturants in gels, such as glyoxal and methylmercury hydroxide, can also affect covalent binding and must be removed by soaking in phosphate buffer.

The stainless steel electrode in the transfer apparatus described here may release black impurities into the buffer with prolonged electrophoresis and may exhibit pitting on the electrode. These do not appear to interfere with the transfer in any way, and the low cost of this electrode justifies its continued use in this system.

[24] Plasmid Screening at High Colony Density

By DOUGLAS HANAHAN and MATTHEW MESELSON

Bacterial plasmid vectors are widely employed in the isolation, amplification, mutagenesis, and analysis of DNA sequences. A number of applications involve locating the products of rare events. Such identification can be considerably facilitated by the ability to screen for specific plasmids at high colony density. This chapter updates and expands upon methodology devised for that purpose.[1-3]

Principle

This procedure involves establishing bacterial colonies directly upon nitrocellulose filters laid on agar plates and replicating the distributions onto other nitrocellulose filters. Maintaining and replicating colonies on a nitrocellulose support allows very large numbers to be readily manipulated (at least 10^5 colonies per 82 mm in diameter filter). Colonies can be established under one set of conditions and then the distribution (or a replica of it) transferred easily to another.

Initially, a primary colony distribution is created. Colony-forming bacteria are spread on a nitrocellulose filter laid on an agar plate. The plate is incubated to establish small colonies, which are then replicated to other nitrocellulose filters; these in turn are incubated to establish duplicate colonies. The replicas can then be replicated again, transferred to plates

[1] M. Grunstein and D. S. Hogness, *Proc. Natl. Acad. Sci. U.S.A.* **72,** 3961 (1975).
[2] D. Hanahan and M. Meselson, *Gene* **10,** 63 (1980).
[3] M. Grunstein and J. Wallis, this series, Vol. 68, p. 379.

ISBN 0-12-182000-9

containing different drugs or inducers, or lysed and hybridized. Probing replicas with radioactive nucleic acids can identify a colony of cells carrying sequences homologous to the probe. Keying back from an autoradiogram to the master plate localizes the colony. A few colonies are removed from the region(s) of hybridization and dispersed in medium, and an appropriate dilution is spread on a fresh nitrocellulose filter, to give 100–200 colonies. This enriched population is replicated and probed, allowing isolation of pure clones of the hybridizing species.

Maintaining colonies on filters also provides for long-term storage of large distributions (banks, libraries). A sandwich comprised of two nitrocellulose filters and a colony array can be stored at −55° to −80° for an indefinite period, thawed, and separated to give two viable replicas.

Procedures

Materials

Media and Solutions. Virtually any bacterial growth medium may be used. A typical rich medium is Luria broth supplemented with magnesium (LM): 1% Bacto-tryptone, 0.5% Bacto yeast extract, 10 m*M* NaCl, 10 m*M* $MgCl_2$, 1.5% Bacto agar. F plates, used in the preparation of frozen replicas, include 5% glycerol. Chloramphenicol plates are supplemented with 170–250 μg of chloramphenicol per milliliter. Tetracycline-HCl (tet) and sodium ampicillin (amp) are employed at minimum concentrations for the particular strain in use, generally 7–17 μg/ml for tet, and 30–100 μg/ml for amp.

SET is 0.15 *M* NaCl, 30 m*M* Tris, pH 8, 1 m*M* EDTA. Denhardt solution[4] is 0.02% Ficoll, 0.02% polyvinylpyrolidone, and 0.02% bovine serum albumin (BSA). Formamide (MCB FX420) is deionized with Bio-Rad AG501-X8 ion-exchange resin and stored frozen.

Filters. Precut nitrocellulose filters are available from a number of suppliers. All bind DNA quite efficiently. Two important criteria bear on this application of nitrocellulose: (*a*) the probability of establishing a single cell into a growing colony (plating efficiency); and (*b*) dimensional stability and integrity through the various treatments (including melting annealed probes off a filter followed by rehybridization with a different probe). In this context, Millipore HATF (Triton-free) filters have proved to be the most reliable and durable[2,5] and are available in 82-mm and 127-mm diameters for use in 100-mm and 150-mm petri dishes, respectively.

Filters may be prepared in advance for plating and replication and then stored indefinitely. The filters are floated on double-distilled water (dd

[4] D. T. Denhardt, *Biochem. Biophys. Res. Commun.* **23,** 641 (1966).

[5] F. G. Grosveld, H. H. M. Dahl, E. de Boer, and R. A. Flavell, *Gene* **13,** 227 (1981).

H_2O), submerged when wetted, and sandwiched between dry Whatman 3 MM filters into a stack, which is wrapped in aluminum foil and autoclaved on liquid cycle for 15 min. After sterilization, the pack of filters is sealed in a plastic bag to maintain humidity.

Alternatively, filters may be wetted by placement on an agar plate just prior to use. This is convenient when only a few plates are to be prepared. The filters are sterile as supplied and may be resterilized by ultraviolet irradiation.

Plating and Growth

A sterile nitrocellulose filter is laid upon an agar plate. The filter has a curl to it, and final placement without air bubbles is easier if the curl of the filter matches the curl up at the edges of the plate. Any remaining bubbles can be nursed out using a bent glass Pasteur pipette.

The cells are applied to the filters (in a 200–400 μl volume for 82-mm, twice this volume for 127-mm filters) and quickly spread with a bent Pasteur pipette (L-shaped) to distribute the cells on the filter. A plating wheel helps to spread the suspension while leaving a blank edge (or rim) on the filter. To assure even distribution, the suspension should not be drying in patches during the spreading process. A uniform film of liquid will be visible (by reflection) immediately upon completion of the spreading. It will dry in a minute or two. If dry patches form during the spreading, use larger volumes of medium.

The plates are then incubated until colonies 0.1 mm in diameter appear. Incubation at 30–32° facilitates control of colony size. The most frequent cause of distortion in colony distributions during replication arises from overgrowth of colonies. For any density, colonies 0.1–0.2 mm in diameter are best. Virtually any visible colony will transfer properly, while colonies greater than 1 mm in diameter can transfer nonuniformly. Heterogeneity in colony size is not a problem if the largest colonies are kept below 0.5 mm in diameter.

An alternative to plating directly on filters is to lift the colonies off an agar plate much in the same manner that phage are lifted off plaques in the method of Benton and Davis.[6] Colonies are established on well dried agar plates, which are then refrigerated for a few hours (to retard smearing). A dry nitrocellulose filter is placed on the colony distribution and allowed to become wet; then the filter is lifted off, carrying the colonies with it. This filter can then either be placed (colonies up) on a fresh plate, replicated immediately, or chloramphenicol amplified, etc.[7,8] Colony lifts work rea-

[6] W. D. Benton and R. W. Davis. *Science* **196,** 180 (1977).
[7] D. Ish-Horowicz and D. F. Burke. *Nucleic Acids Res.* **9,** 2989 (1981).
[8] G. Guild and E. Meyerowitz, personal communication.

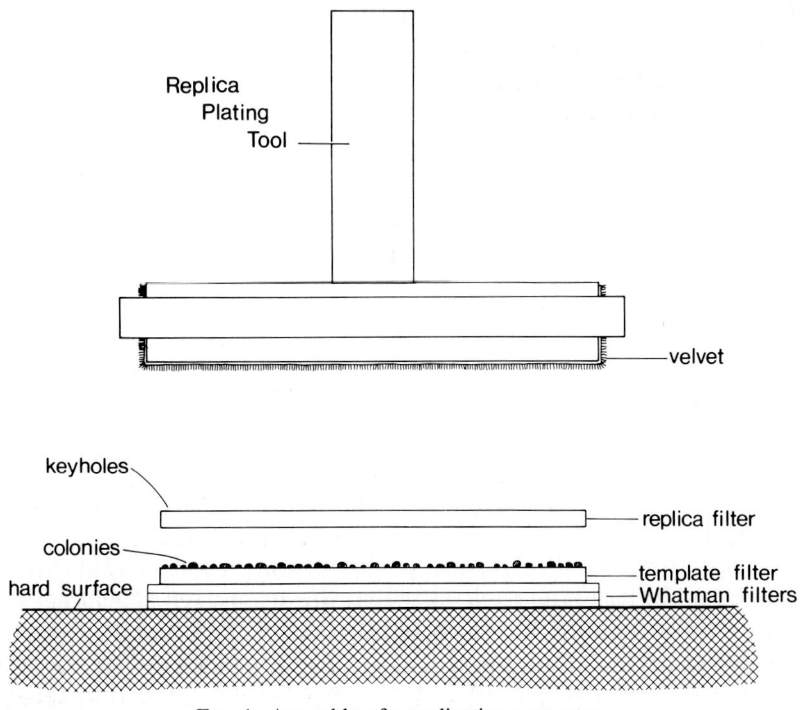

FIG. 1. Assembly of a replication apparatus.

sonably well at moderate densities, but reliability at high density remains uncertain.

Replication

The template filter is peeled off its agar plate and laid, colonies up, on a bed of sterile Whatman paper. A wetted sterile nitrocellulose filter is held between two flat-bladed forceps and laid upon the template filter. The sandwich is pressed firmly together with a velvet-covered replica plating tool (Fig. 1). The filters are keyed to each other by making a characteristic set of holes in the sandwich with a large needle. The replica is peeled off the template and placed on a fresh agar plate.

Additional replicas may be made by placing fresh filters on the template and repeating the process. In this case the sandwich is inverted and the existing holes in the template are used to key the two filters.

Four to five replicas may be made off a template without sacrificing complete transfer. If additional replicas are desired, the template filter can be reincubated to replenish its colonies. The replicas are incubated at 30–

37° until the colonies develop to the desired size, which is generally 0.5–1 mm in diameter for lysis or amplification, but again only 0.1–0.2 mm in diameter for further replication or frozen storage.

The sterile nitrocellulose filters used in the replication step should be moist and supple but free of visible surface moisture. Overly dry filters can be placed briefly on a fresh agar plate just prior to use, which assures the desired degree of wetness.

Various alternatives to the velvet replica plating device may be used: automobile pistons, a pair of thick glass plates,[5] etc. The velvet seems to help even out the force distribution over the occasionally uneven surface of the filter–colony sandwich.

Lysis and Prehybridization

Replica filters to be probed are lysed (colonies up) on several sheets of Whatman 3 MM paper that have been barely saturated with 0.5 *M* NaOH. It is convenient to float the layers of Whatman paper in a tray that is partially filled with liquid, which is then poured off, leaving the saturated paper in the tray. After several minutes, the replica filters are peeled off the lysis pad, blotted momentarily on dry paper towels, and then placed on a second pad saturated with 0.5 *M* NaOH. This is followed by two neutralization steps, first with 1 *M* Tris, pH 8, and then with 1 *M* Tris, pH 8, 1.5 *M* NaCl, each for 3–5 min. The replica filters are then placed (colonies still up) on sheets of dry Whatman 3 MM to blot the surface film of liquid down through the filter. After brief air drying, the filters are sandwiched between Whatman paper and baked under vacuum at 60–80° for 1–3 hr. The filters should still be supple, and not yet bone white when placed in the vacuum oven.

Regardless of the hybridization probe and conditions, prehybridization is important both to occupy nonspecific binding sites and also to clean off cellular debris from the filters. Polypropylene food storage boxes are very useful both in the prehybridization and posthybridization washes. Agitation is very important throughout.

The baked replica filters are floated on double-distilled H_2O, submerging when wetted and then placed in a box containing several hundred milliliters of 5 × Denhardt solution, 0.5% SDS. The filters are incubated with agitation at 68° for at least 2 hr (up to 12 hr). The filters are then removed and placed in the hybridization solution.

Hybridization, Washing, and Visualization

Colony hybridization can be performed under a wide variety of conditions. Those described below have proved to be convenient and give good signal-to-noise ratios. Hybridization in rotating water baths is particularly important for reducing nonspecific background. Filters are placed in heat-sealable freezer bags, the hybridization solution is added, and the bag is sealed and then placed in a water bath of the desired temperature and agitated at 25–100 rpm. A good rule of thumb is to use 7 ml for the first 82-mm filter, and 2 ml for each additional one (2 × for 137-mm filters). Ten large filters can be readily screened in one bag provided that a reasonable volume is used and that the bag is well agitated during the hybridization.

DNA. Standard DNA hybridization conditions are 6 × SET, 1 × Denhardt solution, 0.1% SDS, at 68°. Alternative conditions are 6 × SET, 1 × Denhardt solution, 50% deionized formamide, 0.1% SDS, 42–45°. A number of investigators add single-stranded carrier DNA (50–150 μg/ml) and/or tRNA (100–250 μg/ml) to hybridizations. The usefulness of these additions are dependent on the characteristics of the probe, but are unlikely to be counterproductive even when unnecessary. The hybridization probe is denatured by the addition of 0.1 volume of 1 *N* NaOH for 2 min at room temperature, neutralized by the addition of 0.1 volume of 1 *M* Tris, pH 8, and 0.1 volume of 1 *N* HCl, and then added to the hybridization solution to a final concentration of less than 20 ng/ml.

Several 30-min posthybridization washes are performed with agitation in a few hundred milliliters of 2 × SET, 0.2% SDS, at 68° in large polypropylene boxes. Alternative or additional wash conditions are 0.1 × SET, 0.1% SDS, 53°.

RNA. Standard RNA hybridization conditions are 6 × SET, 1 × Denhardt solution, 50% deionized formamide, 0.1% SDS, 250 μg of tRNA per milliliter, at 42–45° for 10–20 hr. Several washes are performed in 2 × SET, 0.2% SDS, at 68° (with agitation) and/or 0.1 × SET, 0.1% SDS at 53°. The probe is employed in the hybridization at less than 20 ng/ml.

Oligonucleotides.[9–12] Oligonucleotides are hybridized in 6 × SET, 5 × Denhardt solution, 250 μg/ml tRNA, 0.5% Nonidet P-40 (NP-40) (Shell oil) (or 0.1% SDS). Optimal hybridization temperatures depend on the length and nucleotide composition of the oligonucleotide. A good approximation for the hybridization temperature T_H is given by $T_H = T_D - (3°)$, where $T_D = 2° \times$ [the number of A-T base pairs] plus 4°C × [the number of

[9] J. W. Szostak, J. I. Stiles, B.-K. Tye, D. Chiu, F. Sherman, and R. Wu, this series, Vol. 68, p. 419.

[10] R. B. Wallace, M. J. Johnson, T. Hirose, T. Miyake, E. H. Kawashima, and K. Itakura, *Nucleic Acids Res.* **9,** 879 (1981).

[11] R. B. Wallace, M. Schold, M. J. Johnson, D. Dembek, and K. Itakura, *Nucleic Acids Res.* **9,** 3647 (1981).

[12] M. Smith, *in* "Methods of RNA and DNA Sequencing" (S. M. Weissman, ed.), in press. Praeger, New York, 1983.

G-C base pairs].[13] For mixed probes of average GC composition, a reasonable approximation is $T_H = 3° \times$ the length of the oligonucleotide. The oligonucleotides are used at a final concentration of 2 ng/ml. The filters are washed with four changes of 6 × SET, 0.5% SDS at 20° over 20 min, followed by a 1-min wash at T_H.[11] Nonspecific hybridization of oligonucleotides to colonies is reduced significantly by prehybridizing for 6–8 hr at 68° in 5 × Denhardt solution, 0.5% SDS, 10 m*M* EDTA, after which the filters are gently rubbed with a gloved hand, to remove any remaining colonial matter (D. H., unpublished observations).

Mounting and Autoradiography. After posthybridization washes in food storage boxes with gentle agitation, the filters are blotted dry on paper towels, but not allowed to become bone dry and brittle. The filters are sandwiched between two sheets of plastic wrap (which keeps the filters supple) and then mounted on a solid support (e.g., cardboard, old X-ray film). The mounted filters are marked with radioactive ink and subjected to autoradiography, generally using intensifying screens at −70°.

The developed film is aligned to the mounted filters using the radioactive ink marks. The keyholes on the filters can then be marked either directly on the film or on a clear plastic sheet (along with positive hybridization spots). The keyed film can be aligned to the master filter in two ways. The film (or transparent sheet) can be placed on top of the plate. The keyholes are aligned, and the colonies are identified by sighting straight down through the film. An alternative is to put the film on a light box, remove the filter from its plate and lay it on a small square of plastic wrap, and then place the filter (plus plastic wrap) directly on the film, rotating it to align keyholes with their corresponding marks. Circled positives are then readily identified, without parallax problems.

Notes. In contradiction to benefits observed with DNA blots and phage screens, the authors and other investigators have not found dextran sulfate to improve the signal-to-noise ratios in colony hybridization.

Filters may be rehybridized with different probes. The filters are cut out of their plastic wrap mounts (still supple), floated on, and then submerged in several hundred milliliters of 5 × Denhardt solution, 0.5% SDS in a food storage box. The filters are incubated with gentle agitation at 80° for several hours followed by a second wash at 68° with fresh solution. Millipore filters hybridized at 68° can be rehybridized 2–4 times before significant disintegration occurs. Filters hybridized in formamide at 42–45° can be rehybridized considerably more.

[13] S. U. Suggs, T. Hirose, T. Miyake, E. H. Kawashima, M. J. Johnson, K. Itakura, and R. B. Wallace, *ICN–UCLA Symp. Dev. Biol. Using Purified Genes,* VXX, p. 683 (1981).

High levels of nonspecific hybridization (noise or background) may result from several factors: excessive quantities of probe; an inappropriately low hybridization temperature, which can often be raised several degrees without affecting specific hybridization; or a "dirty" probe, which can be filtered quickly through a 0.2-μm filter prior to its addition to the hybridization. A number of investigators routinely filter all hybridization probes.

Freezing

Distributions of colonies on nitrocellulose may be stored indefinitely at -55 to $-80°$. Template filters (which may themselves be replicas) are prepared as usual, except that they are either grown on F plates or transferred to F plates and incubated for a few hours (30–37°). Sterile dry nitrocellulose filters are wetted on F plates just prior to use. The replication is carried through the keying step. The two filters are left together and sandwiched between several dry Whatman filters, plus one wet Whatman filter to maintain humidity. The stack is placed inside a freezer bag, sealed, and stored below $-55°$ (Fig. 2). When needed, the bag is removed and brought to room temperature. The filters are peeled apart, laid on fresh agar plates, and incubated until small distinct colonies appear. As with basic replication, the template colonies usually develop more rapidly than those on the replica; incubation at 30–32° allows more control of the colony size.

Chloramphenicol Amplification

A replica filter may be transferred to an agar plate containing chloramphenicol, which upon incubation at 37° for 12–48 hr will effect substantial amplification of appropriate plasmid vectors (e.g., pBR322). For most healthy *E. coli* K12 stains, chloramphenicol is quite effective at 170–250 μg/ml, although levels ranging from 12 μg/ml to 500 μg/ml have been

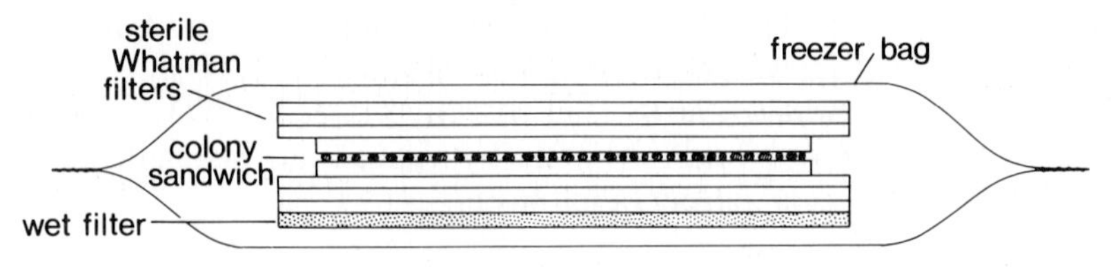

FIG. 2. A colony distribution prepared for frozen storage.

used. Very large plasmids (and cosmids) do not amplify well in some cases, and it is advisable to test the particular host–vector in use to verify the utility of amplification.

Comments

This technique has been employed in the isolation of both cDNA and genomic DNA sequences from a wide variety of organisms. No particular limitations due to fidelity of replication or frozen storage have been observed. Density constraints arise primarily from the quality and specific activity of the hybridization probe. This applies particularly to mixed oligonucleotide probes, where the optimum salt concentration and temperature may be very sensitive to the number and particular sequences of the oligonucleotide mixture, with signal and noise in a delicate balance. It is, however, demonstrably possible to screen a large colony distribution with a mixed oligonucleotide probe.[14]

In another variation, this technique has been applied in probing colony arrays on nitrocellulose filters with radioactive antibodies in order to locate sequences through recognition of their gene products. A cDNA clone encoding part of the structural gene for chicken tropomyosin was isolated from a high-density colony distribution, using an antibody to chicken tropomyosin to identify colonies expressing an antigenic portion of the protein.[15] The pilus protein gene of *Nisgeria gonorrhoeae* was cloned into *E. coli,* and the positive colony was identified through its expression and subsequent recognition by a suitable antibody.[16] As with oligonucleotide probes, hybridization conditions and density limitation are likely to vary considerably with the characteristics of the probe and its complement.

This technique has been applied to the construction and storage of cosmid banks and to the isolation of specific sequences from them.[5,17] Densities have been constrained to about 10^4 colonies per 127-mm filter by the low titers of packaged recombinant cosmids obtained when using higher eukaryotic DNA. This does not represent an intrinsic sensitivity limit, as 10^5 cosmids carrying yeast DNA (50 genome equivalents) have

[14] M. Noda, Y. Furutani, H. Takahashi, M. Toyosato, T. Hirose, S. Inayama, S. Nakanishi, and S. Numa, *Nature (London)* **295,** 202 (1982).

[15] D. M. Helfman, J. R. Feramisco, J. C. Fiddes, G. P. Thomas, and S. H. Hughes, *Proc. Natl. Acad. Sci. U.S.A.* **80,** 31 (1983).

[16] T. F. Meyer, N. Mlawer, and M. So, *Cell* **30,** 489 (1982).

[17] M. Steinmetz, A. Winoto, K. Minard, and L. Hood, *Cell* **48,** 489 (1982).

been screened on one 127-mm filter using a moderate specific activity DNA probe (10^7 dpm/μg) (D. H. and B. Hohn, unpublished). Steinmetz *et al.*[17] reported that if initial platings of cosmids are incubated first on a low concentration of tet (5 μg/ml) for several hours, and then the filters are transferred to higher tet plates (10 μg/ml), three times the number of colonies form as compared to initial plating on the higher concentration of tet (an amp selection was not similarly tested).

There is no clearly defined maximum colony density for plasmid screening when a pure DNA probe of greater than 10^7 dpm/μg is used, and it is likely that 10^6 colonies can be replicated and screened on a single filter.

Acknowledgments

The authors thank many colleagues for comments and suggestions and Patti Barkley for preparing the manuscript. D. H. is a junior fellow of the Harvard Society of Fellows.

[25] Chromogenic Method to Screen Very Large Populations of Bacteriophage Plaques for the Presence of Specific Antigen

By DONALD A. KAPLAN, LAWRENCE GREENFIELD, and R. JOHN COLLIER

Detection of specific macromolecular species associated with individual bacterial colonies or phage plaques is of crucial importance in molecular cloning. Existing methods for detecting either specific nucleic acid sequences or antigens are adequate for many situations, but leave much to be desired when one is searching for the rare clone. Earlier we described an efficient and highly sensitive method to screen very large numbers of bacteriophage microplaques for the presence of specific antigen.[1,2] One to 10 *million* microplaques (and sometimes more) could be screened per standard petri plate with good efficiency of detection and recovery.

[1] D. A. Kaplan, L. Naumovski, and R. J. Collier, *Gene* **13,** 211 (1981).
[2] D. A. Kaplan, L. Naumovski, B. Rothschild, and R. J. Collier, *Gene* **13,** 221 (1981).

ISBN 0-12-182000-9

The system used in these original studies involved detection of diphtheria toxin antigen in plaques of bacteriophage β plated on nonlysogenic *Corynebacterium diphtheriae*. We have applied the principles elucidated in the β phage system to detection of antigens in plaques of coliphage λ; and we have modified or refined many aspects of the detection procedure. The method in its present form should be widely applicable and should permit isolation of rare λ clones that might go undetected by other methods.

Principle of the Method

Horseradish peroxidase (HRP) covalently coupled to specific antibody (Ab) against a phage-borne or cloned antigen is incorporated into the soft agar layer of plaque assay plates. Immunoprecipitation of the HRP–Ab conjugate occurs in plaques in which specific antigen is produced. Once plaque growth has ceased, plates are immersed in buffer to permit residual soluble HRP-Ab to diffuse from the agar. The plates are then removed from the buffer container, and chromogenic HRP substrates are added to the agar surface. Colored insoluble reaction products are deposited at the sites of HRP-containing immunoprecipitate, thereby allowing detection of antigen-positive plaques. Figure 1 shows results obtained when this method was applied to detection of β-galactosidase in plaques formed by a mixture of *lac*-positive and *lac*-negative bacteriophage incubated under standard plating conditions.

This basic detection method may be adapted to screen very large numbers of plaques per unit area of agar surface if one reduces the diameters of both plaques and immunoprecipitate. This may be accomplished by increasing the initial density of indicator cells (D_0) in the soft agar layer. The bacterial density at which cells can no longer sustain phage replication (D_f) is reached much sooner, the period of plaque formation is diminished, and the plaques are therefore smaller. At values of D_0 greater than about 3×10^9 cells/ml, plaques become virtually invisible to the unaided eye. However, the local concentration of HRP-immunoprecipitate remains high and can easily be detected by chromogenic reactions. We have described a model of plaque formation that provides a theoretical basis for these phenomena.[2]

Chromogenic Screening of λ Phage Microplaques

Below we present a protocol to screen for specific antigen (β-galactosidase) in the coliphage λ system. Important concepts for understanding the method are then discussed, together with useful variations.

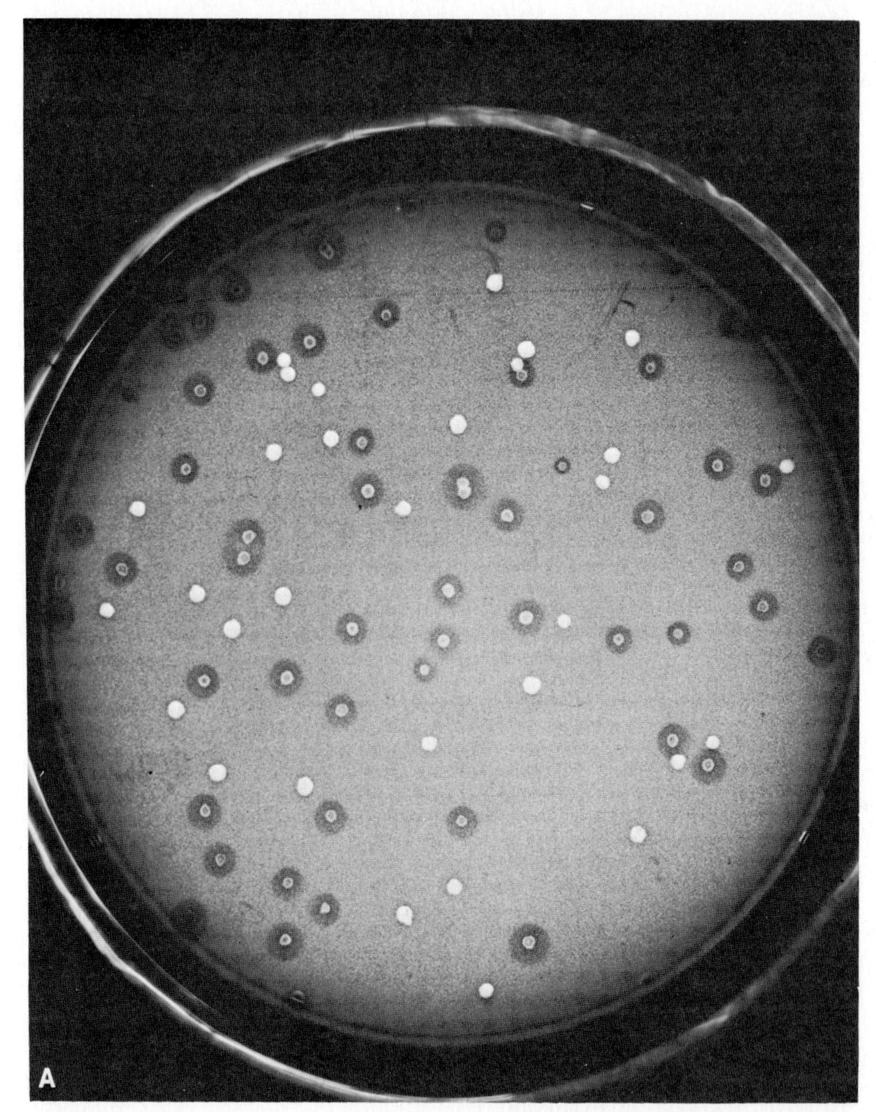

FIG. 1. Detection of β-galactosidase in plaques of Charon 17 bacteriophage. (A) Approximately 65 Charon 17 (λ *lacZ*$^+$) and 46 Charon 30 (λ *lacZ*$^-$) phage were plated in top agar containing 1.7×10^8 cells of *Escherichia coli* LA 108 per milliliter and 300 μg of anti-β-galactosidase-HRP conjugate per milliliter (52-mm plastic petri dish). After incubation at 37° for 12.5 hr, the plates were "dialyzed" for 49 hr and then developed as described in the text. The plate was illuminated from below with light filtered through a Wratten No. 61 green filter

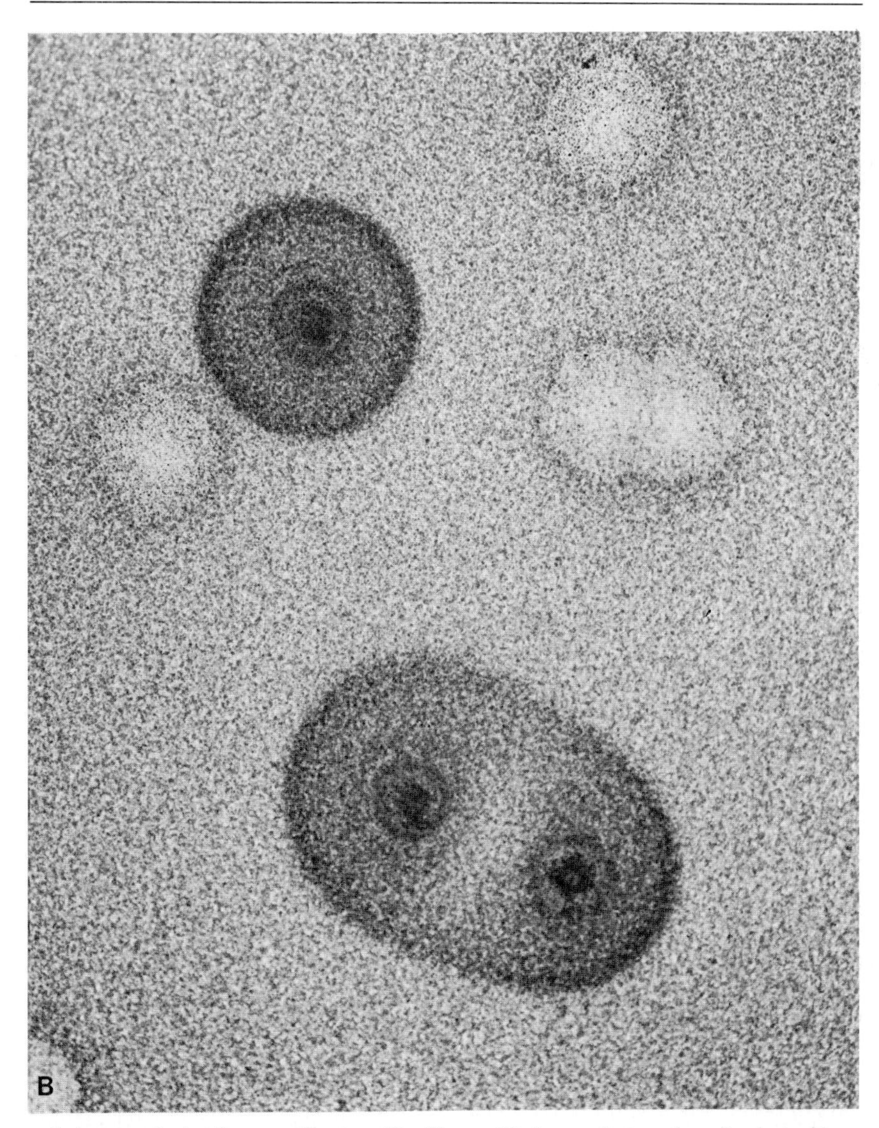

and photographed at 2× magnification. The Charon 17 plaques that produce β-galactosidase are surrounded by rings of pigment whereas the Charon 30 plaques are not. (B) About 71 Charon 17 (λ $lacZ^+$) and 320 Charon 30 (λ $lacZ^-$) were plated in top agar containing 245 μg of conjugate per milliliter and indicator cells (LA108) at a density of 2.2×10^8 cells/ml. The plate was incubated at 37°, dialyzed, and developed as in Fig. 1A. Magnification ×15. Photographed using a Wratten No. 61 green filter.

Buffers

PBS: 8 g of NaCl; 2.17 g of $Na_2HPO_4 \cdot 7\ H_2O$; 0.2 g of KH_2PO_4; 0.2 g of KCl; water to 1 liter

BSP: 10 mM Tris-HCl, pH 7.5; 0.1 M NaCl; 10 mM $MgCl_2$; 0.05% gelatin

λ dilution buffer: 10 mM Tris-HCl, pH 7.5; 10 mM $MgSO_4$

Sodium phosphate, 17.5 mM, pH 6.8: per liter, 10 ml of 0.875 M $Na_2HPO_4 \cdot 7\ H_2O$; 10 ml of 0.875 M $NaH_2PO_4 \cdot H_2O$

Sodium phosphate, 0.1 M, pH 6.8: per 100 ml, 5.83 ml of 0.875 M $NaH_2PO_4 \cdot H_2O$, 5.6 ml of 0.875 M $Na_2HPO_4 \cdot 7\ H_2O$

Carbonate-bicarbonate buffer 1 M: to make 100 ml, 30 ml of 1 M Na_2CO_3, 70 ml of 1 M $NaHCO_3$

Chemicals

DAB: 3,3′-diaminobenzidine (Sigma No. D5637). *Caution: carcinogenic compound*

TMB: 3,3′,5,5′-tetramethylbenzidine (Sigma No. T.2885)

(Diethylaminoethyl) cellulose: Whatman DE-52, preswollen microgranular anion exchanger (Whatman Cat. No. 4057-050)

Sephadex G-25, fine (Pharmacia Fine Chemicals, Item 17-0032-02)

IPTG: isopropyl-β-D-thiogalactopyranoside (Sigma No. I-5502)

Glutaraldehyde: Grade 1 specially purified, 25% aqueous solution (Sigma No. G-5882)

H_2O_2: Superoxol, 30% with 1 ppm $Na_2SnO_3 \cdot 3\ H_2O$ as preservative (Baker 1-2186)

3-Amino-9-ethylcarbazole (Sigma A-5754)

Dimethyl sulfoxide (Sigma D-5879)

Enzymes

β-Galactosidase (gift from A. Fowler)

Horseradish peroxidase: hydrogen-peroxide oxidoreductase (EC 1.11.1.7), type VI salt-free powder (Sigma No. P-8375)

Media

Broth: 10 g of Difco tryptone, 2 g of Difco yeast extract, 5 g of NaCl, 1 liter of water. Sterilize by autoclaving. When cool add 1/100 volume of 10% maltose.

Plates: 10 g of Difco tryptone, 2 g of Difco yeast extract, 5 g of NaCl, 15 g of Difco agar, 1 liter of water. Sterilize by autoclaving. When IPTG is needed, allow liquid to cool to 55° and add 0.12 g of IPTG per liter (0.5 mM final concentration). Fill standard glass petri plates about half full, or add 8 ml to 52 × 15 mm plastic petri plates (Falcon No. 1007). Allow to cool.

Top agar: 10 g of Difco tryptone, 2 g of Difco yeast extract, 5 g of NaCl, 7.5 g of Difco agar, 1 liter of water. Sterilize by autoclaving. Melt agar prior to use and keep molten at 55°. At time of addition of phage and cells, add 10 μl of 50 m*M* IPTG per milliliter of top agar.

Phage

Charon 17 (Williams and Blattner[3]): λ *lac5 sRI3*^{0} *cIam sRI4*^{0} *nin5 sHind*III6^0 *DK1*

Charon 30 (Rimm *et al.*[4]): λ *B1007 KH54 nin5 s*(*Bam*HI *2-3 B1007*)$^+$ *DK1 sR14*^{0}

λ70 (Charnay *et al.*[5]): λ *plac 5-1 UV5 imm*λ *nin5 HBs-1*

cI 857 (Sussman and Jacob[6])

Bacterial strains

LA108: F$^-$ *ton A* Δ(*lac*)×*74 nal supE supF rk*$^-$*mk*$^+$ (Pourcel *et al.*[7]); gift from Christine Pourcel

MC4100: F$^-$ *araD139* Δ(*lac*)*U169 rpsL relA thi;* gift from N. Sternberg

MC4100 F′ *lacIq;* F′ *lacIq araD139* Δ(*lac*)*U169 rpsL relA thi;* gift from N. Sternberg

Preparation of Anti-β-galactosidase Antibody

Six-month-old New Zealand white rabbits were used. β-Galactosidase was stored at −70° in 10 m*M* Tris-HCl, pH 7.5, at a concentration of 10 mg/ml. For the primary injection, 100 μg of enzyme were dissolved in 1.5 ml of PBS, and an equal volume of complete Freund's adjuvant (Difco No. 0638-60) was added. Half the dose was given subcutaneously, and half intramuscularly. This initial injection was followed by bimonthly secondary injections performed in the same manner except that incomplete Freund's adjuvant (Difco No. 639-60) was used in place of complete adjuvant. Rabbits were bled (40–50 ml of whole blood) bimonthly.

Antibody was prepared from the serum by ammonium sulfate precipitation and DEAE-cellulose chromatography in a manner similar to that

[3] B. G. Williams and F. R. Blattner, *J. Virol.* **29,** 555 (1979).

[4] D. L. Rimm, D. Horness, J. Kucera, and F. R. Blattner, *Gene* **12,** 301 (1980).

[5] P. Charnay, M. Gervais, A. Louise, F. Galibert, and P. Tiollais, *Nature* (*London*) **286,** 893 (1980).

[6] R. Sussman and F. Jacob, *C. R. Hebd. Seances Acad. Sci.* **254,** 1517 (1962).

[7] C. Pourcel, C. Marchal, A. Louise, A. Fritsch, and P. Tiollais, *Mol. Gen. Genet.* **170,** 161 (1979).

described by Garvey *et al.*[8] The fraction precipitated at 40% saturation ammonium sulfate was resuspended and dialyzed extensively against 17.5 mM sodium phosphate, pH 6.8. It was then chromatographed on a 2.6 cm × 11 cm DEAE-cellulose column in the same buffer. Immunoreactive fractions were pooled, precipitated with 40% saturated ammonium sulfate, and dialyzed extensively against 0.15 M NaCl. The purified immunoglobulin was then stored at −70°. The yield was ca 5.4 mg of immunoglobin per milliliter of serum.

Conjugation

Antibody was conjugated to horseradish peroxidase by the method of Avrameas and Ternynck[9] with few modifications. Horseradish peroxidase (200 mg) was resuspended in 2 ml of 0.1 M sodium phosphate, pH 6.8, 1.25% glutaraldehyde and incubated in the dark at room temperature for 18–21 hr. It was then chromatographed on a 2.6 cm × 11 cm Sephadex G-25 column equilibrated with 0.15 M NaCl at room temperature. The brown fractions were pooled, and the volume was brought to 20 ml with 0.15 M NaCl. An equal volume of anti-β-galactosidase antibody (70 mg) in 0.15 M NaCl was added, followed by a 1 ml of 1 M carbonate bicarbonate buffer, pH 9.6. After 24 hr of incubation at 4°, lysine was added to 2 M, and the mixture was incubated an additional 2 hr at 4° and then extensively dialyzed against PBS. An equal volume of saturated ammonium sulfate (pH 7.6) was added, and the mixture was slowly stirred for 30 min on ice. After 16 hr at 4° the mixture was centrifuged at 7700 g for 20 min. The pellet was resuspended in 20 ml of PBS, and the ammonium sulfate precipitation step was repeated. This second pellet was resuspended in 4 ml of PBS and extensively dialyzed against PBS. The conjugate preparation was partitioned in 1-ml aliquots and stored at −70°.

Horseradish peroxidase in the conjugate was estimated from the A_{403} (an A_{403} of 1 is equivalent to 0.4 mg of HRP per milliliter; Worthington). The A_{280} contributed by horseradish peroxidase was estimated to be 1/3 × A_{403}. The amount of antibody in each preparation (in mg/ml) was calculated by subtracting from the A_{280}, the contribution by horseradish peroxidase at this wavelength, and dividing by 1.46. The yield with this procedure is usually about 0.23 mg of horseradish peroxidase conjugated per milligram of antibody (0.9 molecule of HRP per antibody molecule).

[8] J. S. Garvey, N. E. Cramer, and D. H. Sussdorf, *in* "Methods in Immunology," p. 215. Benjamin, Reading, Massachusetts, 1977.

[9] S. Avrameas and T. Ternynck, *Immunochemistry* **8,** 1175 (1971).

Phage Lysates

Phage stocks were prepared according to Miller.[10] After two sequential plaque purifications, the phage from a single plaque were resuspended in 0.2 ml of BSP buffer, adsorbed to *E. coli* strain LA108, and plated onto 9-cm plates. After 8 hr of growth at 37°, 4 ml of medium containing 10 m*M* $MgSO_4$ were added, and the plates were incubated at 4° for 16 hr. The liquid and soft agar were collected by scraping the plate, and the soft agar was removed by centrifugation. Phage stocks were stored at 4° over chloroform.

Preparation of Cells

Overnight cultures are prepared by adding 0.1 ml of a −20° glycerol stock to 20 ml of broth and incubating with shaking at 37° overnight. Cultures to be used for infection are initiated by adding 5 ml of the overnight culture to 50 ml of broth in an 250-ml Erlenmeyer flask and incubating for 5–8 hr at 37° on a rotary shaker. The density of the culture is then determined by measuring the optical density at 540 nm using a Beckman Model 25 spectrophotometer (1.0 optical density unit at 540 nm is equivalent to 1.2×10^9 cells/ml), and the cells are used immediately for plating.

Preparation of Membrane Plates

Plating of the conjugate-containing soft agar layer directly onto hard agar has never yielded optimal results; the conjugate diffuses into the hard agar, thereby diminishing conjugate concentration at the site of plaque formation and making removal of unprecipitated conjugate difficult. To circumvent these problems, we have designed a plate (designated membrane plate) in which a dialysis membrane is interposed between the two agar layers (Fig. 2), thereby preventing diffusion of the conjugate into the bottom agar. (Membrane plates are available in disposable plastic versions from Genetic Sciences Inc., 1550 California Street, Suite 6246, San Francisco, CA 94109.)

To prepare the plates, a 12 cm × 12 cm piece of dialysis membrane (Spectrapor 2; molecular weight cutoff 12,000–14,000; dry thickness 0.00218 in.) is moistened, placed between the two halves of the plate, and drawn taut as the two halves are pressed firmly together. Excess membrane is trimmed away, and the assembled plate is placed in a 9-cm glass petri plate and autoclaved for 10 min. Water (5 ml) is added to the upper

[10] J. H. Miller, *in* "Experiments in Molecular Genetics," p. 37. Cold Spring Harbor Laboratory, Cold Spring Harbor, New York, 1972.

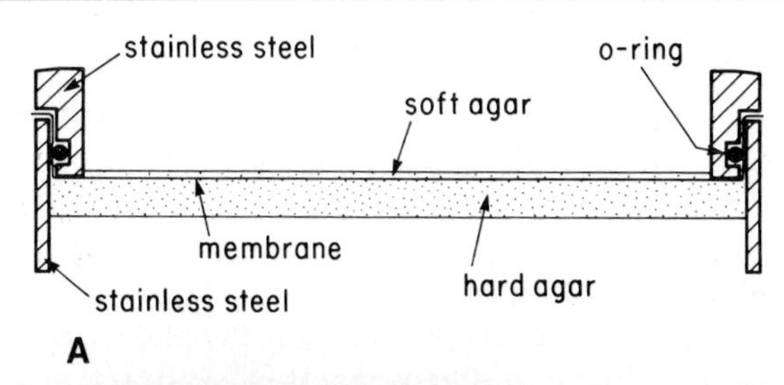

FIG. 2. Membrane plates. (A) The plates are composed of two stainless steel sections. The chambers are separated by a layer of Spectrapor 2 membrane (molecular weight exclusion 12,000–14,000; dry thickness 0.00218 in.). Top agar is added to the upper, smaller chamber, and bottom agar to the lower chamber. (B) The antigen-producing plaques are seen as small, very dark colored rings on an essentially transparent background.

surface of the membrane (shallower compartment) prior to autoclaving to prevent drying and cracking of the membrane. After sterilization, the water is removed, the apparatus is inverted, and 12 ml of molten bottom agar are added to the exposed (deeper) compartment. After the agar has solidified, the plates are inverted and used for plating soft agar as described below. When using membrane plates it is preferable to use enriched media. Our standard broth can be enriched by increasing the yeast extract from 2 g/liter to 7.5 g.

Plating of Phage

The phage adsorption–infection step is performed by adding 0.05 to 0.16 ml of cell suspension, and 0.16 ml of a solution 10 mM in $MgCl_2$ and 10 mM in $CaCl_2$, to 0.16 ml of appropriately diluted phage stock, and incubating the mixture at 37° for 10 min. Then 16 μl of 50 mM IPTG, 20–50 μl of HRP-Ab conjugate, and 1.6 ml of molten top agar (55°) are added, and the mixture is poured directly onto the membrane in the upper chamber of the membrane plates. After the top agar has hardened, the plates are inverted and incubated in 9-cm glass petri plates overnight at 37°.

The number of phage to be plated and screened must be determined from the type of gene bank employed and from estimates of the probability of occurrence of a given clone. In reconstruction experiments we have screened as many as 10 million microplaques per 9 cm diameter membrane plate (1.6×10^5 microplaques per cm^2) with good efficiency of detection of antigen-positives (50% or greater). However, it is probably advisable at the outset to plate phage at lower densities (ca. 100,000 per plate) and increase the densities as one gains experience with the system.

The concentration of indicator cells to be plated is of crucial importance, since excessive intersection (confluence) of plaques may drastically reduce antigen production and hence the efficiency of antigen detection (see discussion below). We recommend that at the outset one examine the sizes of plaques formed with indicator cell densities of 10^7 and 10^8 per milliliter in the soft agar. Plaque size at higher indicator cell densities may then be predicted from the fact that plaque radius varies inversely with the log of initial indicator cell density (D_0).[1]

Knowledge of average plaque size permits one to estimate, in turn, the number of plaques that can be accommodated per unit area of a plate, while retaining a tolerable level of plaque intersection. Either of two approaches may be employed for this estimation. The simpler is to divide the total area of the plate by the average area of a plaque under the plating conditions employed. The number obtained, divided by 3, is in the right range of the number of plaques that can be accommodated without detrimental losses of antigen per plaque. Alternatively, one may use Eq. (1)[1] to predict the fraction of nonintersecting plaques, p, as a function of n, the total number of plaques of average radius r per plate of radius R:

$$p = [1 - 4(r/R)^2]^{n-1} \quad (1)$$

Figure 3 shows curves calculated from this formula for four different values of plaque radius obtained at different initial indicator cell densities. To be conservative, one should initially choose conditions giving values of p greater than 0.5. The degree of intersection tolerable in any given system must in the final analysis be determined empirically.

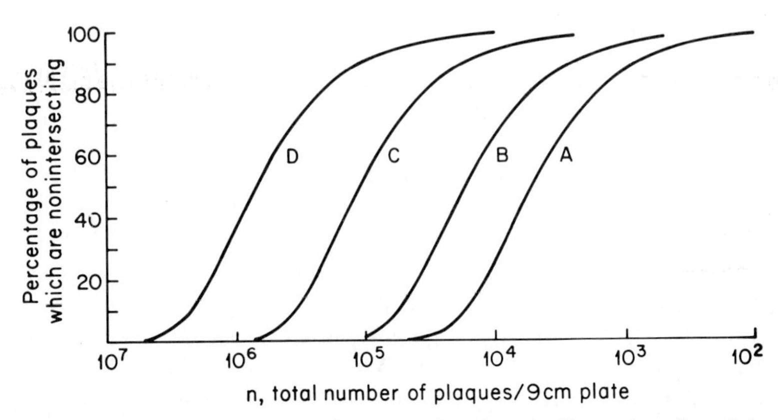

FIG. 3. Fraction of nonintersecting plaques as a function of n (the total number of plaques per plate of 9 cm diameter), and r (the average radius of the plaques). Calculations were made at 4 different values of r: curve A, 0.263 mm; curve B, 0.146 mm; curve C, 0.057 mm; and curve D, 0.023 mm. With Charon 17 bacteriophage plated on *Escherichia coli* strain LA108, these values correspond to initial indicator cell densities of 2.3×10^8, 7.2×10^8, 1.7×10^9, and 2.4×10^9 cells/ml, respectively.

An empirical approach must also be employed to determine the amount of HRP-Ab conjugate required for optimal screening. The optimal concentration depends on several variables, including the specific activity of the conjugate, the amount of antigen produced per plaque, and the avidity of the antibody for the antigen. The ability of the conjugate to form immunoprecipitate is also an important consideration and should be confirmed if possible.

Removal of Unprecipitated Conjugate: "Dialysis"

To remove unprecipitated conjugate, the bottom agar is first removed gently from the lower chamber of the membrane plate with the aid of a small spatula, and the plates are placed in a 27 cm × 32 cm × 12 cm rectangular plastic container containing 4 liters of PBS. To minimize trapping of air in the lower chamber, the plates are initially inserted in a vertical orientation, and are then placed horizontal on the bottom of the container with the soft agar layer uppermost, such that it is bathed in buffer. The container is incubated at 4° for 8–14 hr, while the buffer is kept in gentle motion across plate surfaces either by agitation with a magnetic stirring bar or by slow rocking or rotation of the container. There is little loss of phage or bacteria during this step.

Color Development: the Chromogenic Reaction

The chromogenic reaction is performed immediately after removal of plates from the liquid. To a freshly prepared solution of DAB (0.5 mg/ml in PBS; CAUTION: CARCINOGENIC REAGENT) is added H_2O_2 to a final concentration of 0.015%, and 8 ml of the mixture is added immediately to each plate. Brown HRP reaction product ("stain") is apparent within 2–3 min. Color is fully developed by 10 min; at that time the plates are rinsed to remove substrate.

Retrieval of Lysogens or Phage

The minute "stained" plaques obtained with high indicator cell densities are most easily detected with the aid of a dissecting microscope (magnification 30–60 ×). When phage capable of lysogenization are used, lysogens formed within plaques may be retrieved by inserting a sterile wooden toothpick into the center of the stained area. Material on the tip is resuspended in 1 ml of medium. Plating of 0.2 ml of a 1 : 10 dilution yields 500–2000 colonies, ca 1% of which are antigen-positive. When nontemperate phage are employed, we use a small glass capillary (ca 0.2 mm inside diameter) prepared by drawing out the tip of a Pasteur pipette heated in a flame. The capillary is inserted into the center of the stained area, and the retrieved core of top agar is resuspended in 0.2 ml of BSP buffer. To this are added 0.1 ml of 10 m*M* $MgCl_2$, 10 m*M* $CaCl_2$, and 50–100 μl of cell suspension; the mixture is replated as above. When sampling of antigen-positive plaques was performed on plates containing high densities of antigen-negatives (1.2×10^3 phage/cm^2), the retrieval process usually gave 1 to 2×10^4 isolated plaques with ca 1% showing an antigen-positive reaction. This result is consistent with the fact that bacteriophage λ diffuses slowly from the plaque, either into the dialyzing medium or into the surrounding agar. After 25 hr of dialysis 35% of the original bacteriophage is found in the plaque and only 40% is found within a concentric ring (1.1 mm thick) 2 mm from the plaque center.

Modifications of the Procedure When Conventional Plates Are Used

Although membrane plates give superior results, acceptable results may frequently be obtained with conventional plating methods, if certain precautions are followed.

1. It is important that the thickness of the bottom agar be at least 3 mm in order to obtain maximal growth of the bacterial lawn.
2. If plastic petri plates are used, they must be fixed with double-sided adhesive tape to the bottom of the "dialysis" chamber before the PBS is added.
3. "Dialysis" is usually carried out over 48 hr, with one change of buffer (2 liters per change).

4. Although the background is never as low in conventional plates as in membrane plates, it may be minimized by initiating the chromogenic reaction *immediately* (within seconds) after removal of plates from the "dialysis" medium. Soluble HRP-Ab conjugate is never completely removed from the hard agar layer, but the concentration is minimal in the soft agar immediately adjacent to the "dialysis" fluid. Hence if the chromogenic reaction is performed quickly, diffusion of soluble HRP-Ab conjugate to the surface is minimized.

Factors Affecting Antigen Production in Plaques

Plaque Size and the Number of Cells Lysed per Plaque

The ease with which antigen-positive plaques may be detected is determined by the quantity of antigen produced within individual plaques. Factors affecting the quantity of antigen present include (*a*) the number of cells lysed per plaque; (*b*) the level of gene expression within the phage-infected cell; and (*c*) stability of antigen to denaturation and proteolysis. Our work to date to maximize antigen production has been directed toward understanding the process of plaque formation and the effects of easily manipulated variables of the plating procedure.

As a plaque develops, bacterial cells multiply to form microcolonies. (Although cells are initially distributed evenly throughout the soft agar, little growth occurs in the lower two-thirds of the soft agar, apparently owing to limiting O_2.) Overall growth of cells in the soft agar, as estimated by optical density, appears to be logarithmic (Fig. 4). The final cell density shows relatively little variation over a broad range of initial cell density (D_0) (Fig. 5). During the logarithmic phase of growth, the bacterial density may be described by Eq. (2),

$$D = D_0 e^{\mu t} \tag{2}$$

where D is the density of the cells at time t, D_0 is the density at time zero, and μ is the instantaneous growth constant under the conditions employed.

When the density of phage plated is high, a substantial fraction of the plaques formed will intersect with others during plaque formation. Once a growing plaque has intersected, or contacted, another, the growth of both plaques ceases along the line of intersection, owing to the lack of substrate bacteria.

In order to detect antigen in individual plaques while maximizing the density of phage plated, it is important to understand the effects of intersection of growing plaques. We have shown in the corynephage β sys-

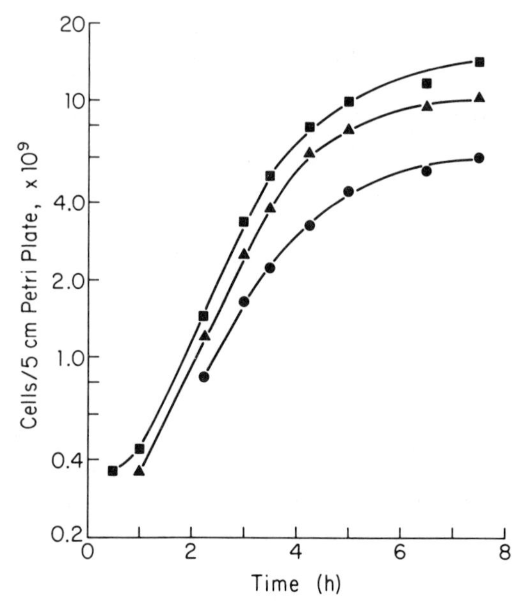

FIG. 4. Effect of media on growth of *Escherichia coli*. *Escherichia coli* strain LA108 was plated at an initial cell density of 7.9×10^7 cells/ml in top agar on 52-mm plastic petri dishes containing different media and incubated at 37°. At various times the plates were scanned in an electrophoresis–TLC densitometer (Quick Scan R and D, Model R4-077; Helena Laboratories) with a 570-nm filter. The number of cells per 5-cm plate was determined from a standard curve derived by measuring the absorbance of plates containing various numbers of cells plated in soft agar. Media: ■, TYE (15 g/liter, Difco Tryptone; 10 g/liter, Difco yeast extract; 5 g/liter NaCl); ▲, hybrid medium (10 g/liter, Difco Tryptone; 2 g/liter Difco yeast extract; 5 g/liter NaCl); and ●, λ medium (media: 10 g/l Difco Tryptone; 5 g/l NaCl).

tem,[2] and assume that the same holds true in the coliphage λ system, that the radius of a plaque increases *linearly* from the time it first becomes visible under a microscope. Hence, $r = kt$, where r is the plaque radius, k is a rate constant, and t is time. We also know that (*a*) uninfected cells are growing logarithmically and (*b*) the number of cells lysed in an increment of time is proportional to the product of the increment in plaque volume and the concentration of cells in this incremental volume. Hence it is not surprising that *the majority of cell lysis in an isolated plaque occurs during the late stages of plaque formation*. Indeed, we have calculated that over 80% of cell lysis occurs during the last 20% of plaque development (Fig. 6).

As a consequence of the foregoing, the number of cells lysed per plaque, and hence the amount of phage-encoded antigen released, would be expected to diminish if the late stages of plaque growth are eliminated

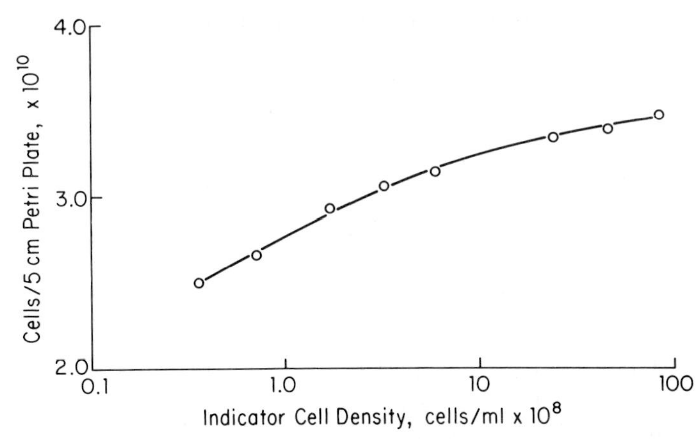

FIG. 5. Final cell density as a function of initial indicator cell density. Different initial indicator cell densities of *Escherichia coli* strain LA108 were plated in top agar on 52-mm plastic petri plates. For the higher concentrations, cells were centrifuged at 7700 *g* for 5 min followed by resuspension in broth. After overnight incubation at 37°, the plates were scanned in an electrophoresis–TLC densitometer (see Fig. 4). The final cell density was estimated from the plate optical density.

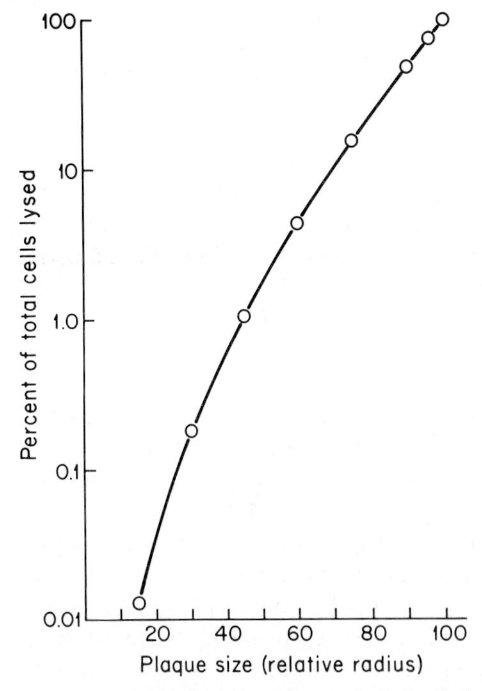

FIG. 6. Cell lysis during plaque development. The number of cells lysed as a function of time was calculated from the equations presented by Kaplan *et al.*[2] Values were normalized to percentage of total cells lysed during plaque development. These values are plotted against the plaque radius rather than time, since it has been shown that the plaque radius is directly proportional to time.[2]

EFFECT OF PLAQUE DENSITY ON THE EFFICIENCY OF DETECTION OF LOW-FREQUENCY ANTIGEN-POSITIVE PLAQUES IN A LARGE POPULATION OF ANTIGEN-NEGATIVES[a]

Number of *lacZ*⁻ phage added	Plaques scored as *lacZ*⁺	Percentage of expected
9.2×10^3	78	100
3.7×10^4	77	99
1.1×10^5	63	81
5.9×10^5	52	67

[a] Approximately 78 Charon 17 (λ *lacZ*⁺) phage were plated with varying numbers of Charon 30 (λ *lacZ*⁻) phage on 52-mm plastic petri plates. The top agar contained 300 μg of conjugate per milliliter and an initial indicator cell density per milliliter of 1.7×10^8 cells of *Escherichia coli* strain LA108. After overnight incubation at 37°, the plates were dialyzed and developed. The number of antigen-positive plaques detected by eye were counted.

by intersection (confluence) of adjacent plaques. In both the corynephage β and coliphage λ systems (see the table) the efficiency of detection of antigen-positive phage in large populations of antigen-negatives decreases under conditions where extensive intersections of plaques occurs. At low plaque densities with minimal intersection, we find large rings of immunoprecipitation (Fig. 7A). When densities of plated phage are increased, rings of immunoprecipitate concentric with the plaques are reduced to slightly irregular patches (Fig. 7B and C); at very high phage densities ("superconfluence"), no immunoprecipitate whatsoever can be detected.

To prevent the deleterious effects of plaque intersection on antigen production at high plaque densities, plaque size must be kept small. By plating phage in the presence of sufficiently high concentrations of indicator cells, the period during which cells are infectable is reduced; hence fewer generations of phage multiplication occur. We have found that with Charon 17 phage, plated on *E. coli* LA108, plaque radius decreased logarithmically as the initial cell density, D_0, was raised from 9×10^6 cells/ml to 9×10^8 cells/ml (Fig. 8). At cell concentrations greater than 3×10^9 per milliliter, plaques were no longer visible under a dissecting microscope at 60×. The logarithmic relationship has also been observed with corynephage β and is predicted by the model of Kaplan *et al.*[1,2]

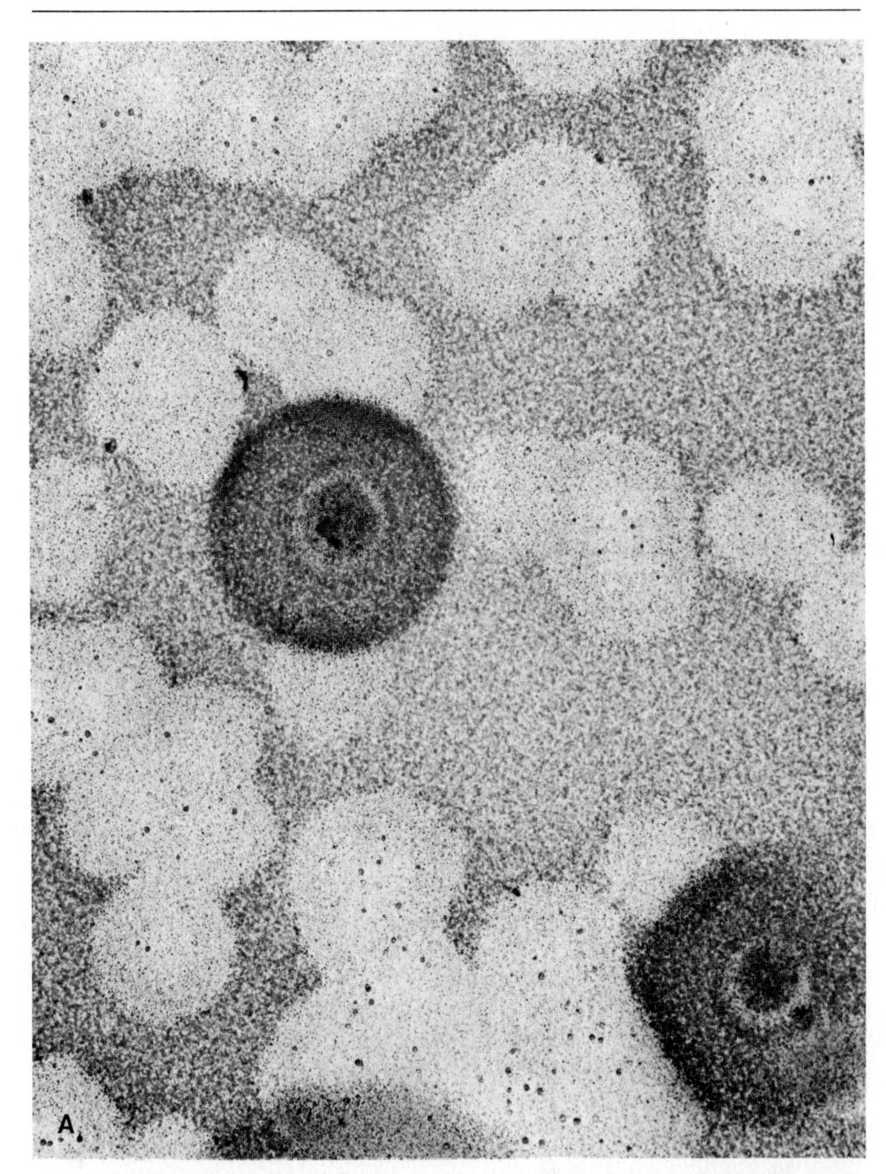

FIG. 7. Effect of crowding on the morphology of the colored immunoprecipitation products. Approximately 75 Charon 17 (λ *lacZ*$^+$) bacteriophage and either (A) 1.6×10^3 (7.5×10^1 per cm^2), or (B and C) 8×10^4 (3.7×10^3 per cm^2) Charon 30 (λ *lacZ*$^-$) bacteriophage were plated on 2.2×10^8 *Escherichia coli* strain LA108 per milliliter in top agar containing 240 μg of conjugate per milliliter. The plates (52-mm plastic petri plates) were incubated at 37° overnight, dialyzed, and developed as described in the text. The plates were photographed

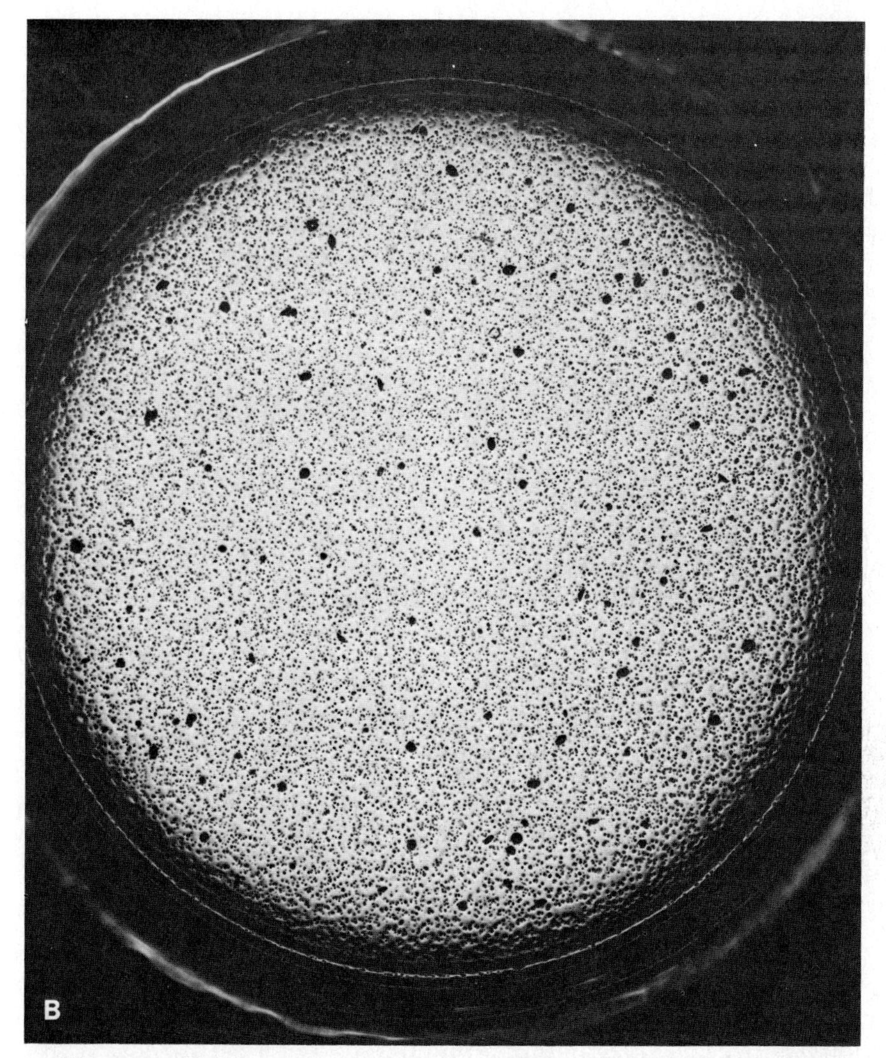

at different magnifications: (A) 15×; (B) 2×; and (C) 15×. The Charon 17 plaques that produce β-galactosidase are marked by either rings (A) or large granules, (B) and (C), of precipitation. The background nonstaining microcolonies represent λ-resistant cells that occur at a frequency of 1 in 10^4 cells of *E. coli* strain LA108.

We have demonstrated in the corynephage β system that antigen-positive plaques as small as 0.03 mm in diameter give dark, pinpoint-size areas of stain that are easily discernible with the unaided eye, and similar results have been obtained with coliphage λ. Although the absolute quan-

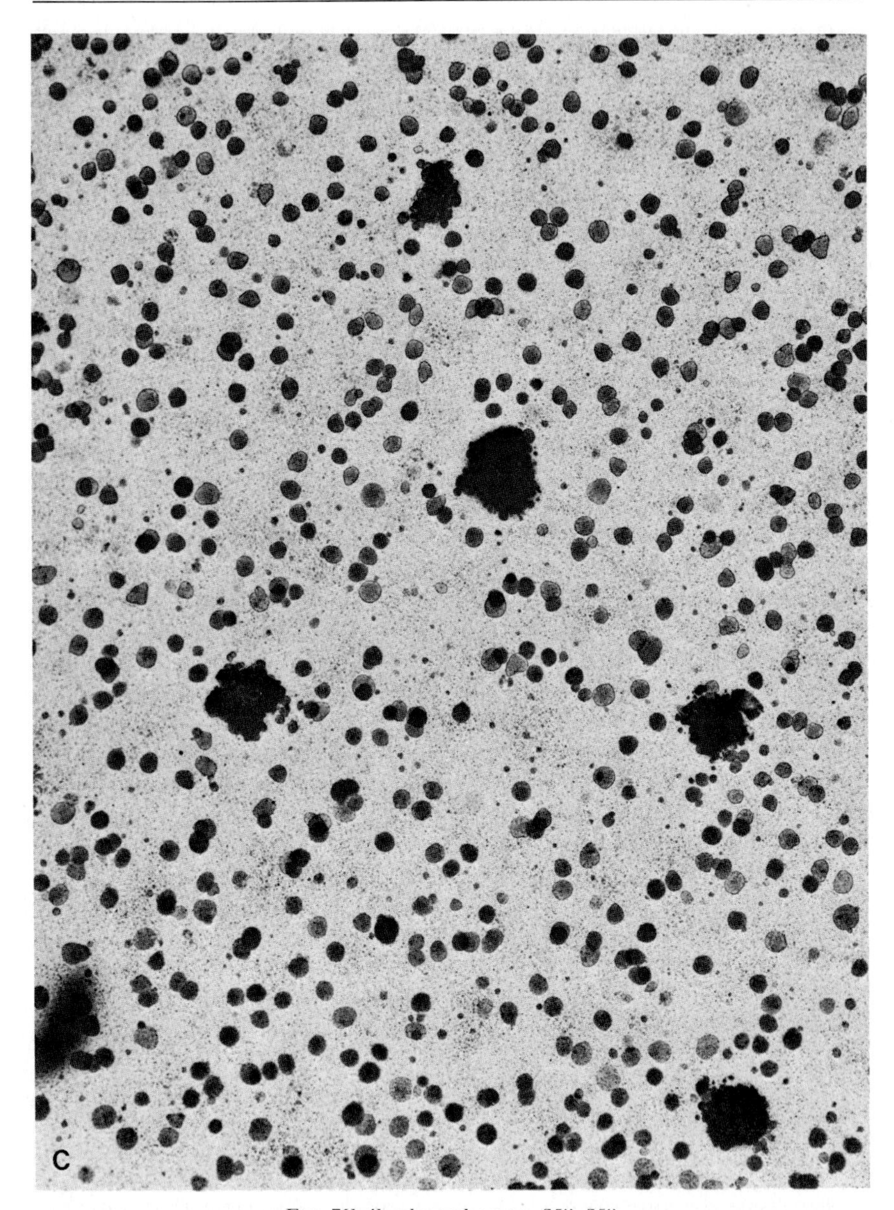

FIG. 7C. See legend on pp. 358–359.

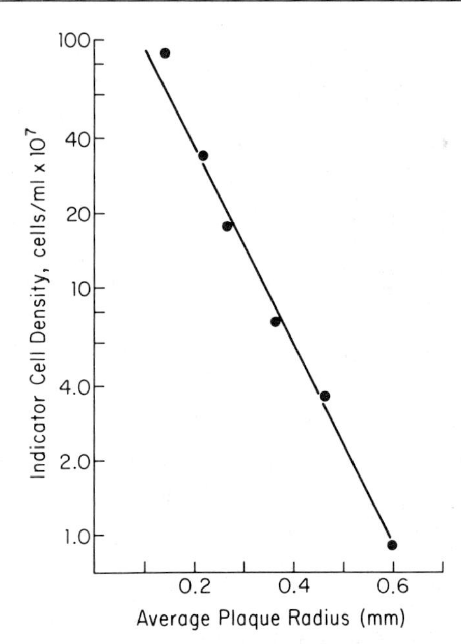

FIG. 8. Average plaque radius as a function of initial indicator cell density. Charon 17 phage were plated on varying initial cell densities of *Escherichia coli* strain LA108. After overnight incubation, at 37°, average plaque size was determined by measuring the radii of 25 randomly chosen plaques from each plate. Measurements were made at 60× magnification under a dissecting microscope equipped with an ocular micrometer.

tity of antigen per plaque decreased as D_0 was increased (Fig. 9), antigen-positive plaques were readily visible. In the corynephage β system the *concentration* of antigen in plaques was found to *increase* with increasing D_0. The same appears to be true in the coliphage λ, although it is difficult to obtain an accurate estimate of the increase.

Plaque size, and hence the amount of antigen produced, is also strongly dependent on the individual strain of phage used. We have observed that under identical plating conditions different λ strains grow to different final plaque radii (e.g., at $D_0 = 7 \times 10^7$ cells/ml, plaque radii of Charon 17, Charon 30, and λ cI857 were 0.36 mm, 0.48 mm, and 0.69 mm, respectively).

The type of medium used can also strongly affect the results obtained. λ medium consistently gave weak plaques and poor antigen production on plates, despite the fact that good results were obtained in liquid culture. With TYE medium we experienced some difficulties in performing the adsorption–infection step, apparently owing to a high proportion of non-viable cells in saturated cultures. A hybrid medium, consisting, per liter,

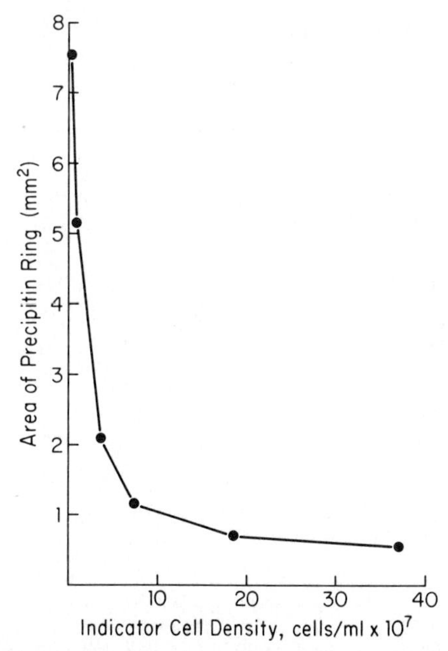

FIG. 9. Variations in quantity of β-galactosidase antigen as a function of initial indicator cell density. Charon 17 bacteriophage were plated (52-mm plastic petri plates) on varying initial cell densities of *E. coli* strain LA108 in top agar containing 280 μg of conjugate per milliliter. After incubation, dialysis and development, average plaque and ring sizes were determined by measuring the diameters of 25 randomly chosen plaques from each plate. Measurements were performed as in Fig. 8. The areas of the precipitin rings were calculated by subtracting the plaque area from the total precipitin ring area.

of 10 g of Difco Tryptone, 2 g of Difco yeast extract, and 5 g of NaCl, proved superior to either λ medium or TYE medium. Figure 4 shows growth curves of *E. coli* LA108 obtained with three different media. Although the instantaneous growth constant does not differ significantly, the lawn density, and therefore the amount of antigen produced per plaque, does. The hybrid medium supports the growth of dense bacterial lawns, gives a short lag period, and allows a relatively long period of exponential growth.

Gene Expression within Phage-Infected Cells and Stability of the Antigenic Product

Expression of translation products from eukaryotic genes cloned into *E. coli* has frequently required fusion to prokaryotic promoters. Thus, the λ P_L promoter has been used to express maize and wheat chloroplast genes for the large subunits of ribulosebisphosphate carboxylase,[11] the major antigen of foot and mouth disease virus,[12] and human fibroblast interferon[13]; similarly the *lac* promoter has been used to express hen ovalbumin[14] and human growth hormone.[15] The *lac* promoter has been employed to construct expression cloning vehicles.[15–17] This promoter is positively controlled by the level of cAMP–CAP complex in the cell, and negatively controlled by the *lac* repressor, a product of the *lacI* gene.

We have examined the effect of the *lac* inducer isopropyl-β-D-thiogalactoside (IPTG) on β-galactosidase production in plaques of Charon 17 phage formed on two *E. coli* strains: MC4100-F′*lacIq,* which produces multiple copies of the *lac* repressor, and MC4100, which lacks repressor. The diameters of stained immunoprecipitin rings were measured and areas calculated. With strain MC4100 a high level of β galactosidase was detected, and the level did not vary within a range of IPTG concentrations from 0 to 5 m*M*. In contrast plaques on strain MC4100-F′*lacIq* contained no detectable β-galactosidase in the absence of IPTG. At 0.1 m*M* IPTG the strain produced 13% of the β-galactosidase produced in the absence of the episome, and at 0.5 m*M* it produced 30% of the level.

Despite proper fusion between eukaryotic genes and bacterial promoters, translational products may be extremely unstable owing to rapid proteolysis within the bacterial cell. Several investigators have resorted to protein fusion to overcome this problem.[5,18] Thus, for example, hepatitis virus surface antigen could be detected in clones when the gene was fused to the C-terminal region of β-galactosidase (a 1005 residue fragment), but was apparently unstable when fused to a short N-terminal fragment (8 amino acids).[5] Fusion to β-galactosidase was also used in producing insulin A and B chains[18] and somatostatin.[19]

[11] A. A. Gatenby, J. A. Castleton, and M. W. Saul, *Nature* (*London*) **291,** 117 (1981).

[12] H. Kupper, W. Keller, C. Kurz, S. Forss, H. Schaller, R. Franze, K. Strohmaier, O. Marquardt, V. G. Zaslavsky, and P. H. Hofschneider, *Nature* (*London*) **289,** 555 (1981).

[13] R. Derynck, E. Remaut, E. Saman, P. Stanssens, E. De Clercq, J. Content, and W. Fiers, *Nature* (*London*) **287,** 193 (1980).

[14] O. Mercereau-Puijalon, A. Royal, B. Cami, A. Garapin, A. Krust, F. Gannon, and P. Kourilsky, *Nature* (*London*) **275,** 505 (1978).

[15] D. V. Goeddel, H. L. Heyneker, T. Hozumi, R. Arentzen, K. Itakura, D. G. Yansura, M. J. Ross, G. Miozzari, R. Crea, and P. H. Seeburg, *Nature* (*London*) **281,** 544 (1979).

[16] P. Charnay, M. Perricaudet, F. Galibert, and P. Tiollais, *Nucleic Acids Res.* **5,** 4479 (1978).

[17] T. M. Roberts and G. D. Lauer, this series, Vol. 68, p. 473.

[18] K. Itakura, T. Hirose, R. Crea, A. D. Riggs, H. L. Heyneker, F. Bolivar, and H. W. Boyer, *Science* **198,** 1056 (1977).

[19] D. V. Goeddel, D. G. Kleid, F. Bolivar, H. L. Heyneker, D. G. Yansura, R. Crea, T. Hirose, A. Kraszewski, K. Itakura, and A. D. Riggs, *Proc. Natl. Acad. Sci. U.S.A.* **76,** 106 (1979).

The chromogenic method described here may facilitate screening for desired determinants from fused genes. It should be possible to use the HRP-Ab in the conventional manner when polyclonal, precipitating antibodies against the target antigen are available and of high enough affinity, and the antigenic determinants on the fusion product are sufficiently exposed. However, if the desired determinants of the fusion product are weak, or if one wishes to use nonprecipitating monoclonal antibodies against these determinants, or when immunoprecipitation does not occur for other reasons, it may be useful to include unconjugated, precipitating polyclonal antibodies against the carrier portion of the fused protein, to immobilize the gene fusion product in the plaque in which it is produced. One is then freed of the constraint of requiring the Ab conjugate also to act as immunoprecipitant, and may include HRP-conjugated monoclonal or polyclonal antibody specific for the target moiety of the fusion product to screen for the expression of a specific antigenic determinant or determinants.

We have obtained preliminary data to support the validity of this approach, using λ70, a clone that produces a hybrid protein consisting of the first 1005 amino acids of β-galactosidase followed by 192 amino acids of the hepatitis B surface antigen. When the soft agar contained only the HRP-conjugate of anti-hepatitis B surface antigen, no chromogenic immunoprecipitate was detected. However, when unconjugated anti-β-galactosidase was included, rings of immunoprecipitate were formed, and the area of the rings correlated with the amount of unlabeled anti-β-galactosidase present. We assume that the anti-β-galactosidase antibody acts as an immunoprecipitant, facilitating detection of weak hepatitis determinants on the hybrid molecule.

Factors Affecting Detection of Antigen

Several experimental variables besides the biological parameters affect antigen detection within a plaque. The sensitivity of the method is dependent on the contrast between the insoluble, intensely colored reaction product and the background. Contrast is affected by the specific activity of the HRP-Ab conjugate, the concentration of the conjugate used, effectiveness of removal of residual soluble HRP-Ab, the conditions (including type of substrate) used for the chromogenic reaction, and the method of visualization of the plates.

Specific Activity of the Conjugate

A major factor affecting specific activity of the conjugate is antibody purity. Optimally the antibody to be labeled with HRP should be affinity-purified or monoclonal antibody. As noted above, if labeled monoclonal antibody is employed, it may be necessary to include unlabeled polyclonal antibody as a means of immunoprecipitating the antigen. The unfractionated immunoglobulin fraction, or even unfractionated antisera, may be substituted for affinity-purified material, but background staining will be higher.

The glutaraldehyde coupling procedure described has routinely given about 0.9 HRP molecule per Ab molecule in our hands. This procedure has been shown to produce some loss of enzymic and antigen-binding activity,[20] and other methods may yield more satisfactory results.[21,22] O'Sullivan *et al.* have reported that with *N,N′,O*-phenylenedimaleimide or *m*-maleimidobenzoyl-*N*-hydroxysuccinimide ester, 80% of added β-galactosidase was conjugated, with 90% retention of both enzymic and antigen-binding activities.[22]

Concentration of the Conjugate

The area of the immunoprecipitate in radial double-diffusion systems (with antigen diffusing from a central point) is inversely proportional to the antibody concentration.[23] Results in the immunodetection system employed here are consistent with this relationship (Fig. 10). At low concentrations of HRP-Ab conjugate, the stained area is large and the intensity weak, therefore giving poor resolution and contrast. As the concentration of conjugate is raised the area becomes smaller, and the staining more intense. However, very high concentrations of conjugate adversely affect detection by increasing background staining, and sometimes diminish the area of stain to the point where detectability is limited. Hence it is advisable to determine an optimal concentration empirically.

Removal of Unprecipitated Ab-HRP Conjugate

Effective removal of unprecipitated conjugate is crucial to the success of the method. Whereas conjugate diffusion is a very slow process from conventional plates, a few hours of "dialysis" of membrane plates gives virtually complete conjugate removal and hence essentially transparent backgrounds. One may calculate that the rate of removal of soluble conjugate should be inversely proportional to the square of the thickness of the agar throughout which it is dissolved. Since the soft agar layer is ca 1/7.4

[20] D. J. Ford, R. Radin, and A. J. Pesce, *Immunochemistry* **15,** 237 (1978).

[21] B. A. L. Hurn and S. M. Chantler, this series, Vol. 70, p. 132.

[22] M. J. O'Sullivan, E. Gnemmi, D. Morris, G. Chieregatti, A. D. Simmonds, M. Simmons, J. W. Bridges, and V. Marks, *Anal. Biochem.* **100,** 100 (1979).

[23] G. Mancini, A. O. Carbonara, and J. F. Heremans, *Immunochemistry* **2,** 235 (1965).

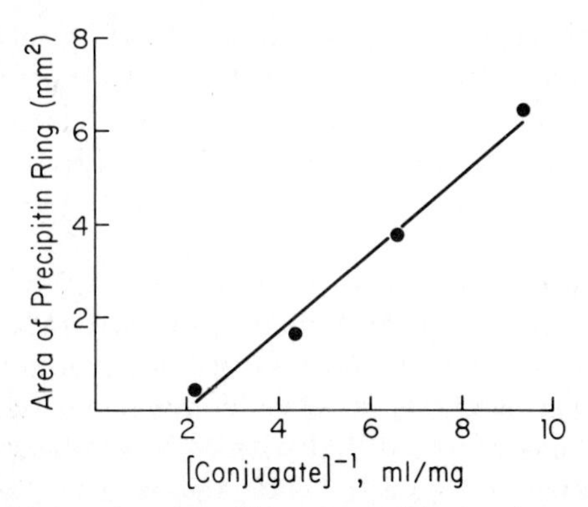

FIG. 10. Relationship between area of immunoprecipitation ring and inverse of conjugate concentration. Charon 17 bacteriophage were plated (52-mm plastic petri plates) on 1.1×10^8 cells of *Escherichia coli* strain LA108 per milliliter in top agar containing from 105 to 431 μg of conjugate per milliliter. After incubation, dialysis, and color development, average ring sizes were determined by measuring the radii of 25 randomly chosen plaques and corresponding rings. Calculations were performed as in Fig. 9.

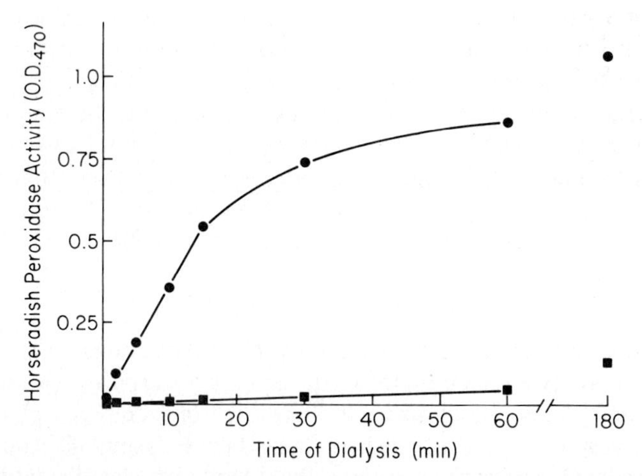

FIG. 11. Rate of conjugate removal during "dialysis." Conjugate was added to top agar at a concentration of 1.5 mg/ml and poured to a thickness of 0.54 mm either onto 8 ml of bottom agar in a 52-mm plastic petri plate or onto a membrane plate. After overnight incubation at 37° to approximate the assay conditions, the plates were placed in 250 ml of PBS and shaken at 24°. At various times, samples were removed, and the amount of conjugate released was assayed by adding 3,3′-diaminobenzidine to 0.5 mg/ml and H_2O_2 to 0.015%. After incubation at 24°, the reaction was stopped by the addition of sodium cyanide to 10 m*M* and placing on ice. The absorption at 470 nm was measured.

of the total, the rate of removal of conjugate from the dialysis membrane plates should be 50–60 times faster than from conventional plates. The data in Fig. 11 confirm this prediction.

A second advantage of the membrane-plate configuration is that HRP-Ab is maintained at a high and more or less constant concentration throughout the experiment, since it is confined to the soft agar layer; thus immunoprecipitation is maximized while conjugate is conserved.

Finally, the use of membrane plates may permit the use of "antibody sandwich" methods in the procedure. Since diffusion is so rapid from the thin agar layer, it should be possible to perform the initial immunoprecipitation with unlabeled or hapten-labeled primary antibody. After removal of residual soluble primary antibody, an HRP-labeled secondary antibody (anti-primary antibody or anti-hapten) could be added, followed by a second "dialysis" step, and staining in the conventional manner. This would have the advantage of minimizing the number of antibody preparations to be conjugated to HRP.

The Chromogenic Reaction

Conditions of the chromogenic reaction also affect the sensitivity of the detection method. HRP, which uses H_2O_2 to oxidize a number of substrates, is commonly employed in antibody conjugates because of its stability and ability to yield colored, insoluble reaction products. Substitution of β-galactosidase, alkaline phosphatase, lactoperoxidase, or other enzymes for HRP may be advantageous under certain circumstances, but we have found HRP to yield excellent results.

Several chromogenic substrates of HRP are available, which have different reaction requirements and color characteristics. We have examined results obtained with 3,3′-diaminobenzidine and two less carcinogenic chromogens, tetramethylbenzidine and 3-aminomethylcarbazole at a variety of pH values ranging from pH 3 to pH 7.5. With 3,3′-diaminobenzidine, optimal results were obtained at pH 7.5; staining was intense and background was minimal. Also, autoxidation of the substrate, which can contribute to the background, is low under these conditions.

Results obtained with tetramethylbenzidine and 3-amino-9-ethylcarbazole were generally less satisfactory. With the former, precipitin rings were most prominent at pH 3, barely detectable at pH 4, and invisible at higher pH levels. The intensity of the rings at pH 3 was much lighter than with DAB at pH 7.5. With 3-amino-9-ethylcarbazole, the intensity of staining was also less intense than that observed with DAB.

Inspection of Plates

Use of a dissecting microscope facilitates detection of antigen-positive microplaques of small radius. Also, a Wratten No. 61 green filter was found to increase contrast between stained areas and the background.

Acknowledgments

This work was supported in part by National Science Foundation grant PCM 79 10684, Cetus Corporation, and Genetic Sciences Inc. We thank P. Charnay and co-workers for phage strain λ70 coding for hepatitis virus surface antigen.

[26] Purification of Nucleic Acids by RPC-5 ANALOG Chromatography: Peristaltic and Gravity-Flow Applications

By J. A. THOMPSON, R. W. BLAKESLEY, K. DORAN, C. J. HOUGH, and ROBERT D. WELLS

RPC-5 column chromatography was established as a powerful technique for the purification (especially preparatively) of nucleic acids.[1] Early versions of RPC column chromatography (RPC-1 to RPC-4) consisted of water-immiscible organic quaternary ammonium compounds coated on diatomaceous earth supports. These were used extensively for the fractionation of transfer RNAs isolated from a wide variety of sources.[2] Later, changing the support to a polychlorotrifluoroethylene resin (RPC-5) led to improved separation of tRNA isoacceptors, shorter analysis times, and easier scale-up for preparative separations.[3] RPC-5 column chromatography has proved to be useful as a tool for both the analytical and preparative fractionation of DNA fragments generated by restriction endonuclease cleavage,[4,5] of complementary strands of specific

[1] R. D. Wells, S. C. Hardies, G. T. Horn, B. Klein, J. E. Larson, S. K. Neuendorf, N. Panayotatos, R. K. Patient, and E. Selsing, this series, Vol. 65, p. 327.

[2] A. D. Kelmers, H. O. Weeren, J. F. Weiss, R. L. Pearson, M. P. Stulberg, and G. D. Novelli, this series, Vol. 20, p. 9.

[3] R. L. Pearson, J. F. Weiss, and A. D. Kelmers, *Biochim. Biophys. Acta* **228,** 770 (1971).

[4] S. C. Hardies and R. D. Wells, *Proc. Natl. Acad. Sci. U.S.A.* **73,** 3317 (1976).

[5] A. Landy, C. Foeller, R. Reszelbach, and B. Dudock, *Nucleic Acids Res.* **3,** 2575 (1976).

ISBN 0-12-182000-9

DNA restriction fragments,[6] of oligomers of single-stranded DNA and RNA,[7-9] and of supercoiled plasmid ColE1 DNA.[10]

Although originally named "reverse-phase chromatography," fractionations were based primarily on ion-exchange mechanisms, where bound nucleic acids eluted from RPC-5 columns predominantly in order of increasing molecular weight.[1,2] However, it also was observed that several species of nucleic acids of similar size (i.e., tRNAs or restriction fragments) were well separated, suggesting that factors other than molecular weight governed elution behavior. Examples of these included (*a*) DNA restriction fragments with unpaired ("sticky") ends eluted at substantially higher salt concentrations than equivalent-sized fragments with paired ("blunt") ends; (*b*) single-stranded DNA eluted at much higher salt concentrations than double-stranded DNA of the same length; and (*c*) AT-rich DNA restriction fragments eluted at higher salt concentrations than fragments of equivalent size that are not AT-rich. At least one other factor influenced the separation of DNA restriction fragments by RPC-5 column chromatography. Some fragments, which eluted later than predicted from their size, contained known genetic regulatory sites. These observations were reviewed elsewhere.[1]

Several difficulties encountered with the originally reported chromatography matrix, however, hindered the widespread use of this technique. The most significant problem was the lack of commercial availability of the matrix. Other difficulties resulted from the composition of the original matrix. First, the active organic layer (quaternary amine) was held loosely on an inert, particulate support. With continued use, this layer was gradually released (i.e., "bleed"), thereby reducing the effective life of the column. Second, the particle size of the original matrix was very heterogeneous, necessitating high operating pressures (200–400 psi) to achieve reasonable flow rates. Finally, preparations of the matrix varied from batch to batch in both the support and the coating, inhibiting development of chromatographic procedures that would give predictable results.

RPC-5 ANALOG Column Chromatography

A new matrix, RPC-5 ANALOG, was developed by us in 1980 to correct those original difficulties described above (i.e., availability and bleeding) in order to take advantage of the potential power of chromato-

[6] H. Eshaghpour and D. M. Crothers, *Nucleic Acids Res.* **5,** 13 (1978).

[7] J. F. Burd and R. D. Wells, *J. Biol. Chem.* **249,** 7094 (1974).

[8] J. B. Dodgson and R. D. Wells, *Biochemistry* **16,** 2367 (1977).

[9] G. C. Walker, O. C. Uhlenbeck, E. Bedows, and R. I. Gumport, *Proc. Natl. Acad. Sci. U.S.A.* **72,** 122 (1975).

[10] A. N. Best, D. P. Allison, and G. D. Novelli, *Anal. Biochem.* **114,** 235 (1981).

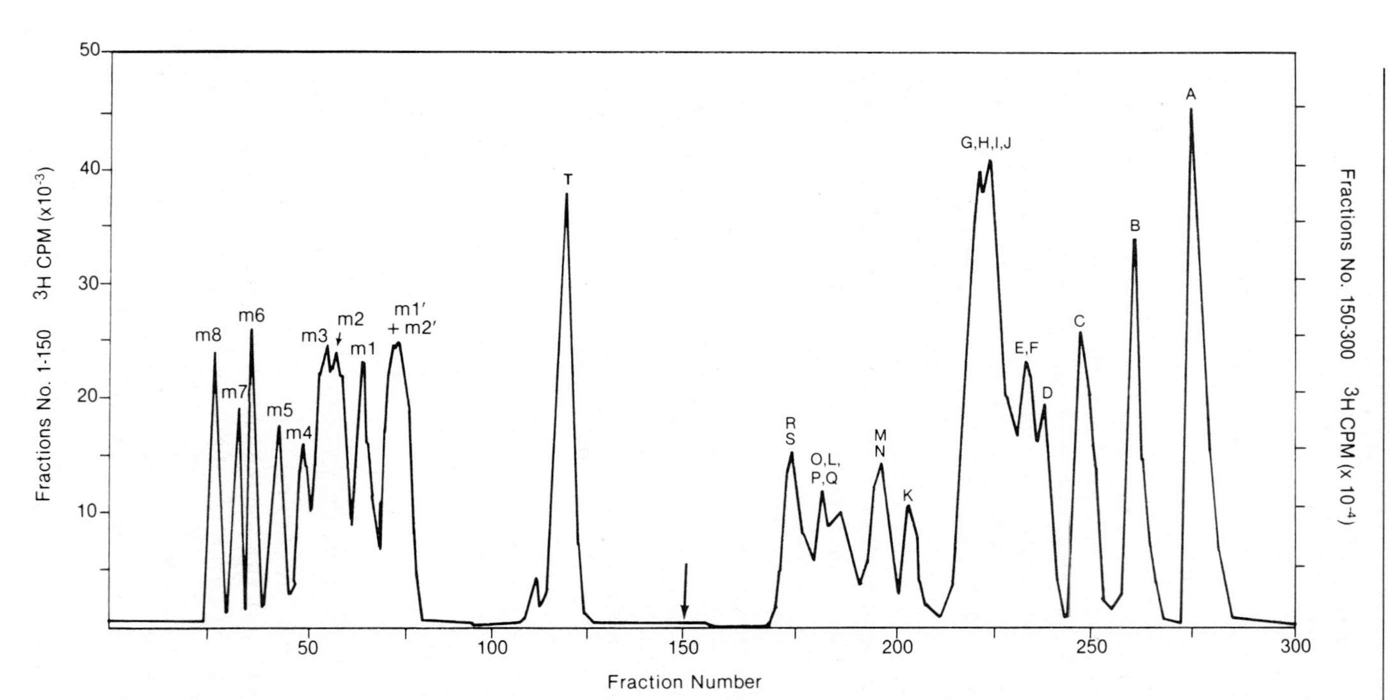

FIG. 1. Column chromatographic purification of DNA restriction fragments. Approximately 10 A_{260} units of form I SV40 [^{3}H]DNA were digested to completion with *Alu*I as determined by agarose gel electrophoresis. The sample was extracted with phenol and ether, dialyzed against 0.2 *M* NaCl in TE buffer, and loaded on an RPC-5 ANALOG column (88 × 0.9 cm) previously equilibrated in the same buffer using a Milton Roy pump (0.5 ml/min) operating at 350 psi. After washing of the bound DNA with the same buffer, a linear (1000 ml) gradient of 0.45 *M* to 0.65 *M* NaCl in TE buffer was applied to the column (0.5 ml/min). Fractions (6 min) were collected, and eluted [^{3}H]DNA was determined by liquid scintillation. The arrow (fraction 150) shows the point at which the *Y* axis scale of ^{3}H counts per minute was reduced by 10-fold. Appropriate peak fractions were pooled, dialyzed into water, and lyophilized to dryness. Aliquots (1 μg) of resuspended DNA fractions were analyzed by polyacrylamide (10%) gel electrophoresis and identified as indicated. More than 95% of the [^{3}H]DNA applied was recovered from the column.

graphic purification of nucleic acids. RPC-5 ANALOG is readily available commercially (Bethesda Research Laboratories, Inc.), is reproducible from batch to batch, and is significantly reduced in bleeding. Therefore, once a specific chromatographic condition has been established, reproducible results can be expected. This matrix, named ANALOG, directly substitutes in essentially all the types of nucleic acid chromatographic purifications originally described for RPC-5.[1,2]

RPC-5 ANALOG column chromatography was used for improved separations of tRNA isoacceptors.[11,12] Furthermore, preparative fractionations of DNA fragments generated by restriction endonuclease cleavage readily were achieved. These include an *Alu*I digest of SV40 DNA (Fig. 1), a *Hae*III digest of ϕX174 RF DNA, a *Hae*III digest of pRZ2 DNA, and a *Taq*I digest of pBR325 DNA (data not shown). Additionally, 50-mg samples of eukaryotic chromosomal DNA, digested with *Eco*RI, routinely were fractionated by RPC-5 ANALOG column chromatography. This fractionation was followed by hybridization to specific viral probes, similar to experiments previously described.[13] RPC-5 ANALOG also was used both to separate complementary strands of DNA restriction fragments (Fig. 2) and to purify milligram quantities of supercoiled pBR322 plasmid DNA (data not shown). Table I shows some specific examples of purification of nucleic acids by RPC-5 ANALOG column chromatography that include both analytical and preparative applications. This list of applications (Table I) is by no means exhaustive; rather it is intended as a guide to some general chromatography conditions for applications more frequently encountered.

Another major impediment to the widespread use of both RPC-5 and RPC-5 ANALOG column chromatography was the requirement for high operating pressures (>100 psi) in order to achieve reasonable flow rates. The operating characteristics of RPC-5 ANALOG were dramatically improved by size classification of the matrix into a narrow, discrete particle size distribution. The particle size was defined empirically by the desirability of gravity flow or peristaltic pump chromatography and by the resolution requirements of several specific nucleic acid purifications. Achievement of such a size-classified ANALOG allowed us to develop both preparative and analytical purification methods for two important types of nucleic acids, supercoiled DNAs and synthetic oligodeoxyribonucleotides. These methods based upon improved RPC-5 ANALOG are described below.

[11] G. D. Novelli, personal communication.

[12] R. Modali, personal communication.

[13] S. M. Tilghman, D. C. Tiemeier, F. Polsky, M. H. Edgell, J. G. Seidman, A. Leder, L. W. Enquist, B. Norman, and P. Leder, *Proc. Natl. Acad. Sci. U.S.A.* **74,** 4406 (1977).

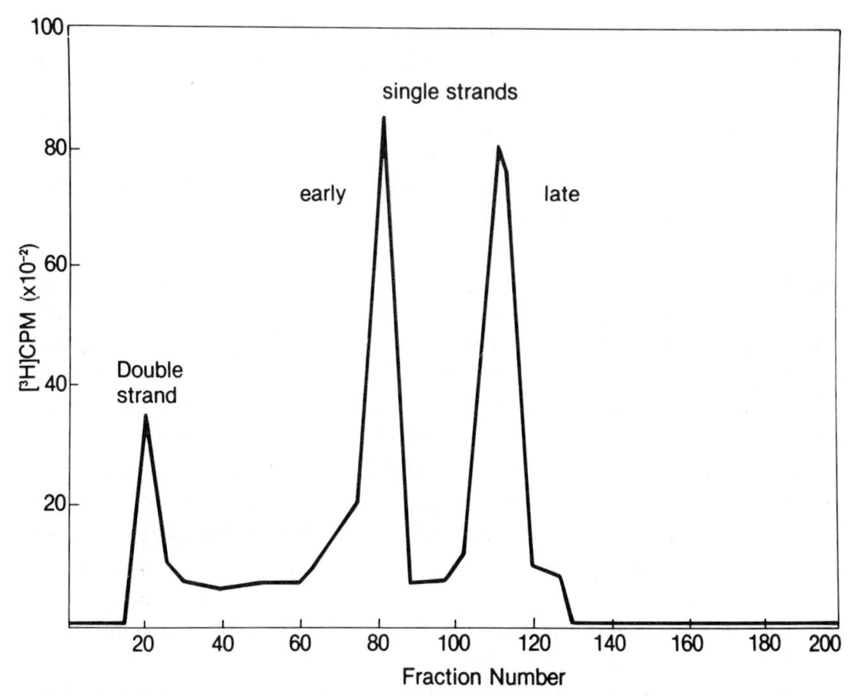

FIG. 2. Column chromatographic purification of separated strands of a DNA restriction fragment. Approximately 10 μg of the [^{3}H]*Alu*C fragment (330 bp) (see Fig. 1) were suspended in 0.5 *M* NaCl, 12 m*M* NaOH (pH 12), then loaded on an RPC-5 ANALOG column (45 × 0.2 cm) previously equilibrated with the same buffer using a Milton Roy pump (0.25 ml/min) operating at 300 psi. After washing the bound DNA with the same buffer (5 ml), a linear (50 ml) gradient of 0.9 *M* to 0.95 *M* NaCl, 12 m*M* NaOH (pH 12) was applied to the column (0.25 ml/min). Fractions (1 min) were collected, and eluted [^{3}H]DNA was determined by liquid scintillation spectrometry. Appropriate peak fractions were pooled, dialyzed against H_2O, and lyophilized to dryness. Aliquots of resuspended DNA were analyzed by both agarose (1%) gel electrophoresis and blot hybridization, then identified as indicated. More than 90% of the [^{3}H]DNA applied was recovered from the column.

General Principles

Operating Mechanisms

Separations on RPC-5 ANALOG apparently are based primarily on ion-exchange mechanisms. RPC-5 ANALOG is composed of a thin film of trialkylmethylammonium chloride covering a solid, nonporous, noncompressible, inert microparticulate resin. The tightly bound coating functions as an anion exchanger with the strongly acidic phosphate residues in nucleic acids. Once bound to RPC-5 ANALOG, nucleic acids are eluted by applying a salt solution of ionic strength that is higher than the ionic strength found for binding. Typically, fractionations of complex mixtures of nucleic acids are executed by applying a linear salt gradient of increasing ionic strength to a column of appropriate size. Lower molecular weight nucleic acids elute before higher molecular weight owing to the presence of fewer total available negative phosphate groups. Hence, separations are usually according to the length (number of bases) of the nucleic acid molecules. However, the available charge for a given nucleic acid molecule may be different from that calculated for its length owing to either base sequence or secondary and tertiary structure. Thus, such changes in a nucleic acid molecule occasionally may interfere with a separation owing to an unusual elution behavior, as previously observed.[14] On the other hand, this unique behavior can be utilized advantageously. Amplification or suppression of these unique charge properties (e.g., by altering chromatographic conditions such as pH or temperature) can resolve otherwise difficult separations.

Preparation of RPC-5 ANALOG

To prepare RPC-5 ANALOG (Bethesda Research Laboratories, Inc.) for use, 1 g of the matrix is suspended in 10 ml of 2.0 *M* NaCl buffered with 10 m*M* Tris-HCl (pH 7.2), 1 m*M* Na_2 EDTA (TE buffer) and stirred gently for 1 hr. The matrix is allowed to settle (10 min), and the supernatant along with any fines is removed by aspiration (do not decant). This process is repeated several times until the supernatant, after settling, remains clear. The suspension is then placed at 4° for 16 hr (overnight) with gentle stirring to ensure complete hydration. After this final hydration, the matrix is allowed to settle and the supernatant is removed by aspiration. The settled matrix is resuspended (10 ml/g) in fresh, degassed 2 *M* NaCl in TE buffer and stored at 4° until use.

Chromatography Conditions

Both preparative and analytical columns are prepared essentially as previously described.[1] The size of the column depends on the amount of RPC-5 ANALOG to be used and the method by which it is packed into the

[14] R. K. Patient, S. C. Hardies, J. E. Larson, R. B. Inman, L. E. Maquat, and R. D. Wells, *J. Biol. Chem.* **254**, 5548 (1979).

TABLE I
EXAMPLES OF NUCLEIC ACID PURIFICATIONS BY RPC-5 ANALOG COLUMN CHROMATOGRAPHY

Applications	Example	Chromatography conditions			
		Amount of nucleic acid	Column dimensions (cm)	Load buffer	Gradient elution
DNA restriction fragments	*Hae*III digest of ϕX174 RF DNA	1.0 mg	0.9 × 25	0.2 *M* NaCl in TE buffer (pH 7.2)	500 ml; 0.45 to 0.65 *M* NaCl in TE buffer (pH 7.2)
Insert DNA	30 bp insert DNA from pSP14 vector	10.0 mg	0.9 × 88	0.2 *M* NaCl in TE buffer (pH 7.2)	500 ml; 0.4 to 0.6 *M* NaCl in TE buffer (pH 7.2)
Digested genomic DNA	*Eco*RI digest rat liver DNA	30 mg	0.9 × 88	0.5 *M* NaCl in TE buffer (pH 7.2)	1000 ml; 0.5 to 0.8 *M* NaCl, in TE buffer (pH 7.2)
Oligodeoxyribonucleotides (homopolymers)	Mixture of $p(dA)_n$, where $n = 4$ to 40	5 μg	0.9 × 5	0.1 *M* NaCl, 12 m*M* NaOH (pH 12)	1000 ml; 0.1 to 0.7 *M* NaCl, 12 m*M* NaOH (pH 12)

Synthetic DNA polymers	*Eco*RI linker (8-mer)	250 μg	0.9 × 5	0.1 *M* NaCl, 12 m*M* NaOH (pH 12)	200 ml; 0.2 to 0.4 *M* NaCl, 12 m*M* NaOH (pH 12)
Supercoiled DNA	φX174 RF DNA	1.0 mg	0.9 × 20	0.5 *M* NaCl in TE buffer (pH 7.2)	300 ml; 0.5 to 0.8 *M* NaCl in TE buffer (pH 7.2)
Transfer RNA	[^{14}C]Leucine tRNA isoacceptors	60,000 CPM	0.63 × 22	0.2 *M* NaCl in TE buffer (pH 4.5)	300 ml; 0.4 to 0.9 *M* NaCl in TE buffer (pH 4.5)
Ribosomal RNA	Ribosomal RNA (4 S, 18 S, and 28 S) from rabbit reticulocyte	2.0 mg	0.6 × 25	0.5 *M* NaCl in TE buffer (pH 4.5)	300 ml; 0.5 to 1.3 *M* NaCl in TE buffer (pH 4.5)
Separated strands of DNA restriction fragments	SV40 *Alu*C restriction fragment (330 bp)	10 μg	0.2 × 45	0.5 *M* NaCl, 12 m*M* NaOH (pH 12)	50 ml; 0.9 to 0.95 *M* NaCl, 12 m*M* NaOH (pH 12)
Large molecular weight single-stranded DNA	Resolution of M13 "+" from RF form of viral DNA	200 μg	0.9 × 25	0.5 *M* NaCl in TE buffer (pH 7.2)	500 ml; 0.5 to 1.0 *M* NaCl in TE buffer (pH 7.2)

column (i.e., gravity or peristaltic pump). In general, the functional capacity of an RPC-5 ANALOG column is 0.5 mg of nucleic acid per gram of matrix (in batch method chromatography; 0.2 mg of nucleic acid per gram of matrix). The geometry of the column (diameter and height) is related to the particle size of the matrix, the capacity of the specific matrix, the type of nucleic acid being resolved, the resolution requirements, subsequent dilutions of recovered nucleic acids, and other principles of liquid chromatography that will not be reviewed here. Recommendations for these parameters relative to specific applications are available on request from one of us (J. A. Thompson). The column is placed in a total chromatography system as previously described,[1] with modifications included below.

Sample Preparation

As with RPC-5 chromatography,[1] sample preparation is of key importance for the accurate, reproducible resolution of nucleic acids by RPC-5 ANALOG. Preliminary purifications should be employed to remove contaminating molecules, e.g., proteins, organic solvents, blocking groups, polyvalent and divalent cations (i.e., spermine and Mg^{2+}), especially when studying complex biological systems. Proper sample preparation is essential to eliminate contaminants that may bind either to the ANALOG matrix, which reduces both column performance and life, or to the nucleic acid, which prevents binding to the matrix. In our experience, frequently failures of RPC-5 ANALOG to perform can be attributed to improper sample preparation.

Recovery of Sample

Recovery of nucleic acids from the RPC-5 ANALOG matrix is routinely greater than 95%. When RPC-5 ANALOG is used in a total chromatography system, in order to maintain this level of recovery, care must be taken to ensure that all the hardware components are compatible with the type of nucleic acid being purified. Glass columns with Teflon tubing and connectors are recommended.

The recovery of nucleic acid molecules from the ANALOG matrix is determined to a great extent by the salt concentration at which binding occurs. Large molecular weight (>100 bases) single-stranded RNA and DNA molecules cannot be eluted from RPC-5 ANALOG unless these nucleic acids are applied to the column at a relatively high salt concentration (0.5 *M* NaCl). In fact, better recoveries of high molecular weight (>1000 bp) double-stranded DNA can be achieved by binding in this same salt concentration. Excellent recovery of lower molecular weight single-stranded DNA and RNA (<100 bases) and double-stranded DNA (<1000 bp) are obtained when bound in buffers of lower salt concentration (0.1 *M* to 0.2 *M* NaCl). The degree of secondary structure within an mRNA molecule determines the salt concentration for binding and recovery of this nucleic acid. A preliminary analysis of mRNA binding to an RPC-5 ANALOG minicolumn is recommended (see Other Purifications).

The volume in which a specific nucleic acid is eluted from RPC-5 ANALOG is dictated by the extent and size of the gradient relative to the column volume. Eluted nucleic acids are conveniently desalted and concentrated when necessary by a variety of commonly used techniques: precipitation (ethanol, polyethylene glycol, spermine), dialysis, lyophilization, or minicolumn chromatography (RPC-5 ANALOG, DEAE-cellulose, Sephadex).

We have not observed cross-contamination with previous samples, coating contamination, or degradation of nucleic acids subjected to RPC-5 ANALOG chromatography. Post-column biological functionality of these purified nucleic acids was determined by the following tests: digestion with restriction enzymes (including *Acc*I, *Alu*I, *Bam*HI, *Bgl*I, *Eco*RI, *Hin*dIII, *Hin*dII, *Pst*I, *Taq*I, *Sal*I, *Hae*III, and *Hpa*I); labeling with T4 polynucleotide kinase, *E. coli* DNA polymerase, or calf thymus terminal deoxynucleotidyltransferase; solution or filter hybridization; chemical sequencing reactions; priming and template activity for dideoxy sequencing with either *E. coli* DNA polymerase or AMV reverse transcriptase; and reactions with T4 DNA ligase together with subsequent cloning and transformation efficiencies. In most cases, nucleic acids highly purified by RPC-5 ANALOG maintain a biological activity the same as or higher than that of nucleic acids purified by other methods.

Specific Applications

The availability of a discrete particle size of RPC-5 ANALOG permitted the development of methods for the purification of specific nucleic acids by liquid chromatography using a peristaltic pump. Below we discuss two of these purification methods using specific examples, one for supercoiled DNA molecules and the other for synthetic oligodeoxyribonucleotides. Both of these applications represent contemporary procedures for molecular biology.[1]

Purification of Supercoiled DNA

Sample Preparation

Growth of Bacterial Cells

Escherichia coli (HB101) containing pBR322 were grown in shaking culture at 37° in M9 media containing ampicillin (0.01%, w/v). Plasmid amplification was achieved using chloramphenicol by the following method.

Media. The quantities given are for 1 liter of medium. All media should be kept sterile.

LB medium: 5 g of yeast extract, 10 g of Bacto-tryptone, 10 g of NaCl, 1 g of glucose

1 X M-9 medium: 100 ml of 10X stock M-9 salts, 835 ml of distilled H_2O, 10 ml of 0.1 M $MgSO_4 \cdot 7 H_2O$, 10 ml of 0.01 M $CaCl_2$, 25 ml of 20% w/v, casamino acids, 20 ml of 20% (w/v) glucose, 0.2 ml of thiamin (10 mg/ml), appropriate amino acids as required, and appropriate antibiotic if desired

10 X stock M-9 salts: 10 g of NH_4Cl, 60 g of Na_2HPO_4, 30 g of KH_2PO_4, 5 g of NaCl

Procedure

1. Grow overnight culture (10–50 ml) of selected *Escherichia coli* strain (in this case HB101 harboring pBR322) in LB (or any other "rich" medium) and required antibiotics at 37° with constant shaking (125 rpm).
2. Prewarm for 30 min 1 liter of the 1 × M-9 medium to 37° in a sterile 2-liter flask. Remove 1.0 ml as a blank for A_{550} determinations.
3. Inoculate the 1 × M-9 medium with 10 ml of the overnight culture and incubate at 37° with vigorous shaking (250 rpm). Remove 1.0 ml of the culture and determine the A_{550}. Repeat this every 30–60 min and plot the log A_{550} versus time (growth curve) (A_{550} starts ~0.03).
4. When $A_{550} = 0.1$ (~2 hr after beginning incubation), add 10 ml of sterile uridine (100 mg/ml); continue shaking.
5. When $A_{550} = 0.7$ (~1.5 hr after adding uridine), add solid chloramphenicol to final concentration of 100 mg/liter and continue shaking for 17–20 hr at 37°.
6. Centrifuge cells (9000 g) and rinse pellet (*do not resuspend*) once with 100 ml of cold (4°) 25 mM Tris-HCl (pH 8.0), 10 mM Na_2EDTA.

7. Collected cells (usually 1–2 g wet weight) can immediately be lysed, or resuspended in lysis buffer and stored at −20°.

This protocol is our most reliable method for growing and obtaining high yields of plasmid DNA.

Bacterial Lysis and Sample Preparation (1-Liter Culture)

METHOD A

Cleared Lysate

1. Resuspend cells (1–2 g) in 70 ml: 50 m*M* Tris-HCl (pH 8), 50 m*M* Na_2EDTA, 15% (w/v) sucrose. Equilibrate for 15 min at room temperature.
2. Add 70 mg of powdered lysozyme; let stand for 15–30 min at room temperature.
3. Add 3 ml of 10% sodium dodecyl sulfate (SDS); *gently* invert to mix.
4. Add 6 ml of 3 *M* potassium acetate; *gently* invert to mix, and place on ice for 30 min.
5. Centrifuge for 30 min at 12,000 *g*. Remove supernatant to a new tube being careful to avoid contamination from viscous pellet.
6. Add *exactly* two volumes of cold (−20°) 95% ethanol; mix well and let stand for 20 min at 4°. Centrifuge 12,000 *g* for 20 min.
7. Resuspend pellet in 10 ml of TE buffer; add 1.93 g of solid ammonium acetate; mix well and let stand for 20 min at 4°. Centrifuge for 30 min at 12,000 *g*; discard pellet, and save supernatant. Add 2 volumes of cold (−20°) 95% ethanol to supernatant; mix well, and let stand for 20 min at 4°.
8. Centrifuge for 30 min at 12,000 *g*; rinse pellet with cold (−20°) 80% ethanol.
9. Carefully remove supernatant and dry pellet under vacuum (5 min). Resuspend pellet in 4 ml of 0.05 *M* NaCl buffered with 10 m*M* Tris-HCl (pH 7.2), 1 m*M* Na_2EDTA (TE buffer).
   Optional enzymic digestion of RNA
   a. Add 1 unit of RNase T_1 per each unit of A_{260}.
   b. Incubate for 15 min at 37°.
10. Add 1.2 ml of 2.0 *M* NaCl in TE buffer.
11. Add 5.0 ml of a 49 : 49 : 2 mixture of phenol (saturated with 0.5 *M* NaCl in TE buffer)–chloroform–isoamyl alcohol. Mix for 15–30 min at ambient temperature.

12. Centrifuge for 10 min at 12,000 *g*; remove upper aqueous phase to new tube.
13. Add exactly *two volumes* of cold (−20°) 95% ethanol; mix well and let stand for 20 min at 4°.
14. Centrifuge for 30 min at 12,000 *g*; remove supernatant and dry pellet under vacuum (5 min).
15. Resuspend pellet in 5 ml of TE buffer.
16. Add 5 ml of 1.0 *M* NaCl in TE buffer. Load sample on column equilibrated in 0.5 *M* NaCl in TE buffer.

METHOD B

Base/Acid Extraction

1. Resuspend cells (1–2 g) in 20 ml of 25 m*M* Tris-HCl (pH 8)–50 m*M* Na_2EDTA, 1% (w/v) glucose.
2. Add 40 ml of 0.2 *N* NaOH, 1% SDS (made fresh weekly); gently mix, then incubate for 10 min on ice.
3. Add 30 ml of cold (4°) 3 *M* potassium acetate (pH 4.8). Gently mix, then incubate on ice for 5 min.
4. Centrifuge 30 min at 27,000 *g*. Remove supernatant to new tube; discard pellets (to avoid contamination from pellet filter through cheesecloth).
5. Add exactly *2 volumes* of cold (−20°) 95% ethanol; mix well, and let stand for 20 min at 4°; centrifuge at 12,000 *g* for 20 min; decant supernatant.
6. Resuspend pellet in 10 ml of TE buffer; add 1.93 g of solid ammonium acetate; mix well, and let stand for 20 min at 4°.
7. Centrifuge for 30 min at 12,000 *g*; discard pellet, and *save* supernatant.
8. Add 20 ml of cold (−20°) 95% ethanol to supernatant; mix well, and let stand for 20 min at 4°.
9. Centrifuge for 20 min at 12,000 *g*; decant supernatant and dry pellet under vacuum (5 min). Resuspend pellet in 4 ml 0.05 *M* NaCl in TE buffer.
   Optional enzymatic digestion of RNA
   a. Add 1 unit of RNase T_1 per each unit A_{260}.
   b. Incubate for 15 min at 37°.
10. Add 1.2 ml of 2.0 *M* NaCl in TE buffer.
11. Add 5 ml of a 49:49:2 mixture of phenol (saturated with 0.5 *M* NaCl in TE buffer)–chloroform–isoamyl alcohol. Mix for 15–30 min at ambient temperature.

12. Centrifuge for 10 min at 12,000 g; remove upper aqueous phase to new tube.
13. Add 10 ml of cold (−20°) 95% ethanol; mix well and let stand for 20 min at 4°.
14. Centrifuge for 30 min at 12,000 g; remove supernatant and dry pellet under vacuum (5 min).
15. Resuspend pellet in 5 ml of TE buffer, then add 5 ml of 1.0 M NaCl in TE buffer.
16. Load sample on ANALOG column equilibrated in 0.5 M NaCl in TE buffer.

Chromatography

Preparative RPC-5 ANALOG Column Chromatography System for Supercoiled DNA Purification (1-Liter Culture)

1. Prepared RPC-5 ANALOG (12 g) is packed in a column (1.5 × 6.0 cm) with a peristaltic pump using 2.0 M NaCl in TE buffer (2.0 ml/min).
2. The column is placed in a chromatography system containing gradient maker and fraction collector. The A_{260} of the effluent is monitored either in-line with a flow cell coupled to a strip-chart recorder or off-line by analyzing the collected fractions.
3. Equilibrate the column with 50–100 ml of 0.5 M NaCl in TE buffer (1.0 ml/min).
4. When the conductivity of the effluent equals the influent, the column is ready for loading. A nucleic acid sample resuspended in 0.5 M NaCl in TE buffer is applied to the column through the pump (0.5 ml/min); start collecting fractions.
5. Wash the column extensively (0.5–1.0 ml/min) with 0.5 M NaCl in TE buffer until <0.01 A_{260} appears in the effluent.
6. Apply 300-ml linear salt gradient (0.5 to 0.8 M NaCl in TE buffer) at 0.5–1.0 ml/min (size and range of gradient can be altered to fit experimental requirements).
7. Analyze fractions containing A_{260} peaks by agarose gel electrophoresis. Pool appropriate fractions; concentrate and desalt by ethanol precipitation or dialysis and lyophilization.
8. To prepare for re-use, wash column with 50–100 ml of 2 M NaCl in TE buffer (1–2 ml/min); store in this buffer at ambient temperature until further use.

Preparative Batch Method Purification of Supercoiled DNA (1-Liter Culture)

1. Prepare RPC-5 ANALOG (5 g) and suspend in 50 ml of 2 *M* NaCl in TE buffer.
2. Centrifuge for 5 min at 3000 *g*; discard supernatant.
3. Resuspend pellet in 50 ml of 0.5 *M* NaCl in TE buffer with gentle mixing (10 min).
4. Centrifuge for 5 min at 3000 *g*; discard supernatant.
5. Repeat steps 3 and 4 *twice*.
6. The plasmid preparation above, suspended in 30 ml of 0.5 *M* NaCl in TE buffer, is gently mixed (15–30 min) with the equilibrated ANALOG.
7. Centrifuge for 5 min at 3000 *g*; *save* supernatant.
8. Resuspend matrix in 50 ml of 0.5 *M* NaCl in TE buffer with gentle mixing (10 min).
9. Centrifuge for 5 min at 3000 *g*; discard supernatant.
10. Repeat steps 8 and 9 *twice*.
11. Resuspend matrix in 10 ml of 0.65 *M* NaCl in TE buffer and mix gently (15–30 min).
12. Centrifuge 5 min at 3000 *g*; *save* supernatant.
13. Repeat steps 11 and 12 *twice*. Each supernatant is saved separately.
14. Resuspend matrix in 30 ml of 1.0 *M* NaCl in TE buffer and mix gently (15–30 min); centrifuge for 5 min at 3000 *g*; *save* supernatant.
15. Analyze supernatants (steps, 7, 12, and 13) by agarose gel electrophoresis.
16. Pool appropriate supernatants; concentrate and desalt by ethanol precipitation or dialysis and lyophilization.
17. To prepare matrix for re-use, resuspend and wash pelleted matrix 3 times (50 ml each) with 2 *M* NaCl in TE buffer; resuspend matrix in 50 ml of 2 *M* NaCl in TE buffer and store at 4° until further use.

Note: A preliminary analytical batch method procedure is recommended to determine the exact salt concentration necessary to elute a specific plasmid DNA (see Fig. 4). The above procedure can be scaled to any size preparation as long as the ratio of nucleic acids to be bound to matrix remains constant (20 units A_{260}; 5 g matrix). The recovered plasmid DNA can be concentrated by minicolumn techniques (see Table III) and conveniently ethanol precipitated.

Discussion

A highly purified preparation of plasmid DNA should be composed only of the supercoiled closed circular duplex form of the molecule (form I). Other forms of the plasmid molecule [nicked circular (form II) and linear (form III)], cellular DNA and RNA, and proteins should not be present. Historically, the amount of these impurities tolerated in a purified plasmid preparation has depended on the intended experimental use of the DNA. To date, chromatography on RPC-5 ANALOG appears to be the best method for conveniently obtaining high purity preparations of plasmid DNAs.

To use RPC-5 ANALOG for the purification of nucleic acids from a biological extract, a major emphasis is placed on sample preparation. In order to purify supercoiled plasmid DNA molecules, a minimal amount of sample preparation is necessary to remove most of the high molecular weight cellular DNA and RNA, contaminating proteins, and any interfering cations. Any of several cell lysis procedures may be incorporated into the purification of plasmid DNA by RPC-5 ANALOG as long as it reflects these considerations of sample preparation. Some alternative procedures include selective extraction in acid phenol,[15] direct boiling of the bacterial suspension,[16] and differential alkaline denaturation.[17,18]

Most lysis methods routinely remove most of the high molecular weight cellular DNA. We find that short-incubation ethanol precipitations (10 min, 4°) of supercoiled DNAs in high salt (2.5 *M* ammonium acetate) selectively eliminates high molecular weight DNA as well as interfering polyvalent cations (e.g., spermine). General lysis procedures remove most contaminating proteins; however, to increase the predictable performance and life of the ANALOG column, phenol extraction is recommended during sample preparation. Further, some high molecular weight RNA will coelute with the supercoiled DNA molecules under the described chromatographic conditions (Fig. 3). Removal of this high molecular weight RNA should be included in the sample preparation strategy by employing either enzymic (RNase T_1) or chemical (selective precipitation with ammonium acetate or alkaline lysis) methods. If treatment with RNase is, for some reason, undesirable, the resulting supercoiled plasmid DNA purified by RPC-5 ANALOG may contain up to 1% contaminating RNA molecules depending on both the lysis method and the strain of *E. coli* cells. Small molecular weight RNA molecules (i.e. oligonucleotides

[15] M. Zasloff, G. D. Ginder, and G. Felsenfeld, *Nucleic Acids Res.* **5,** 1139 (1978).

[16] D. S. Holmes and M. Quigley, *Anal. Biochem.* **114,** 193 (1981).

[17] T. C. Currier and F. W. Nester, *Anal. Biochem.* **70,** 431 (1976).

[18] H. C. Birnboim and J. Doly, *Nucleic Acids Res.* **7,** 1513 (1979).

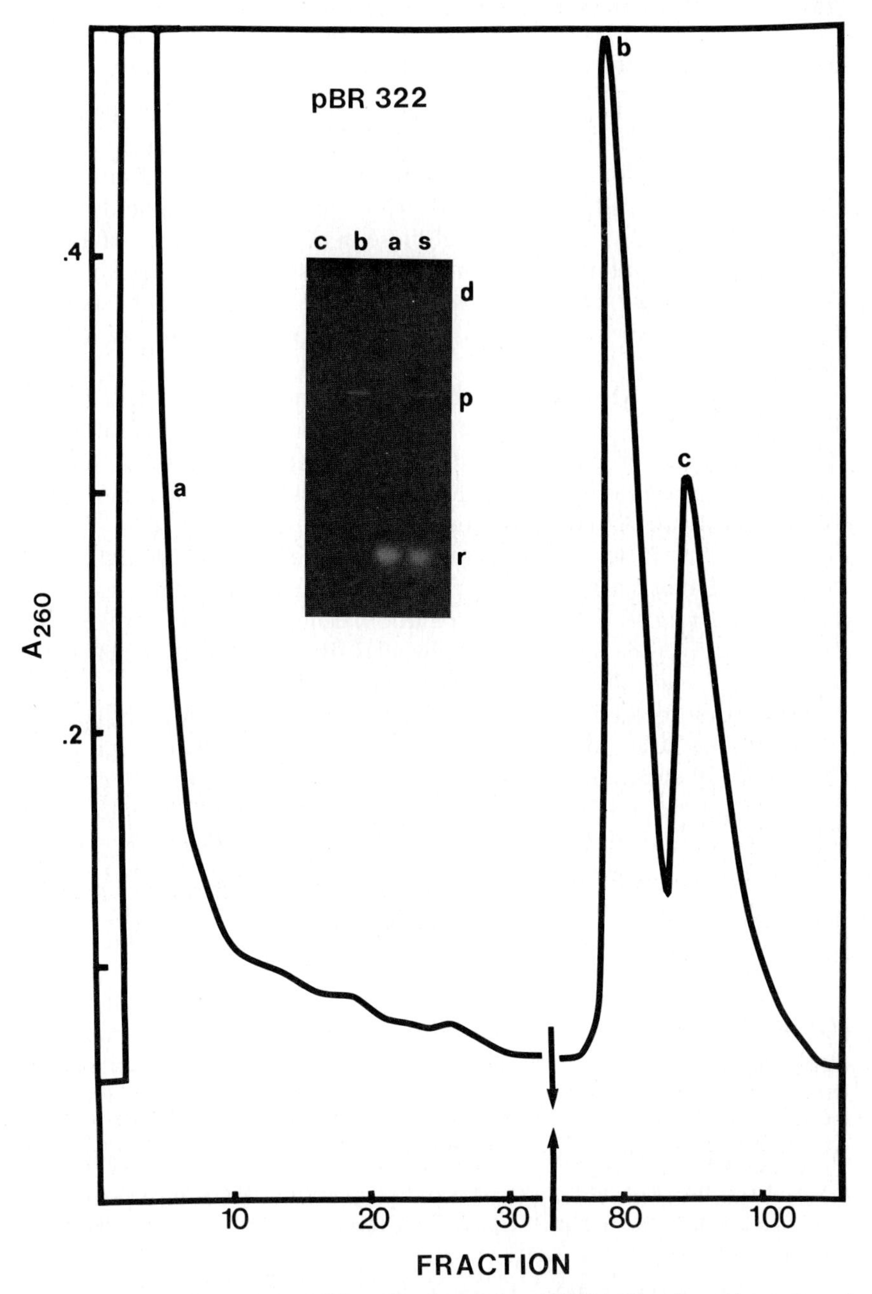

FIG. 3. Column chromatographic purification of plasmid DNA. The cleared lysate from 2 liters of *Escherichia coli* cells harboring the amplified plasmid, pBR322, was treated with RNase T_1 according to Sample Preparation, Method A. The resuspended sample was loaded on an RPC-5 ANALOG column (1.5 × 6 cm) previously equilibrated with 0.5 *M* NaCl in TE

and tRNA) wash through the column upon loading the extracted nucleic acid sample (peak a, Fig. 3).

Since most of the A_{260} material in the extracted sample does not bind to the matrix at a high salt concentration (0.5 *M* NaCl) and since the matrix has a high binding capacity at this salt concentration ($\geq$10 units of A_{260} per gram of matrix), small chromatography columns (1.5 $\times$ 6 cm) containing a minimal amount of RPC-5 ANALOG (12 g) can be used conveniently to purify rapidly milligram quantities of plasmid DNA (from 2–3 liters of *E. coli* cells). On the other hand, plasmid DNA from only 100 ml of *E. coli* cells can be purified conveniently on the same column without creating subsequent dilution problems. All multimeric (i.e., monomers, dimers, trimers) forms of the supercoiled DNA elute at a salt concentration of 0.6–0.7 *M* NaCl (peak b, Fig. 3). With the specific gradient chosen in Fig. 3, peak c contained all other forms (II and III) of the plasmid DNA and the contaminating cellular DNA. The components of peaks b and c can be further resolved by changing the conditions of chromatography.

The chromatographic separation obtained for plasmid DNA purification (Fig. 3) was characteristic for supercoiled molecules in general (3 kb to 49 kb) regardless of the size we studied. Further, preliminary evidence suggests that supercoiled DNA molecules up to 140 kb can be recovered from the RPC-5 ANALOG chromatography matrix. In addition to numerous types of plasmids, RPC-5 ANALOG column chromatography was used to purify many types of supercoiled viral DNA molecules, including polyoma, SV40, and the replicative forms of M13mp7, 8, and 9, and ϕX174. This matrix should also be useful for purifying supercoiled mitochondrial DNA molecules. All supercoiled DNA molecules purified by the ANALOG matrix have maintained a biological activity the same as or better than nucleic acids purified by other methods.

---

buffer using a peristaltic pump (1.0 ml/min). Fractions (4 min) were collected and detected by monitoring (flow cell) the absorbance at 260 nm. The column was washed with 130 ml of the same buffer; peak a represents material not bound to the matrix. A linear (300 ml) gradient of 0.5 *M* to 0.8 *M* NaCl in TE buffer was applied to the column (1.0 ml/min). The A_{260} of the first 70 fractions (2 ml each) of the gradient elution profile, the A_{260} after fraction 110 through the completion of the gradient and the A_{260} through a 100-ml postgradient wash of 2 *M* NaCl in TE buffer remained at baseline (data not shown). Fractions containing peaks a, b, and c were pooled, dialyzed against water, and lyophilized. The dried samples were resuspended in 10, 5, and 5 ml, respectively, of TE buffer. Aliquots (1 μg) of peaks a, b, and c were analyzed by 1% agarose gel electrophoresis, in parallel with 10 μg of the column load (s), containing RNA (r), plasmid DNA (p), and cellular DNA (d); 2.8 mg of supercoiled pBR322 (peak b) was recovered.

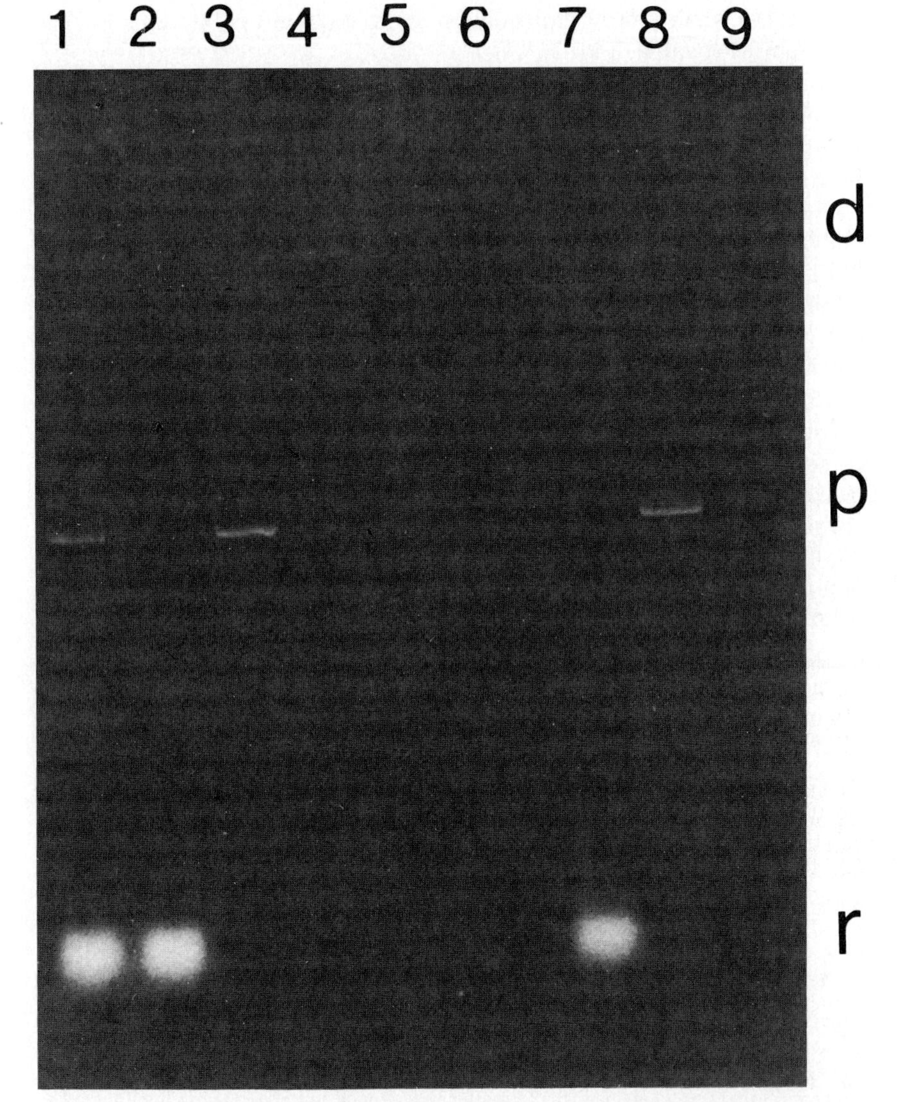

FIG. 4. Batch method purification of plasmid DNA. Agarose (1%) gel electrophoresis was used to analyze nucleic acids recovered by batch method RPC-5 ANALOG chromatography. The total nucleic acid extract from 1 liter of *Escherichia coli* cells harboring the amplified plasmid, pBR322, was treated with RNase T_1 and prepared according to Sample Preparation, Method A (lane 1). To test the plasmid preparation for a batch elution, 2.0 A_{260} units were resuspended in 2.0 ml of 0.2 *M* NaCl in TE buffer were mixed with 0.5 g of preequilibrated ANALOG matrix for 30 min. After centrifugation the pelleted matrix was washed with 2.0 ml of 0.2 *M* NaCl in TE buffer. Analysis of the supernatants from the binding (lane 5) and washing (lane 6) steps showed that all the nucleic acid material bound to

The batch method offers a convenient noncolumn alternative to obtain milligram quantities of supercoiled DNAs (Fig. 4). This method is particularly useful when simultaneously purifying several different supercoiled DNA molecules. Frequently several different plasmid DNAs are purified by isopycnic banding in parallel cesium chloride gradients containing ethidium bromide.[19,20] However, this procedure must be followed by laborious extraction and chromatography methods to remove contaminating reagents as well as residual nucleic acids. In fact, a comparison of purifying plasmid DNA by double banding in cesium chloride–ethidium bromide gradients or by using RPC-5 ANALOG (batch or column) revealed some striking differences (Table II). About the same amount of time was required for sample preparation (lysis and extraction); however, the separation time was significantly shorter, and sample recovery was easier, using RPC-5 ANALOG chromatography. Table II also shows that a greater yield of plasmid DNA of a higher degree of purity was obtained by RPC-5 ANALOG techniques.

Finally, analytical use of the batch method for purification strategies or determining eluting salt concentrations can be somewhat tedious. The minicolumn technique described (see Other Purifications below) is recommended for analytical use, especially when screening specific clones (Fig. 9) or purifying less than milligram quantities of plasmid DNA.

[19] R. Radloff, W. Bauer, and J. Vinograd, *Proc. Natl. Acad. Sci. U.S.A.* **57,** 1514 (1967).
[20] D. B. Clewell and D. R. Helinski, *Proc. Natl. Acad. Sci. U.S.A.* **62,** 1159 (1969).

---

the matrix and none eluted during the washing procedure. Nucleic acids bound to the matrix were eluted in a stepwise manner using buffers of increasing salt concentration (similar to Preparative Batch Method Purification). The supernatants from the elutions were dialyzed against water, lyophilized, resuspended in 0.1 ml of TE buffer, and aliquots (1 μg) were analyzed on gels. The supernatant from elutions with 0.5 *M* NaCl in TE contained small RNA molecules (lane 7); the supernatant from elutions with 0.6 *M* NaCl in TE buffer contained supercoiled plasmid DNA (lane 8); and, the supernatant from elutions with 0.7 *M* NaCl in TE buffer contained high molecular weight DNA (lane 9). No nucleic acid material was found in the supernatant after washing with 2.0 ml of 2.0 *M* NaCl in TE buffer. This preliminary analytical batch procedure permitted establishment of the salt concentrations to use for the preparative purification below. The remaining material from the plasmid preparation was resuspended in 0.5 *M* NaCl in TE buffer and treated with 5 g of RPC-5 ANALOG exactly as described in Preparative Batch Method Purification. One-microgram aliquots from the supernatants of steps 7, 12, and 15 were analyzed by agarose gel electrophoresis. Under these conditions, small RNA molecules did not bind to the matrix (lane 2). Supercoiled plasmid DNA molecules eluted with 0.65 *M* NaCl in TE buffer (lane 3). Elutions with 1.0 *M* NaCl in TE buffer recovered high molecular weight DNA (lane 4). Approximately 0.9 mg of purified supercoiled plasmid DNA was recovered by this preparative batch procedure from 1 liter of *E. coli* cells. r = RNA, p = form I plasmid DNA, d = high molecular weight DNA.

TABLE II
COMPARISON OF PLASMID DNA PURIFICATION METHODS[a]

	CsCl–EtBr gradients	RPC-5 ANALOG column	RPC-5 ANALOG batch method
Yield of supercoiled DNA (mg)	0.9	1.4	1.2
Purity (%)	90	>99	>99
$A_{260\ nm}/A_{280\ nm}$	1.9	1.9	1.9
Purification time	4–5 days	8 Hr	8 Hr

[a] An amplified plasmid, pBR322, was purified from a 1-liter culture of *Escherichia coli* cells (lysis and extraction by Method A) using either RPC-5 ANALOG chromatography (batch or column as described) or isopycnic banding in cesium chloride gradients containing ethidium bromide.[19,20] Concentrations of purified plasmid DNA were determined by UV absorbance at 260 nm (1 mg/ml = 20 A_{260} units). Purity was determined by loading the individual purified plasmid preparations on an RPC-5 ANALOG column under conditions listed in the legend to Fig. 3. Any A_{260} material that did not elute in a single peak (Fig. 3, peak b) was measured, analyzed by agarose gel electrophoresis, and used to determine the percentage of contamination present (within the optical limits of detection of the UV flow cell).

Purification of Synthetic Oligodeoxyribonucleotides

Sample Preparation

1. Construct a "minicolumn" from a 1.0-ml plastic pipette tip (Rainin) plugged at the end with siliconized glass wool. Load DEAE-cellulose (suspended in 2 *M* NaCl in TE buffer) into the column, and allow to drain under gravity flow. Note that typically a 2- to 4-mm bed height is used for these "minicolumns."
2. Wash the minicolumn with 3–5 ml of 0.05 *M* NaCl in TE buffer.
3. Resuspend the final lyophilized extraction product from the chemical synthesis reaction oligodeoxyribonucleotides in 1.0 ml of 0.05 *M* NaCl in TE buffer and load on the minicolumn.
4. Wash the minicolumn with 3–5 ml of 0.05 *M* NaCl in TE buffer.
5. Wash the minicolumn with 3–5 ml of water; continue washing with water until <0.002 A_{260} is eluted from the column.
6. Elute bound oligonucleotides with three 0.2-ml portions of 30% (v/v) triethylamine. Allow the first two elutions to proceed under gravity flow; the final elution should be forced through by air pressure (syringe bulb). Collect effluent in a single 1.5-ml microfuge tube (Eppendorf).
7. Lyophilize effluent to dryness.

8. Resuspend oligodeoxyribonucleotides in 1–2 ml of 0.1 *M* NaCl, 12 m*M* NaOH (pH 12), and load on column.

Preparative RPC-5 ANALOG Column Chromatography System for Synthetic Oligodeoxyribonucleotide Purification

1. Prepared RPC-5 ANALOG, 5 g, is packed in a column (0.9 × 5 cm) with 2 *M* NaCl in TE buffer using a peristaltic pump (2.0 ml/min).
2. The column is placed in a chromatography system containing gradient maker and fraction collector. Either the A_{260} (preparative) or the presence of radioactivity (analytical) is monitored in the effluent buffer. Either on-line measurements with an appropriate flow cell coupled to a strip chart recorder or off-line measurements of collected fractions can be used.
3. Equilibrate column with 50–100 ml of 0.1 *M* NaCl, 12 m*M* NaOH (pH 12) at 1.0 ml/min.
4. When the pH of the effluent equals that of influent (i.e., pH 12 ± 0.5), the oligonucleotide sample from above, resuspended in 1–2 ml of 0.1 *M* NaCl–12 m*M* NaOH (pH 12), is loaded onto the column through the pump (0.5 ml/min); start collecting fractions.
5. Wash the column extensively with 0.1 *M* NaCl–12 m*M* NaOH (pH 12) until the effluent is at baseline (<0.002 A_{260} or $<0.01\%$ of initial radioactivity).
6. Apply a 100-ml gradient of increasing salt concentration (use standard curve for limits, Fig. 6) in 12 m*M* NaOH (pH 12); flow rate 0.3–0.5 ml/min.
7. Pool appropriate fractions, neutralize with concentrated HCl, and dilute the salt concentration to ≤ 0.1 *M* with water.
8. Concentrate sample by minicolumn DEAE-cellulose chromatography (see Sample Preparation).
9. Lyophilized oligonucleotides are resuspended in TE buffer and identified by polyacrylamide gel electrophoresis and modified chemical sequencing reactions.[21]
10. To prepare column for re-use, wash with 50–100 ml of 2 *M* NaCl in TE buffer (1–2 ml/min), and store column in this buffer at ambient temperature until further use.

[21] A. M. Maxam and W. Gilbert, this series, Vol. 65, p. 499 as modified by K. E. Rushlow, personal communication.

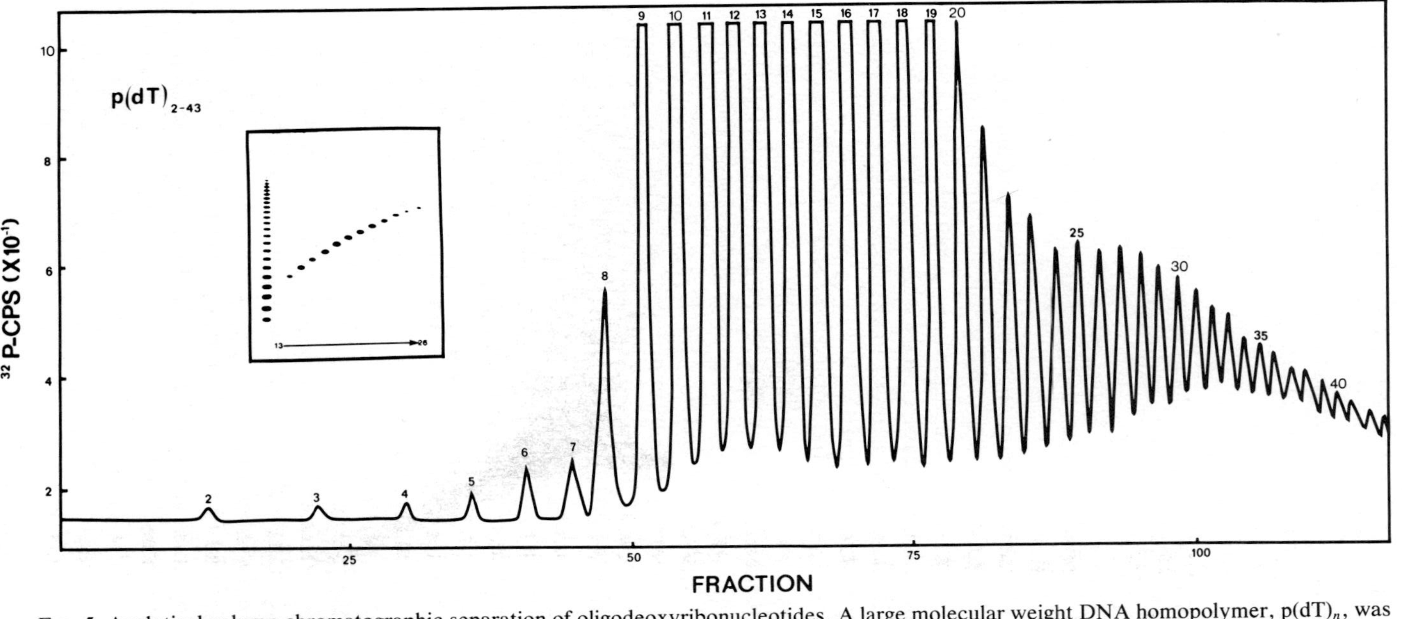

FIG. 5. Analytical column chromatographic separation of oligodeoxyribonucleotides. A large molecular weight DNA homopolymer, $p(dT)_n$, was partially digested with pancreatic DNase I; 5 μg of a mixture of $p(dT)_m$ oligonucleotides (m = 2 to 50) was 5′-end-labeled with [γ-^{32}P]ATP and T4 polynucleotide kinase (see K. L. Berkner and W. R. Folk, this series, Vol. 65, p. 28). Unreacted [γ-^{32}P]ATP was removed by a DEAE-cellulose minicolumn (see Sample Preparation). Recovered, labeled oligonucleotides were resuspended in 1.0 ml of 0.1 *M* NaCl–12 m*M* NaOH (pH 12). The entire volume (5×10^9 cpm) was loaded on an RPC-5 ANALOG column (0.9×5 cm), previously equilibrated with 0.1 *M* NaCl–12 m*M* NaOH, using a peristaltic pump (0.5 ml/min). The effluent was continuously monitored with a radiometric flow cell. The column was washed with the same buffer until the effluent was <0.01% of the initial radioactivity, about 30 ml (data not shown). A linear gradient (1000 ml of 0.1 *M* to 1.1 *M* NaCl in 12 m*M* NaOH, pH 12) was applied to the column (0.4 ml/min), and 4.8-ml fractions were collected. No ^{32}P-labeled material could be detected after fraction 125 through the completion of the gradient and through a 100-ml postgradient wash of 2 *M* NaCl in 12 m*M* NaOH (pH 12) (data not shown). Aliquots (10 μl) of the peak fractions (labeled 13–26) were analyzed directly by 20% polyacrylamide gel electrophoresis (inset). Oligonucleotide sizes of each peak (as numbered in the figure) were determined further by modified chemical sequencing reactions.[21]

Discussion

Ideally, to purify a specific oligodeoxyribonucleotide, a preparative column should reproducibly resolve DNA polymers differing by a single monomeric unit over a defined size range. Furthermore, oligonucleotides should elute from the column at a predictable eluent concentration. RPC-5 ANALOG chromatography approaches this ideal, as seen in Figs. 5 and 6.

Calibration of the RPC-5 ANALOG column was initially determined by several analytical fractionations of different homologous series of oligonucleotides, e.g., $p(dT)_n$ (Fig. 5). Single nucleotide differences in the oligonucleotides of each series were verified by polyacrylamide gel electrophoresis (e.g., see inset, Fig. 5) and chemical sequencing reactions.[21] The exact salt concentration necessary to elute a specific size (number of bases) of oligonucleotide was determined by measuring the conductivity of the effluent fractions following analytical fractionation of specific DNA polymers. These calculations were used to generate the standard curves in Fig. 6.

The results from the analytical purification (Figs. 5 and 6) allowed us to determine the size and range of a salt gradient necessary for preparative purification of synthetic oligodeoxyribonucleotides of defined length (Fig. 7). Most other purification schemes, including reverse-phase (C_{18} columns) and anion-exchange (DEAE) chromatography,[22] do not possess the resolution capabilities of RPC-5 ANALOG. Alternatively, time-consuming gel extraction methods do not routinely yield the same recovery obtained with RPC-5 ANALOG chromatography (>98%).

Exact nucleotide sequences of purified oligonucleotides are readily determined chemically owing to the high degree of purity obtained from ANALOG chromatography (Fig. 8). No inhibition of biological activity was observed for ANALOG-purified oligonucleotides. In fact, subsequent ligation reactions and primer directed sequencing and cloning experiments are more efficient using purified oligodeoxyribonucleotides.

In our as yet limited experience with RPC-5 ANALOG purification of mixtures of synthetic oligodeoxyribonucleotides, no unusual elution behavior was observed. Within a given sample, DNA polymers differing by a single monomeric unit were resolved with an increasing linear salt gradient. These separations were obtained regardless of the sequence or size of the specific oligonucleotide. Potential secondary structures as a result of base sequences within a specific single-stranded polymer were eliminated by operating all chromatography columns at pH 12. In fact, it has been shown that alkaline conditions are a requirement to resolve $p(dC\text{-}dI)_n$ and

[22] H. G. Khorana and W. J. Conners, *Biochem. Prep.* **11,** 113 (1966).

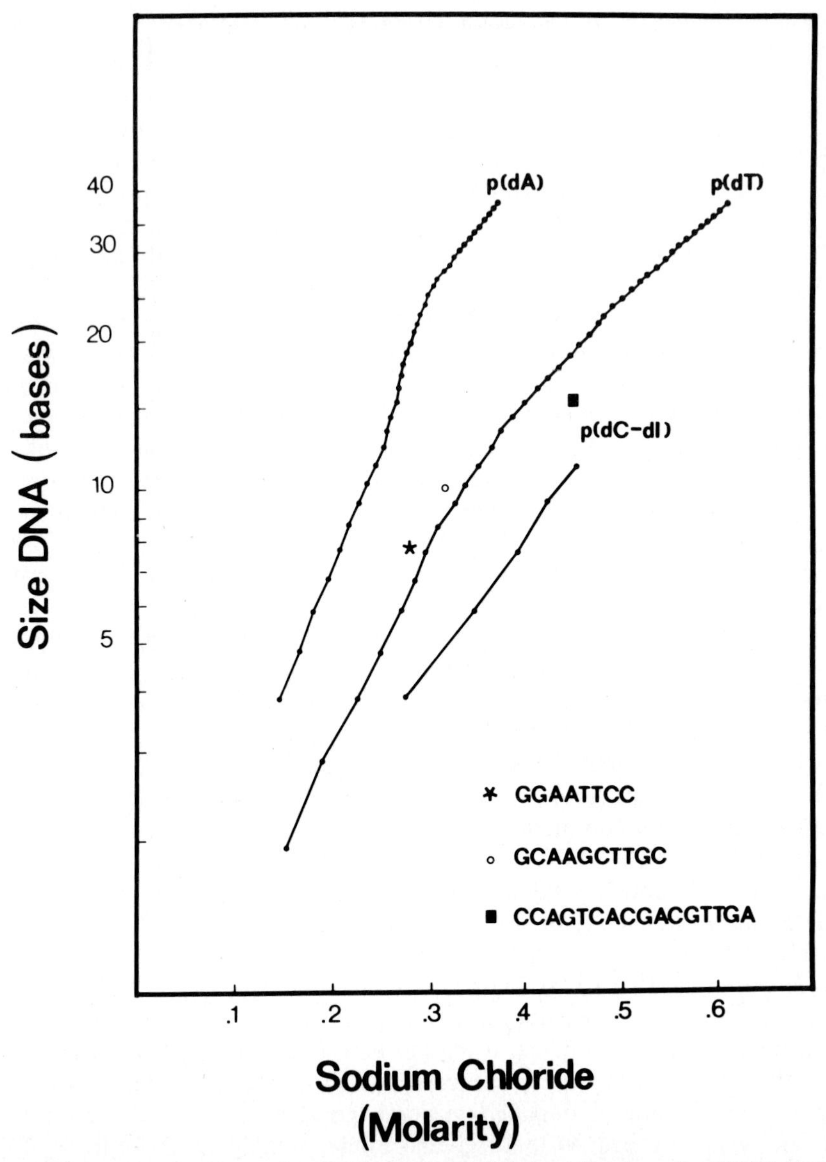

FIG. 6. Calibration curves for oligodeoxyribonucleotide purification using RPC-5 ANALOG column chromatography. Analytical fractionations of several complex mixtures of single-stranded DNA polymers were separately performed on the same RPC-5 ANALOG column (0.9 × 5 cm). The size of the DNA polymer is plotted versus the salt concentration at which the peak elution occurred as determined by direct conductivity measurements. Several different experiments are summarized to determine the typical behavior of the ANALOG matrix in fractionating single-stranded DNA oligomers. Although specific details varied among experiments, the chromatography conditions were essentially those described for Fig. 5.

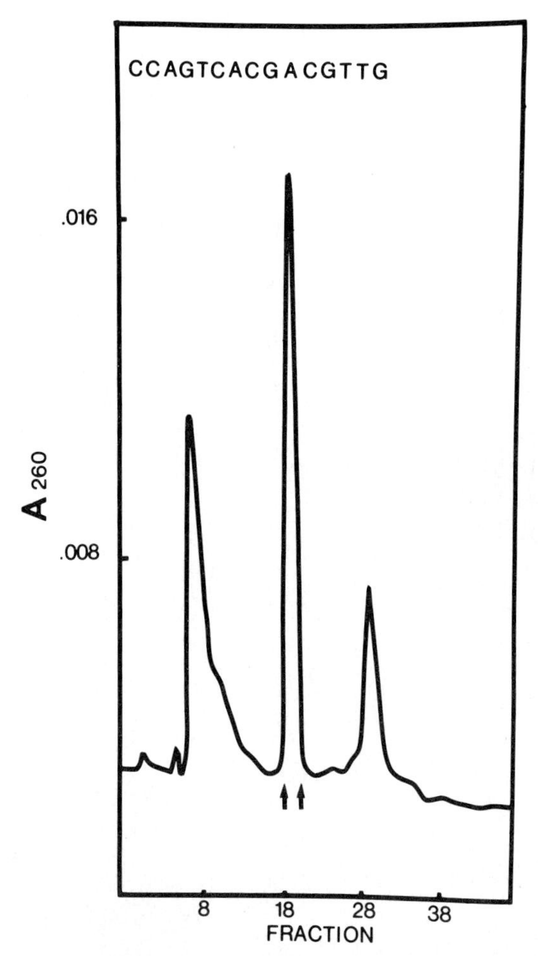

FIG. 7. Preparative purification of synthetic oligodeoxyribonucleotides by RPC-5 ANALOG column chromatography. Of the M13 dideoxy sequencing primer (15-mer) synthesized by solution phase triester chemistry (S. A. Narang, R. Brousseau, H. M. Hsiung, and J. J. Michniewicz, this series, Vol. 65, p. 610), 250 μg were prepared as described using a DEAE-cellulose minicolumn. One microgram was removed for subsequent analytical analysis. The remainder of the oligonucleotide mixture, resuspended in 0.1 *M* NaCl in 12 m*M* NaOH (pH 12), was loaded on an RPC-5 ANALOG column (0.9 × 5 cm) previously equilibrated with 0.1 *M* NaCl in 12 m*M* NaOH (pH 12) using a peristaltic pump (1.0 ml/min). The effluent was continuously monitored at 260 nm. The column effluent was essentially to baseline (<0.002 A_{260}). A linear gradient (100 ml) of 0.4 *M* to 0.6 *M* NaCl (based on calibrated curve, Fig. 6) in 12 m*M* NaOH (pH 12) was applied to the column (0.3 ml/min), and fractions (1.8 ml) were collected. The A_{260} remained at baseline after fraction 50 through the completion of the gradient, and no A_{260} was detected during a 100-ml postgradient wash of 2 *M* NaCl in 12 m*M* NaOH (data not shown). Fractions containing the purified 15-mer (indicated by arrows) were concentrated and desalted as described in Sample Recovery. The recovered oligonucleotide was resuspended in 100 μl of TE buffer, and the exact chemical sequence (as indicated) was determined by modified chemical sequencing methods.[21]

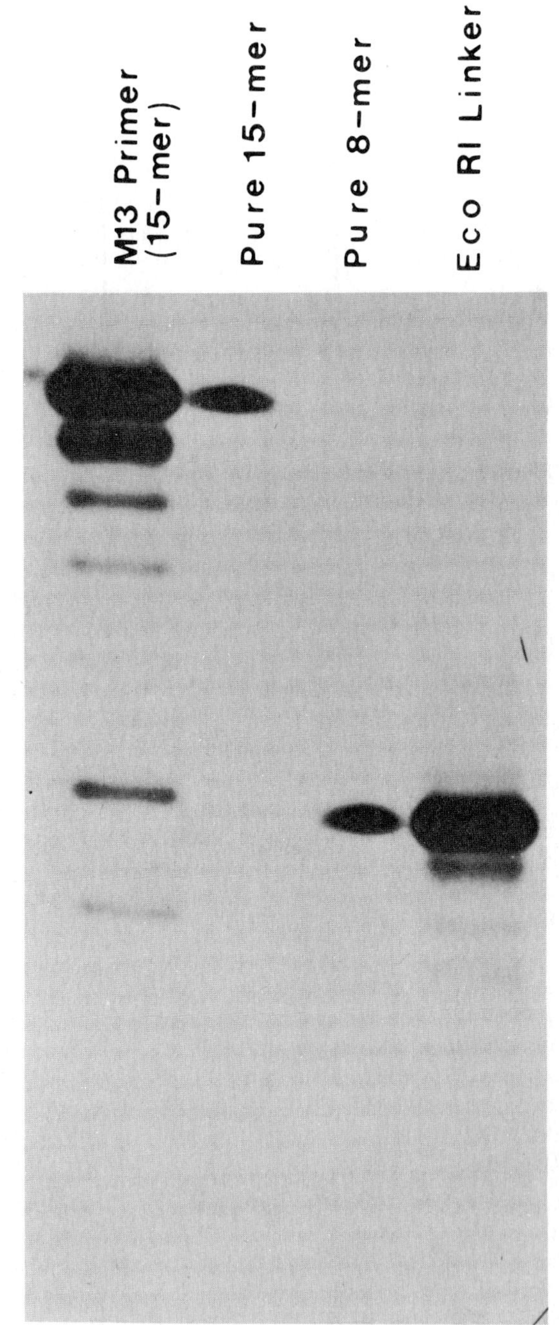

FIG. 8. Analytical analysis of preparatively purified synthetic oligodeoxyribonucleotides. Both the M13 dideoxy sequencing primer (15-mer) and the *Eco*RI linker (8-mer) were purified on a preparative scale essentially as described in Fig. 5. One microgram each of the purified oligodeoxyribonucleotides and the unpurified mixture of oligonucleotides were 5′-end labeled with [γ-^{32}P]ATP and T4 polynucleotide kinase. Unreacted [γ-^{32}P]ATP was removed from each sample with DEAE-cellulose minicolumns. Radioactively labeled oligonucleotides were analyzed by 20% polyacrylamide gel electrophoresis. Shown here is a 16-hr exposure to XR-5 film. Even after a 3-day exposure to the same film with an intensifying screen (Lightning Plus), the purified oligonucleotide samples appeared to be homogeneous (<99%).

$p(dG)_n$ polymers.[1] Eventually, with continued use, a standard curve will be generated to predict reliably the salt concentration needed for elution of a given length of DNA polymer from RPC-5 ANALOG.

Other Purifications

RPC-5 ANALOG chromatography can solve a wide variety of purification problems in nucleic acid biochemistry. Table III lists a number of these analytical applications. Besides providing a rapid, efficient mechanism to obtain preparative amounts of plasmid and oligonucleotide DNAs of high purity, RPC-5 ANALOG is used in our laboratory for "clean-up" of nucleic acids following manipulations commonly used in the study of DNAs and RNAs. The following skeletal chromatography procedure can be implemented for each of the applications mentioned by substituting, where appropriate, the proper buffer concentrations from Table III.

Chromatography

1. Construct a minicolumn from either a 1-ml plastic pipette tip (Rainin) or a 1-ml disposable syringe plugged at the end with siliconized glass wool. Load 1 ml (0.1 g) of suspended, prepared RPC-5 ANALOG into the column, and allow to drain under gravity flow.

Note: Typically, 0.1 g of ANALOG is used for 50 μg of binding A_{260} material. Up to 1.0 g of ANALOG (for 500 μg of A_{260} material) can be used successfully with this protocol.

2. Wash the minicolumn with 30–50 bed volumes of buffer A (see Table III). Allow buffer to drain out completely.
3. Load nucleic acid sample ($>$1.0 μg/ml) in buffer A onto the matrix. Drain.
4. Wash minicolumn with 30–50 bed volumes of buffer A. For this 0.1 g of RPC-5 ANALOG minicolumn, the wash volume is 3–5 ml of buffer A.
5. Drain buffer completely. To reduce dilution of recovered nucleic acid in the next step, remove residual buffer in the matrix by air pressure from a syringe bulb on the top of the minicolumn. Dry matrix presents no harm to the matrix or bound nucleic acid.
6. Elute nucleic acid with one bed volume of buffer B (3 times). Collected effluent fractions are pooled appropriately, and the nucleic acid is processed further.

TABLE III
PURIFICATIONS USING RPC-5 MINICOLUMNS

Applications	Load/wash buffer A[a]	Elution buffer B[a]	Comments
Clone screening by rapid lysis[b]	0.5 M NaCl	1.0 M NaCl	Removes interfering host RNA/DNA for unobstructed gel observation and removes inhibitors of restriction enzyme digestion.
Purification of inserts from vectors	0.2	0.6	For inserts approximately ≤800 bp; use gel extraction for larger molecular weight insert DNA
Extraction from low melting point agarose gel electrophoresis[c]			Excise band nucleic acid from gel; melt to 65°; dilute with 3–4 volumes of buffer A and place at 42°; equilibrate minicolumn with 42° buffer A; load/wash sample with 42° buffer A; elute with buffer B at room temperature. EtBr does not interfere; removes inhibitors of restriction enzymes. mRNA partition between low and high salt binding is dependent on secondary structure.
ss-DNA (>100 bases)	0.5	2.0	
ds-DNA(<1000 bp)	0.2	0.7	
ds-DNA(>1000 bp)	0.5	2.0	
rRNA	0.5	2.0	
tRNA	0.2	0.7	
Concentration of dilute DNA solutions	0.2	1.0	Concentrate dilute (≥1.0 μg/ml) samples.
Restriction fragment–crude size classification	0.5	2.0	Flow-through contains ≤100 bp; elution >100 bp. Use to increase ligation efficiency of larger fragments for shotgun cloning.
Removal of linker from ligation reactions	0.5	2.0	Improve ligation efficiency for subsequent steps after linker addition.
Removal of small molecules from DNA solutions	0.2–0.5	2.0	Load conditions depending upon size of nucleic acid: e.g., remove unreacted labeled nucleotides from end-labeling or fill-in reaction; change buffers between consecutive enzyme reactions.

TABLE III (*continued*)

Applications	Load/wash buffer A[a]	Elution buffer B[a]	Comments
RNA—crude size classification	0.5	2.0	Achieve simple classification; tRNA from rRNA; mRNA partition dependent on secondary structure.
Single-stranded viral DNA isolation	0.5	2.0	Useful for template isolation in M13 dideoxy sequencing method.

[a] The buffers used in the minicolumn procedure are TE buffer (10 m*M* Tris-HCl, pH 7.2; 1 m*M* Na_2EDTA) containing the indicated molarity of NaCl.

[b] R. W. Davis, M. Thomas, J. Cameron, T. P. St. John, S. Scherer, and R. A. Padgett, this series, Vol. 65, p. 404.

[c] DNA and RNA molecules can be purified by electroelution from either agarose or polyacrylamide gels directly to RPC-5 ANALOG; alternatively, RPC-5 ANALOG minicolumns can be used to purify nucleic acids electroeluted into dialysis tubing.

Discussion

The RPC-5 ANALOG minicolumn approach is a rapid, efficient means to purify nucleic acids. ANALOG chromatography is a generalized technique that can be applied to a wide variety of purification problems. The list of applications in Table III is by no means exhaustive, rather is intended as a guide to the more frequently used procedures in a molecular biology laboratory. In addition, Table III should indicate the versatility of the matrix; that is, it should not be thought of as a specialized reagent with limited applications. For example, the results of using minicolumns to screen two different colonies of *E. coli* cells (1 ml of culture) harboring a pBR322 plasmid DNA containing inserted SV40 DNA restriction fragments are shown in Fig. 9. In order to identify the inserted DNA by agarose gel electrophoresis, contaminating RNA molecules, which ordinarily would mask visualization of the insert by migration to approximately the same position on the agarose gel, were eliminated from the plasmid preparation by RPC-5 ANALOG minicolumns (Fig. 9, lanes 6 and 7). After digestion of the RNA-free recombinant plasmids with the appropriate restriction endonucleases, the proper size of the insert DNA fragment (Fig. 9, lanes 4 and 5) was easily identified. Purification of these specific insert SV40 DNA fragments from the plasmid vector also was achieved by RPC-5 ANALOG minicolumns (Fig. 9, lanes 2 and 3). Further mapping of these purified SV40 DNA fragments with appropriate restriction endonucleases (data not shown) clearly identified each of the

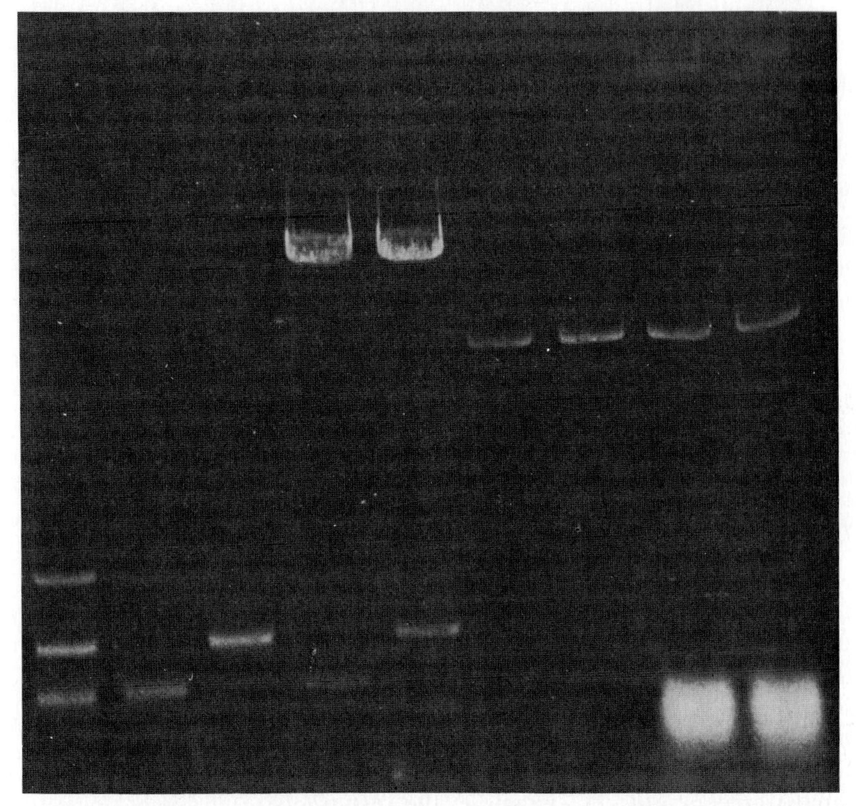

FIG. 9. Minicolumn purification of nucleic acids. The purified [^{3}H]*Alu*B and [^{3}H]*Alu*C fragments (see Fig. 1) were ligated into pBR322 using *Bam* linkers and transformed into *Escherichia coli* (HB101). The cleared lysate (Method A) from a 1-ml overnight culture of both *Alu*C- and *Alu*B-selected clones were resuspended in TE buffer (0.1 ml) and analyzed by agarose (1%) gel electrophoresis (lanes 8 and 9, respectively). Both nucleic acid samples were found to contain contaminating cellular RNA and DNA. One-tenth milliliter of 1.0 *M* NaCl in TE buffer was added to each sample and loaded on an RPC-5 ANALOG minicolumn equilibrated in 0.5 *M* NaCl in TE buffer. Bound nucleic acids were washed with the same buffer essentially as described in the text and were eluted with 0.7 *M* NaCl in TE buffer (0.2 ml) and precipitated with 0.4 ml of 95% ethanol. The pelleted nucleic acid was resuspended in 0.1 ml of TE buffer and analyzed on the same gel. The plasmid DNA from both the *Alu*C and *Alu*B clones were purified free of contaminating cellular nucleic acids (lanes 6 and 7, respectively). Both preparations were digested with *Bam*HI, analyzed by agarose (1%) gel electrophoresis (*Alu*C, lane 4; *Alu*B, lane 5), phenol extracted, ethanol-precipitated, and resuspended in 0.2 *M* NaCl in TE buffer (0.1 ml). These samples were loaded on RPC-5 ANALOG minicolumns and washed with the same buffer. The *Alu*C and *Alu*B fragments were eluted with 0.2 ml of 0.6 *M* NaCl in TE buffer (lanes 2 and 3, respectively) and identified by agarose (1%) gel electrophoresis by comparison to a standard mixture (lane 1) of the purified *Alu*A, *Alu*B, and *Alu*C fragments (see Fig. 1).

cloned DNA fragments. These fragments were used to generate primers for sequencing early SV40 mRNA as previously described.[23]

We can screen up to 48 separate clones in 8 hrs, beginning with the cell pellet and terminating with the gel analysis, using RPC-5 ANALOG minicolumns in between to remove contaminating reagents and residual nucleic acids that may interfere with gel analysis. Sample preparation is not as important a consideration in using minicolumns as with the other ANALOG purification methods described. In fact, the minicolumn in most cases is used to clean up a nucleic acid preparation for a subsequent manipulation. Contaminants from these procedures, such as specific enzymes and their corresponding buffers, can be tolerated by the matrix, as long as the final ionic strength of the sample is equal to buffer A. Dilution to achieve the proper salt concentration is recommended provided that the nucleic acid concentration remains ≥ 1.0 μg/ml. Elution with a small volume will concentrate the sample. As pointed out earlier though, these contaminants will reduce column performance and life. Successful performance can be achieved with minicolumns by following the chromatography procedure outlined above. However, after use the minicolumn must be discarded. This ensures that the performance expected from the matrix will be obtained.

Appendix

We have noticed that the base–acid extraction procedure (Method B) results in poor yields of plasmid (viral) DNA when the nucleic acids are isolated from certain bacterial strains. A preliminary analytical extraction is recommended to determine the applicability of this technique.

Defined particle size distributions of RPC-5 ANALOG have been made available commercially by Bethesda Research Laboratories, Inc. under the trademark name *NACS* (Nucleic Acid Chromatography System). Besides size ranges specific for the peristaltic and gravity flow chromatography systems mentioned here, a smaller discrete particle size range for HPLC applications is also available.

[23] J. A. Thompson, M. F. Radonovich, and N. P. Salzman, *J. Virol.* **31,** 437 (1979).

[27] Hybrid Selection of Specific RNAs Using DNA Covalently Coupled to Macroporous Supports

By HANS BÜNEMANN and PETER WESTHOFF

For immobilization of sonicated denatured DNA, macroporous supports of the Sepharose CL and Sephacryl S type can be used advantageously instead of fine-grained cellulose or Sephadex.[1,2] Direct comparison of parameters important for sensitive hybridization reactions with sonicated DNAs or RNAs clearly proves the macroporous supports to be equivalent or superior to their solid counterparts.

Coupling via diazotization reaction[3–6] or via cyanogen bromide[7,8] immobilizes initially up to 500 μg of sonicated DNA per gram of wet Sephacryl S-500 or Sepharose CL-6B. After preincubation for 2–3 days under hybridization conditions, about 50–70% of the initial amount is retained on the support as "stably bound DNA." This DNA has a leakage rate of about 1% per day under hybridization conditions. As long as overactivation is strictly avoided, the immobilized DNAs are 90–100% accessible during hybridization reactions with sonicated denatured DNA.[1,2] The amount of base substitutions generated by the various coupling procedures is estimated to be 0–1.5% by mismatch measurements for materials with optimally accessible DNA and 1.5–2.9% for those with partially inaccessible nucleic acids. No substantial difference is found for the rate of hybridization reaction for DNAs immobilized to macroporous or solid supports as long as the hybridization is performed with sonicated DNA.[9] In general, this heterogeneous type of hybridization reaction proceeds

[1] H. Bünemann, P. Westhoff, and R. G. Herrmann, *Nucleic Acids Res.* **10,** 7163 (1982).

[2] H. Bünemann, *Nucleic Acids Res.* **10,** 7181 (1982).

[3] B. E. Noyes and G. R. Stark, *Cell* **5,** 301 (1975).

[4] M. L. Goldberg, R. P. Lifton, G. R. Stark, and J. G. Williams, this series, Vol. 68, p. 206.

[5] J. C. Alwine, D. J. Kemp, B. A. Parker, J. Reiser, J. Renart, G. R. Stark, and G. M. Wahl, this series, Vol. 68, p. 220.

[6] B. Seed, *Nucleic Acids Res.* **10,** 1799 (1982).

[7] D. J. Arndt-Jovin, T. M. Jovin, W. Bähr, A. M. Frischauf, and M. Marquardt, *Eur. J. Biochem.* **54,** 411 (1975).

[8] S. G. Siddell, *Eur. J. Biochem.* **92,** 621 (1978).

[9] If *Eco*RI-digested λ DNA is probed under otherwise identical conditions, the rate of hybridization is about one to two orders of magnitude slower than for sonicated DNA in the case of macroporous S-500.

ISBN 0-12-182000-9

about one order of magnitude slower for all supports tested so far than for the comparable renaturation reaction in homogeneous solution.[2]

Based on these results, we recommend DPTE[10]-Sephacryl S-500, DBM-Sephacryl S-500, and BrCN-Sephacryl S-500 as favorable supports for immobilization of DNA. Other Sephacryl S and Sepharose CL materials can be used similarly. Normally they give substantially lower yields of DNA coupling, reduced accessibilities, and enhanced effects of mismatch.[1,2] When Sephacryl S-500 is activated by the standard procedures described below, the immobilized DNAs enable the sensitive selection of minor RNA species as well as the enrichment of DNA sequences specific for certain chromosomes.

The recommended S-500 materials can be characterized by the following criteria.

DPTE-Sephacryl S-500. The material is synthesized by two simple "one-bottle" overnight reactions at room temperature by the aid of inexpensive chemicals.[6] The extended spacer of 1,4-butanediol diglycidyl ether between support and DNA minimizes mismatch effects. If the particles are handled carefully, abrasion effects can be neglected.

DBM-Sephacryl S-500. The synthesis is practically identical to the well known procedure for preparation of DBM-cellulose[4] or DBM-paper,[5] but substantially quicker because the finely grained cellulose[3] need not be prepared. Sephacryl S-500 is also easier to handle because it can be collected by mild centrifugation or by using normal filtering devices. Its stability is comparable to that of its DPTE analog. The observed mismatch in all cases exceeds that of DPTE S-500. CL Sepharoses cannot be used for the DBM procedure because they are destroyed irreversibly in the course of NBPC treatment at 135°.

BrCN-Sephacryl S-500. Use of the poisonous BrCN and the sensitivity of the activated material to impurities of the DNA solution are the main disadvantages of this otherwise simple coupling procedure.[7] Furthermore, the activation reaction generates cationic groups on the support.[8] Therefore, elution of hybridized nucleic acids cannot be achieved with distilled water alone but demands salt concentrations on the order of 0.2–0.5 *M*. Perhaps owing to this ion exchange character, hybridization reactions are about two times faster with BrCN-activated S-500 than with the diazo-activated homologs.[2] This rate enhancement by preconcentration of the nucleic acid in the vicinity of the support makes BrCN-activated S-500 preferable for those experiments that are critical for kinetic reasons.

[10] The abbreviations used are: ABM, aminobenzyloxymethyl; DBM, diazobenzyloxymethyl; NBPC, 1-[(*m*-nitrobenzyloxy)methyl]pyridinium chloride; APTE, 2-aminophenylthioether; DPTE, 2-diazophenylthioether.

Materials and Methods

Sephacryl S-500 is purchased from Pharmacia. All reagents are analytical grade: cyanogen bromide (Merck or Serva), NBPC (BDH or Pierce), 1,4-butanediol diglycidyl ether (Aldrich), and 2-aminothiophenol (Aldrich).

Preparation of DNA. Plasmid DNAs are isolated by standard procedures. The DNA should be free from impurities that may react with the activated materials. All DNAs are brought to a molecular weight of 3–5 × 10^5 by sonication of 1-ml volumes of ice-cold DNA solution for 4 × 30 sec (we use a Braun Labosonic 1510, equipped with a microtip at 100 W setting). The output energy is limited to 50 scale units. (For smaller volumes, e.g., 100 μl of DNA solution, DNA fragments of a similar size are obtained by sonication for 8 × 5 sec in an Eppendorf tube under otherwise identical conditions.)

Quantitative Determination of Immobilized DNA by Nuclease Digestion. Aliquots of 0.1–0.2 g of wet material are washed successively on a sintered-glass filter with H_2O and 0.1 *M* sodium borate, pH 8.8. The wet particles are transferred into a Sorvall tube, suspended in a mixture of 2 ml of 0.05 *M* sodium borate, pH 8.8, and 0.05 ml of 0.1 *M* $CaCl_2$. *Staphylococcus aureus* nuclease is added to a final concentration of 300 units/ml, and the mixture is incubated at 37° overnight. After centrifugation, the amount of immobilized DNA is calculated from the absorbance of the supernatant at 260 nm. One optical density unit of native DNA is equivalent to 1.6 OD units of nuclease-digested DNA.

Coupling of DNA via Diazotization[3,6]

Both diazo-coupling procedures consist of the covalent attachment of an aromatic amino compound to the matrix in a first step. In the second step, the amino function is diazotized and coupled with denatured DNA. Once the different amino derivatives are synthesized, all further steps are identical.

Preparation of APTE-Sephacryl S-500. Fifty milliliters of S-500 are washed in a Büchner funnel with 1 liter of distilled water; 35 g of the moist cake are placed into a 100-ml Sovirel bottle or any bottle with a tight fitting cover. Then 1 *M* NaOH (35 ml) and 1,4-butanediol diglycidyl ether (3 ml) are added, and the sealed bottle is shaken at room temperature overnight. The slurry is transferred to a Büchner funnel. The aqueous solution is sucked off as completely as possible by pressing the material with a broad spatula before the moist cake is returned to the bottle. Acetone (40 ml) and 2-aminothiophenol (5 ml) are added before the bottle is sealed, and the mixture is shaken at room temperature overnight. The bottle is opened in a hood, and its contents are washed in a Büchner funnel successively with 200 ml each of acetone, 0.1 *M* HCl, water, 0.1 *M* HCl, and water again. The APTE material, suspended in H_2O and stored in a stoppered bottle at 4°, is stable for months.

Preparation of ABM-Sephacryl S-500. The commercially available Sephacryl S-500 is obtained in aqueous solution. Contrary to the recommendations of the manufacturer the particles are dried: the slurry is transferred into a Büchner funnel, washed with acetone, spread out in a glass pan, and dried to constant weight at 60° in an oven. (Since the dried and shrunken material recovers its original volume within some minutes after resuspension in aqueous solutions, pore sizes and other parameters of the particles are apparently not altered appreciably by the drying procedure.) The dry material is wetted with a solution of 1 g of NBPC and 0.3 g of sodium acetate dihydrate in 9 ml of H_2O. For homogeneous swelling and moistening of the material, 8.5 ml of the aqueous solution are sufficient per 1 g of dried material. The wet material is spread out in a glass pan and dried for 30 min at 60° in an oven. Then the incubation temperature is raised to 135° for an additional 1 hr. The dry brownish cake is homogenized by careful grinding in a porcelain mortar, taking care to avoid mechanical destruction of the Sephacryl particles. The finely divided NBM material is washed in a glass filter with 100 ml of toluene and again dried at 60°. For reduction the NBM material is incubated for about 10 min at 60° with 50 ml of freshly prepared 20% (w/v) solution of sodium dithionite in a sealed bottle. Finally, the suspension of the ABM material is transferred to a sintered-glass filter and washed with 100 ml each of water, 30% acetic acid, and water. In a stoppered bottle the suspension of the ABM material is stable for months when stored at 4°.

Preparation of DBM- and DPTE-Sephacryl S-500. About 3 g of moist ABM or APTE material from the stock are washed in a sintered-glass filter with 50 ml of distilled water and suspended in a mixture of 3 ml of water and 10 ml of 1.8 *M* HCl. The suspension is kept on ice, and 100-μl amounts of a freshly prepared solution of 10 mg/ml sodium nitrite in water are added successively; the progress of the reaction is followed with starch iodide paper. In the case of APTE S-500, the particles turn yellow immediately. No color change is observed for ABM S-500. Addition of $NaNO_2$ is finished after about 30 min, when the color reaction (dark blue) remains positive. When the activation reactions are formed under the standard conditions described above, the following volumes of nitrite solution are required for 3 g of wet amino material: 0.5 ml for ABM S-500 and 0.3 ml for APTE S-500. The slurry is quickly transferred to an ice-cold sintered-glass filter and washed immediately with 100 ml of ice-cold water and finally with about 15 ml of an ice-cold mixture of 25 m*M* sodium phosphate, pH 6.0, and analytical grade DMSO (Merck) (20:80, v/v).

Alternatively sodium phosphate can be replaced by 0.05 *M* sodium borate, pH 8.0, and the coupling procedure can be performed at 4° instead of room temperature. In the case of APTE S-500, the coupling efficiency at room temperature is exactly the same for both buffers.

Both materials slowly turn darker during the last washing step. The washing is stopped as soon as the surface of the cake becomes dry. As quickly as possible, 2 g of the wet DBM or DPTE material are placed in a 10-ml Sovirel glass tube and mixed with 1 ml of DNA solution freshly prepared in the following way. About 1 mg of DNA is sonicated, and precipitated by ethanol. The dried pellet is dissolved in 0.2 ml of 25 m*M* sodium phosphate, pH 8.0, or 0.05 *M* sodium borate, pH 8.0, *before* 0.8 ml of analytical grade DMSO is added. (Denaturation by boiling is not necessary because it takes place instantaneously with the addition of DMSO, owing to the enthalpy of the mixture.) When a phosphate buffer is used, the reaction mixture is kept at room temperature overnight or, with better results, for 2 days. For borate buffer the suspension is incubated at 4° overnight. The suspension, which contains many bubbles of nitrogen, is poured into a sintered-glass filter and washed at 40° successively with water and 0.4 *M* NaOH. After a final washing with 100 ml of 10 m*M* Tris-HCl, pH 8.0, the materials are ready for preincubation and nuclease digestion.

Coupling of DNA via Cyanogen Bromide Activation[7]

Preparation of BrCN-Sephacryl S-500. Five grams of wet material are washed, using a Büchner funnel or a sintered-glass filter, with an excess of distilled water before the wet cake is transferred into a 25-ml Erlenmeyer flask equipped with a magnetic stirrer and a pH electrode. (To avoid the destruction of the particles, S-500 should be stirred with a blade stirrer.) Ice-cold water (20 ml) is added, and the suspension is stirred rigorously and kept below 15° by occasional addition of pieces of ice. Two grams of solid cyanogen bromide are added, and the pH is maintained between 10.5 and 11.5 with 5 *M* NaOH for about 20 min. After activation, the material is washed quickly with 5 volumes of ice-cold water, followed by 1 volume of ice-cold 10 m*M* potassium phosphate, pH 8.0. Generally, 2 g of activated wet material are mixed with 1 ml of ice-cold solution of sonicated denatured DNA. (The solution of about 1 mg of DNA in 1 ml of 10 m*M* potassium phosphate, pH 8.0, is boiled and stored on ice immediately before use.) The reaction mixture is maintained at room temperature overnight. Then the material is washed on a fritted-glass funnel with the following solutions: 10 m*M* potassium phosphate, pH 8.0; 1 *M* potassium phosphate, pH 8.0; 1 *M* potassium chloride; 0.1 *M* sodium hydroxide; and water.[7] After a final washing with 100 ml of 10 m*M* Tris-HCl, pH 8.0, the material is ready for preincubation and nuclease digestion.

Coupling of Small Amounts of DNA. All methods of DNA immobilization on Sephacryl S-500, described above for immobilization of about 0.5 mg of DNA on about 2 g of wet material, are large-scale preparations. With reduced amounts of activated support, e.g., 0.1–0.2 g of wet material, 5–50 μg of sonicated plasmid DNA can be coupled without substantial loss of quality of the hybrid selection reaction as proved by selection of specific transcripts of the chloroplast genome.

Results

Selection and Isolation of Specific mRNAs by Hybridization to DNA Immobilized on S-500

Purified and sonicated restriction fragments of spinach plastid DNA are immobilized on DPTE S-500 as described above. After preincubation for 2–3 days at 42° in hybridization medium[11] (50%, v/v deionized formamide; 1 m*M* EDTA, 0.6 *M* NaCl; 0.2% SDS; 20 m*M* PIPES, pH 6.4), the material is washed twice with 5 ml of hot water (90–100°) and stored in hybridization medium at 4°.

Immediately before addition of RNAs, samples of 0.2 g of wet material are washed three times with 1 ml of hybridization buffer at room temperature. The material is drained, transferred to a 1.5-ml tube (Cryotube, Nunc), mixed with 100 μl of RNA in hybridization buffer [about 60–120 μg of total plastid RNA or total cellular poly(A)$^-$ RNA], and incubated for about 4 hr at 42° under occasional or, preferably, continuous gentle shaking. The suspension is transferred to a small plastic column (Bio-Rad Econo column) and washed quickly at 42° (four times) with 5 ml of 50% formamide, 2 × SSC, 0.2% SDS; and at 60° (once) with 5 ml of 10 m*M* Tris, pH 8.0, 2 m*M* EDTA. The solutions are added and immediately drained by suction or pressure. For mRNA elution, the particles are suspended again in 1 ml of Tris–EDTA and transferred to an Eppendorf tube with a disposable pipette. After brief centrifugation, the supernatant is discarded, 500 μl of water heated to 90° are added, and the Eppendorf tube is incubated for 90 sec at 90°. The suspension is immediately returned to the Econo column; the liquid is drained from the particles by pressure and collected in a siliconized ice-cold 15-ml Corex tube containing 75 μg of calf liver carrier tRNA (Boehringer) as well as 80 μl of 4 *M* sodium acetate, pH 6.0. Two subsequent washings, each with 350 μl of

[11] Preincubation for 4 hr at 70° gives practically identical results.

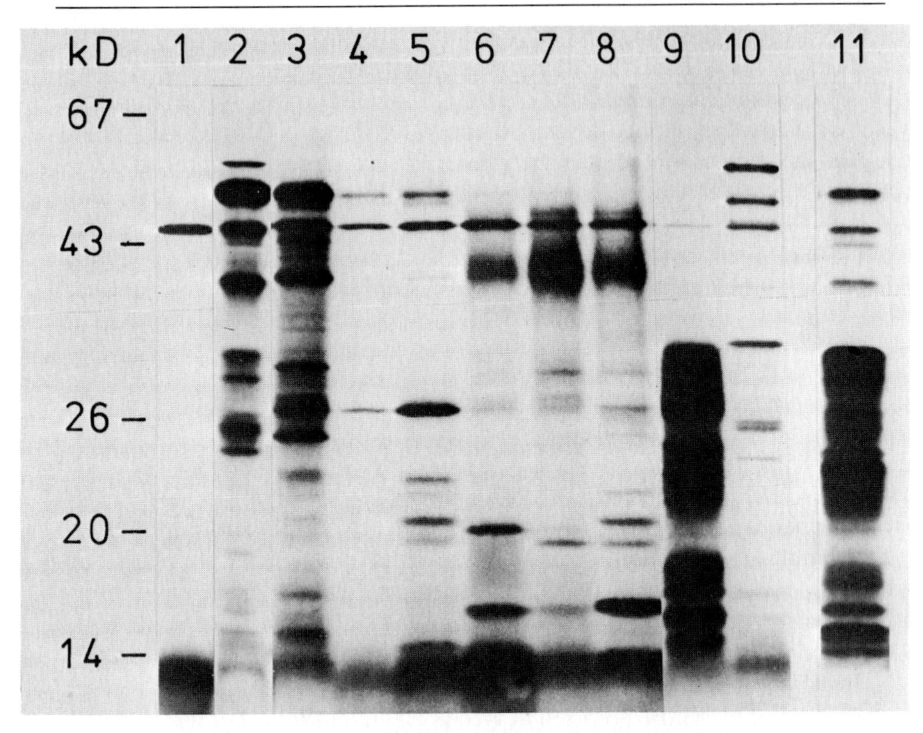

FIG. 1. Fluorography of a 10–15% sodium dodecyl sulfate–polyacrylamide gel of translation products of hybrid selected mRNAs complementary to immobilized spinach plastid DNA restriction fragments. Lanes 1 and 2 contain the *in vitro* translated products without any mRNA and total plastid RNA, respectively. The different protein pattern in lanes 3–11 represent subsets of the profiles obtained with the mixed large *Sal* fragments S-2 plus S-3 of spinach plastid DNA. Lanes 4 and 5 display subsets of these protein patterns, since their messengers were selected by subfragments B-21 (lane 4) and B-7 (lane 5) derived from S-3. The proteins shown in lanes 6–8 are translated from mRNAs complementary to three adjacent and partially overlapping DNA fragments B-4 (lane 6), S-7 (lane 7), and S-8 (lane 8). The highly specific enrichment of mRNA species by S-6 is documented in lane 9 (cf. lane 2). Lane 10 again represents proteins coded by fragment S-10. For the position of restriction fragments and genes on the physical map of spinach plastid DNA, see P. Westhoff, N. Nelson, H. Bünemann, and R. G. Herrmann, *Curr. Genet.* **4,** 109 (1981); and R. G. Herrmann and J. V. Possingham, *in* "Results and Problems in Cell Differentiation" (J. Reinert, ed.), Vol. 10, p. 45. Springer-Verlag, Berlin and New York, 1980. For lanes 9 and 11, see text.

hot water, are also drained into the Corex tube, and 4 ml of ethanol are added to precipitate the RNA at $-70°$ for 30–45 min or at 20° overnight.

The RNA precipitates are collected by centrifugation, washed four times in 70% ethanol, and dried briefly at 0° under vacuum. Finally, they are dissolved in 4 μl of sterile double-distilled water and stored frozen or are immediately translated in a rabbit reticulocyte cell-free system.[12] After incubation at 30° for 60 min, EDTA and pancreatic RNase are added to final concentrations of 10 m*M* and 20 μg/ml, respectively, and digestion is performed at 30° for 15 min.

For electrophoresis on SDS–polyacrylamide gels, aliquots of the translation assays are adjusted to 2% (w/v) SDS, 5% (v/v) 2-mercaptoethanol, 10% (w/v) glycerol, 125 m*M* Tris-HCl, pH 6.8. Before loading onto the gel the samples are heated at 100° for 1 min or—if heat-sensitive proteins have to be analyzed—at 70–80° for 30 sec (see Fig. 1).

It appears that mRNAs that encode proteins of different sizes can be hybridized equally well with high efficiency to macroporous S-500. The background, caused by unspecific adsorption of RNA to the macroporous support, is especially low for S-500. This is evident from comparison of the protein patterns between lanes 9 and 11. Both are derived from mRNAs hybrid-selected with the same DNA restriction fragment, immobilized on DPTE S-500 (lane 9) or on BrCN G-25 superfine[8] (lane 11), respectively.

General Comments

The rate of hybrid formation between a large excess of immobilized plasmid DNA of low complexity and minor amounts of RNA species within a complex mixture of many other RNAs is governed exclusively by the concentration of the accessible part of the immobilized DNA. Therefore overactivation of materials generally is to be avoided because it may cause inaccessibility of 70–80% of the immobilized DNA. Often, reduced yields of particular coupling reactions are more than compensated by excellent accessibilities.

Furthermore, a harsh preincubation procedure helps to increase the sensitivity of the hybrid selection reaction, because it eliminates the 30–50% of the immobilized DNAs that are bound only loosely. Since the hybridization reaction in homogeneous solution proceeds one order of magnitude faster than the reaction with the immobilized DNA, the detachment of 10% of the coupled DNA can cause the failure of an experiment by trapping the RNA of interest. Therefore we recommend an intensive preincubation of all materials before their use for hybrid selection reactions.

[12] H. R. B. Pelham and R. J. Jackson, *Eur. J. Biochem.* **67,** 247 (1976).

Section V

Analytical Methods for Gene Products

[28] Quantitative Two-Dimensional Gel Electrophoresis of Proteins

By JAMES I. GARRELS

Methods of high-resolution two-dimensional gel electrophoresis were introduced by O'Farrell,[1] and important improvements and variant techniques were described subsequently by Garrels and Gibson,[2] O'Farrell *et al.*,[3] Garrels,[4] and Anderson and Anderson.[5,6] The fundamental method for two-dimensional separation of complex mixtures of proteins, as first used by O'Farrell, involves the separation of proteins in the first dimension by charge, using isoelectric focusing in high concentrations of urea and nonionic detergents, followed by separation in the second dimension by molecular mass using sodium dodecyl sulfate (SDS) electrophoresis. The method has been improved for resolution, reproducibility, and convenience since the original procedure. Some progress has been made in the methods and equipment for routine production of large numbers of two-dimensional gels and in methods of computerized data analysis. A special issue of *Clinical Chemistry* provides an excellent overview of the field.[7]

The methods described here are used in the author's laboratory for the resolution and quantitation of radiolabeled proteins from cultured cells. The gel system has been optimized for resolution and reproducibility in order to facilitate computer quantitation and eventual collection of highly standardized data. Aspects of quantitation, including sample counting and film calibration, will be stressed in this report. The details of the electrophoresis procedures have not been changed substantially since an earlier publication.[4] Much progress has been made in the development of semiautomated equipment for routine production of two-dimensional gels and in computer software for quantitation and comparison of two-dimensional gel patterns, but detailed discussion of these subjects is beyond the scope of this chapter. More detailed descriptions of equipment and documented computer software are available from the author on request.

[1] P. H. O'Farrell, *J. Biol. Chem.* **250,** 4007 (1975).
[2] J. I. Garrels and W. Gibson, *Cell* **9,** 793 (1976).
[3] P. Z. O'Farrell, H. M. Goodman, and P. H. O'Farrell, *Cell* **12,** 1193 (1977).
[4] J. I. Garrels, *J. Biol. Chem.* **254,** 7961 (1979).
[5] N. G. Anderson and N. L. Anderson, *Anal. Biochem.* **85,** 331 (1978).
[6] N. L. Anderson and N. G. Anderson, *Anal. Biochem.* **85,** 341 (1978).
[7] *Clinical Chemistry* **28,** 737–1092 (1982).

ISBN 0-12-182000-9

Sample Preparation Procedures

Preparation of samples should be done quickly in the cold. Cells are harvested in a dilute buffer of 20 m*M* Tris, 2 m*M* $CaCl_2$, pH 8.8, and dispersed by passage through a small needle. Viscosity due to broken cells can be reduced by optional inclusion of 50 μg of staphylococcal nuclease per milliliter. A cold 10× solution of SDS and β-mercaptoethanol is added immediately after cell harvest to bring the final concentrations to 0.3% and 1.0%, respectively. It is also acceptable to heat samples in SDS to 100° for 2–3 min, or even to add hot SDS directly to the cell monolayer to inactivate proteases, phosphatases, or other enzymes that might alter the protein size or charge. A 10× concentrated nuclease solution is then added to yield a final concentration per milliliter of 100 μg of DNase I, 50 μg of RNase A. The 10× nuclease solution should contain 0.5 *M* Tris-HCl, 50 m*M* $MgCl_2$, pH 7.0, for optimal enzyme activity during digestion. The nucleases are allowed to act briefly in the cold, while mixing with a pipette or syringe. When the viscosity has disappeared, the sample is quick-frozen in liquid nitrogen and lyophilized. The lyophilized sample is finally dissolved in two-dimensional gel sample buffer containing 9.95 *M* urea, 4% NP-40, 2% ampholytes, and 100 m*M* dithiothreitol. Samples can be stored at −70° for at least several months.

Preparation of Isoelectric Focusing Gels

Materials

Acrylamide and bisacrylamide (Bio-Rad)
Ultra-pure urea (Schwarz-Mann)
Ampholytes (LKB)
Nonidet P-40 (NP-40; Particle Data Laboratories) or Triton X-100 (Calbiochem)
Glass tubes, 21 cm long, 0.047 inch i.d., 0.065 inch o.d. (Glass Company of America, Millville, New Jersey)

Method

The isoelectric focusing gels are composed of 2.7% acrylamide, 0.135% bisacrylamide, 9.5 *M* urea, and 2% NP-40. This composition assures a high-porosity gel for rapid protein migration and easy equilibration, yet the gel is strong enough to be handled without breakage. Ampholytes of the desired pH range are included at a concentration of 2%. The gel solution is dispensed into 5-ml aliquots and stored frozen at −70°.

Polymerization is initiated by adding 15–22 μl of 10% ammonium persulfate to a 5-ml aliquot of gel solution that has been warmed to 37°. The solution is degassed and the gels are promptly cast in the apparatus described below. The gels are kept at a temperature of at least 30° for 1 hr after casting to ensure complete polymerization. Weakness of the gel, especially those containing basic ampholytes, is often caused by incomplete polymerization or insufficient ammonium persulfate.

A automatic casting device for isoelectric focusing (IF) gels can be easily constructed from two Lucite cylinders and a funnel as diagrammed in Fig. 1. The large cylinder is affixed to a square base so that it will stand upright and will contain water. The large end of the funnel is joined to the smaller cylinder to form the filling chamber. A small (1/32 inch) hole should be drilled in the stem of the funnel, and a larger (1/8 inch) hole should be drilled in the cylinder about 1 inch above the junction with the funnel. This completes the assembly of the filling apparatus.

Use of the isoelectric focusing casting apparatus requires that isoelectric focusing solutions be prepared in volumes of 2–5 ml in the specified plastic vials. These vials can be attached by a push-fit over the stem of the funnel. After attachment, the desired number of glass tubes is dropped through the funnel into the vial, and the entire assembly is lowered slowly into the larger cylinder filled with water. The water will first enter the small hole in the funnel, layering over the dense gel solution without mixing. As the funnel lowers, eventually water enters through the larger hole and filling proceeds rapidly. As water enters the filling chamber (funnel plus attached cylinder), the gel solution is forced upward into the gel tubes. If the water level is adjusted to a preset level, the tubes will be filled to the maximal level just as the funnel comes to rest on the bottom of the water reservoir. This procedure can progress unattended and assures that all gel tubes are filled to the same level without leakage. No overlay of the top of the isoelectric focusing gels is necessary during polymerization.

Isoelectric Focusing

The isoelectric focusing gel tubes are mounted into a vertical electrophoresis stand with an upper and a lower electrode chamber. Silicone rubber grommets convenient for holding the gel tubes can be purchased from BoLab (part No. BB1012B). The upper chamber is filled with 0.1 *M* NaOH and the lower chamber is filled with 0.01 *M* H_3PO_4. An optional prefocusing step is usually performed prior to sample loading. For prefocusing, the gel is covered with 3 μl of overlay solution and electrophoresed at a maximum current of 100 μA per gel for about 45 min until the voltage reaches 1000 V.

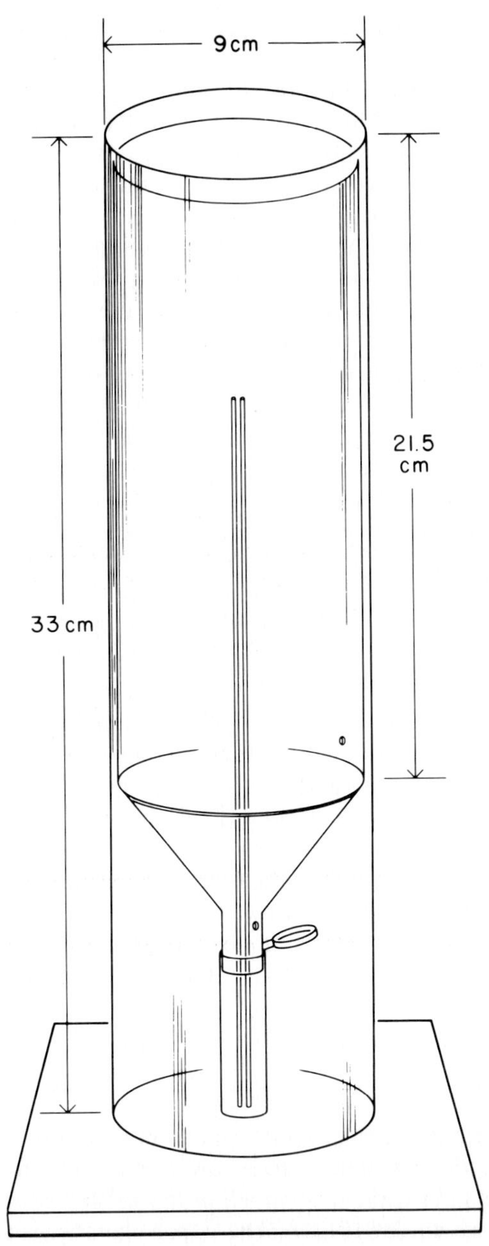

FIG. 1. Isoelectric focusing gel casting apparatus. The outer and inner cylinders are each made of 1/8-inch thick Lucite. The inner, removable, cylinder is attached to a funnel (Nalgene PF60). For casting, a vial (BelArt F-17574) containing isoelectric-focusing gel solution is pushed over the end of the funnel, the glass tubes are inserted, and the assembly is lowered into the outer cylinder containing water. As water enters slowly, first through the small hole in the stem of the funnel and later through the larger hole above the funnel, the gel solution is displaced upward into the glass tubes.

Samples are applied to the top of the isoelectric focusing gels without removing the electrode buffer or the overlay solution. There should be enough space in the glass tubes above the gel for 15–20 μl of sample so that a standard load of 10 μl can be applied without fear of overflow. The upper electrode buffer should be prewarmed to 30°, and the samples should be prewarmed to 37° to prevent precipitation of urea during sample loading.

The isoelectric focusing gels are run in an incubator maintained at 30°. The duration of focusing for 20 cm gels of any pH range should be 19,000 V-hr. Usually this is done overnight at 1100–1200 V.

Equilibration

After the isoelectric focusing step, the gels are pushed from the glass tubes by controlled air pressure. A pressure in excess of 20 psi may be needed, but the gels can be extruded slowly and carefully if the air flow rate can be controlled during extrusion. The urea in the gels need not be precipitated as previously described.[4]

The extruded gels are placed directly into equilibration buffer and agitated at room temperature for 3–10 min. Little difference is seen in gels equilibrated for different times, although if equilibration is too short, the NP-40 remaining in the gel can interfere with SDS binding resulting in lack of resolution for the small acidic proteins.

Slab Gel Electrophoresis

The composition of the slab gels and the electrode buffer are the same as described by Laemmli[8] and O'Farrell.[1]

For convenience and enhanced reproducibility, wide slab gels are cast, each of which can accommodate four standard isoelectric focusing gels. These wide slabs of acrylamide are 1 mm thick, 19 cm high, and 100 cm wide. The glass plates are assembled as shown in Fig. 2 with a length of rubber tubing stretched between the plates across the bottom and up each side. Rigid external clamps compress the tubing and maintain an accurate spacing between the plates. Additional clamps are placed across the top of the plates to prevent spreading from the weight of the gel. One of the clamps on the top should have a small hole that holds a blunt-ended needle through which the gel solution is pumped. The gel solution should run smoothly down the side of one plate without forming bubbles. After the gel solution is pumped in to within 1–2 mm of the top, an overlay of water-saturated isobutanol is added. The slab gels are left to polymerize overnight.

[8] U. K. Laemmli, *Nature (London)* **227,** 680 (1970).

FIG. 2. Slab gel-casting apparatus. Large glass plates (10 × 38 1/4 × 3/16 inches) are sealed with a length of amber latex tubing between the plates along three sides. The rigid Delrin clamps compress the tubing and maintain accurate spacing between the plates. Thin Lucite strips slightly less than 1 mm thick are glued to the end of each notched slab. These do not serve as spacers, but merely as guides for the rubber tubing. A small amount of 1% agarose, 0.1% SDS is applied at each end to form a seal between the glass and the Lucite strip that will retain electrode buffer when the slab is mounted for electrophoresis. The gel solution is pumped in through needles held in a small hole in one of the upper clamps. The vertical lines on the glass plates are etched guide lines demarcating the boundaries of the four isoelectric focusing gels to be loaded onto each slab gel.

For electrophoresis, the slab gel plates are clamped to an electrode apparatus as diagrammed in Fig. 3. The top and bottom electrode chambers are filled with electrode buffer, and the first-dimension gels are applied beneath the buffer. Using a forceps and a blunt-ended spatula, it is not difficult to push the long, thin first-dimension gel between the glass plates seating it smoothly on the top of the slab. A constant power of 60 W is applied to the gels until the bromophenol blue tracking dye has reached the bottom (about 4.5 hr).

Postelectrophoresis Processing

After electrophoresis in the second dimension, the glass plates are separated and the wide slabs are cut into four individual gels. Guidelines scratched onto the outside of the glass plates are helpful, both during isoelectric focusing gel application and during cutting of the slab.

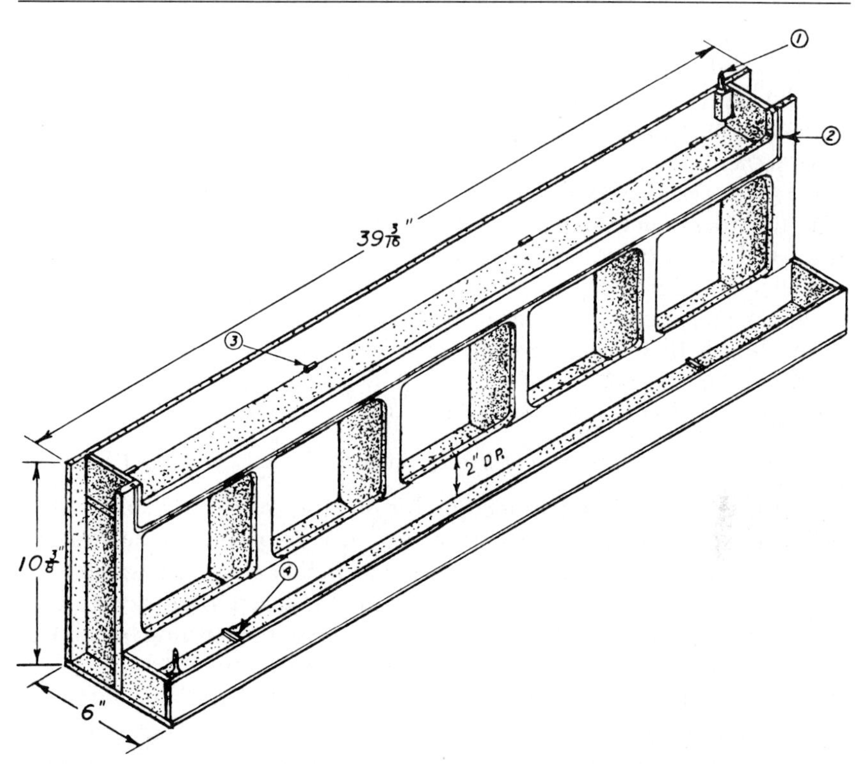

FIG. 3. Slab gel electrophoresis apparatus. The glass plate assemblies are clamped against the front of this chamber forming a seal with the upper electrode chamber. The platinum wire runs through cylindrical guides (3) and must extend the full length of each chamber. The electrode posts (1) are at opposite ends to assure an equal voltage drop along all parts of the gel. A length of 3/16-inch in diameter silicone sponge cord (Ja-Bar Silicone Corp, Andover, New Jersey) is placed in a notch (2) to seal the upper chamber against the notched glass plate. Small spacers (4) hold the gel plates off the bottom of the lower electrode chamber.

Calibration strips (see below) are thawed, and one strip is processed with each gel during subsequent fluorography and drying. For fluorographic detection of ^{35}S, the gels are dipped into two successive dimethylsulfoxide (DMSO) baths for 30 min each followed by a 90-min dip in a solution of 8–10% (w/w) PPO. If detection of tritium is necessary, higher concentrations of PPO (16%) should be used, as described by Bonner and Laskey.[9] After the PPO bath, the gels are agitated in cold water overnight.

After fluorography, the gels and calibration strips are placed on blotting paper and dried under heat and vacuum. Gels should not be in high

[9] W. M. Bonner and R. A. Laskey, *Eur. J. Biochem.* **46,** 83 (1974).

vacuum for more than 30–60 min after drying to hardness. Excess drying causes brittleness and fracturing of the gels.

The gels should not be excessively curled when exposed to film. Low humidity causes rapid curling of the gels, but an hour or two in a humid environment, such as a cold room, leads to flattening of the gels. The dried gels should be exposed to X-ray film, such as Kodak XAR, XS, or XRP (in order of decreasing sensitivity and cost) at a temperature of −70°. If desired, the films can be prefogged to increase sensitivity and linearity for detection of photon emissions. In the author's laboratory, this step is omitted, since it creates a higher and less uniform background, and since the nonlinear film response curve can be accurately measured for each exposure (see below).

Assay for Trichloroacetic Acid-Precipitable Radioactivity in Gel Samples

A rapid and quantitative method for determination of trichloroacetic acid-precipitable counts in gel samples is necessary for routine work involving quantitative analysis. Usually the radioactivity determined for individual protein spots is related to the total radioactivity loaded on the gel to give a fraction of total incorporation during the labeling period into each protein.

A carrier solution is prepared containing per milliliter, 0.2 mg of bovine serum albumin and 1 mg of DL-methionine. This carrier is supplemented with 0.1 part of sample buffer (see formula above) and with 0.005 part of a 0.05% Coomassie Blue staining solution. The sample buffer supplies excess NP-40, which improves the consistency of the pellet, and the stain adds color to the pellet for more accurate aspiration of the supernatant.

Samples prepared for two-dimensional gel electrophoresis are diluted into the carrier solution in volumes of 3–10 μl per 300-μl aliquot of carrier. The dilutions can be stored at 4° for later assay. The protein radioactivity is precipitated by addition of 150 μl of 50% trichloroacetic acid followed by a 2-min centrifugation in a microcentrifuge. The supernatant is removed carefully from the pellet. A second wash with 10% trichloroacetic acid can be carried out if desired, but is usually not necessary. The pellet is then resuspended in 1 ml of NCS solubilizing solution (85% NCS solubilizer, 15% water), and the entire microcentrifuge tube is placed into a scintillation vial. Counting is done in 10 ml of toluene-Liquifluor. Since the entire procedure is done in the same tube, losses due to transfers are minimized. The presence of a plastic tube does not interfere with scintillation counting. (The tubes should not be labeled with marking pens, since ink in the scintillation fluid will greatly reduce sensitivity.)

Preparation of Calibration Gels

The sensitivity for detection of radioactive proteins in dried gels by autoradiography or fluorography involves a large number of possible variables. These include the composition and thickness of the gel, the energy of the isotope, the concentration of the fluorescent enhancer (for fluorography), the exposure time and temperature, the characteristics of the film, and the conditions of film development. To calibrate and standardize each step would be impractical. A better and more reliable method is to prepare calibration gel strips that contain various known levels of radioactive protein and to expose one of these strips with each gel to be calibrated. Making and storing large numbers of the calibration strips is relatively easy by the method described below.

Overview

The calibration process should provide the relationship between protein radioactivity per square millimeter in the gel and optical density on the film. Because film response is usually nonlinear, the calibration should cover a range of radioactive protein concentration sufficient to span the usable range of optical density on the film. Our approach is to prepare calibration gels of the thickness and polymer concentration of the gels to be quantitated, and to incorporate known amounts of radioactivity into the calibration gel during polymerization. By casting the calibration gels in layers (see below), it is possible to prepare gels with many segments of different radioactive protein concentration. Once cast these gels are cut into long, thin strips and frozen for storage. One calibration strip is thawed for each two-dimensional gel to be calibrated. If the two-dimensional gel is processed for fluorography, the calibration gel is processed through the same baths. The two gels are dried on the same paper and exposed to the same piece of film.

Procedures

Preparation of Radioactive Protein. Radioactive protein is prepared biosynthetically by addition of a labeled amino acid to the medium of a convenient cell culture. For labeling of confluent mammalian cells in tissue culture, we add 400 μCi of [^{35}S]methionine per milliliter in methionine-free medium with normal serum and incubate for at least 24 hr. The cells are harvested by our normal gel sample preparation procedure involving SDS and nucleases, except that the sample is not frozen and lyophilized. Instead the sample is heated to 100° for 3 min to inactivate any proteases and then dialyzed extensively to 0.015 *M* Tris, 0.1% SDS, pH 8.8 to remove nonprotein radioactivity. After dialysis the sample is collected, a 1 : 100 dilution is counted in triplicate by the standard trichloroacetic acid precipitation assay, and the sample is stored frozen at −70°.

Preparation of Gel for Calibration Strips. The gel solution used for the calibration gels should be as close as possible in composition to the gels to be quantitated, except that the calibration gel should contain 0.01 mg/ml BSA as carrier for the radioactive protein, and it should not contain SDS (in order to avoid bubbles between the layers during casting). We prepare separate calibration gels for each concentration of acrylamide used in our second-dimension slab gels, and we do not attempt to calibrate gradient gels. A total of 800 ml of gel solution is prepared and aliquoted into 24 50-ml plastic centrifuge tubes with 32 ml per tube.

Addition of Radioactive Protein. We prepare our calibration gels in 24 segments consisting of three radioactive segments followed by a nonradioactive blank. We use a computer program to calculate the dilution of the radioactive protein sample into each aliquot of gel to give us a geometric progression of radioactive protein concentrations ranging from 900,000 dpm/ml down to 300 dpm/ml. The original protein sample should be serially diluted with unpolymerized acrylamide solution so that the volume added to each gel aliquot will be greater than 10 μl and not more than 100 μl. After addition of radioactive protein, each tube is mixed well and two aliquots of 200 μl are removed for counting by the trichloroacetic acid precipitation assay.

Overview of Casting. The gel aliquots for calibration gels are poured into our usual slab gel casting apparatus, which makes gels 36 inches wide. Each aliquot of gel with radioactive protein will be poured into four such slabs making a gel layer about 7 mm high and 0.1 mm thick. With the concentration of TEMED and ammonium persulfate used, this layer will polymerize within 3 min. With a constant overlay of nitrogen gas, this layer will polymerize completely and smoothly. No overlay solution is needed, and no mixing of layers occurs. The layers are poured in a repetitive cycle 3 min apart until completion. The least radioactive layers are poured first.

After the 24 layers are poured and polymerized, the plates are separated and the gels are cut vertically. For storage, the calibration gels are cut into strips 4 cm wide (each will be cut to a width of 1 cm before use). Thus, a typical preparation yields over 360 individual calibration gels.

Slab Apparatus. The glass plates are assembled as usual (see Fig. 2) except that extra clamps are placed along the top to form a nearly complete seal. Small spacers are placed between the plates at the top to hold them apart. (Since the gel will be poured in short segments, the weight of the solution is not sufficient to hold the plates tightly against the upper clamps.) The nitrogen gas is admitted through a hole in one of the top clamps and escapes in the small openings at the top corners. An initial flush followed by a very slow flow is sufficient to maintain the oxygen-free atmosphere above the gels.

Polymerization of Calibration Gels. The TEMED can be added to each of the aliquots at the beginning of the casting period. Each gel aliquot is polymerized by addition of 100 μl of 10% ammonium persulfate followed by thorough mixing on a vortex mixer. The gel is administered to each slab casting unit by pipetting it between the plates from the top at each edge. Four slabs are cast at once, using 7 ml of solution per slab, except for the bottom layer, which receives 8 ml (some of the bottom layer will not polymerize next to the sealing tube). The least radioactive layers are poured first to avoid contamination. Each layer will polymerize within 3 min with a smooth surface and with no unpolymerized liquid remaining above the gel. Using a timer, this cycle is repeated each 3 min until all layers have been polymerized.

Freezing of Calibration Gels. The glass plates are first marked with a marking pen to indicate where the slab is to be cut. The plates are then separated, and the slab is cut vertically (perpendicular to the gel layers) every 4 cm. The separate layers of the gel will be firmly bonded together. The gels must be individually quick-frozen by a method that leaves them flat for convenient storage. The best method we know is to prepare a Dry Ice–methanol bath containing a large metal plate partially submerged in the bath. Each strip is laid on the metal plate for a few seconds until it has frozen solid and is then placed in a precooled storage box. To prevent sticking of the gel to the metal, the plate should be partially wet with methanol, and any liquid methanol should be carefully wiped from each gel before storage in a freezer. There is no practical limit to the length of storage, except for the decay of short-lived isotopes.

Results and Strategy for Computerized Analysis

A typical separation of proteins on three different pH gradients is shown in Fig. 4. These three gradients are used routinely; often each sample is applied to gels of all three types. It can be seen from the figure that the pH 5–7 and the pH 6–8 ranges are complementary. The broad range gel is useful because it resolves proteins more basic than the pH 6–8 gel, it resolves some of the small acidic proteins better than the pH 5–7 gel (best seen on 12.5% slabs, not shown), and it causes some of the spots in the middle range to be focused more tightly.

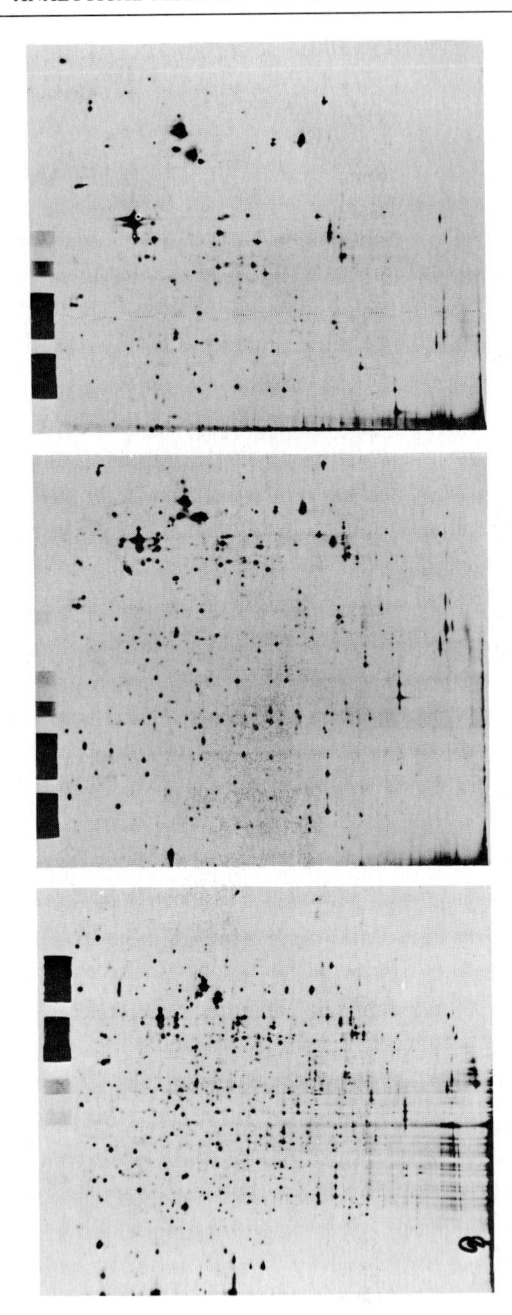

Our approach for computerized analysis is to run gels of several pH ranges and several acrylamide concentrations. The use of separate pH ranges, as shown here, greatly enhances the resolution. The use of separate acrylamide concentrations in the second dimension also enhances resolution. More important, the ability to detect each protein on several different patterns improves the potential for correct identification of the protein and matching to standard patterns. Protein spots that happen to overlap on a 7.5% gel probably will not overlap on a 10% or 12.5% gel due to small differences in the relationship between mobility and pore size for each protein.

Computer analysis requires the availability of a scanner, such as the Optronics rotating drum scanner, which is often used in X-ray crystallography, a minicomputer, and sufficient disk storage to hold at least several images. An interactive graphics system is a virtual necessity because user interaction with the images is required, especially when matching one image to another. Several laboratories have published preliminary computer programs for computerized gel analysis (4,10–12). In the author's laboratory, programs for automatic spot detection and integration and for pattern matching have been developed for a PDP-11 computer. These are available on magnetic tape upon request.

[10] K.-P. Vo, M. J. Miller, E. P. Geiduschek, C. Nielson, A. Olson, and N. H. Xuong, *Anal. Biochem.* **112,** 258 (1981).

[11] J. Taylor, N. L. Anderson, and N. G. Anderson, *in* "Electrophoresis '81" (R. Allen and P. Arnaud, eds.), p. 383. de Gruyter, Berlin, 1981.

[12] P. Lemkin and L. Lipkin, *Comp. Biomed. Res.* **14,** 272 (1981).

FIG. 4. Electrophoretic separation of proteins from HeLa cells. HeLa cells were labeled with [^{35}S]methionine for 48 hr in complete medium. The samples prepared from these cells were run on two-dimensional gels containing either pH 5–7 (left), pH 6–8 (center), or pH 3.5–10 (right) ampholytes. The second-dimension gels contained 7.5% acrylamide. Approximately 440,000 dpm in a volume of 10 μl were applied to each gel. The gels and calibration strips were impregnated with PPO, dried, and exposed to Kodak X-RP film for 6 days. For each panel, the acidic proteins are on the left. The lowest molecular weight resolved on these gels is about 35,000. The pH range on the pH 5–7 gel goes down to approximately 4.0, and on the broad range gel it extends up to approximately 8.0.

The segments in the calibration strips are arranged in triplets separated by nonradioactive gel. The triplets are not in order, but contain a code that can be interpreted by the scanning program to determine the batch number and the radioactivity data for each calibration gel.

[29] Peptide Mapping in Gels

By STUART G. FISCHER

Protease mapping is a convenient method of identifying proteins and polypeptides, requiring no special equipment or technical facility beyond that necessary for sodium dodecyl sulfate (SDS)–polyacrylamide slab gel electrophoresis. In its most popular application, a gel slice containing a stained polypeptide band is cut from an SDS gel, equilibrated with stacking gel buffer (0.125 *M* Tris-HCl at pH 6.8–0.1% SDS), and set into a sample well of a second SDS gel. The gel slice is overlayered with protease in sample buffer, and electrophoresis is performed in a slab gel apparatus[1] in the discontinuous system described by Laemmli[2] except that, to facilitate controlled incubation, the power is turned off for a time when the protease and polypeptide are compressed into a band in the stacking gel. Proteolytic cleavage of the polypeptide substrate generates a pattern of bands whose intensity and size can be compared with a similarly treated reference sample in an adjacent slot. The banding pattern varies with site specificity of the protease.

A more burdensome alternative to digestion within the gel, but which generally produces sharper bands, allows the use of more enzymes, and may be advised where substrate is less difficult to obtain, is digestion in sample buffer prior to electrophoresis. A new procedure permits peptide mapping of polypeptides that are not resolvable in stained gels, such as isotopically labeled polypeptides in a background of densely staining bands. In this modification a channel from the first acrylamide gel is set horizontally across the top of the protease gel, protease in sample buffer is overlayered across the whole strip, and electrophoresis is conducted in the same way as for a single band. Cleavage generates a column of spots visualized by autoradiography, below the original position of the polypeptide band of interest.

Protease Digestion in Sample Buffer

The method of peptide mapping by digestion in sample buffer is shown in Fig. 1. In this experiment a commercial preparation of bovine serum albumin (A-4378) from Sigma was digested with *Staphylococcus aureus*

[1] F. W. Studier, *J. Mol. Biol.* **79,** 237 (1973).
[2] U. K. Laemmli, *Nature (London)* **227,** 680 (1970).

ISBN 0-12-182000-9

V8 protease (36-900-1) from Miles Laboratories. Five hundred micrograms of BSA was dissolved by boiling for 2 min in 1 ml of modified Laemmli sample buffer containing 0.125 *M* Tris-HCl at pH 6.8, 0.5% SDS, 10% glycerol, and 0.0001% bromophenol blue. Aliquots (100 μl) containing 50 μg of BSA are divided into six test tubes. A seventh tube contains 100 μl of sample buffer with no BSA; to this tube protease only is added.

Twenty microliters of sample buffer containing varying amounts of *S. aureus* protease are added and incubated at 37°. After 1 hr, 10 μl of 10% SDS and 7 μl of 2-mercaptoethanol are added to each sample, bringing the final concentrations of SDS and 2-mercaptoethanol to 1% and 5%, respectively, and boiled for 2 min. Fifty microliters of each sample, containing 18 μg of BSA and 0–14 μg of protease, are run into a 15% acrylamide slab gel (30 : 0.4, acrylamide–bisacrylamide). The gel is stained with 0.1% Coomassie Brilliant Blue in 50% methanol, 10% acetic acid and is destained in 5% methanol, 10% acetic acid. The stained gel shows that with increasing amounts of protease (a $\rightarrow$ f) BSA is cleaved into more than 20 smaller polypeptides. The total of their molecular weights greatly exceeds that of BSA, and many must, therefore, be only partially digested. After much longer incubation with protease, they will be cleaved to very small polypeptides that migrate near the dye front. However, we see here that the pattern of partial polypeptides is remarkably stable over a wide range of enzyme concentrations, with major differences in intensity rather than size. Protease itself, incubated under the same conditions, also appears as a set of bands that under close inspection can be seen between the BSA bands in channel f. For this reason, it is important to include a sample of protease with no sample and thereby avoid interpreting the appearance of common bands between two samples as proof of their relatedness, if indeed they come from the protease itself.

Similar intensity (and pattern) changes are generated by varying digestion time at constant enzyme concentration. When Coomassie Brilliant Blue stain is used, the large number of bands generally requires at least 10 μg of substrate applied in each well for clear visualization. Smaller polypeptide substrates usually generate fewer bands and proportionately lower loads are sufficient. With more sensitive stains, or radioactively labeled sample much less can be used.

This procedure may also be used on proteins that have been electrophoretically eluted from a stained gel. However, precaution should be taken not to let the gel from which the bands are to be eluted stain and destain longer than necessary. Prolonged exposure to acetic acid at room temperature may hydrolyze the polypeptide in the gel, causing the peptide bands to smear.

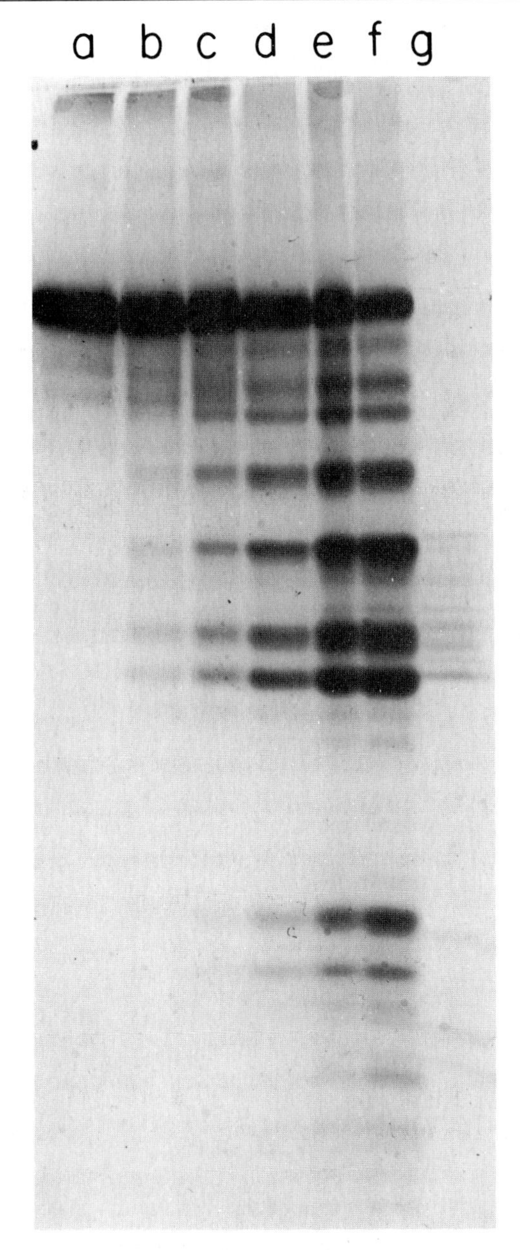

FIG. 1. Peptide maps of bovine serum albumin (BSA) digested in sample buffer with *Staphylococcus aureus* protease. The BSA at a concentration of 0.5 mg/ml in 0.125 *M* Tris-

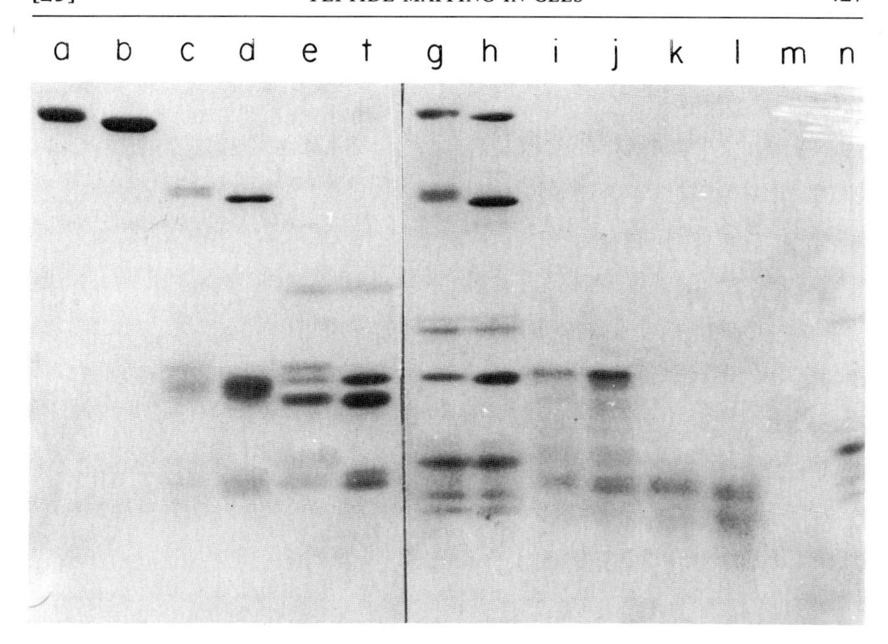

FIG. 2. Peptide maps of *Physarum polycephalum* H1 cut from a sodium dodecyl sulfate (SDS) gel and reelectrophoresed with protease. Lanes: (a) PO_4H1; (b) H1; (c) PO_4H1 and 0.02 μg of *Staphylococcus aureus* protease; (d) H1 and 0.02 μg of *S. aureus* protease; (e) PO_4H1 and 0.6 μg of *S. aureus* protease; (f) H1 and 0.6 μg of *S. aureus* protease; (g) PO_4H1 and 10 μg of chymotrypsin; (h) H1 and 10 μg of chymotrypsin; (i) PO_4H1 and 0.1 μg of subtilisin; (j) H1 and 0.1 μg of subtilisin; (k) PO_4H1 and 2 μg of subtilisin; (l) H1 and 2 μg of subtilisin; (m) 2 μg of subtilisin; (n) 10 μg of chymotrypsin.

Protease Digestion in Gels

Peptide mapping by partial proteolysis in gel slices is shown in Fig. 2. In this experiment histone H1 from *Physarum polycephalum* and a phosphorylated variant with reduced electrophoretic mobility were compared. Extracts containing about 5 μg of each H1 per channel were size-fractionated on a 15% acrylamide–SDS gel, stained for 1 hr at room temperature with 0.1% Coomassie Brilliant Blue in 50% methanol, 10% acetic acid, and destained for 2 hr at room temperature with 5% methanol, 10% acetic acid. The histone bands, visualized over a light box, were cut from the gel

HCl at pH 6.8, 0.5% sodium dodecyl sulfate (SDS), 10% glycerol, 0.0001% bromophenol blue was incubated for 1 hr at 37° with increasing amounts of *S. aureus* protease, then brought to 1% SDS and 5% 2-mercaptoethanol, boiled for 2 min, and analyzed on a 15% polyacrylamide–SDS gel. The *S. aureus* protease (in milligrams per 50-μl sample load) is (a) 0, (b) 0.7, (c) 1.4, (d) 3.5, (e) 7, (f) 14, (g) 14; no BSA.

with a clean razor blade. Since the band below each well typically broadens to nearly twice the width of the well, a longitudinal cut through the center of each band generates a slice that will fit into the same sized well of a second gel without distorting the well walls. The small gel pieces from each H1 fraction were soaked for 10–15 min with occasional swirling in test tubes containing 20 ml (about 10 times the gel volume) of 0.1% SDS, 0.125 *M* Tris-HCl pH 6.8. This process was repeated two more times with fresh Tris-SDS until the pH of the buffer remained neutral.

A second 15% polyacrylamide–SDS gel (30 : 0.4, acrylamide–bisacrylamide) with a 3-cm-long 6% acrylamide stacking gel was prepared, and the gel pieces were set in the same orientation into the bottoms of the sample wells of the second gel. This is done most easily if the sample wells are first filled with 0.1% SDS. The gel pieces sink to the bottoms of the wells with gentle prodding, trapping no air beneath them, even if the two gels are the same thickness. Prodding is done with minimal damage to the gel slices by sliding a microspatula tip between the gel slice and either glass face and applying brief, gentle, downward pressure, then removing the microspatula and pushing slowly on top of the slice with an acrylic strip slightly less thick and wide as the sample well. This process is usually repeated two or three times until the gel slice rests against the bottom of the sample well. Each sample slice is then overlayered with 20 μl of sample buffer (0.125 *M* Tris-HCl at pH 6.8, 0.1% SDS, 10% glycerol) containing different amounts of protease. The protease sample volume ought to be large enough to keep roughly equal amounts of protease passing through all parts of the sample band.

The gel is run at low voltage (3 V/cm) until the protease catches up with the polypeptide substrate near the lower edge of the stacking gel, as indicated by the fusion and sharp focusing of the bromophenol blue and Coomassie Brilliant Blue bands. Power is turned off for 30 min (although longer incubations may be necessary for weak enzymes) and then run at higher voltage (10 V/cm) until the dye tracks reach the bottom of the gel. Staining and destaining of the gel is performed in the usual way for SDS gels.

Figure 2 shows that with each of three proteases, the banding patterns of the phosphorylated and unphosphorylated histone are similar except for slight upward displacement of some phosphohistone bands. The uppermost band in slots e and f is *S. aureus* protease, and several of the common bands in slots g and h are chymotrypsin. Enzyme activity in the gel is sometimes difficult to predict and therefore it is recommended that each sample be digested over a 20-fold range of enzyme in order to obtain patterns with a useful number of peptide bands.

Peptide Mapping of Gel Strips

A typical problem for which peptide mapping of gel strips in two dimensions is useful is shown in Fig. 3. In this experiment, radioisotopically labeled polypeptides made *in vivo* early during *B. subtilis* infection by SPO1 were compared with those produced *in vitro* from a purified restriction fragment. Bacterial proteins *in vivo* obscure the phage bands, which are visualized by autoradiography, and the small amount of protein made *in vitro* also can be detected only by autoradiography. Although the principal polypeptide product made *in vitro* has the same electrophoretic mobility in SDS–gel electrophoresis as a polypeptide made *in vivo,* their identity could be more firmly established by peptide mapping. However, the precise location of the radiolabeled polypeptide band of interest in each sample is difficult to determine, and it is even more difficult to remove from the gel free of contamination from bands just above and below it. These problems can be avoided by removing strips from the first

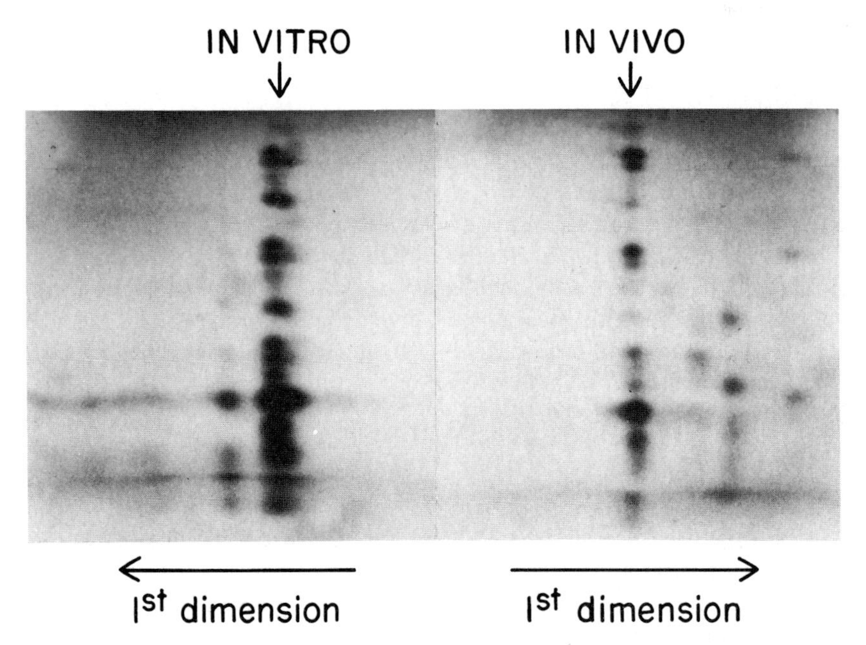

FIG. 3. Two-dimensional peptide maps of SPO1 proteins. Autoradiogram of 20% acrylamide–sodium dodecyl sulfate gel. *Left side: Staphylococcus aureus* protease digestion products of *in vitro* products of SPO1 restriction fragment. *Right side: S. aureus* protease digestion products of *in vivo* products of early SPO1 infection. The small arrows indicate the major polypeptide bands of interest.

gel containing the region where the polypeptide is known to migrate. The strips are equilibrated with several changes of stacking gel buffer (0.125 *M* Tris-HCl at pH 6.8, 0.1% SDS) and laid across the top of a 20% acrylamide–SDS gel with a 3 cm high 6% acrylamide stacking gel. To avoid lateral diffusion of the bands as they pass between the gels, it may be helpful to underlay the strips with a thin layer of 1% agarose in the same buffer as a sealant. The agarose should also fill the space beside the strips and to their upper edges. Twenty micrograms of *S. aureus* protease in 1 ml of sample buffer (0.125 *M* Tris-HCl at pH 6.8, 0.5% SDS, 10% glycerol, 0.0001% bromophenol blue) is then layered above the strips, and electrophoresis is performed in the same way as with individual bands. The autoradiogram shows that the major polypeptide made *in vitro* has the same proteolytic banding pattern as that made *in vivo*.

The same technique can also be used to compare several polypeptides from a single channel of an SDS gel without the excision of individual bands. Moreover, staining the first gel can be omitted, saving time and labor and avoiding acid hydrolysis of the substrate.

Concluding Remarks

Several common proteases are useful in peptide mapping, including chymotrypsin, *Staphylococcus aureus* protease, papain, subtilisin, *Streptomyces griseus* protease, ficin, and elastase. Of these, papain requires 2-mercaptoethanol, which inhibits acrylamide polymerization and is therefore not useful for proteolysis during reelectrophoresis. Cofactors, such as EDTA, required by some proteases, can, however, be included in the gel solutions. Some enzymes, notably trypsin and Pronase, often generate smeared bands and may be less useful. In addition, some substrates easily and clearly cleaved in solution are poor substrates during reelectrophoresis for reasons not understood.

Acknowledgments

I wish to thank Marion Perkus, Nathaniel Heintz, and David Shub for permission to use their SPO1 peptide mapping experiment prior to publication.

[30] Methods of RNA Sequence Analysis

By ALAN DIAMOND and BERNARD DUDOCK

Rapid methods for the determination of the nucleotide sequence of DNA by base-specific chemical cleavage[1] and chain termination by dideoxynucleotides[2] are in wide use. Similar techniques have been developed for RNA sequence analysis using base specific cleavage by both chemical[3] and enzymic methods[4-8] as well as chain termination with dideoxynucleotides using reverse transcriptase.[9] While these procedures have proved to be sufficient for the determination of the primary structure of messenger and ribosomal RNAs, they have not been sufficient in the sequencing of RNAs that are rich in modified residues, such as tRNAs. The above techniques cannot distinguish the numerous modified nucleotides present in tRNAs. Furthermore, the stable secondary structure of tRNAs[10] often contain regions of inaccessibility which are resistant to cleavage.

A method has been developed by Stanley and Vassilenko[11] that is particularly well suited for the sequence analysis of tRNAs and other RNAs rich in modified residues. This method requires as little as 1 μg of a purified tRNA. Partial digestion of the tRNA under conditions that yield on the average one break per molecule generates fragments with newly produced 5′-hydroxyl ends, which can then be radioactively labeled with [γ-^{32}P]ATP and T4 polynucleotide kinase. The labeled fragments are then separated by gel electrophoresis, excised, eluted, digested to [5′-^{32}P]nucleoside diphosphates and analyzed by thin-layer chromatography. This procedure usually yields between 80 and 100% of the entire sequence of the molecule. The remainder of the sequence, as well as confirmation of the entire sequence, can be determined by mobility shift analysis and base-specific cleavage of ^{32}P-end labeled tRNA and tRNA fragments.

[1] A. M. Maxam and W. Gilbert, this series, Vol. 65, p. 499.
[2] F. Sanger, S. Nicklen, and A. R. Coulson, *Proc. Natl. Acad. Sci. U.S.A.* **74,** 5463 (1977).
[3] D. A. Peattie, *Proc. Natl. Acad. Sci. U.S.A.* **76,** 1760 (1979).
[4] H. Donis-Keller, A. M. Maxam, and W. Gilbert, *Nucleic Acids Res.* **4,** 2527 (1977).
[5] A. Simoncsits, G. G. Brownlee, R. S. Brown, J. R. Rubin, and H. Guilley, *Nature (London)* **269,** 833 (1977).
[6] H. Donis-Keller, *Nucleic Acids Res.* **7,** 179 (1977).
[7] H. Donis-Keller, *Nucleic Acids Res.* **8,** 3133 (1980).
[8] M. S. Boguski, P. A. Hieter, and C. C. Levy, *J. Biol. Chem.* **255,** 2160 (1980).
[9] P. H. Hamlyn, M. J. Gait, and C. Milstein, *Nucleic Acids Res.* **9,** 4485 (1981).
[10] A. Rich and U. RajBhandary, *Annu. Rev. Biochem.* **45,** 805 (1976).
[11] J. Stanley and S. Vassilenko, *Nature (London)* **274,** 87 (1978).

METHODS IN ENZYMOLOGY, VOL. 100

ISBN 0-12-182000-9

Principle

Stanley and Vassilenko[11] have described a procedure for the sequence analysis of RNA. This procedure involves partial digestion of the RNA with deionized formamide at 100° for short periods of time (0.1–10 mins). This digestion results in an average of one break or "hit" per molecule. In order to obtain only one break per molecule, it is presumed that most of the molecules remain intact and the relative occurrence of multiple hits is insignificant. The position of the break within the molecule should be random, and this is facilitated by the denaturing effects of both high temperature and formamide. In addition, the presence of 0.4 m*M* $MgCl_2$ appears to enhance the probability of obtaining random cleavage. A similar procedure, using water instead of formamide to digest the RNA, has been described by Gupta and Randerath.[12] Water digestion has the advantage that it can be used directly for subsequent steps whereas 6–12 hr are usually required to remove the formamide under vacuum before proceeding.

The products of digestion of tRNA with either formamide or water are two fragments, one containing a 5′-phosphate and a 3′-phosphate and the other having a newly generated 5′-hydroxyl as well as a 3′-hydroxyl termini. Only the newly generated 5′-hydroxyl end is a substrate for end labeling with polynucleotide kinase and [γ-^{32}P]ATP. The uncut molecules, which comprise most of the tRNA population, cannot be labeled in this manner, as they already have 5′-phosphates. Some labeling of uncut molecules does occur, however, presumably due to a phosphate exchange reaction between the 5′-terminal phosphate of the RNA molecule and the γ-phosphate of [^{32}P]ATP catalyzed by T4 polynucleotide kinase.[13]

The entire array of end labeled fragments can be separated on denaturing polyacrylamide gels essentially on the basis of size. Thin gels[14] 0.035–0.05 cm are used because (*a*) less sample is required; (*b*) gels can be run at elevated temperature; (*c*) the time required for electrophoresis is shorter; and (*d*) the ease with which labeled RNA can be eluted from the gel is greatly enhanced. Autoradiography of such a gel should reveal a uniformly labeled ladder of bands, each band representing a 5′-end-labeled oligonucleotide which is one residue shorter than the band immediately above it. Each species of RNA analyzed in this manner will have an optimum time of digestion, which results in the production of a uniform ladder due to obtaining single "hit" kinetics. It is therefore advisable that a number of different digestion times be tried and the one that gives the

[12] R. C. Gupta and K. Randerath, *Nucleic Acids Res.* **6,** 3443 (1979).

[13] J. H. van de Sande, K. Kleppe, and H. G. Khorana, *Biochemistry* **12,** 5050 (1973).

[14] F. Sanger and A. R. Coulson, *FEBS Lett.* **87,** 107 (1978).

most uniformly labeled ladder be used for the rest of the procedure. A common problem is the appearance of a compression effect, regions where the distance between successive bands on the autoradiograph are small compared with other regions of the ladder. This phenomenon makes the determination of the sequence in the region of the compression very difficult. Compressions arise when complementary regions of the RNA fold back on themselves. This can be minimized by heating and rapidly cooling the sample before application to the gel and by using denaturing conditions during gel electrophoresis, such as elevated temperature and including 7 *M* urea in the gel and loading buffer.

Each band in the ladder can be excised by lining up an autoradiograph with the gel, making use of radioactive marker spots on the gel. Slices are made as thin as possible to avoid contamination with adjacent bands. Radioactive oligonucleotides can then be eluted by diffusion and concentrated by ethanol precipitation. An aliquot of each eluted oligonucleotide is digested with a mixture of enzymes to produce a [5′-^{32}P]pXp, where X is the terminal nucleoside of the particular oligonucleotide. Identification of the radioactive nucleoside 5′,3′-diphosphate and determination of the nucleotide sequence is then made by thin-layer chromatography on polyethyleneimine (PEI)-cellulose sheets using ammonium sulfate and ammonium formate as solvents.[12] These solvent systems are particularly useful as the relative mobility of modified nucleoside 5′,3′-diphosphates and ribose-methylated dinucleoside triphosphates common to tRNAs have been determined.

Residues suspected of being modified can be digested to liberate the terminal 5′-^{32}P-labeled nucleotides using nuclease P_1 and analyzed by the two-dimensional thin-layer system described by Silberklang *et al.*[15] This procedure can be used to confirm the identity of most of the modified residues found in tRNA, and indeed some investigators[16] prefer to substitute this procedure entirely for PEI-cellulose chromatography.

Modified versions of the above procedure have been described that circumvent the need to cut the bands from the gel.[12] Partial hydrolysis, end-labeling, and gel electrophoresis are performed essentially as above, but the ladder is directly transferred to PEI-cellulose by contact-transfer, and radioactive oligonucleotides are then digested with RNase *in situ*. The resulting nucleoside 5′,3′-diphosphates are then identified as described above for the ammonium sulfate and ammonium formate solvent systems.[12] Another modification of this technique has been described[17] where contact transfer is made to DEAE-cellulose thin-layer plates fol-

[15] M. Silberklang, A. M. Gillum, and U. L. RajBhandary, this series, Vol. 59, p. 58.
[16] Y. Kuchino, S. Watanabe, F. Harada, and S. Nishimura, *Biochemistry* **19**, 2085 (1980).
[17] Y. Tanaka, T. A. Dyer, and G. G. Brownlee, *Nucleic Acids Res.* **8**, 1259 (1980).

lowed by *in situ* digestion. The nucleotide sequence is then determined by electrophoresis at pH 2.3. Contact-transfer methods have the advantage of by-passing the tedious task of cutting bands but also suffer disadvantages. Compressions in the ladder are also transferred and can make interpretation of the chromatogram difficult. Moreover, cutting bands affords the opportunity to use eluted labeled oligonucleotides for numerous purposes other than one-dimensional analysis of the radioactive termini. As mentioned above, aliquots may be digested with nuclease P_1 and analyzed by two-dimensional chromatography. In addition, labeled oligonucleotides, especially those derived from regions of the ladder showing severe compression effects or other anomalies, can be purified and used for mobility shift analysis.[15,18,19] This procedure is particularly useful in determining the sequence of regions showing unusually stable secondary structure (see below).

Sequence determination of tRNAs by formamide fragment analysis usually yields approximately 80–100% of the entire sequence. On occasion the 5′-terminal nucleotide will not be labeled efficiently by this method. In this case, identification can be made by removing the terminal phosphate with alkaline phosphatase and labeling the 5′ end with polynucleotide kinase and [γ-^{32}P]ATP.[15] The 5′-end-labeled RNA is then digested with RNase P_1 and the 5′ terminus determined by two-dimensional thin-layer chromatography. The 3′-terminal nucleotide may be determined by 3′-end-labeling the tRNA with [5′-^{32}P]pCp[20] and digesting the labeled tRNA with RNase T_2, followed by two-dimensional thin-layer chromatography.

Confirmation of the entire sequence as well as determination of the 3′- and 5′-terminal nucleotides can be accomplished by analysis of 5′- and 3′-end-labeled tRNA and tRNA fragments by cleavage with base-specific RNases and by mobility shifts. Both procedures have been described previously in this series.[15] In the mobility shift procedure, a partial digest of end-labeled RNA is separated on cellulose acetate (on the basis of charge) in the first dimension and by homochromatography on DEAE-cellulose thin-layer chromatography (TLC) at 65° (on the basis of size) in the second dimension. The nucleotide sequence of an oligonucleotide can then be determined by noting the relative "shift" between successive spots on an autoradiogram. Since the second dimension is performed at 65° in the presence of 7 *M* urea, this technique helps to overcome the adverse affects of secondary structure. RNA sequence gels displaying the

[18] F. Sanger, J. Donelson, A. R. Coulson, and D. Fisher, *Proc. Natl. Acad. Sci. U.S.A.* **70,** 1209 (1973).

[19] R. Pirtle, I. Pirtle, and M. Inouye, *J. Biol. Chem.* **255,** 199 (1980).

[20] T. E. England and O. Uhlenbeck, *Nature* (*London*) **275,** 560 (1978).

products of partial cleavage with base-specific ribonucleases[4-8] are particularly useful in confirming the sequence of an RNA molecule. The commonly used ribonucleases are RNase T_1, which cleaves at G residues, RNase U_2, which cleaves at A residues, and an enzyme from *Bacillus cereus*,[21,22] which can be used to cleave at pyrimidines. Traditionally, the difficulty with sequencing RNA by cleavage with base-specific enzymes has been discriminating between pyrimidines. Some enzymes have been described that help to overcome this problem. An RNase from *Physarum polycephalum* (*Phy*1)[23] has the specificity to cleave $U > G \sim A > C$ and RNase CL3[24,25] from chicken liver, which cleaves $C \gg A > U$, and therefore these enzymes can be helpful for pyrimidine determination. These enzymes tend to be influenced by adjacent nucleotides, somewhat limiting their usefulness. We prefer to use a ribonuclease activity present in relatively unfractionated preparations from *Physarum* made essentially the same way as *Phy*1, which is denoted *Phy*M.[7] This ribonuclease, now commercially available from P-L Biochemicals, cleaves specifically at U and A residues when the RNA is digested under denaturing conditions (7 *M* urea, 50°, pH 5.0). Used in conjunction with the other base-specific enzymes, *Phy*M can be used to distinguish a C from U residue.

Materials and Reagents

[γ-^{32}P]ATP was either purchased from New England Nuclear (1000–3000 Ci/mmol) or synthesized from ADP and inorganic [^{32}P]phosphate (~7000 Ci/mmol) by the methods of Walseth and Johnson[26] with the required enzymes purchased from either Boehringer Mannheim or Sigma. [5′-^{32}P]cytidine 5′,3′-diphosphate was from Amersham (2000–3000 Ci/mmol). Calf intestine alkaline phosphatase, Nuclease P_1 and T4 polynucleotide kinase were from Boehringer Mannheim. Analytical grade acrylamide was from Serva and *N,N*′-methylethylenediamine (TEMED) was purchased from Bio-Rad Laboratories. Formamide was from Fisher Scientific Company and deionized immediately before use with analytical grade mixed bed resin, [AG 501-X8(D)], purchased from BioRad Laboratories.

[21] E. C. Koper-Zwarthoff, R. E. Lockard, B. Alzner-DeWeerd, U. L. RajBhandary, and J. F. Bol, *Proc. Natl. Acad. Sci. U.S.A.* **74,** 5504 (1977).

[22] R. E. Lockard, B. Alzner-DeWeerd, J. E. Heckman, M. W. Tabor, J. MacGee, and U. L. RajBhandary, *Nucleic Acids Res.* **5,** 37 (1978).

[23] D. Pilly, A. Niemeyer, M. Schmidt, and J. P. Bargetzi, *J. Biol. Chem.* **253,** 437 (1978).

[24] C. C. Levy and T. P. Karpetsky, *J. Biol. Chem.* **255,** 2153 (1980).

[25] M. S. Bogusky, P. A. Hieter, and C. C. Levy, *J. Biol. Chem.* **255,** 2160 (1980).

[26] T. F. Walseth and R. A. Johnson, *Cyclic Nucleotide Res.* **9,** 771 (1978).

RNases T_1, T_2, and U_2 were from Calbiochem-Behring Corp., RNase from *B. cereus* was a gift from J. Dunn, and *Phy*M was a gift of H. Donis-Keller. Both are now commercially available from P-L Biochemicals. In addition, RNases *Phy*I and CL3 are available from Bethesda Research Laboratories.

Cellulose-precoated TLC plastic sheets (0.10 mm layer thickness without fluorescent indicator) were from EM Reagents, and Macherey-Nagel polyethyleneimine-coated plastic thin-layer sheets (0.10 mm layer) were from Brinkmann.

Radioactive gels and plates were autoradiographed with Kodak XAR-5 or BB5 film using a lightning-plus intensifying screen from DuPont as needed.

Formamide Fragment Analysis

The tRNA sample used should be at least 95% pure, since this technique is particularly sensitive to even low levels of contaminating RNA. As a final step in the purification of a tRNA, we usually use preparative gel electrophoresis (20% polyacrylamide gel run in the presence of 7 *M* urea), using UV shadowing to detect the bands and eluting the RNA as described by Donis-Keller *et al.*[6] Considerable care must be taken to remove residual acrylamide, since it will interfere with subsequent steps. This can be accomplished by two extractions of the eluted RNA with phenol–chloroform (1 : 1) followed by an ether extraction and ethanol precipitation. It is often difficult to precipitate small quantities of RNA quantitatively using ethanol. It may therefore be necessary to concentrate the eluted RNA by 2-butanol extractions. Repeated extractions with equal volumes of 2-butanol can quickly and effectively increase the RNA concentration so as to ensure efficient precipitation.

One microgram of tRNA is digested for 0.1–10 min in the presence of 11 μl of freshly deionized formamide containing 0.4 m*M* $MgCl_2$. The digested sample is chilled on ice and dried over P_2O_5 in a vacuum desiccator overnight. The dried sample is then dissolved in 10 μl of H_2O and used as a substrate for labeling with polynucleotide kinase and [γ-^{32}P]ATP. This reaction is done in a solution containing 10% glycerol, 50 m*M* Tris-HCl, pH 8.0, 10 m*M* $MgCl_2$, 10 m*M* dithiothreitol (DTT), 10 μg of bovine serum albumin (BSA) per microliter, 100 pmol of [γ-^{32}P]ATP, and 5 units of T4 polynucleotide kinase in a 100-μl volume. The reaction is allowed to proceed for 30 min at 37°. The reaction is quenched by adding 0.5 ml of H_2O, 50 μl of 20% sodium acetate, pH 5.2, 4 μl of 50 m*M* ATP, and 4 μl of carrier RNA (2 mg/ml). It is necessary to use as carrier an RNA that is larger than the undigested sample so as not to interfere with the resolution of the radioactive oligonucleotides on the gel. Two volumes of ethanol are added and the sample is placed at −70° for 1 hr and centrifuged (3000 *g* for 30 min at 4°); the supernatant is decanted, and the pellet is dried under vacuum.

The pellet is dissolved in 20 μl of H_2O, and a 5-μl aliquot is dried and dissolved in 2–3 μl of gel loading buffer containing 7 *M* urea, 0.02 *M* sodium citrate, pH 5.0, 1 m*M* EDTA, and 0.1% bromophenol blue (BPB) and xylene cyanole (XC). The sample is heated for 90 sec at 90°, immediately placed on ice, and loaded onto a 20% polyacrylamide gel that was prerun at 1500 V for 1 hr. The gel is run at constant power such that the glass plate of the gel is hot to the touch (~45°). Electrophoresis is terminated when the BPB dye has migrated one-third the length of the gel. One of the glass plates is then carefully removed, and the gel is covered with plastic wrap. Radioactive dye markers are placed at numerous positions on the covered gel to permit alignment of the autoradiograph and the gel. Autoradiography usually takes between 5 and 45 min; if a longer time is required there is probably insufficient radioactivity for further analysis. An example of a typical formamide-generated ladder is shown in Fig. 1.

The autoradiograph is placed on a light box and the gel is positioned on top of it, using the radioactive markers for alignment. Bands are cut out of the gel with a razor blade and placed in 1 ml of elution solution containing 0.3 *M* NaCl, 0.1% SDS, and 0.15 OD_{260} unit[27] per milliliter of carrier RNA. The RNA is eluted from the gel by incubating the sample at 37° overnight with gentle shaking. The solution is then pipetted off using a Pasteur pipette, leaving the intact gel slice behind, and the RNA is precipitated with two volumes of ethanol. The 5′-end-labeled oligonucleotides are collected by centrifugation, dried under vacuum, and dissolved in 10 μl of "digestion mix" containing 0.3 *M* ammonium acetate, pH 4.5, and 6 units/ml RNase T_2, 1000 units/ml RNase T_1, and 1.5 mg/ml RNase A. The digestion is allowed to proceed overnight at 37°. One-microliter aliquots from each sample are applied to a 20 × 40 cm PEI-cellulose thin-layer plate that had been predeveloped overnight with H_2O and dried. As many as 60 samples can be run per plate. After the application of the sample the plate is dried in a gentle stream of cool air, washed with gentle agitation in absolute methanol for 7 min to remove residual salts, and then air-dried. A 3-cm wick of Whatman 3 MM paper is attached by stapling to the top of the plate, and the plate is developed in water until the water just passes the origin line (1–2 cm from the bottom of the plate).

[27] OD_{260} unit: one absorbance (OD_{260}) unit is defined as that amount of material per milliliter of solution that produces an absorbance of 1 in a 1.0 cm light path at a wavelength of 260 nm.

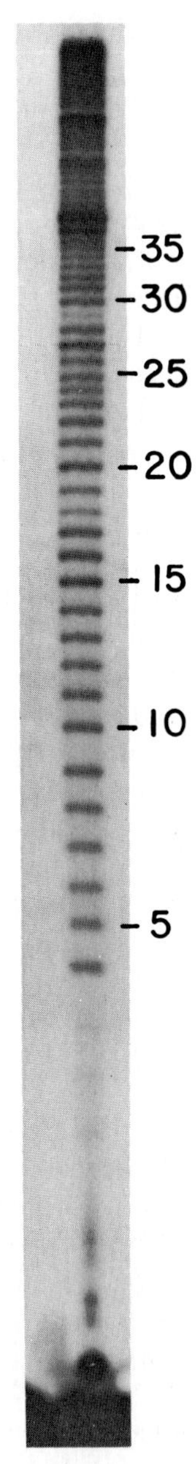

FIG. 1. Fractionation of [5′-^{32}P]oligonucleotides derived by partial formamide hydrolysis (2.5 min at 100°) of spinach chloroplast tRNA$_f^{Met}$ on a 20% polyacrylamide gel.

The plate is then placed, without drying, in a solution of 0.55 *M* ammonium sulfate at 4°. Chromatography is allowed to proceed at 4° until the solvent front has migrated half way up the wick. Ascending chromatography in ammonium formate is performed in the same manner except that the plate is washed in 0.1 *M* ammonium formate for 7 min prior to methanol washing, and chromatography is in 1.75 *M* ammonium formate at room temperature. Autoradiography usually takes 1 day using an intensifying screen and Kodak XAR-5 film at −70°. A sample chromatogram is shown in Fig. 2. The identity of each ribonucleotide is deduced from its relative mobility in both solvent systems (see Fig. 3). The entire procedure can be repeated using 15% and 12% polyacrylamide gels to expand regions of the ladder corresponding to the middle and 5′ portions of the RNA molecule.

Sometimes a gap will appear in the radioactive ladders generated by this technique where a single band seems to be missing. This is usually due to the presence of a 2′-*O*-methyl modification in the tRNA sequence. Treatment of a 5′-end-labeled oligonucleotide containing a 2′-*O*-methylated residue at its 5′ end will result in a labeled [5′-^{32}P]dinucleoside triphosphate, owing to the inability of formamide to cleave such a modification. Identification of the dinucleoside triphosphate may be determined by its mobility on PEI-cellulose (see Fig. 3). In addition, the identity of the ribose-modified nucleotide may also be determined by two-dimensional TLC following digestion of the [5′-^{32}P]dinucleoside triphosphate with nuclease P_1. The nucleotide 3′ to the 2′-*O*-methylated residue can be determined by partial cleavage of end-labeled RNA with base-specific ribonucleases (see below).

Two-Dimensional Thin-Layer Chromatography of Modified and Terminal Nucleotides

In the course of the formamide fragment analysis procedure, most, if not all, of the modified residues in the tRNA can usually be identified. These assignments are confirmed by two-dimensional TLC. This is performed by digesting [5′-^{32}P]nucleoside 5′,3′-diphosphates (see above) with RNase P_1 in a mixture containing 3 μl of sample with 0.25 μl of 1.0 *M* ammonium acetate, pH 4.5, 0.25 μl of RNase P_1 (5 mg/ml), and 0.5 μl of H_2O. The reaction is incubated for 3 hr at 37°. An aliquot of the digested sample (1 μl) is chromatographed together with 0.3 OD_{260} units each of pA, pU, pC, and pG standards on 10 × 10 cm cellulose thin-layer plates in solvents A and C of Silberklang *et al.*[15] Cold standards are visualized by UV light and compared to the position of the radioactive nucleotide determined by autoradiography (see Fig. 4). A similar approach can be used to

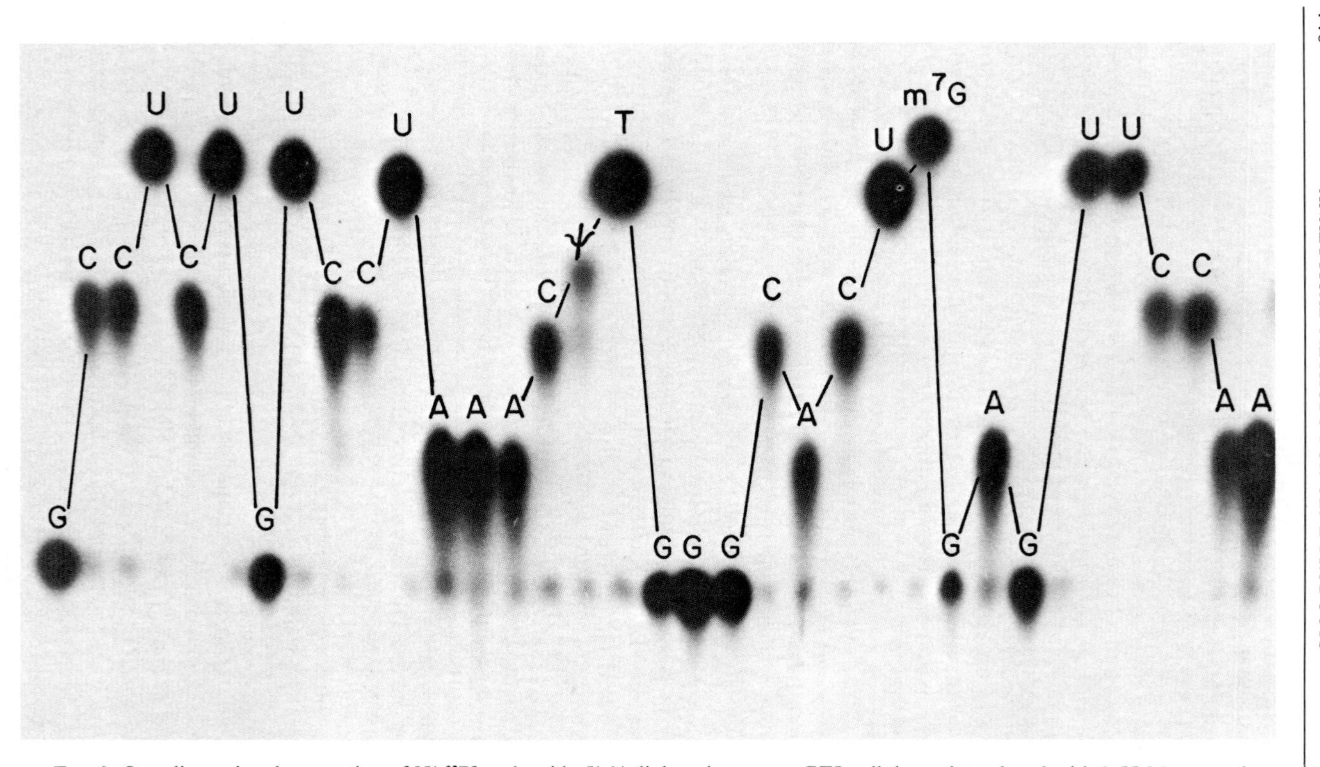

FIG. 2. One-dimensional separation of [5′-^{32}P]nucleoside 5′,3′-diphosphates on a PEI-cellulose plate eluted with 0.55 *M* ammonium sulfate. These residues are from bands 4–37 of the ladder shown in Fig. 1 and correspond to nucleotides 38–71 in spinach chloroplast $tRNA_f^{Met}$.

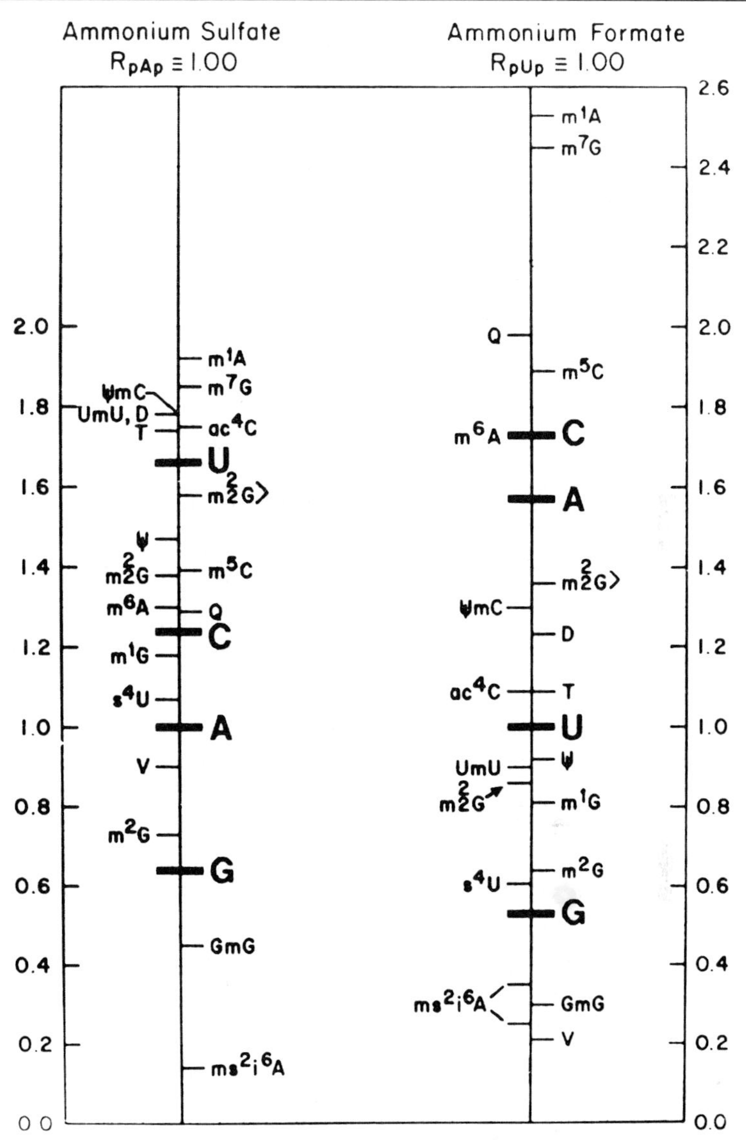

FIG. 3. Relative R_f values of nucleoside 5′,3′-diphosphates and ribose-methylated dinucleoside triphosphates in the ammonium sulfate and ammonium formate solvent systems.[12] V, Q, and ms^2i^6A denote nucleoside 5′,3′-diphosphates of uridine-5-oxyacetic acid, 7-(4,5-*cis*-dihydroxy-1-cyclopenten-3-ylaminomethyl)-7-deazaguanosine, and 2-methylthio-N^6-isopentenyladenosine, respectively. From R. C. Gupta and K. Randerath, *Nucleic Acids Res.* **6,** 3443 (1979).

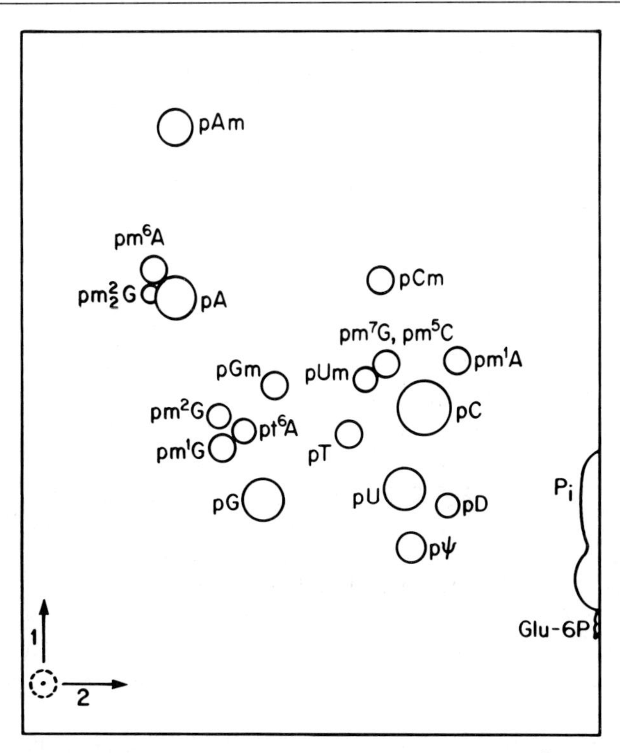

FIG. 4. Schematic diagram showing the mobilities of nucleoside [5′-^{32}P]monophosphates upon two-dimensional thin-layer chromatography on cellulose plates. First dimension (solvent a) isobutyric acid–concentrated NH_4OH–H_2O (66 : 1 : 33, v/v/v); second dimension (solvent c) 0.1 *M* sodium phosphate, pH 6.8–ammonium sulfate–*n*-propanol (100 : 60 : 2, v/w/v). P_i is inorganic phosphate, and Glu-6P is glucose 6-phosphate. From M. Silberklang, A. M. Gillum and U. L. RajBhandary, this series, Vol. 59, p. 58.

identify the terminal nucleotides of a tRNA. The tRNA is labeled at its 5′ end by treatment with calf intestine alkaline phosphatase and end-labeled with [γ-^{32}P]ATP and polynucleotide kinase, digested with RNase P_1 and chromatographed as above, to identify the 5′-terminal nucleotide. The 3′-terminal nucleotide is determined by 3′-end-labeling the tRNA with [5′-^{32}P]pCp and RNA ligase. The end-labeled molecule is treated with "digestion mix" (see above) and chromatographed as above using Ap, Up, Cp, and Gp standards as UV markers.

RNA Sequence Analysis Using Mobility Shifts and Base-Specific Ribonucleases

The 5′ and 3′ termini of the molecule, as well as confirmation of the entire sequence, can be accomplished by mobility-shift analysis and by base-specific cleavage of ^{32}P end-labeled tRNA and tRNA fragments. Both of these techniques have been described,[15] and therefore the details will not be repeated here. Purified RNA (40 pmol) is labeled at its 5′ end by removing the 5′-phosphate with calf intestine alkaline phosphatase using conditions previously described.[28] The sample is extracted three times with phenol–chloroform (1 : 1) followed by three extractions with diethyl ether and dried under vacuum. Kination is accomplished in 50 m*M* Tris-HCl (pH 8.0), 10 m*M* $MgCl_2$, 10 m*M* DTT, 10 μg of BSA per milliliter, and 10% glycerol. In general, a fivefold excess of [γ-^{32}P]ATP over the 5′ end of RNA is used together with 5 units of T4 polynucleotide kinase for 30 min at 37°. The reaction is quenched by addition of 0.5 ml of 20% sodium acetate (pH 5.2) and 1 OD_{260} unit of carrier RNA (wheat germ 5 S RNA), precipitated with two volumes of ethanol; the pellet is applied to a 20% polyacrylamide gel, which is run and eluted and the sample recovered as described above for the purification of tRNA by gel electrophoresis. The RNA can be labeled at its 3′ terminus by incubating 40 pmol of RNA in 50 m*M* *N*-2-hydroxyethylpiperazine-*N*′-2-ethanesulfonic acid (HEPES)–KOH (pH 7.5), 15 m*M* $MgCl_2$, 3.3 m*M* DTT, 10 μg of BSA per milliliter, 15% dimethyl sulfoxide, 500 pmol of ATP, 33 pmol of [5′-^{32}P]pCp, and 3 units of T4 RNA ligase at 4° overnight. Quenching the reaction and preparative gel electrophoresis are as described above for 5′-end labeling.

For mobility-shift analysis either 5′- or 3′-^{32}P end-labeled tRNA is digested with formamide (see above) for 30 min at 100°, dried overnight under vacuum, and dissolved in 2–3 μl of H_2O. This sample is then electrophoresed on a strip of cellulose acetate paper for 45 min at 5000 V using an electrophoresis buffer consisting of 5% acetic acid, 5 m*M* EDTA, and 7 *M* urea. The sample is then transferred to a DEAE-cellulose plate and chromatographed using 50 m*M* KOH homomix at 65° for 2 hr. The plate is then dried and autoradiographed. Examples of such autoradiographs are shown in Fig. 5. Separation in the first dimension at pH 3.5 is essentially by charge, whereas the second dimension at pH 8.3 separates oligonucleotides by size. This procedure can usually resolve 8–15 nucleotides from the labeled end. An autoradiograph of a mobility shift performed by two-dimensional gel electrophoresis on a sample prepared as described above is shown in Fig. 6. The characteristic shifts in mobility between oligonucleotides are essentially the same in both systems. Two-dimensional gel electrophoresis has the advantage of being able to resolve more nucleotides from the labeled end (15–40 nucleotides), but suffers the disadvantages of being a more tedious and difficult procedure requiring approximately 3 days to complete a single run.

[28] R. Pirtle, I. Pirtle, and M. Inouye, *Proc. Natl. Acad. Sci. U.S.A.* **75,** 2190 (1978).

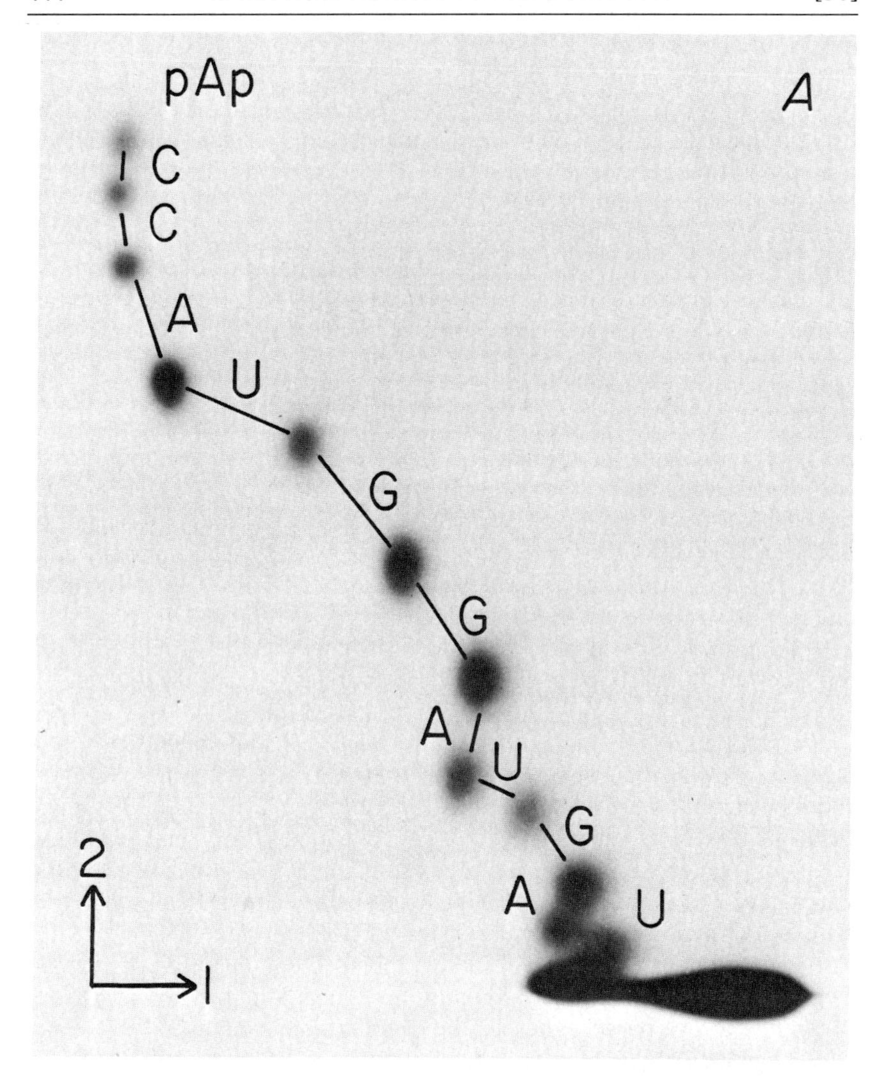

FIG. 5. Mobility-shift analysis by two-dimensional electrophoresis–homochromatography. The oligonucleotides for the analysis are (A) 3′-^{32}P-labeled spinach chloroplast $tRNA^{Met}_{m}$; (B) 3′-^{32}P-labeled spinach chloroplast $tRNA^{Thr}_{3}$; (C) 5′-^{32}P-labeled chloroplast $tRNA^{Met}_{m}$; (D) 5′-^{32}P-labeled spinach chloroplast $tRNA^{Thr}_{3}$; (E) a 5′-^{32}P-labeled oligonucleotide from bovine liver $tRNA^{Ser}_{CmCA}$; (F) a 5′-^{32}P-labeled oligonucleotide from spinach chloroplast $tRNA^{Ile}_{I}$.

Mobility-shift analysis is a most suitable procedure to determine the nucleotide sequence of a region of a molecule that shows compression effects in the formamide fragment analysis and/or RNA sequence gel procedures. It is therefore a most valuable technique to determine the

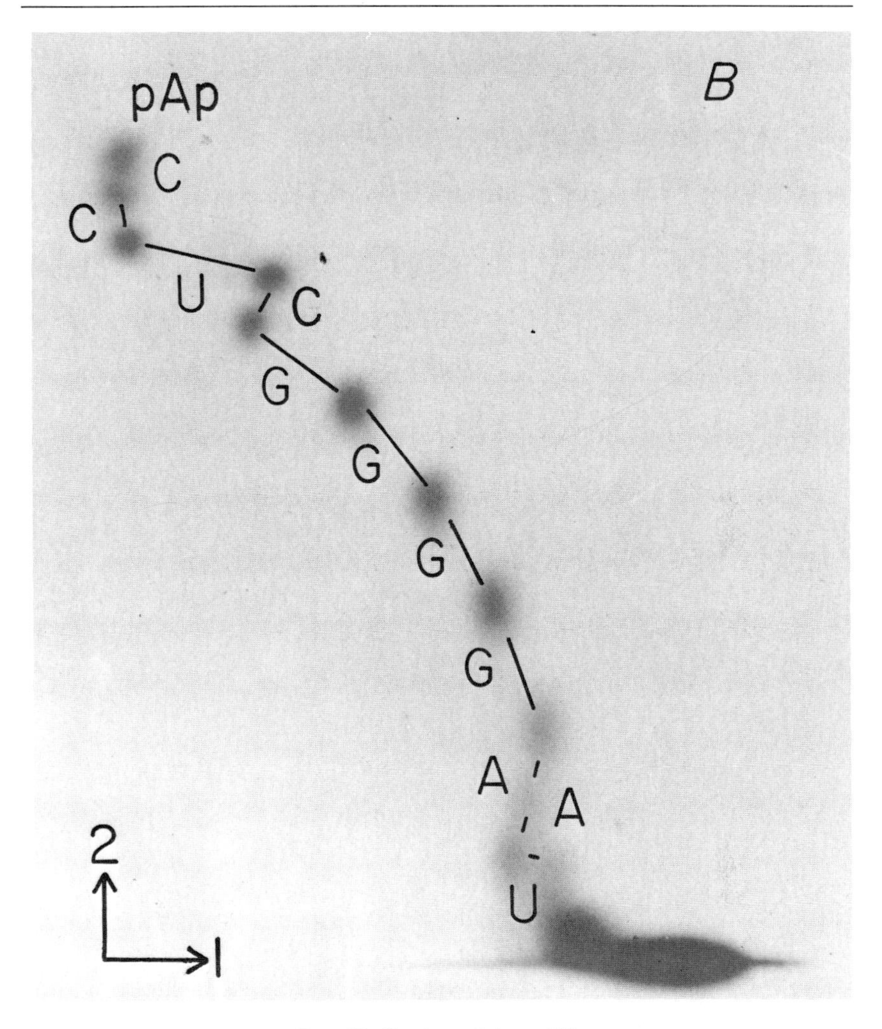

FIG. 5B. See legend on p. 444.

sequence of a "difficult" region of an RNA molecule. In addition to 5'- and 3'-^{32}P end-labeled tRNA there are two other relatively simple procedures to obtain end labeled tRNA fragments suitable for mobility-shift analysis. The first method is to use bands eluted from the formamide generated ladder. When mobility shift analysis is to be performed on bands from the ladder, the formamide fragment analysis can be carried out on a preparative scale. This involves formamide digesting 1–3 μg of tRNA. Sometimes the digest is carried out for a longer time so that more fragments are available for labeling. Occasionally bands from a for-

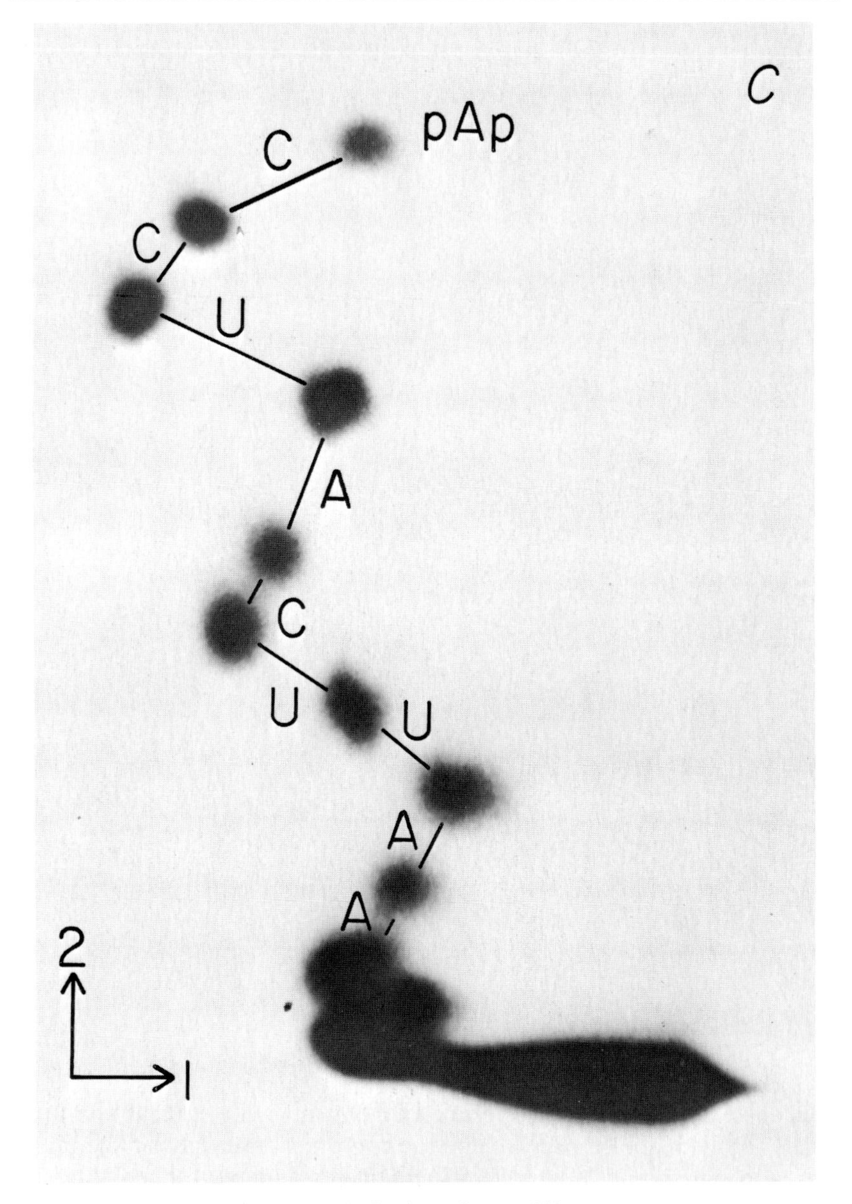

FIG. 5C. See legend on p. 444.

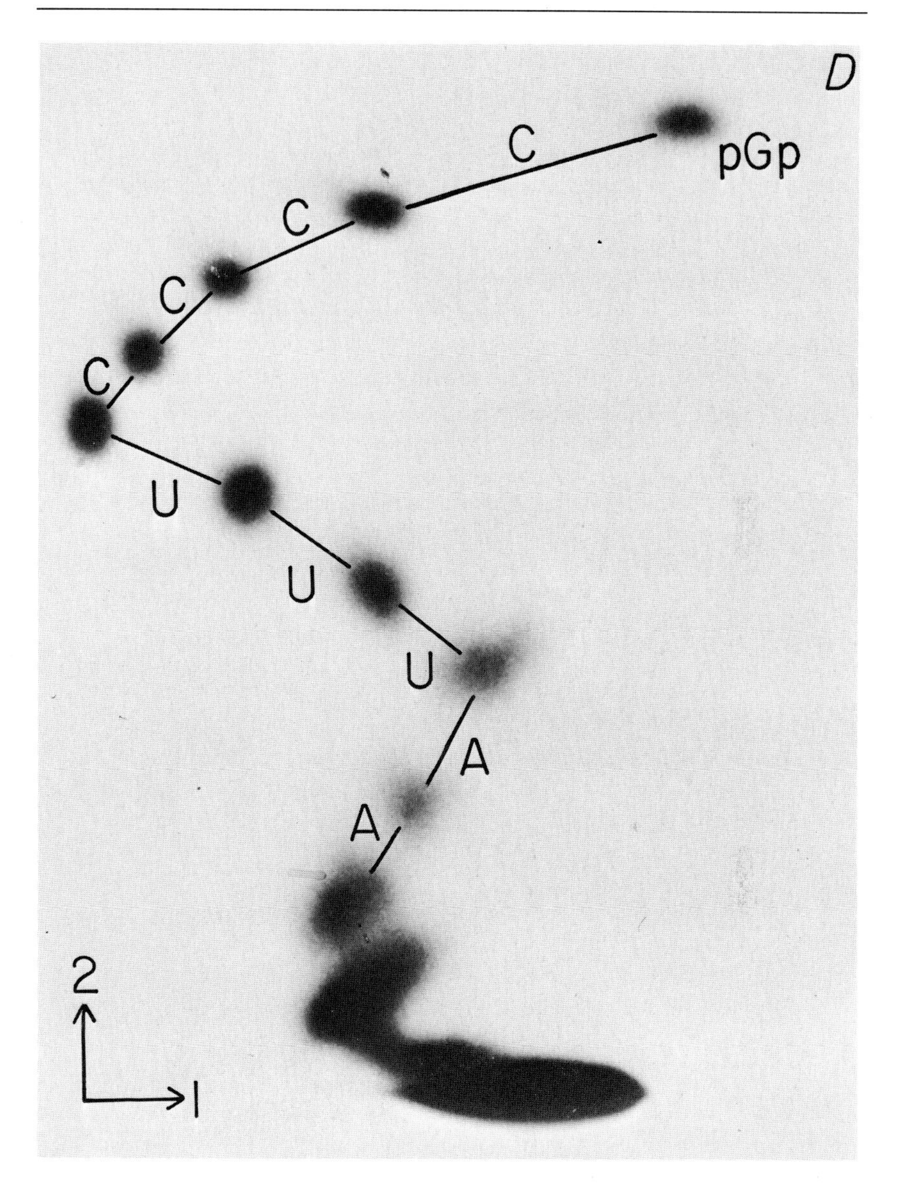

FIG. 5D. See legend on p. 444.

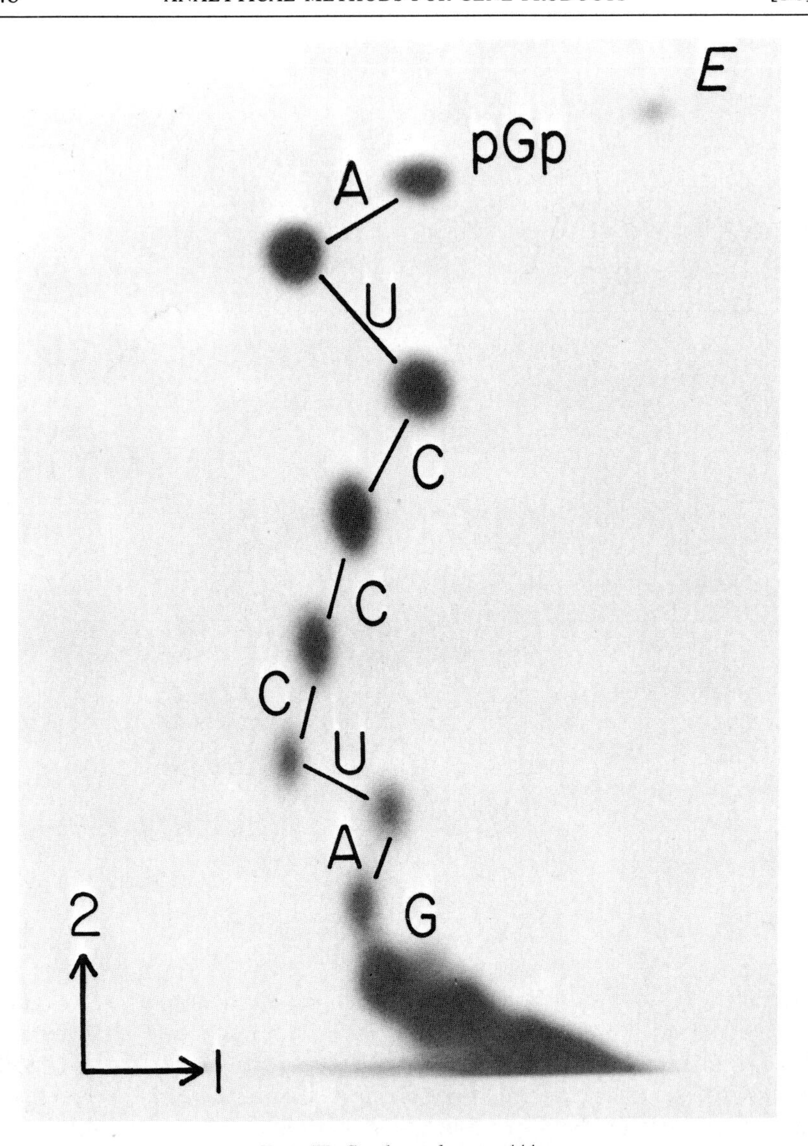

FIG. 5E. See legend on p. 444.

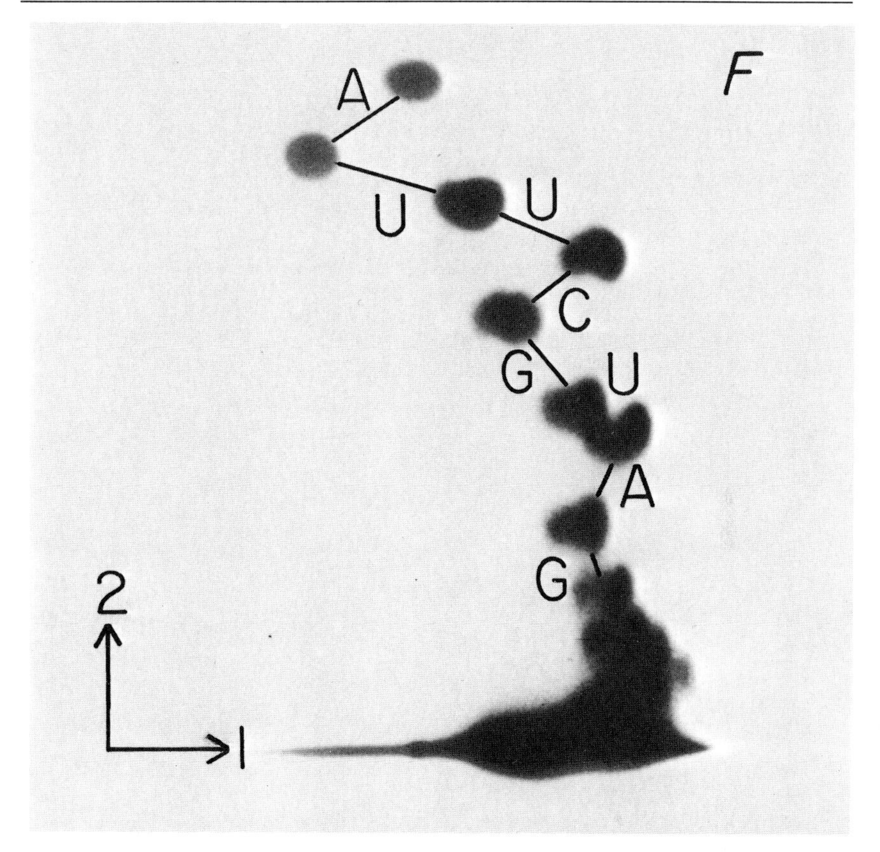

FIG. 5F. See legend on p. 444.

mamide ladder are not pure enough to give a clear and unambiguous mobility shift without further purification. This purification can be accomplished by gel electrophoresis. With this procedure many end-labeled RNA fragments can be readily obtained that are suitable for mobility-shift analysis. Another procedure to obtain such end-labeled RNA fragments is as a by-product of the 5′- or 3′-end-labeling of RNA. Apparently in the course of the end-labeling reaction some of the RNA is partially cleaved and some of these cleavage products are suitable substrates for the end-labeling reaction. These labeled fragments are readily obtained from the polyacrylamide gel used to purify the 5′- or 3′-end-labeled RNA.

The conditions used for partial cleavage of end-labeled tRNA with RNase T_1 (G residues), RNase U_2 (A residues), BC RNase (U + C resi-

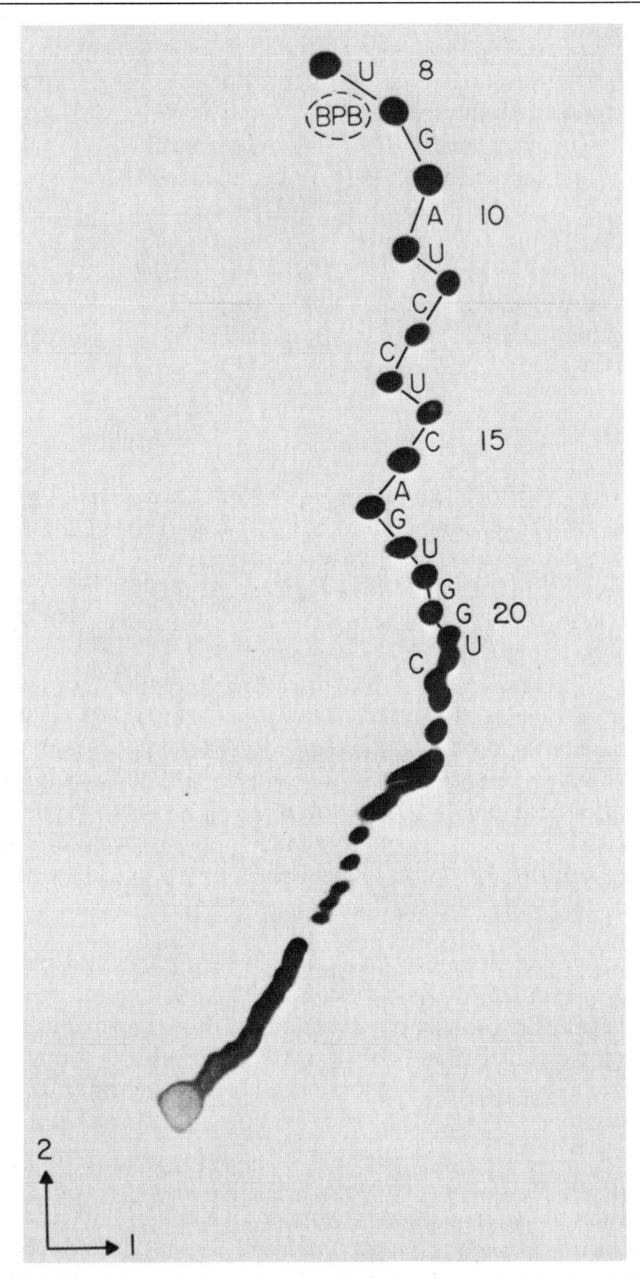

FIG. 6. Mobility-shift analysis by the two-dimensional gel vectoring method. A 5′-^{32}P-labeled fragment of bovine liver $tRNA^{Ser}_{CmCA}$ produced during the end-labeling reaction was partially digested by treatment with formamide for 30 min at 100° and subjected to electro-

dues), and RNase *Phy*M (U + A residues) have been described.[4-8,14] As can be seen in Fig. 7, cleavage by these ribonucleases does not result in the appearance of bands corresponding to every residue of the molecule. This is due to inaccessability of regions of stable secondary structure and the presence of modified residues that are not substrates for the enzymes. It is therefore essential that one lane be included on the gel showing the position of every residue (a ladder). This is accomplished by digesting a dried sample of end-labeled RNA with formamide for 30 min at 100°. The reaction is terminated by chilling and adding 1 μl of 3% xylene cyanole and bromophenol blue. RNA sequence gels are useful in confirming the identity of most of the residues of the RNA molecule.

Comments

Although it is also important for mobility shifts and analysis by partial cleavage with base-specific ribonucleases, it is essential for formamide fragment analysis that the sample being sequenced is at least 95% pure. Contamination with other RNA species will appear as an unequal distribution of radioactivity in the bands of the ladder generated by formamide digestion. Indeed, one can sometimes see multiple superimposed ladders with impure RNA samples. Subsequent elution and digestion of end-labeled oligonucleotides from such a ladder will characteristically result in the appearance of spots corresponding to all four major nucleoside 5′,3′-diphosphates (pAp, pCp, pUp, and pGp) after PEI-cellulose chromatography. The same results will also be observed if the RNA is overdigested with formamide, resulting in multiple "hits" per molecule. An additional indication that an RNA molecule has been overdigested is that the bands at the 5′ end of the ladder are lighter than the bands at the 3′ end of the ladder.

These procedures should be suitable for sequencing an RNA molecule up to 150–200 nucleotides in length. For RNA molecules much larger than this, it may be prudent to explore alternative sequencing procedures.

---

phoresis at 4° on a 10% polyacrylamide gel (pH 3.5) at 300 V until the bromophenol blue dye marker (BPB) traveled 80% of the length of the gel. A strip containing the radioactive material was cut from the gel and soaked in Tris-borate buffer at pH 7.6.[4] This strip was then placed at 90° between glass plates, and a 20% polyacrylamide gel (pH 7.6) without urea was poured until the strip was just covered. The second-dimension gel (20% polyacrylamide containing 7 *M* urea) was then poured on top of the plug and subjected to electrophoreses at 600 V until the BPB dye marker migrated 80% of the length of the gel. From A. Diamond, B. Dudock, and D. Hatfield, *Cell* **25,** 497 (1981).

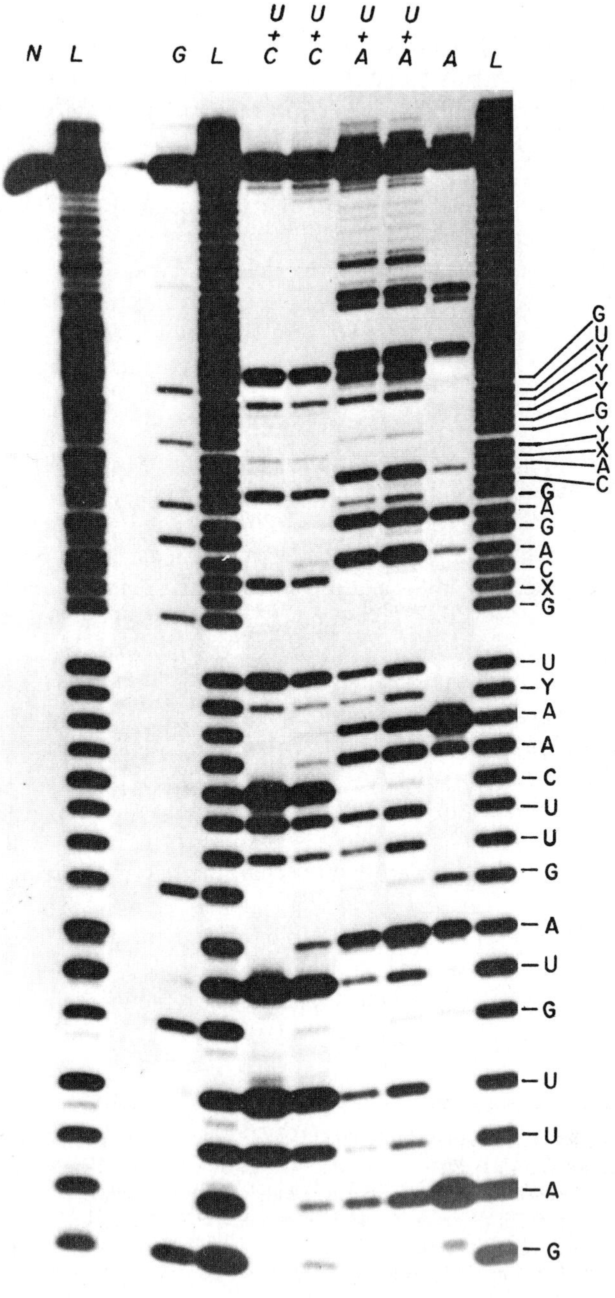

N L G L U+C U+C U+A U+A A L

G
U
Y
Y
Y
G
Y
X
A
C
G
A
G
A
C
X
G
U
Y
A
A
C
U
U
G
A
U
G
U
U
A
G

Acknowledgments

The authors would like to express their gratitude to M. Kashdan, M. Francis, J. W. Straus, K. Kelly, and R. Pirtle for the application of these techniques in our laboratory, for the use of their unpublished data, and for helpful discussions. This work was supported by Grants GM-25254 from the National Institutes of Health and PCM-7922751 from the National Science Foundation.

FIG. 7. An RNA sequence gel of oligonucleotides from a partial digestion of a 5′-^{32}P-labeled spinach chloroplast tRNA. The base specific cleavages are described in the text. The "ladder" on the left was produced by partial alkaline digestion, and the other two "ladders" were obtained by partial formamide digestions. Formamide digestion produces a clearer "ladder" for nucleotide sequence analysis. N, untreated control. The products of these digestions were fractionated by electrophoresis on a 20% polyacrylamide gel (40 × 33 × 0.05 cm).

Section VI

Mutagenesis: *In Vitro* and *in Vivo*

[31] Directed Mutagenesis with Sodium Bisulfite

By DAVID SHORTLE and DAVID BOTSTEIN

With the development of recombinant DNA technology and the many analytic techniques based on restriction endonucleases, it has become relatively routine to isolate a segment of genomic DNA carrying a gene of interest, to define the limits of the gene within this DNA segment, and then to determine the gene's complete nucleotide sequence. In this way, a number of structural questions can be directly answered. To extend the analysis of a gene further, particularly with regard to functional and regulatory phenomena, requires the isolation and systematic study of a sizable collection of mutant alleles of the gene. In the initial stages of such a mutational analysis, deletion and insertion mutations are often useful to identify important functional elements of the gene, particularly those at the 5′ and 3′ ends. When base substitution mutations are required to extend the analysis, directed mutagenesis with sodium bisulfite can provide an *in vitro* method for efficiently inducing C to T transition mutations at sites in a DNA molecule specified in advance by the experimenter.

Principles

The mutagen sodium bisulfite catalyzes the deamination of cytosine to form uracil under mild conditions of temperature and pH.[1] Cytosine residues in single-stranded DNA react at nearly the same rate as the mononucleotide. However, because of the stereochemistry involved in the bisulfite ion's attack on the cytosine ring, residues embedded within the Watson–Crick helix of double-stranded DNA are essentially unreactive. From the data available, the rate of cytosine deamination in duplex DNA appears to be less than 0.1% of the rate in single-stranded DNA.[1,2] Consequently, sodium bisulfite is, in effect, a single-strand specific mutagen; and a particular nucleotide sequence can be "targeted" for bisulfite mutagenesis by exposing it in a stretch of single-stranded DNA.

Segments of a circular, duplex DNA can be converted to a single-stranded, unpaired form in one of two ways. One approach, described in detail below, is to introduce a single nick at a specific site with an endonu-

[1] H. Hayatsu, *Prog. Nucleic Acid Res. Mol. Biol.* **16,** 75 (1976).
[2] D. Shortle and D. Nathans, *Proc. Natl. Acad. Sci. U.S.A.* **75,** 2170 (1978).

ISBN 0-12-182000-9

clease and then to convert the nick into a short gap by exonucleolytic removal of a limited number of nucleotides.[2-4] In addition to serving as targets for bisulfite mutagenesis, short single-stranded gaps can be used as specific sites for construction of small deletions with S1 nuclease[5] or as sites for mutagenesis by nucleotide misincorporation.[6] Alternatively, single-stranded components of a circular DNA can be annealed to generate a duplex molecule in which a specific segment of one strand remains unpaired. Examples of this second approach that have been used to construct unique targets for bisulfite mutagenesis are (*a*) reannealing of separated strands of full-length linear plasmid DNA with separated strands of a large restriction fragment[7]; (*b*) annealing of plus strand DNA of an M13 phage carrying a DNA fragment cloned into a unique restriction site with denatured linears generated by cleavage of RFI DNA from the parental M13 (without the insert) at the same restriction site (William Folk, personal communication); (*c*) construction of a unique deletion loop by heteroduplex formation between wild-type DNA and DNA from a deletion mutant of known sequence[7a,7b]; and (*d*) formation of a displacement loop, or D loop, by annealing a unique single-stranded restriction fragment to a covalently closed circular DNA (Maria Jason and Paul Schimmel, personal communication). It should be noted that, when the resultant single-stranded region is relatively long, care must be taken to reduce the level of bisulfite mutagenesis in order to avoid multiple, widely separated C to T mutations.

Nicking Reactions

Restriction Endonuclease plus Ethidium Bromide

When incubated with duplex DNA containing one or more restriction sites, type II restriction endonucleases cleave at, or near, their recognition sequences by generating two nicks, one in each strand. With some type II enzymes, the cleavage reaction on negatively supercoiled circular

[3] D. Shortle, D. Koshland, G. M. Weinstock, and D. Botstein, *Proc. Natl. Acad. Sci. U.S.A.* **77,** 5375 (1980).

[4] D. Shortle and D. Botstein, *in* "Molecular and Cellular Mechanisms of Mutagenesis" (J. F. Lemontt and W. M. Generosa, eds.), p. 147. Plenum, New York, 1982.

[5] D. Shortle, J. Pipas, S. Lazarowitz, D. DiMaio, and D. Nathans, *in* "Genetic Engineering" (J. K. Setlow and A. Hollaender, eds.), Vol. 1, p. 73. Plenum, New York, 1979.

[6] D. Shortle, P. Grisafi, S. J. Benkovic, and D. Botstein, *Proc. Natl. Acad. Sci. U.S.A.* **79,** 1588 (1982).

[7] P. E. Giza, D. M. Schmit, and B. L. Murr, *Gene* **15,** 331 (1981).

[7a] D. Kalderon, B. A. Oostra, B. K. Ely, and A. Smith, *Nucleic Acids Res.* **17,** 5161 (1982).

[7b] K. W. C. Peden and D. Nathans, *Proc. Natl. Acad. Sci. U.S.A.* **79,** 7214 (1982).

DNA can be inhibited after only one nick has been induced by including in the reaction mixture the intercalating agent ethidium bromide.[2,8] Using purified supercoiled DNA and ethidium bromide at an optimal concentration, it is sometimes possible to nick 50–90% of the input DNA, yielding open circular molecules with a single nick at the restriction enzyme's normal cleavage site. Since open circular DNA binds a greater amount of ethidium bromide than does the negatively supercoiled form, inhibition of double-strand cleavage presumably results from the increased binding of ethidium bromide to the restriction site after the enzyme has made the first single-strand break.

Procedure. The concentration of ethidium bromide that will stimulate a maximal level of nicking with a given restriction enzyme must first be determined by titrating the amount of ethidium bromide (over a range from 20 to 200 μg/ml) added to a series of 10-μl reactions containing 0.5–1.0 μg of circular DNA in the enzyme's standard buffer. After adding an amount of restriction enzyme sufficient to cleave completely the DNA in 30–60 min in the absence of ethidium bromide, the reaction mixtures are incubated at room temperature for 2–4 hr; then the reaction is stopped by addition of 10 μl of 10% sucrose–25 m*M* EDTA, pH 8.0–0.4% sodium dodecyl sulfate (SDS)–bromophenol blue. The relative rate of nicking versus double-strand cleavage is estimated by electrophoresis of this reaction mixture on an agarose gel system that provides efficient separation of open circular DNA from closed circular and linear DNA forms. Figure 1 illustrates the results of titration of the ethidium bromide concentration on the cleavage of pBR322 by *Cla*I. Above a concentration of 75–100 μg/ml, the ratio of open circular to linear forms does not change significantly, although the overall reaction proceeds more slowly. This titration behavior is typical for enzymes that display the nicking reaction in the presence of ethidium bromide; for those enzymes that do not, little or no open circular DNA is generated at any ethidium bromide concentration.

Once the optimal conditions are determined, a large-scale reaction is carried out to prepare the necessary quantity of nicked circular DNA. It is important that the concentration of input DNA be kept close to the value used in the titration. To terminate the nicking reaction and remove the ethidium bromide and restriction enzyme, the mixture is made 25 m*M* EDTA, pH 8.0–0.25 *M* NaCl and extracted with phenol, followed by ethanol precipitation.

Comments. Two parameters that may slightly increase the efficiency of nicking are lower temperature and higher salt concentration. In the

[8] R. C. Parker, R. M. Watson, and J. Vinograd, *Proc. Natl. Acad. Sci. U.S.A.* **74,** 851 (1977).

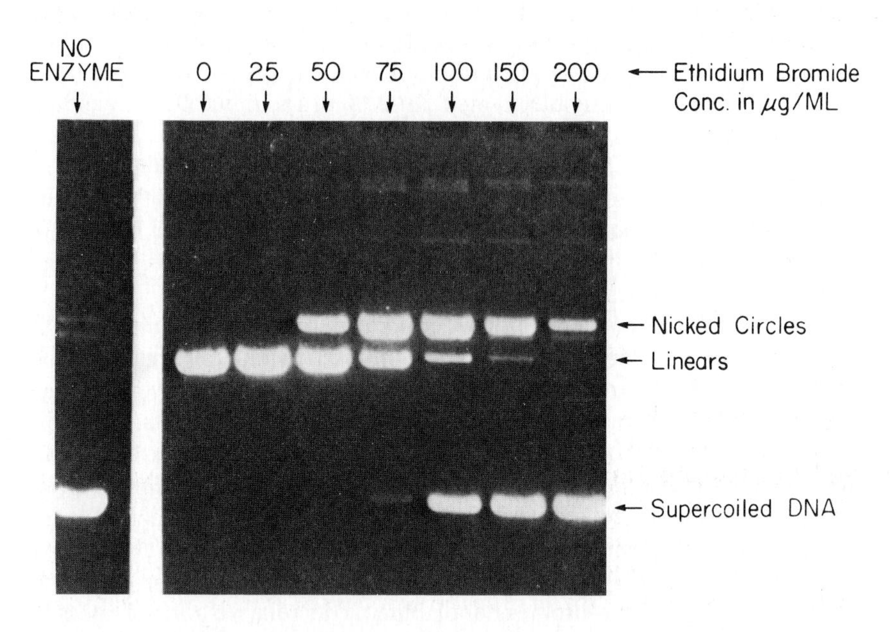

FIG. 1. Assay of the *Cla*I plus ethidium bromide-catalyzed nicking of covalently closed circular pBR322 DNA.

authors' experience, however, at least 5–10% of the input DNA is linearized under the best of circumstances. A few of the restriction enzymes that display an efficient nicking reaction are *Bgl*I, *Hpa*II, *Eco*RI, *Hin*dIII, *Bam*HI, *Cla*I, *Hinc*II, and *Bst*NI. However, with the conditions described above, little or no nicking was observed with the enzymes *Kpn*I, *Pvu*I, *Pvu*II, *Sal*I, or *Taq*I. With restriction enzymes that cleave at more than one site on a circular DNA, nicking may occur preferentially at a subset of sites.[9] To evaluate the extent of possible deviation from randomness, restriction enzyme-generated nicks can be labeled by carrying out a limited nick translation with an [α-^{32}P]dNTP, and the labeled sites can be localized by cleavage with one or more restriction enzymes, followed by autoradiography of gel-separated fragments.[3] In any event, a mutation

[9] K. W. C. Peden, J. M. Pipas, S. Pearson-White, and D. Nathans, *Science* **209,** 1392 (1980).

induced in one of the recognition sites of a multicut enzyme can be readily mapped by digestion with the cognate enzyme and identification of the two missing bands.

DNase I plus Ethidium Bromide

When the objective is to obtain mutations anywhere within a circular DNA, a simple approach is to generate a single bisulfite-sensitive site per molecule by nicking with DNase I in the presence of ethidium bromide.[10] Since this reaction induces one nick essentially at random per DNA circle, the percentage of molecules that acquire a nick in a given DNA segment will be equal to the percentage length of the DNA circle comprised by that segment. In principle, it should be possible to induce a C to T mutation at every cytosine residue without the limitation of mutational hot spots observed with other methods of random mutagenesis. Obviously, this approach will be most useful when an easily scorable phenotype is available to detect mutants of interest.

Procedure. To a solution of 50 m*M* Tris-HCl, pH 7.2–5 m*M* $MgCl_2$–0.01% gelatin–100 μg/ml ethidium bromide–50–100 μg/ml purified superhelical DNA is added an amount of pancreatic DNase I (typically 20–200 ng/ml) sufficient to convert 95–99% of the input DNA to an open circular form upon incubation at room temperature for 60 min. The minimal concentration of DNase I should be determined by titration with a series of small reactions and assaying the extent of nicking by agarose gel electrophoresis, as described earlier. For large-scale preparations, the reaction is stopped by adding EDTA to a final concentration of 20 m*M*, phenol extraction, and ethanol precipitation.

Comments. If a large excess of DNase I is used in this reaction, more than one nick will be induced per DNA circle. One semiquantitative method for assessing the average number of nicks per molecule is to denature the open circular DNA recovered from a reaction and then determine the ratio of single-stranded circles to single-stranded linears, either by gel electrophoresis or by counting these two forms by electron microscopy.[11] Finally, it should be noted that a small fraction of input DNA is linearized in this reaction. Presumably, these molecules have undergone a double-strand cleavage by DNase I in a "single hit" event[12]; therefore, their formation is not indicative of a high level of multiple nicks per molecule.

[10] L. Greenfield, L. Simpson, and D. Kaplan, *Biochim. Biophys. Acta* **407**, 365 (1975).
[11] R. W. Davis, M. N. Simon, and N. Davidson, this series, Vol. 21, p. 413.
[12] E. Melgar and D. A. Goldthwait, *J. Biol. Chem.* **243**, 4409 (1969).

Segment-Specific Nicking Procedure

Single nicks can be induced in defined segments of a circular DNA molecule with a two step, segment-specific nicking procedure.[3] In the first step, the *recA* protein of *E. coli* is used to catalyze the annealing of a unique, single-stranded DNA fragment to its complementary sequence on a covalently closed circular DNA.[13,14] The annealed fragment displaces one strand of the circular DNA, forming a single-stranded displacement loop, or D loop. Since one negative superhelical turn is removed for every 10 nucleotides displaced, the D loop is quite stable under the conditions of reaction. In the second step, a small amount of the single-strand specific S1 nuclease is used to nick the displaced strand, resulting in rapid breakdown of the D loop by spontaneous displacement of the fragment and termination of S1 nuclease action. The final product is an open circular DNA molecule with a nick located within the segment defined by the single-stranded fragment.

The reaction in which specific D loops are generated to provide a target for S1 nuclease nicking presents two nontrivial experimental tasks. First of all, a moderate quantity (0.5 to 2 μg) of a unique restriction fragment must be obtained in a single-stranded form. One of several possible solutions to this problem is described below. Second, *recA* protein catalyzed D-loop formation is an intrinsically complex reaction, one in which the stoichiometric ratios of the reactants (circular DNA, single-stranded fragment, and *recA* protein) must be carefully controlled. For a detailed discussion of the kinetics and probable mechanism of this reaction, the references cited in footnotes 13 and 14 can—and should—be consulted.

Exonuclease III Digestion of Isolated Restriction Fragments. A purified duplex DNA fragment can be converted to a heterogeneous mixture of single-stranded subfragments by limited digestion with exonuclease III. In this reaction, 1–2 μg of a restriction fragment is dissolved in 25 μl of 50 m*M* Tris-HCl, pH 8.0–0.5 m*M* $MgCl_2$–1 m*M* 2-mercaptoethanol–100 μg of gelatin per milliliter. After a 5-min preincubation of this solution at 45°, approximately 5 units of exonuclease III (New England BioLabs) is added and incubation is continued at 45° for 1 min per 70 base pairs of fragment length. The reaction is then terminated by adding 25 μl of 0.5 *M* NaCl–10 m*M* EDTA, pH 8.0, heating at 65° for 15 min, followed by phenol extraction and ethanol precipitation with 4 μg of carrier tRNA. The pellet is

[13] T. Shibata, C. DasGupta, R. P. Cunningham, and C. M. Radding, *Proc. Natl. Acad. Sci. U.S.A.* **76,** 1638 (1979).

[14] K. McEntee, G. M. Weinstock, and I. R. Lehman, *Proc. Natl. Acad. Sci. U.S.A.* **76,** 2615 (1979).

dissolved in a small volume of 2 m*M* Tris-HCl, pH 8.0–0.2 m*M* EDTA, and the DNA concentration is calculated on the assumption of 50% hydrolysis of the restriction fragment. Prior to use in a D-loop reaction, this DNA mixture is heated by immersion in a boiling water bath for 30 sec.

Although exonuclease III is a double-strand specific exonuclease that degrades each strand in the 5′ to 3′ direction, it will slowly hydrolyze single-stranded DNA.[15] Therefore, it may be advisable to confirm that hydrolysis is not proceeding significantly beyond 50% by monitoring the release of acid-soluble counts from a radiolabeled fragment.

Alternatively, one specific strand can be completely degraded with exonuclease III by first incorporating an α-thiophosphate nucleotide (which is resistant to hydrolysis by this enzyme) onto one end of the fragment in a reaction with DNA polymerase.[16] With some restriction fragments, the two DNA strands can be separated electrophoretically and recovered from the gel matrix.[17]

recA Protein Catalyzed D-Loop Reaction. The stoichiometry of substrates must be controlled if this reaction is to be used reproducibly. One way to simplify this end is to hold constant the concentrations of circular DNA and single-stranded fragment and to make adjustments primarily in the *recA* protein concentration. For example, in a 30-μl volume of 20 m*M* Tris-HCl, pH 8.0–10 m*M* $MgCl_2$–1 m*M* dithiothreitol–1.3 m*M* ATP, efficient D-loop formation occurs between 150 ng of pBR322 and 30–50 ng of single-stranded fragment when 20–100 pmol of *recA* protein are added. Incubations are carried out in polypropylene tubes at 37° for 30 min. The fraction of circular DNA molecules that acquire a D-loop structure (typically between 40 and 90%) can be assayed either by nitrocellulose filter binding[13] or by electrophoresis of the reaction products (after addition of excess EDTA) on a standard agarose gel[3] *without* ethidium bromide (which destabilizes D-loop structures). This second assay is based on the fact that the circular DNA loses one negative superhelical turn for every 10 nucleotides of fragment annealed. Consequently, the electrophoretic mobility of D-looped molecules will depend on the size of the annealed fragment relative to the size of the circular DNA. Above a fragment size that eliminates all superhelical turns, the mobility of the circular DNA will remain constant at a value near that of open circular DNA. However, with fragments below a certain size, the mobility will not be detectably different from unreacted circular DNA; and therefore, a different assay

[15] B. Weiss, *in* "The Enzymes" (P. D. Boyer, ed.), 3rd ed., Vol. 14. Academic Press, New York, 1981.

[16] S. D. Putney, S. J. Benkovic, and P. R. Schimmel, *Proc. Natl. Acad. Sci. U.S.A.* **78,** 7350 (1981).

[17] A. M. Maxam and W. Gilbert, this series, Vol. 65, p. 499.

may be necessary to detect D-loop formation (e.g., appearance of S1 nuclease-sensitive sites). The minimal fragment size that can form a stable D loop by *recA* protein catalysis with a small circular DNA such as pBR322 has not been determined.

The amount of *recA* protein required to catalyze D-loop formation is primarily a function of the amount of single-stranded DNA in the reaction mixture. Below a threshold *recA* protein level corresponding to approximately 1 enzyme molecule per 5 nucleotides, no reaction occurs.[13] Furthermore, at excessive levels of *recA* protein, the yield of D-looped molecules declines. Therefore, a titration series of reactions must usually be carried out to determine the enzyme's activity per nanogram of single-stranded DNA. Once this value has been established, it applies to all circular DNA molecules and homologous fragment substrates, provided that their ratios are held constant.

S1 Nuclease Reaction. In this reaction, the minimal amount of S1 nuclease necessary to convert 95–99% of D-looped molecules to a nicked circular form is used (see Fig. 2). The reaction is carried out by diluting 1 volume of the D-loop reaction mixture directly (i.e., without adding EDTA) into 9 volumes of solution (preheated at 45° for 10 min) containing 55 m*M* sodium cacodylate, pH 6.4–1.1 m*M* $ZnSO_4$–110 m*M* NaCl–0.44%

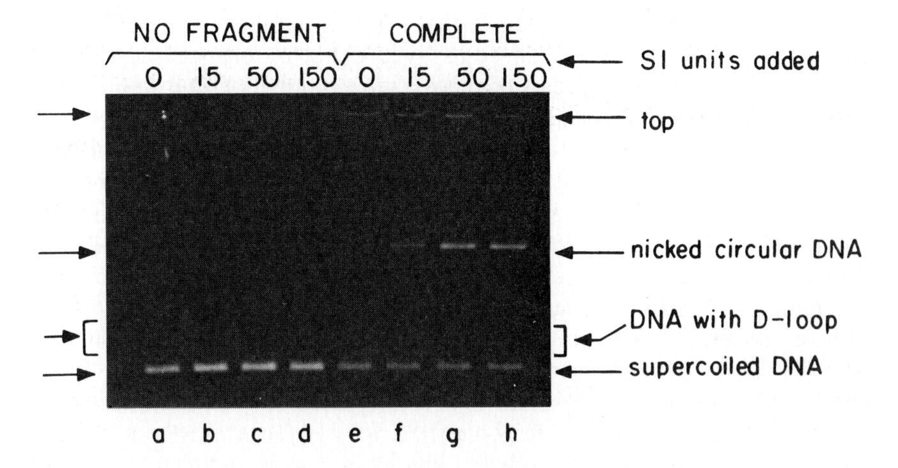

FIG. 2. Assay of the *recA* protein-catalyzed conversion of covalently closed circular pBR322 DNA to D-looped structures and their nicking by S1 nuclease. The single-stranded fragment, a 128 base-pair *Pvu*I-*Pst*I fragment digested with exonuclease III, was omitted from reactions a–d.

SDS–20–500 units of S1 nuclease (Miles Laboratories) per 100 ng of circular DNA. After incubation at 45° for 2 hr, S1 nuclease is inactivated by making the solution 150 m*M* Tris-HCl, pH 8.9–20 m*M* EDTA and incubating at 45° for an additional 30 min. The DNA is recovered by ethanol precipitation in the presence of carrier tRNA, followed by phenol extraction and a second ethanol precipitation. The extent of nicking of covalently closed circular DNA without D-loop structures should be assessed by running a control in which the single-stranded fragment has been omitted from the D-loop reaction (see Fig. 2). If the level of this side reaction is significant, the concentration of NaCl can be increased to 220 m*M*. If necessary, nicked circular molecules can be purified free of unreacted circular DNA by either agarose gel electrophoresis or acridine yellow polyacrylamide bead chromatography.[3]

Conversion of Nicks into Short Single-Stranded Gaps

After open circular DNA with a single, site-specific nick has been prepared, the next step is to convert the nick into a short gap by exonucleolytic removal of a small number of nucleotides. Of the exonucleases that initiate hydrolysis of one strand starting from a nick, two—*Micrococcus luteus* DNA polymerase I and T4 DNA polymerase—have properties that allow the extent of hydrolysis to be controlled. The *M. luteus* enzyme possesses both the 3′ → 5′- and 5′ → 3′-exonuclease functions typical of this class of bacterial polymerase.[18] However, both of these activities are relatively weak, particularly the 3′ → 5′-exonuclease. In the time-controlled reaction described below, an average of 5–6 nucleotides can be removed in predominantly the 5′ to 3′ direction from nicks,[19] although single-stranded gaps of 20 nucleotides or more in length may occasionally be generated. Alternatively, the exonuclease activity of T4 DNA polymerase can be used to remove nucleotides in the 3′ to 5′ direction from a nick.[20] This gapping reaction has the advantage that the extent of hydrolysis is controllable by addition of one or more deoxyribonucleoside triphosphates to the reaction mixture.[21]

In a 25-μl volume of 70 m*M* Tris-HCl, pH 8.0–7 m*M* $MgCl_2$–1 m*M* 2-mercaptoethanol, 0.5–5 μg of nicked circular DNA is incubated at room temperature for 60 min with 0.5 unit of *M. luteus* DNA polymerase I (Miles Laboratories) per microgram of DNA. (Note: Contrary to previous protocols,[2,3] the addition of one deoxyribonucleoside triphosphate to this

[18] L. K. Miller and R. D. Wells, *J. Biol. Chem.* **247,** 2667 (1972).

[19] D. Shortle, Ph.D. Dissertation, The Johns Hopkins University, Baltimore, Maryland, 1979.

[20] D. R. Rawlins and N. Muzyczka, *J. Virol.* **36,** 611 (1980).

[21] P. T. Englund, S. S. Price, and P. M. Weigel, this series, Vol. 29, p. 273.

reaction is not recommended.) The reaction is stopped by adding an equal volume of 0.5 *M* NaCl–25 m*M* EDTA, pH 8.0, followed by phenol extraction and ethanol precipitation. A simple assay to determine the yield of gapped molecules is to allow an aliquot of 100–200 ng of gapped DNA to react with T4 DNA ligase and ATP overnight at 0°. When electrophoresed on a standard agarose gel containing 0.5 μg of ethidium bromide per milliliter, molecules closed by ligation (i.e., ungapped) move as a band with a higher mobility than open circular molecules (i.e., gapped). Therefore, by running as controls on the same gel equal amounts of gapped, unligated DNA and nicked, ligated DNA, the percentage of total molecules which have acquired a single-stranded gap can be estimated.

Mutagenesis with Sodium Bisulfite

The reaction in which sodium bisulfite catalyzes the deamination of cytosine residues occurs in three steps. The bisulfite ion first adds to the double bond of cytosine via a covalent linkage to C-6. The resulting intermediate 5,6-dihydrocytosine 6-sulfonate then, in the rate-limiting step, undergoes hydrolytic loss of the amino group at C-4 to generate 5,6-dihydrouracil 6-sulfonate. Although this compound is stable at the pH of 6.0 used in the first two steps, it undergoes elimination of the bisulfite moiety at slightly alkaline pH to form uracil. At high concentrations (1–3 *M*) of sodium bisulfite, the initial kinetics of cytosine deamination are linear with time and approximately first order with respect to bisulfite concentration.[1] Consequently, the percentage of cytosine residues deaminated can most readily be controlled by adjustments in one or both of these two variables, with pH and temperature held constant.

Procedure. A solution of 4 *M* sodium bisulfite, pH 6.0, is prepared immediately prior to use by dissolving 156 mg of sodium bisulfite ($NaHSO_3$) plus 64 mg of sodium sulfite (Na_2SO_3) in 0.43 ml of distilled water. To obtain levels of cytosine deamination greater than 5–10%, three volumes of this solution are added to one volume of DNA solution (10–50 μg/ml in 15 m*M* NaCl–1.5 m*M* sodium citrate, pH 7.0) in a 6 × 50 mm or 7 × 70 mm glass tube, followed by 0.04 volume of a freshly prepared solution of 50 m*M* hydroquinone. After thorough mixing, 100 μl of paraffin oil are layered above the mixture, and the reaction is carried out at 37° in the dark. Under these conditions of 3 *M* sodium bisulfite, pH 6.0 at 37°, the rate of reaction is approximately 8–10% of susceptible cytosine residues deaminated per hour for the first 2–4 hr. To terminate the reaction and remove the sodium bisulfite, the mixture is transferred to a dialysis bag and dialyzed against the following sequence of five buffers: (*a*) 1000 volumes or more of 5 m*M* potassium phosphate, pH 6.8–0.5 m*M* hydroquinone at 0° (on ice) for 2 hr; (*b*) a repeat of (*a*); (*c*) 1000 volumes or more of 5 m*M* potassium phosphate, pH 6.8 without hydroquinone at 0° for 4 hr. To promote elimination of bisulfite ion from the 5,6-dihydrouracil 6-sulfonate present at this stage, dialysis is continued against (*d*) 1000 volumes or more of 0.2 *M* Tris-HCl, pH 9.2–50 m*M* NaCl–2 m*M* EDTA at 37° for 16–24 hr; (*e*) 1000 volumes of 10 m*M* Tris-HCl, pH 8.0–1 m*M* EDTA at 4° for 4 hr. At this point the DNA is recovered and concentrated by ethanol precipitation.

Bisulfite ion is readily oxidized by dissolved O_2 gas to generate free radicals.[1] Particularly at low concentrations of bisulfite and in the presence of divalent cations, this side reaction can cause extensive damage to DNA. During the incubation with 3 *M* bisulfite and the early stages of dialysis, a low concentration of hydroquinone serves as a free-radical scavenger. Nevertheless, degassing of the water used to make the dialysis solutions by 10 min of vigorous boiling is recommended as a precaution.

Recovery of Mutants

DNA molecules that have undergone cytosine deamination within a single-stranded region contain the nonphysiological base uracil. Before such molecules can give rise to pure mutant clones, the uracil base must function as template for incorporation of a complementary A residue during synthesis of the opposite strand, either *in vitro* or *in vivo* after DNA transformation. However, most if not all cells possess one or more repair enzymes that excise uracil from DNA, the enzyme DNA uracil *N*-glycosylase being the major repair activity in *E. coli*.[22] Not surprisingly, the transformation efficiency of plasmid DNA molecules with large single-stranded gaps drops appreciably after bisulfite treatment when an *Escherichia coli* strain with this enzyme (*ung*$^+$) is used as host.[7] In addition, single CG base-pair deletions have been identified at sites mutagenized with bisulfite among pBR322 mutants recovered on transformation of an *ung*$^+$ *E. coli* strain.[4] *In vitro* repair of mutagenized gaps before transformation by using DNA polymerase and T4 DNA ligase increased both the transformation efficiency and the efficiency of mutagenesis[7] and may reduce the incidence of deletions in the *E. coli* system.[4] Therefore, it is advisable to repair enzymically single-stranded gaps before transformation; and if using *E. coli*, an *ung*$^-$ strain should be considered as a recipient for transformation.

Finally, the authors would like to point out that the degree of single-strand specificity of bisulfite mutagenesis—under the conditions described in this chapter—is not yet known. At least one instance of a

[22] T. Lindahl, *Prog. Nucleic Acids Mol. Biol.* **22,** 135 (1979).

bisulfite-induced mutation occurring outside the single-stranded target site has been reported.[23] Therefore, the appearance of a mutant phenotype in an isolate recovered after directed mutagenesis with sodium bisulfite is only circumstantial evidence that the phenotypic change results from a mutation identified at the target site. Before this conclusion can be safely drawn, genetic mapping must be employed to confirm that the mutation responsible for the mutant phenotype maps to the expected region.

[23] R. Rothstein and R. Wu, *Gene* **15,** 167 (1981).

[32] Oligonucleotide-Directed Mutagenesis of DNA Fragments Cloned into M13 Vectors

By MARK J. ZOLLER and MICHAEL SMITH

The isolation and sequencing of genes has been a major focus of biological research for almost a decade. The emphasis has now turned toward identification of functional regions encoded within DNA sequences. A variety of *in vivo*[1] and *in vitro*[2-4] mutagenesis methods have been developed to identify regions of functional importance. These techniques provide useful information pertaining to the location and boundaries of a particular function. Once this has been established, a method is required to define the specific sequences involved by precisely directing mutations within a target site. Oligonucleotide-directed *in vitro* mutagenesis provides a means to alter a defined site within a region of cloned DNA.[5] This powerful technique has many far-reaching applications, from the definition of functional DNA sequences to the construction of new proteins.

[1] J. W. Drake and R. H. Baltz, *Annu. Rev. Biochem.* **45,** 11 (1976).
[2] W. Muller, H. Weber, F. Meyer, and C. Weissmann, *J. Mol. Biol.* **124,** 343 (1978).
[3] D. Shortle, D. DiMaio, and D. Nathans, *Annu. Rev. Genet.* **15,** 265 (1981).
[4] C. Weissmann, S. Nagata, T. Taniguichi, T. Weber, and F. Meyer, *in* "Genetic Engineering," (J. K. Setlow and A. Hollaender, eds.), Vol. 1, p. 133. 1979.
[5] M. Smith and S. Gillam, *in* "Genetic Engineering," (J. K. Setlow and A. Hollaender, eds.), Vol. 3, p. 1. 1981.

METHODS IN ENZYMOLOGY, VOL. 100

ISBN 0-12-182000-9

The method stemmed from the combination of a number of recent discoveries and observations about nucleic acids: (*a*) marker rescue of mutations in ϕX174 by restriction fragments[6,7]; (*b*) the stability of DNA duplexes containing mismatches[8-11]; and (*c*) the ability of *Escherichia coli* DNA polymerase to extend oligonucleotide primers hybridized to single-stranded DNA templates.[12,13] The basic principle involves the enzymic extension by *E. coli* DNA polymerase of an oligonucleotide primer hybridized to a single-stranded circular template. The oligonucleotide is completely complementary to a region of the template except for a mismatch that directs the mutation. Closed circular double-stranded molecules are formed by ligation of the newly synthesized strand with T4 DNA ligase. Upon transformation of cells with the *in vitro* synthesized closed circular DNA (CC-DNA), a population of mutant and wild-type molecules are obtained. Mutant molecules are distinguished from wild type by one of a number of screening procedures.

In their initial studies, Hutchison *et al.*[14] and Gillam and Smith[15] created a number of mutations in the single-stranded phage ϕX174. Razin *et al.* reported a similar experiment using ϕX174.[16] Wallace and co-workers applied the basic principle to create mutations in genes cloned into pBR322.[17,18] Wasylyk *et al.* demonstrated the first example of oligonucleotide mutagenesis of a cloned fragment in a vector derived from a single-stranded phage, fd.[19] More recently, phage M13 and M13 derived vectors have been used in a number of oligonucleotide mutagenesis ex-

[6] P. J. Weisbeek and J. H. van de Pol, *Biochim. Biophys. Acta* **224,** 328 (1970).

[7] C. A. Hutchison III and M. H. Edgell, *J. Virol.* **8,** 181 (1971).

[8] C. R. Astell and M. Smith, *J. Biol. Chem.* **246,** 1944 (1971).

[9] C. R. Astell and M. Smith, *Biochemistry* **11,** 4114 (1972).

[10] C. R. Astell, M. T. Doel, P. A. Jahnke, and M. Smith, *Biochemistry* **12,** 5068 (1973).

[11] S. Gillam, K. Waterman, and M. Smith, *Nucleic Acids Res.* **2,** 625 (1975).

[12] M. Goulian, A. Kornberg, and R. L. Sinsheimer, *Proc. Natl. Acad. Sci. U.S.A.* **58,** 2321 (1967).

[13] M. Goulian, S. H. Goulian, E. E. Codd, and A. Z. Blumenfield, *Biochemistry* **12,** 2893 (1973).

[14] C. A. Hutchison III, S. Phillips, M. H. Edgell, S. Gillam, P. A. Jahnke, and M. Smith, *J. Biol. Chem.* **253,** 6551 (1978).

[15] S. Gillam and M. Smith, *Gene* **8,** 81 (1979).

[16] A. Razin, T. Hirose, K. Itakura, and A. Riggs, *Proc. Natl. Acad. Sci. U.S.A.* **75,** 4268 (1978).

[17] R. B. Wallace, M. Schold, M. J. Johnson, P. Dembek, and K. Itakura, *Nucleic Acids Res.* **9,** 3647 (1981).

[18] R. B. Wallace, P. F. Johnson, S. Tanaka, M. Schold, K. Itakura, and J. Abelson, *Science* **209,** 1396 (1980).

[19] B. Wasylyk, R. Derbyshire, A. Guy, D. Molko, A. Roget, R. Teoule, and P. Chambon, *Proc. Natl. Acad. Sci. U.S.A.* **77,** 7024 (1980).

periments.[20-25] Vectors derived from single-stranded phage, such as M13 and fd, are more convenient for oligonucleotide-directed mutagenesis than double-stranded vectors, because isolation of pure single-stranded circular template DNA in these systems is quite simple.[26-28] However, the use of a double-stranded vector may be required if cloning of a particular fragment into M13 proves to be difficult or in the creation or destruction of a particular restriction site in the vector itself. Table I summarizes the targets and vehicles utilized in a number of recent mutagenesis experiments.

This chapter describes a mutagenesis procedure that is simple, versatile, and efficient. Three criteria guided the development of the methodology: (*a*) that a mutation could be produced at any position in a cloned fragment of known sequence; (*b*) that the efficiency of mutagenesis be sufficiently high to facilitate screening; and (*c*) that identification of a mutant could be made without biological selection or direct sequencing. As an example, a single G to A transition in the *MATa1* gene of the yeast *Saccharomyces cerevisiae* has been produced as part of a study into the role of an inframe UGA codon within the coding region of the gene. The chapter is divided into two parts. The first presents a step-by-step description of the mutagenesis experiment, from the design of the mutagenic oligonucleotide to the screening procedures and verification of the mutant by sequence determination. The second part presents a detailed protocol.

Experimental Rationale

The mating type of the yeast *S. cerevisiae* is determined by the specific gene, *a* or α, present at the *MAT* locus on chromosome III.[29,30] These

[20] P. D. Baas, W. R. Teertstra, A. D. M. van Mansfield, H. S. Janz, G. A. van der Marel, G. H. Veeneman, and J. H. van Boom, *J. Mol. Biol.* **152,** 615 (1981).

[21] G. F. M. Simons, G. H. Veeneman, R. N. H. Konigs, J. H. van Boom, and J. G. G. Schoenmakers, *Nucleic Acids Res.* **10,** 821 (1982).

[22] I. Kudo, M. Leineweber, and U. RajBhandary, *Proc. Natl. Acad. Sci. U.S.A.* **78,** 4753 (1981).

[23] C. G. Miyada, X. Soberon, K. Itakura, and G. Wilcox, *Gene* **17,** 167 (1982).

[24] C. Montell, E. F. Fisher, M. H. Caruthers, and A. J. Berk, *Nature (London)* **295,** 380 (1982).

[25] G. F. Temple, A. M. Dozy, K. L. Roy, and Y. W. Kan, *Nature (London)* **296,** 537 (1982).

[26] J. Messing, this series, Vol. 101 [2].

[27] G. Winter and S. Fields, *Nucleic Acids Res.* **8,** 1965 (1980).

[28] P. H. Schreier and R. A. Cortese, *J. Mol. Biol.* **129,** 169 (1979).

[29] R. K. Mortimer and D. C. Hawthorne, *in* "The Yeasts" (A. H. Rose and J. S. Harrison, eds.), Vol. 1, p. 385. Academic Press, New York, 1969.

[30] I. Herskowitz and Y. Oshima, *in* "The Molecular Biology of the Yeast *Saccharomyces*" (J. N. Strathern, E. Jones, and J. R. Broach, eds.), p. 181. Cold Spring Harbor Laboratory, Cold Spring Harbor, New York, 1982.

TABLE I
EXAMPLES OF OLIGONUCLEOTIDE-DIRECTED MUTAGENESIS EXPERIMENTS

Target	Alteration	Oligonucleotide length	Vehicle	Selection or screening procedure	% Mutants
ϕX174 gene *E*[a]	A → G	12	ϕX174	Biological	15
	G → A	12	ϕX174	Biological	15
ϕX174 gene *E*[b]	A → G	14	ϕX174	Biological	1.9
	A → G	17	ϕX174	Biological	13.9
ϕX174 gene *E*[c]	A → G	11	ϕX174	Biological	39
	G → A	10	ϕX174	Biological	23
gene *A*	T → G	10	ϕX174	Biological	22
	G → T	11	ϕX174	Biological	13
ϕX174 ribosome binding site[d]	del T	10	ϕX174	*In vitro* selection and DNA sequencing	100[n]
Yeast $tRNA^{Tyr}$ intervening sequence[e]	del 14 bases	21	pBR322	Hybridization	4
Chicken conalbumin promoter (TATA)[f]	T → G	11	fd103	New restriction site	4
Human β-globin[g]	T → A	19	pBR322	Hybridization	0.5
E. coli $tRNA^{Tyr}$[h]	*T* → A	10	M13mp3	Biological	7–10
ϕX174 origin[i]	*T* → C	16	ϕX174	Biological	—
M13 gene *IX*[j]	T → G	13	M13	Biological	1.9
E. coli araBAD operon[k]	del 3 bases	18	M13mp2	Hybridization	7.5
E. coli CRP-cAMP binding site[k]	del 3 bases	19	M13mp2	Hybridization	2.4
Human adenovirus splice junction[l]	T → G	12	M13Goril	*In vitro* selection and DNA sequencing	13[n]
Human $tRNA^{Lys}$[m]	AAA → TAG	15	M13mp7	*In vitro* selection and DNA sequencing	100[n]

[a] Hutchison *et al.*[14] [b] Razin *et al.*[16] [c] Gillam and Smith.[15] [d] Gillam *et al.*[40] [e] Wallace *et al.*[18] [f] Wasylyk *et al.*[19] [g] Wallace *et al.*[17] [h] Kudo *et al.*[22] [i] Baas *et al.*[20] [j] Simons *et al.*[21] [k] Miyada *et al.*[23] [l] C. Montell *et al.*[24] [m] Temple *et al.*[25] [n] Enrichment by multiple rounds of mutagenesis. See Gillam *et al.*[40]

genes are thought to code for regulatory proteins that control the expression of mating type specific genes that act in mating and sporulation.[31,32] Genetic analysis has revealed that *MATα* consists of two complementation groups (α1 and α2), and that *MATa* consists of only one (a1).[32,33] *MATα1* acts to turn on the expression of α specific genes. *MATα2* functions in the repression of a-specific genes and is also active in sporulation. *MATa1* is required for sporulation and for the maintenance of the a/α diploid cell phenotype, but functions only in the presence of α2. In the absence of α2, *a*-specific genes are constitutively expressed. Both *MATa* and *MATα* have been cloned[34,35] and sequenced.[36] Each codes for two transcripts, whose position and direction of synthesis have been physically mapped.[37] The putative protein sequences of a1, α1, and α2 can be predicted because in each case there is only one extended open reading frame. Inspection of these sequences revealed that the *MATa1* gene codes for a UGA stop codon 43 codons from the AUG initiator codon. The open reading frame continues for another 103 codons, terminating with UAA. *In vitro* constructed mutants on either side of this in-frame UGA codon resulted in the loss of a1 function.[38] This suggested that the UGA must be read-through. In order to understand the role of this in-frame UGA codon, two mutants have been constructed: (*a*) the UGA changed to UAA, a strong stop codon; (*b*) the UGA changed to UGG, coding for tryptophan. A 4.2 kb *Hin*dIII fragment containing the *MATa* gene was cloned into the vector M13mp5, derived from the single-stranded phage M13. Single-stranded circular DNA, containing the + strand of M13mp5 and the noncoding strand of the *MATa* gene, was isolated and used as template for oligonucleotide-directed mutagenesis.

Design of the Mutagenic Oligonucleotide

Once the experimental rationale and the desired changes have been formulated, the next step is the design of the oligonucleotide to direct the mutagenesis. Three factors should be considered in this regard: (*a*) the

[31] V. L. Mackay and T. R. Manney, *Genetics* **76,** 273 (1974).

[32] J. N. Strathern, J. Hicks, and I. Herskowitz, *J. Mol. Biol.* **147,** 357 (1981).

[33] G. F. Sprague, J. Rine, and I. Herskowitz, *J. Mol. Biol.* **153,** 323 (1981).

[34] J. B. Hicks, J. N. Strathern, and A. J. S. Klar, *Nature* (*London*) **282,** 478 (1979).

[35] K. A. Nasmyth and K. Tatchell, *Cell* **19,** 753 (1980).

[36] C. R. Astell, L. Ahlstrom-Jonasson, M. Smith, K. Tatchell, K. A. Nasmyth, and B. D. Hall, *Cell* **27,** 15 (1981).

[37] K. A. Nasmyth, K. Tatchell, B. D. Hall, C. R. Astell, and M. Smith, *Nature* (*London*) **289,** 244 (1981).

[38] K. Tatchell, K. A. Nasmyth, B. D. Hall, C. R. Astell, and M. Smith, *Cell* **27,** 25 (1981).

synthetic capability; (*b*) placement of the mismatch within the oligonucleotide; and (*c*) competing sites within the M13 vector or in the cloned fragment.

Studies by Gillam and Smith demonstrated the efficacy of short oligonucleotides 8–12 long to direct mutagenesis in ϕX174.[15,39,40] For mutagenesis of cloned fragments in M13, we have chosen to utilize oligonucleotides ranging in length from 14–21 nucleotides. Oligomers in this range serve as primers for DNA synthesis at room temperature and above. In addition, they are more likely to recognize the unique target in the M13-recombinant DNA than are shorter oligonucleotides. The synthesis of oligonucleotides 14–21 long is well within the capability of the chemical methods presently available.[41–46]

The oligonucleotide is designed so that the mismatch(es) is located near the middle of the molecule. This is especially important if the oligomer will be used as a probe to screen for the mutant. Placement of the mismatch in the middle yields the greatest binding differential between a perfectly matched duplex and a mismatched duplex.[47] A second consideration concerns protection of the mismatch from exonuclease activity of DNA polymerase. *Escherichia coli* DNA polymerase I contains intrinsic 5′ → 3′- and 3′ → 5′-exonuclease activities. The "large fragment" (Klenow derivative) lacks the 5′ → 3′-exonuclease activity, and therefore is used in oligonucleotide-directed mutagenesis to prevent correction of the mismatch. Placement of the mismatched nucleotide near the 3′ end might result in repair of the mismatch by the 3′ → 5′-exonuclease activity still present in DNA polymerase (large fragment). Gillam and Smith have demonstrated that three nucleotides following the mismatch are enough to protect against this problem.[15] The final step in designing the mutagenic oligonucleotide is to conduct a computer analysis on the DNA sequences of the M13 vector and the cloned fragment for regions of partial complementarity with the mutagenic oligonucleotide. The purpose of this step is

[39] S. Gillam and M. Smith, *Gene* **8,** 99 (1979).

[40] S. Gillam, C. R. Astell, and M. Smith, *Gene* **12,** 129 (1981).

[41] M. Edge, A. M. Greene, G. R. Heathcliffe, P. A. Meacock, W. Schuch, D. B. Scanlon, T. C. Atkinson, C. R. Newton, and A. F. Markham, *Nature* (*London*) **292,** 756 (1981).

[42] K. Itakura and A. Riggs, *Science* **209,** 1401 (1979).

[43] M. L. Duckworth, M. J. Gait, P. Goelet, G. F. Hong, M. Singh, and R. C. Titmas, *Nucleic Acids Res.* **9,** 1691 (1981).

[44] F. Chow, T. Kempe, and G. Palm, *Nucleic Acids Res.* **9,** 2807 (1981).

[45] G. Alvarado-Urbina, G. M. Sathe, W-C. Liu, M. F. Gillen, P. D. Duck, R. Bender, and K. K. Ogilvie, *Science* **214,** 270 (1981).

[46] M. D. Matteucci and M. H. Caruthers, *J. Am. Chem. Soc.* **103,** 3185 (1981).

[47] M. Smith, *in* "Methods of RNA and DNA Sequencing" (S. M. Weissmann, ed.). Praeger, New York, 1983 (in press).

```
                H   F   K   D   S   L   *   I   N
          5'-TTTCATTTCAAGGATAGCCTTTGAATCAATTTA-3'        coding strand
                                       Hinf I

mutagenic             5'-AAGGATAGCCTTTAAATC-3'           MS-1 (UAA stop)

oligonucleotides      5'-AAGGATAGCCTTTGGATC-3'           MS-2 (UGG TRP)
```

FIG. 1. The DNA sequence of the coding strand of *MATa1* in the region containing the in-frame UGA codon. Below are shown the sequences of two oligonucleotides synthesized to create point mutations. Each mutation destroys the *Hin*fI site in this region (underlined).

to identify competing sites from which priming might also occur. Such analysis might indicate the need to extend the oligomer, or whether one strand of the cloned fragment is a better template than the other. Generally, a competing target site with a predominance of matches with the 3′ end of the oligonucleotide will be a more efficient priming site than one that exhibits complementarity to the 5′ end of the oligomer.[47]

Figure 1 shows the DNA sequence of the coding strand of *MATa1* in the region of the UGA codon.[36] The octadecanucleotide 5′-AAGGA-

TABLE II
SEQUENCES IN *MATa* AND M13mp5 PARTIALLY COMPLEMENTARY TO OLIGONUCLEOTIDE MS-1[a,b]

Sequence	Position[c]	Matches/18
MATa		
5′-GAATTAAGGGATATATTA-3′	(1355–1338)	12
5′-GGAATTAAGGCTTTGCTT-3′	(1870–1853)	12
M13mp5		
5′-AATTAAAACGCGATATTT-3′	(339–356)	12
5′-CAATTAAAGGCTCCTTTT-3′	(1533–1550)	12
5′-GCTTTAATGAGGATCCAT-3′	(2210–2237)	12
5′-GATGTAAAAGGTACTGTT-3′	(4370–4387)	12
5′-TTTTTAATGGCGATGTTT-3′	(4962–4979)	12
5′-TGTTTTAGGGCTATCAGT-3′	(4975–4992)	12
5′-GCTTTACACTTTATGCTT-3′	(6141–6158)	12
5′-GATTGACATGCTAGTTTT-3′	(6839–6856)	12

[a] Computer search for DNA sequences in *MATa* 4.2 kb *Hin*dIII fragment (noncoding strand) and in M13mp5 (+ strand) that exhibit partial complementarity to oligonucleotide MS-1 (5′-AAGGATAGCCTTTAAATC-3′). Positions that form matched base pairs are underlined.

[b] Computer program for DNA sequence analysis was developed by A. Delaney and described in *Nucleic Acids Res.* **10,** 61 (1982).

[c] The sequences in *MATa* are numbered according to Astell *et al.*[36] The sequences in M13 are numbered according to P. M. G. F. van Wezenbeek, T. J. M. Hulsebos, and J. G. G. Schoenmakers, *Gene* **11,** 129 (1980).

TAGCCTTTAAATC-3′ (MS-1) was designed to change the UGA to UAA, a strong stop codon in yeast. The second oligonucleotide (MS-2) was designed to change the UGA to UGG, coding for tryptophan. Table II summarizes the results of a computer analysis for sequences in the + strand of the vector M13mp5 and *MATa* that are complementary to the oligonucleotide. A potential competing site is the sequence in M13 spanning nucleotides 1533–1550, which has nine consecutive matches with the oligonucleotide. Originally we synthesized a 10-long oligonucleotide that spanned this region. Although oligonucleotides 10–12 long have proved to be very efficient mutagens,[15] preliminary experiments indicated that the 10-mer primed from multiple sites in both the *MATa* insert and M13 vector. To avoid priming from these competing sites, the oligonucleotide was extended. The results of the computer analysis shown in Table II suggested that an 18-mer would prime specifically from the desired site in *MATa*. As demonstrated in a subsequent section, the 18-mer specifically primed from the desired site. This example demonstrated the importance of conducting a thorough computer search before synthesizing the oligonucleotide.

Synthesis of the Oligonucleotide

Short oligonucleotides can be synthesized *in vitro* by either enzymic or chemical means. The enzymic methods provide relatively low yields of final products and are dependent on slow empirical reactions.[48,49] The chemical procedures based on phosphotriester[41–43] or phosphite[44–46] chemistry are the most widely used. The trend has turned toward the development of solid-phase procedures that are amenable to either manual or automatic operation. The amount of oligonucleotide required for one mutagenesis experiment is small (less than 250 pmol) compared to the amount one obtains from the chemical procedures (5–50 nmol).

The octadecanucleotide 5′-AAGGATAGCCTTTAAATC-3′ for mutagenesis of *MATa1* was synthesized by the solid-phase phosphotriester procedure.[41] Purification was accomplished using ion-exchange and C_{18} reverse-phase HPLC.

Determination of the Sequence of the Oligonucleotide

Once the oligonucleotide has been synthesized, it should be sequenced before preceding with the mutagenesis experiment. Two meth-

[48] S. Gillam and M. Smith, this series, Vol. 65 [65].
[49] D. A. Hinton, C. A. Brennan, and R. I. Gumport, *Nucleic Acids Res.* **10**, 1877 (1982).

ods are available: (*a*) two-dimensional homochromotography[50]; (*b*) a modified Maxam and Gilbert procedure designed for oligonucleotides.[51,52] We have found the later procedure to be easier; however, the homochromatography method might be used in order to determine whether two different nucleotides were present at the same position, as would be the case in the deliberate synthesis of a mixture of oligonucleotides. The oligonucleotide is phosphorylated at the 5′ end with [γ-^{32}P]ATP and polynucleotide kinase[44] or at the 3′ end with terminal transferase,[53,54] then subjected to a modified Maxam and Gilbert procedure. Figure 2 shows an autoradiogram of a sequencing gel verifying the sequence of oligomer MS-1.

Preparation of Template DNA

There are a number of vectors available that are derived from the single-stranded phage M13[26,55] or fd.[19] The application of these phage vectors to oligonucleotide-directed site-specific mutagenesis provides for simple and rapid isolation of single-stranded template DNA and makes screening for the mutant clone easy. We have used the M13 vectors developed by Messing and co-workers.[26] These contain a number of useful cloning sites, usually exhibit the "blue to white" plaque coloration upon insertion of a DNA fragment, and can be used in conjunction with a number of commercially available sequencing primers. The size of the insert is for the most part one of convenience. We have conducted successful mutagenesis experiments on inserts ranging from 200 bases to 4.2 kilobases.

The basic strategy for cloning a fragment into an M13 vector (e.g., M13mp8) is depicted in Fig. 3. For mutagenesis of *MATa1,* a 4.2 kb *Hin*dIII fragment containing the entire *MATa* gene was inserted into the *Hin*dIII site of the vector M13mp5. We have also conducted experiments using M13mp7, M13mp8, and M13mp9 (unpublished). For a detailed discussion of M13 cloning see Messing.[26] A number of white plaques were picked, and single-stranded DNA was prepared from 1-ml cultures. The orientation of the insert was determined by sequencing a number of clones into the gene from the *Hin*dIII cloning site using an universal M13

[50] E. Jay, R. Bambara, R. Padmanbhan, and R. Wu, *Nucleic Acids Res.* **1,** 331 (1973).

[51] A. Maxam and W. Gilbert, this series, Vol. 65 [57].

[52] A. M. Banuszuk, K. V. Deugau, and B. R. Glick, unpublished. Biologicals Incorporated, Toronto.

[53] G. Chacoras and J. H. van de Sande, this series, Vol. 65 [10].

[54] R. Roychoudhury and R. Wu, this series, Vol. 65 [7].

[55] J. C. Hines and D. Ray, *Gene* **11,** 207 (1980).

sequencing primer. The sequence of the 4.2 kb *MATa* fragment near each *Hin*dIII site had been previously determined by C. Astell in our laboratory (unpublished results). One clone with the desired orientation was chosen and used as a source of single-stranded template DNA for the mutagenesis experiment.

A single mutagenesis experiment requires only 1–2 μg of single-stranded template DNA. Thus, a 2–3 ml culture, yielding 6–15 μg of single-stranded DNA,[28] should provide enough template DNA to carry out preliminary *in vitro* tests and the actual mutagenesis experiment.

Preliminary *in Vitro* Tests To Determine Specific Targeting by the Mutagenic Oligonucleotide

Once the oligonucleotide has been synthesized and sequenced and the template DNA has been isolated, it is important to carry out a number of preliminary experiments. The purpose of these tests is to demonstrate that the oligomer efficiently primes from the desired site. In addition, priming from this site is optimized by testing several primer : template ratios and priming temperatures. Two experiments can be done:

Primer Extension and Restriction Endonuclease Digestion. In this experiment, 5′-^{32}P-labeled oligonucleotide is annealed with the template, then extended by addition of deoxyribonucleotide triphosphates and DNA polymerase (large fragment). After extension, duplex DNA is cleaved with a restriction endonuclease that recognizes a site downstream from the desired priming site. The resulting fragments are denatured and electrophoresed on a polyacrylamide gel under denaturing conditions; the gel is autoradiographed. The resulting autoradiogram will indicate how many priming sites exist and which one is the major site.

Chain Terminator Sequencing. This test utilizes the mutagenic oligonucleotide as a primer in a chain-terminator sequencing reaction.[56,57] The resulting sequence is compared with sequences downstream from the priming site to determine whether the desired site is recognized. The specificity of priming can be assessed by the clarity of the sequencing ladder and the absence of a background pattern.

These experiments should be done using several priming temperatures and primer : template ratios. Figure 4 shows an example of the use of MS-1 and MS-2 as sequencing primers. The template : oligomer ratio was 1 : 20. Annealing was carried out by heating the mixture to 55° for 5 min,

[56] F. Sanger, A. R. Coulson, B. G. Barrell, A. J. H. Smith, and B. A. Roe, *J. Mol. Biol.* **143,** 161 (1981).

[57] F. Sanger, S. Nicklen, and A. R. Coulson, *Proc. Natl. Acad. Sci. U.S.A.* **74,** 5463 (1977).

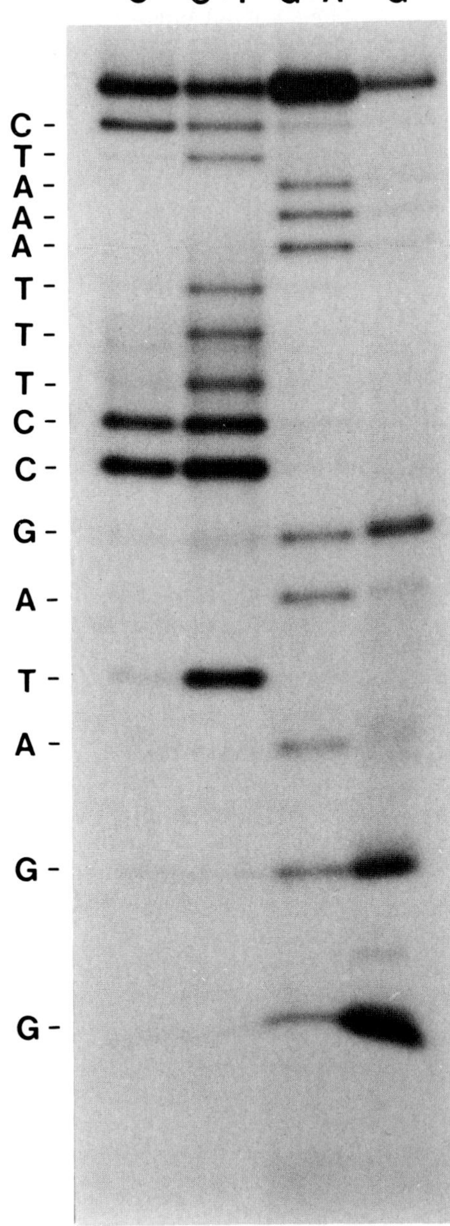
C
C+T
G+A
G
C
T
A
A
A
T
T
T
C
C
G
A
T
A
G
G

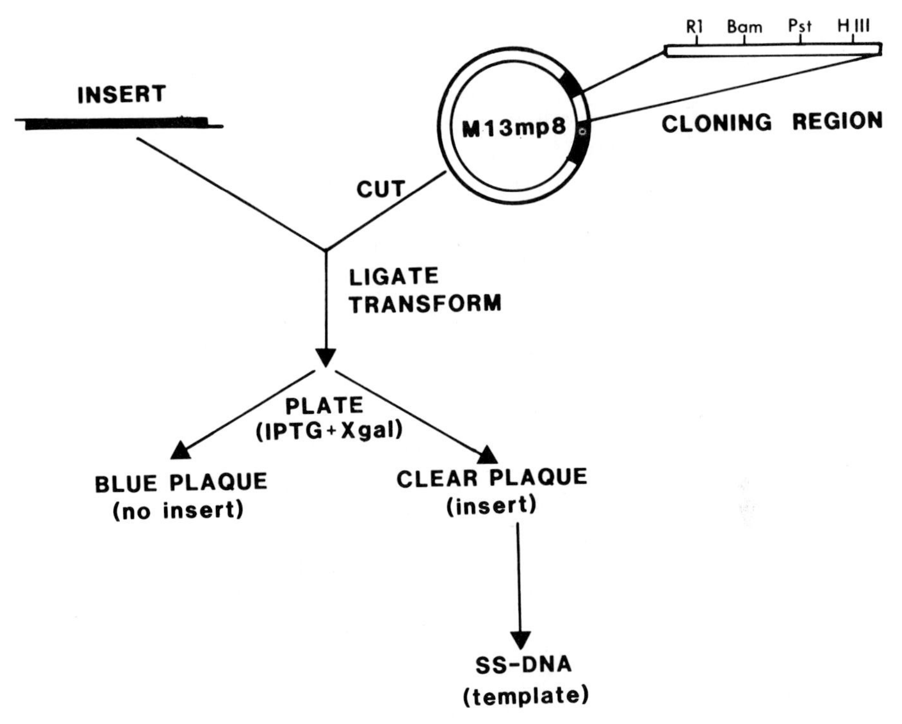

FIG. 3. General strategy for cloning a DNA fragment in the vector M13mp8.[26]

then cooling to room temperature. Extension was carried out at room temperature (23°) as described by Sanger *et al.*[56,57] These are the conditions that generally apply to oligonucleotides from 14 to 18 long directing a one- or two-base change. The specific conditions for each experiment should be determined to ensure success.

The sequence in Fig. 4 corresponds to the sequence in *MATa1* approximately 30 bases downstream from the UGA.[36] The clarity of the pattern demonstrates that priming occurred specifically at the desired site.

FIG. 2. DNA sequence determination of the oligonucleotide MS-1. Autoradiograph of a 20% polyacrylamide gel. A 5′-^{32}P-labeled oligonucleotide (500,000 cpm, 20 pmol) was sequenced according to the procedure of Maxam and Gilbert, modified for short oligonucleotides.[51,52] 10,000 ^{32}P cpm from the "C" and "G" reactions and 20,000 ^{32}P cpm from the "C + T" and "A + G" reactions were loaded into their respective lanes. Electrophoresis was carried out at 1400 V until the bromophenol blue dye migrated half way down the gel.

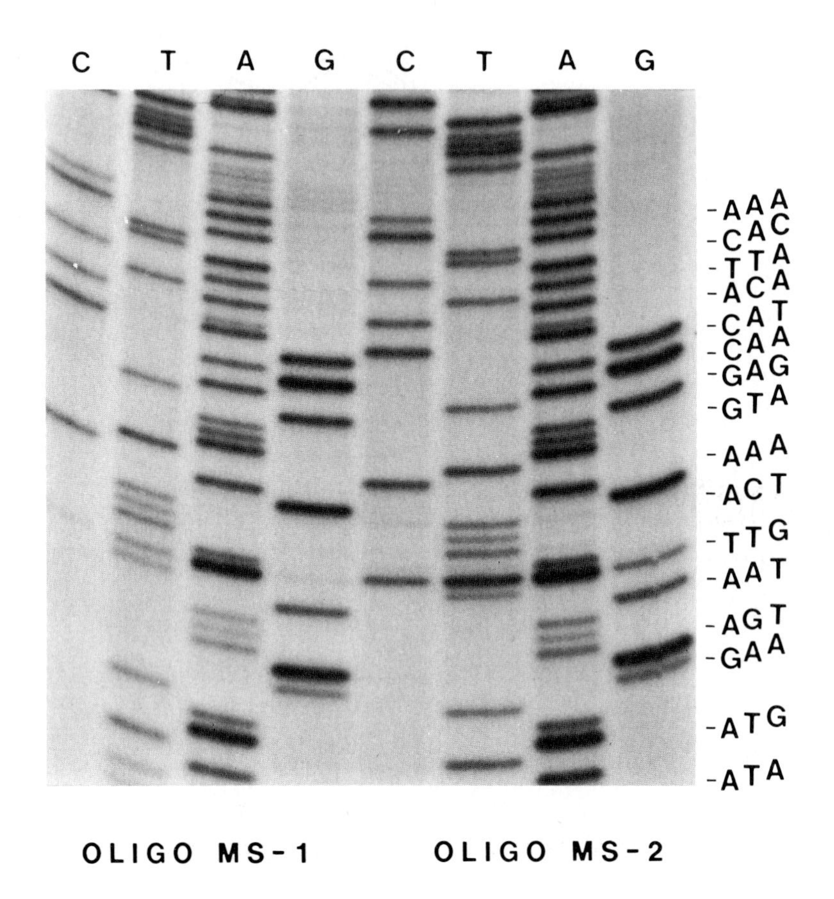

FIG. 4. Preliminary *in vitro* test for specific priming by the mutagenic oligonucleotide. Autoradiograph of a chain-terminator sequencing gel. Single-stranded M13 recombinant DNA containing the noncoding strand of the *MATa1* gene was subjected to chain-terminator sequencing using each mutagenic oligonucleotide as a primer.[56,57] In each reaction, the primer : template ratio was 20 : 1 and enzymic primer extension was carried out at 23°. The sequence corresponds to the region of the *MATa1* gene adjacent to the desired priming site. The clarity of the pattern indicates the high specificity of priming.

Procedure for Oligonucleotide-Directed Mutagenesis

Figure 5 shows the general scheme for oligonucleotide-directed site-specific mutagenesis. The mutagenic oligonucleotide and single-stranded template DNA are annealed and extended by DNA polymerase (large fragment) using the conditions determined in the preliminary tests. The

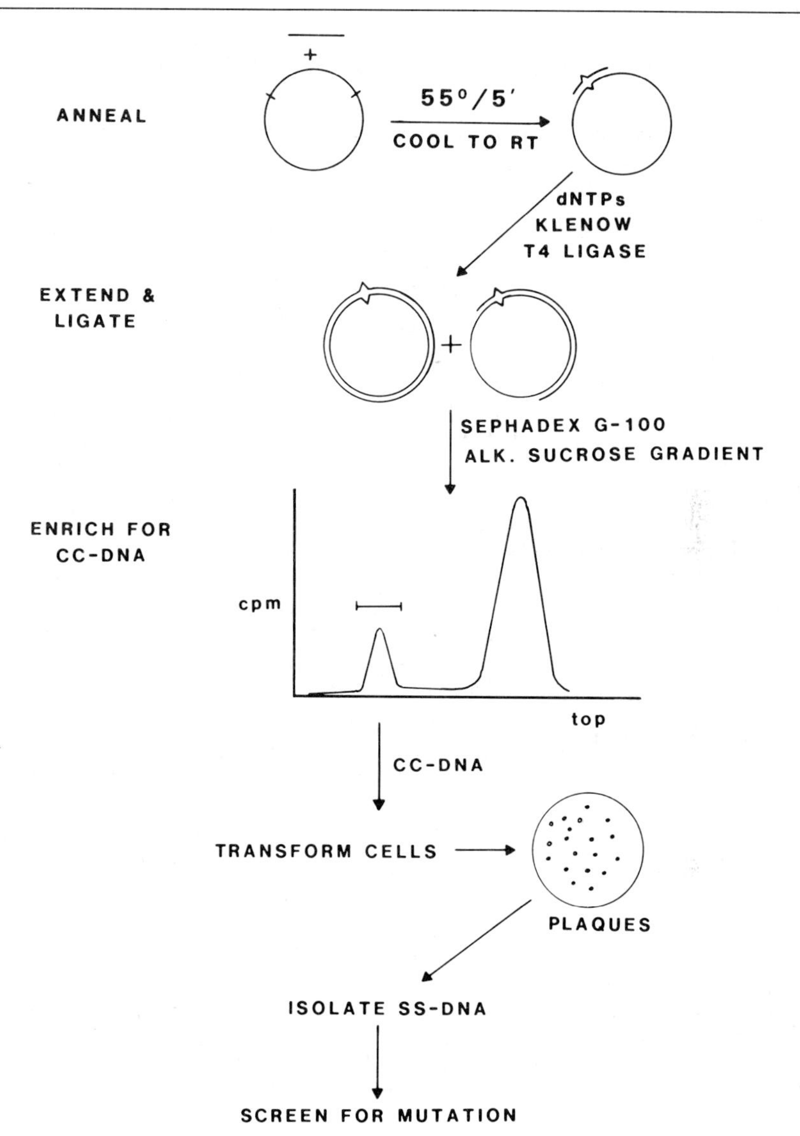

FIG. 5. General scheme for oligonucleotide-directed mutagenesis. CC-DNA, closed circular DNA; SS-DNA, single-stranded DNA.

newly synthesized strand is radioactively labeled by incorporation of [α-^{32}P]dATP during the initial period of extension. The labeled DNA can thus be followed in subsequent steps. Extension continues at 15° for 12–20 hr producing a population of covalently closed molecules. During this period, the primer is extended around the circular template, and the two ends of the newly synthesized strand are ligated by T4 DNA ligase. We have observed that the conversion of single-stranded circular molecules into closed double-stranded molecules is incomplete. Closed circular DNA (CC-DNA) is separated from unligated and incompletely extended molecules by alkaline sucrose gradient centrifugation. The CC-DNA fraction is pooled and used to transform $CaCl_2$-treated JM101 cells. Phage are prepared from 36 individual plaques and are screened for the desired mutant.

A major problem in oligonucleotide-directed mutagenesis has been the background of wild-type molecules resulting from inefficient conversion of single-stranded template DNA to CC-DNA. This increased the number of recombinants that must be screened to obtain a mutant. The work of Gillam and Smith demonstrated the substantial increase in mutagenesis efficiency by enrichment for closed circular molecules by treatment with S1 nuclease.[39] Other useful methods include binding single-stranded DNA to nitrocellulose,[14] agarose gel electrophoresis,[20] CsCl density gradient centrifugation,[21] and alkaline sucrose gradient centrifugation.[22] Alkaline sucrose gradient centrifugation is utilized in the present method because it effectively enriches for CC-DNA, it provides a diagnostic indication of the efficiency and kinetics of the extension and ligation reactions, and it is an exceedingly simple procedure.

Figure 6 shows the gradient profiles from two experiments. After extension and ligation, the DNA was precipitated by addition of a polyethylene glycol (PEG)–NaCl solution to remove ^{32}P-labeled triphosphate, then layered onto a 5 to 20% alkaline sucrose gradient. Centrifugation was carried out at 37,000 rpm for 2 hr using a SW 50.1 rotor. Aliquots were collected by puncturing the bottom of the tube. ^{32}P cpm in each fraction was determined by scintillation counting (Cerenkov). In the first experiment, extension and ligation proceeded for 5 hr at 15°. No CC-DNA was observed. In the second experiment, the reaction proceeded for 20 hr at 15°. In contrast to the first experiment, a peak of radioactivity representing CC-DNA was observed in the gradient from the second experiment. The CC-DNA fraction (tubes 3–8) was pooled, neutralized with 1 *M* Tris-citrate (pH 5), and used to transform $CaCl_2$-treated JM101 cells.

Transformation and Phage Isolation

Aliquots of neutralized CC-DNA are used to transform $CaCl_2$-treated *E. coli* JM101 cells as described by Sanger *et al.*[56] Single-stranded phage are isolated from 36 individual plaques and subjected to the appropriate screening procedure.

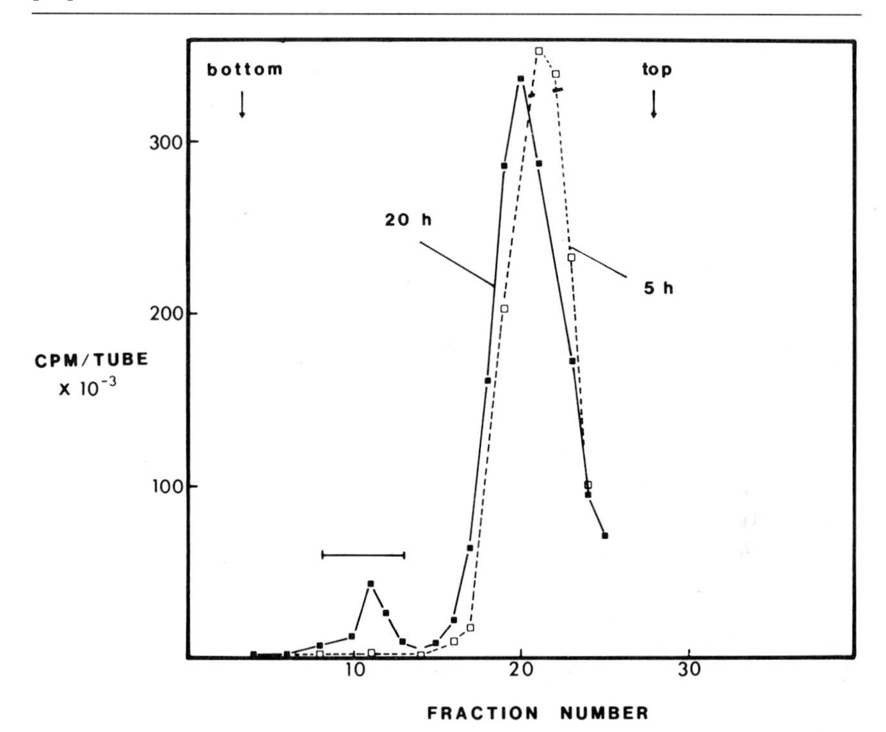

FIG. 6. Profile of ^{32}P-labeled DNA molecules separated by alkaline sucrose gradient centrifugation. After *in vitro* synthesis of closed circular DNA molecules, the sample was denatured with NaOH and centrifuged through a 5 to 20% alkaline sucrose gradient. Centrifugation was carried out at 37,000 rpm for 2 hr at 4° in an SW50.1 rotor. Fractions (175 μl) were collected by puncturing the bottom of the tube, and ^{32}P-cpm in each fraction was measured by scintillation counting. Extension and ligation proceeded at 15° for 5 hr (----) and 20 hr (——).

Screening for the Mutant

The CC-DNA obtained from the sucrose gradient is a heteroduplex consisting of one wild-type strand and one mutant strand. Transformation with this DNA results in a mixture of wild-type and mutant phage. These are distinguished from each other by a selection or screening procedure. Biological selection has been utilized in a number of reports. However, to create mutations in any fragment of DNA, a screening procedure that does not rely on a biological method is required. Three screening procedures are presented; two of them require isolation of single-stranded DNA, and the third is carried out using an aliquot of phage containing supernatant from a 1-ml culture. The choice of which method to use is dependent on the location of the mutation with respect to the cloning site, the availability of another sequencing primer, or whether the mutation deletes or produces a restriction endonuclease site.

Single-Track Sequencing. Single-stranded DNA from each clone is subjected to single-channel chain-terminator sequencing using the dideoxynucleotide that would differentiate wild-type from mutant clones.[56] A universal M13 sequencing primer is used in the case where the mutation is close to the cloning site. Otherwise, another synthetic oligonucleotide or a restriction fragment that primes adjacent to the mutation site can be used.

Restriction Site Screening. This method can be used if the desired mutation creates or destroys a restriction endonuclease site. Single-stranded DNA from each clone is annealed to either an M13 sequencing primer or a restriction fragment primer. Enzymic extension is carried out by DNA polymerase in the presence of deoxyribonucleotide triphosphates, one of which is α-^{32}P-labeled. The DNA is cleaved with the appropriate restriction enzyme, and the resulting fragments are separated on a polyacrylamide gel, then visualized by autoradiography. A mutant clone will exhibit a different restriction pattern from the wild type.

Hybridization Screening. The simplest and most versatile method is the hybridization method, in which the labeled mutagenic oligonucleotide is used as a probe for mutant clones. The effect of mismatches on the thermal stability of heteroduplexes formed between short oligonucleotides and poly(dT) or poly(dA) cellulose was investigated by Astell *et al.*[10] and by Gillam *et al.*[11] These studies demonstrated that a mismatch dramatically destabilized a heteroduplex compared with a perfectly complementary duplex. Wallace and co-workers showed that this principle could be utilized in a screening procedure to differentiate mutant from wild-type molecules produced by site-specific mutagenesis.[58,59] The method presented here is independent of the position of the mutation within the cloned fragment, is carried out on an aliquot of intact phage, and identifies mutant phage in a single day. The principle behind this procedure is that the mutagenic oligonucleotide will form a more stable duplex with a mutant clone, with perfect match, than with a wild-type clone bearing a mismatch. The mutant is detected by forming a duplex at a low temperature where both mutant and wild-type DNA interact, then carrying out washes at increasingly higher temperatures until only mutant molecules hybridize. Phage from each clone are spotted directly onto a

[58] R. B. Wallace, M. J. Johnson, T. Hirose, T. Miyake, E. H. Kawashima, and K. Itakura, *Nucleic Acids Res.* **9,** 879 (1981).

[59] R. B. Wallace, J. Shaffer, R. F. Murphy, J. Bonner, T. Hirose, and K. Itakura, *Nucleic Acids Res.* **6,** 3543 (1979).

nitrocellulose filter, then baked at 80° for 2 hr. There is no need to carry out the denaturation steps as in other *in situ* hybridization procedures.[60,61] The probe solution is added to the filter after a prehybridization step. Hybridization proceeds for 1 hr at a temperature at which both mutant and wild-type molecules hybridize with the probe. The hybridization solution is removed, and the filter is washed, then autoradiographed. The same filter is washed again at a higher temperature, then autoradiographed. This process is repeated several times, each time increasing the wash temperature, until only mutant molecules exhibit a hybridization signal.

Comments. In the example with *MATa*, the mutation lies approximately 2.7 kb downstream from the cloning site. This position cannot be sequenced using a standard M13 primer. However, the desired change destroys a restriction site for the enzyme *Hin*fI (GAATC to AAATC). Figure 7 shows an example of 6 out of 36 clones screened by this method. Based on the wild-type sequence[37] the loss of the *Hin*fI site around position 1667 would yield a new fragment of 605 bp (320 + 285 bp) and would result in the loss of the two smaller fragments. Clones 7, 32, and 34 exhibited a new fragment at approximately 650 bp (see arrow). Clone 21 (not shown) also exhibited a 650 bp *Hin*fI fragment.

The hybridization method offers a versatile and simple way to identify mutant clones. Figure 8 shows an example of 36 clones screened by hybridization to ^{32}P-labeled MS-1 oligonucleotide. The four clones that exhibited hybridization signals at 47° corresponded directly with the clones that showed the restriction site alteration. Thus, approximately 11% of the isolated phage were mutant.

Plaque Purification of a Suspected Mutant

We have observed that the suspected mutant clones can contain up to 50% wild-type molecules. In a hybridization screen, these clones exhibit signals of lower intensity compared with signals from homogeneous mutant clones. Heterogeneous mutants arise because transformation and plating were carried out without a cycle of replication and reinfection. Any single infected cell should contain a mixture of wild-type and mutant phage. However, we have observed that phage isolated from a single phage comprise predominantly one type. Nevertheless, it is important to replate the suspected mutant phage and isolate homogeneous mutant phage.

[60] M. Grunstein and D. S. Hogness, *Proc. Natl. Acad. Sci. U.S.A.* **72,** 3961 (1975).
[61] W. D. Benton and R. W. Davis, *Science* **196,** 180 (1977).

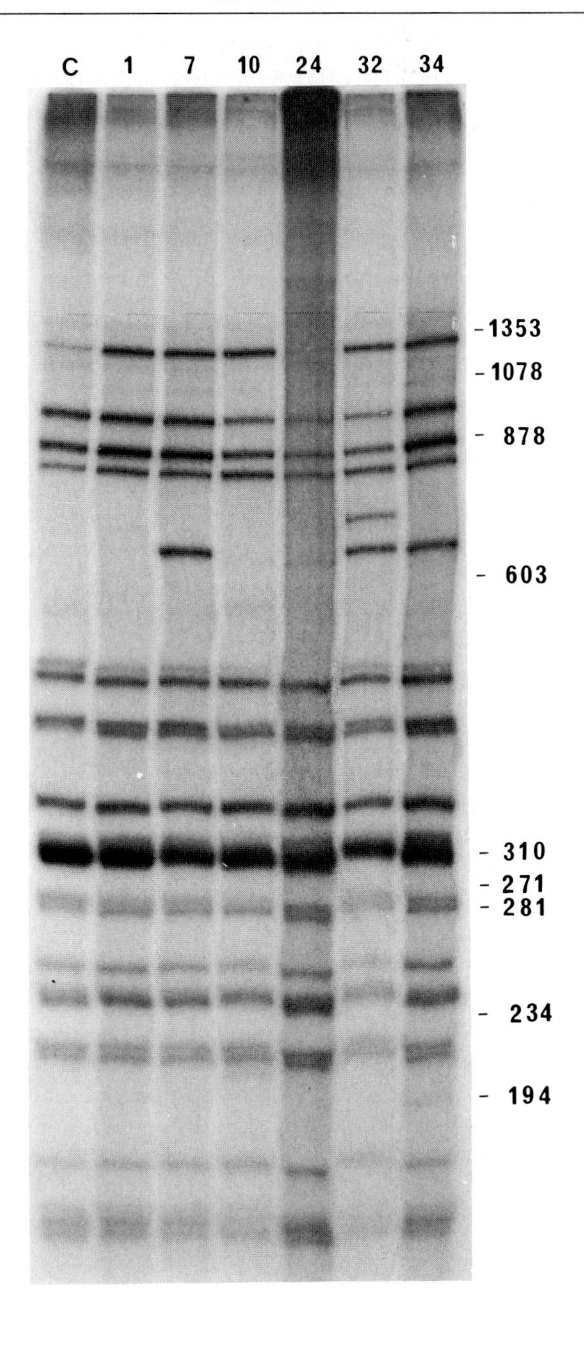
C
1
7
10
24
32
34
-1353
-1078
- 878
- 603
- 310
- 271
- 281
- 234
- 194

This was done for one of the suspected *MATa1* mutants. Phage from this clone were replated, and single-stranded DNA was isolated from 16 individual plaques, then screened by the restriction site method. In all cases, the new fragment of 650 bp was observed. In other experiments in our laboratory, isolates of phage from a suspected mutant contained 0–50% wild-type molecules determined by dot-blot hybridization analysis (unpublished).

Sequencing the Suspected Mutant

The final step of this procedure is to verify that the desired mutation has been produced. This can be accomplished by chain-terminator sequencing on the single-stranded DNA[56,57] or by the procedure of Maxam and Gilbert using isolated double-stranded DNA.[51]

Double-stranded phage DNA was prepared from one plaque-purified mutant phage[62] and sequenced in the region containing the in-frame UGA codon according to the procedure of Maxam and Gilbert. Figure 9 shows the sequence of mutant compared with the wild-type sequence around this site. The only difference between the two sequences is a C to T transition at the desired site.

MS-2 UGA to UGG Mutant

Construction of this mutant was carried out as described above. The oligonucleotide used to direct the change is shown in Fig. 1 and was synthesized according to the procedure of Edge *et al.*[41] The yield of mutant phage was 16.6%, identified by the restriction enzyme screening method. The DNA sequence within the region of the mutation was determined from DNA isolated from a plaque-purified phage by the procedure of Maxam and Gilbert.[51] A single T to C transition was observed within this region (data not shown).

[62] D. B. Clewell and D. R. Helinski, *Proc. Natl. Acad. Sci. U.S.A.* **62,** 1159 (1969).

FIG. 7. Screening for mutants by restriction fragment analysis. Autoradiograph of a nondenaturing 5% polyacrylamide gel separating restriction fragments of potential mutants. The autoradiograph shows an example of six recombinants screened by the restriction site method for the loss of a particular *Hin*fI site. Double-stranded ^{32}P-labeled DNA was synthesized *in vitro* by enzymic primer extension, cleaved with the restriction enzyme *Hin*fI, and electrophoresed. Three mutants are shown that exhibited a new fragment of approximately 650 bp (arrow).

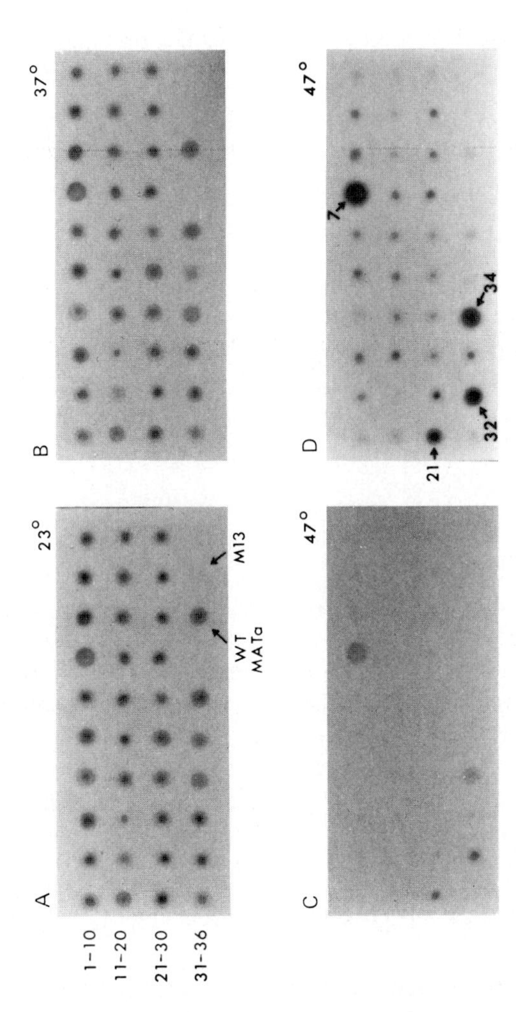

FIG. 8. Dot-blot hybridization screening for mutants using ^{32}P-labeled mutagenic oligonucleotide MS-1. Single-stranded DNA from 36 phage was bound to nitrocellulose and hybridized to ^{32}P-labeled oligomer MS-1. Hybridization was carried out at 23° for 1 hr. The filter was washed in 6 × SSC at 23°, 37°, and 47°, respectively. After each wash, the filter was autoradiographed for 1 hr (panels A–C) or 8 hr (panel D) using Kodak NS-5T film.

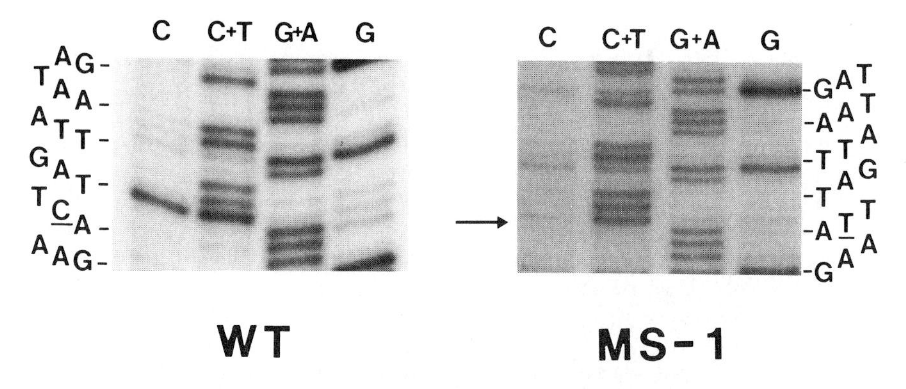

FIG. 9. Comparison of mutant and wild-type (WT) DNA sequences. An autoradiograph of two 12% polyacrylamide gels comparing the sequences of the wild-type and mutant (MS-1) DNA in the region of the *MATα1* gene containing the in-frame UGA codon. A 420 bp *Dde*I–*Bgl*II fragment was 3′-^{32}P-labeled at the *Dde*I site and sequenced by the procedure of Maxam and Gilbert.[51] The only difference between these two sequences is the C to T change in the mutant DNA at the desired position (arrow).

Materials and Methods

Escherichia coli JM101 and M13mp5 were the gift of J. Messing (University of Minnesota). *Escherichia coli* DNA polymerase (large fragment) and T4 DNA ligase were obtained from Bethesda Research Laboratories. T4 polynucleotide kinase was purchased from New England BioLabs. [α-^{32}P]ATP were obtained from New England Nuclear. Deoxyribonucleotide triphosphates were purchased from P-L Biochemicals. RiboATP, Trizma-base, and ultrapure sucrose were obtained from Sigma. The octadecanucleotide 5′-AAGGATAGCCTTTAAATC-3′ was the gift of A. Markham and T. Atkinson (ICI). Hydrazine (99%) and acrylamide were purchased from Kodak Chemicals. Dimethyl sulfate was obtained from BDH. Formic acid (91%), liquefied phenol, $CaCl_2$, EDTA, and PEG-6000 were purchased from Fisher. Nitrocellulose filters were obtained from Schleicher & Schuell.

The mutagenesis reactions are carried out in 0.5-ml siliconized Eppendorf tubes. Mix by gentle vortex or hand agitation. Additions of 5 μl or less are made directly into the tube using siliconized 5 μl graduated micropipettes. Keep the contents of the tube at the bottom by a short spin (1–2 sec) in an Eppendorf centrifuge. For a convenient siliconizing procedure, see Roychoudhury and Wu.[54]

Solutions

Solution A: 0.2 *M* Tris-HCl (pH 7.5, 23°), 0.1 *M* $MgCl_2$, 0.5 *M* NaCl, 0.01 *M* dithiothreitol (DTT)

Solution B: 0.2 *M* Tris-HCl (pH 7.5, 23°), 0.1 *M* $MgCl_2$, 0.1 *M* DTT

Solution C: 1 μl of solution B, 1 μl of 10 m*M* dCTP, 1 μl of 10 m*M* dTTP, 1 μl of 10 m*M* dGTP, 0.5 μl of 0.1 m*M* dATP, 1 μl of 10 m*M* ribo-ATP, 1.5 μl of [α-^{32}P]dATP (7 μCi/μl), 1.5 μl of T4 DNA ligase (2 U/μl), 2 μl of H_2O. Prepare in a 0.5-ml vial before adding to annealed DNA. Keep on ice.

Alkaline sucrose gradient stock solutions: *X*% sucrose, 1 *M* NaCl, 0.2 *M* NaOH, 2 m*M* EDTA; *X* = 20, 17.5, 15, 10, and 5. Autoclave, store at 4°.

Sucrose–dye–EDTA solution (for nondenaturing gels): 60% sucrose, 0.02% (w/v) bromophenol blue, 0.02% (w/v) xylene cyanole FF, 0.025 *M* EDTA

Formamide–dye–EDTA solution (for denaturing gels): 0.02% bromophenol blue, 0.02% xylene cyanole FF, 0.025 *M* EDTA; in 90% formamide (deionized)

20× SSC (concentrated stock solution): 3 *M* NaCl, 0.3 *M* sodium citrate, 0.01 *M* EDTA. Adjust pH to 7.2 with HCl.

100× Denhardt's (concentrated stock solution): 2% bovine serum albumin, 2% poly(vinylpyrrolidone), 2% Ficoll

Media

Supplemented M9: Prepare the following solutions:

1. 10× M9 salts (per liter): $Na_2HPO_4 \cdot 2\ H_2O$, 70 g, KH_2PO_4, 30 g, NaCl, 5 g, NH_4Cl, 10 g.
2. 20% glucose
3. 10 m*M* $CaCl_2$
4. 1 *M* $MgSO_4$
5. 0.1% thiamin (filter sterilize)

Prepare 50 ml of 1× M9 salts in a 200-ml Ehrlenmeyer flask using the 10× stock solution. Autoclave. After solution has cooled add: 0.5 ml of 20% glucose, 0.5 ml of 10 m*M* $CaCl_2$, 50 μl of 1 *M* $MgSO_4$, 50 μl of 0.1% thiamin.

YT medium: Per liter: 8 g of tryptone, 5 g of yeast extract, 5 g of NaCl. For YT plates add 15 g of agar per liter; for YT top agar add 8 g of agar per liter.

2× YT medium: Per liter: 16 g of tryptone, 10 g of yeast extract, 5 g of NaCl

Schedule for a Typical Mutagenesis Experiment

Day 1
- 5′ phosphorylate oligonucleotide (procedure IA)
- Start *in vitro* mutagenesis reaction (procedure IV)
- Prepare $CaCl_2$ treated JM101 cells (procedure VA)
- Prepare alkaline sucrose step gradient (procedure IVD)

Day 2
- PEG–NaCl precipitate DNA (procedure IVC)
- Alkaline sucrose gradient centrifugation (procedure IVD)
- Transform JM101 with CC-DNA (procedure VB)
- Start an overnight culture of JM101 (procedure VA,1)

Day 3
- Prepare phage from 36 individual plaques (procedure VI)

Day 4
- Prepare hybridization probe (procedure IB)
- Screen by dot-blot hybridization (procedure VIIC)
- Plate out suspected mutant phage to purify

Day 5
- Prepare phage from 6 individual plaques of suspected mutants (procedure VI)

Day 6
- Screen by dot-blot hybridization (procedure VIIC)

Procedures

Procedure I. 5′ Phosphorylation of the oligonucleotide[53]

The oligonucleotide should be lyophilized after synthesis, then resuspended in water. A convenient working concentration is 10–20 pmol/μl.

A. For mutagenesis
1. Add to a 0.5-ml siliconized Eppendorf tube: 200 pmol oligonucleotide (dry down if necessary), 3 μl of 1 *M* Tris-HCl (pH 8), 1.5 μl of 0.2 *M* $MgCl_2$, 1.5 μl of 0.1 *M* DTT, 3 μl of 1 m*M* riboATP, H_2O to 30 μl total volume.
2. Mix

3. Add 4.5 units of T4 polynucleotide kinase.
4. Mix and incubate at 37° for 45 min.
5. Stop by heating at 65° for 10 min.

The oligonucleotide must bear a 5'-phosphate group in order to ligate. An aliquot from the reaction vial is used directly for the mutagenesis reaction. To check that the kinase reaction has gone to completion, end label and chromatograph an aliquot as discussed below. This preparation of kinased oligonucleotide can serve as a stock for a number of experiments. Store at −20°.

B. For hybridization screening
1. Dry down 20 μCi of [γ-^{32}P]ATP (2000 Ci/mmol) in a 0.5 ml Eppendorf tube.
2. Add 20 pmol of oligonucleotide, 3 μl of 1 *M* Tris-HCl (pH 8), 1.5 μl of 0.2 *M* $MgCl_2$, 1.5 μl of 0.1 *M* DTT, H_2O to 30 μl total volume.
3. Follow steps 2–5 as in A.
4. Chromatograph on Sephadex G-25 to remove unincorporated [α-^{32}P]dATP. (Alternatively, see Wallace *et al.*[17])

The column (6 × 200 mm) should be made of either plastic or siliconized glass. Chromatograph the sample in 50 m*M* ammonium bicarbonate (pH 7.8). Collect 100-μl aliquots in 1.5-ml Eppendorf tubes, and measure the radioactivity in each fraction by scintillation counting (Cerenkov). The first radioactive peak contains the labeled oligonucleotide. Pool the major fractions.

The purity of the ^{32}P-labeled oligonucleotide can be checked by chromatographing an aliquot of the reaction mixture on a strip of Whatman DE-81 paper using 0.3 *M* ammonium formate (pH 8). The oligonucleotide remains at the origin and [γ-^{32}P]ATP migrates near the solvent front. This analysis can be used to assess either reaction A or B. To check reaction A, add 10 μCi of [γ-^{32}P]ATP to the reaction solution, and chromatograph 1 μl.

Incorporation can be measured qualitatively using a hand-held Geiger counter or quantitatively by cutting out the origin and counting it in a scintillation counter. Generally, procedure IB results in the incorporation of 2 to 6 × 10^6 ^{32}P cpm (Cerenkov) per 20 pmol of oligonucleotide. The 5' ^{32}P-labeled oligonucleotide from reaction IB is also used for sequencing (procedure II) and in the preliminary tests (procedure IVA).

Procedure II. Sequence determination of the mutagenic oligonucleotide.

End-labeled oligonucleotide is subjected to a modified Maxam and Gilbert procedure.[51,52]

1. Dry down 5×10^5 cpm ^{32}P-labeled oligonucleotide in a siliconized Eppendorf tube.
2. Resuspend in 32 μl of H_2O.
3. Add 5 μl to the "C" and "G" tubes. Add 10 μl to the "C + T" and "A + G" tubes.
4. Add 2 μl of 1 mg/ml denatured calf thymus DNA to each tube.
5. Add 20 μl of 5 *M* NaCl to the "C" tube, 15 μl of H_2O to the "C + T" tube, 10 μl of H_2O to the "A + G" tube, 300 μl of cacodylate buffer to the "G" tube.[51]
6. Mix.
7. Add 30 μl of hydrazine to the "C" and "C + T" tubes. Mix and incubate at 45° for 20 min and 5 min, respectively.
8. Add 3 μl of 50% (v/v) formic acid to the "A + G" tube. Mix and incubate at 37° for 20 min.
9. Add 2 μl of dimethyl sulfate to the "G" tube. Mix and incubate at 37° for 3 min.
10. Stop all reactions as described.[51]
11. Carry out subsequent precipitation and piperidine cleavage steps as described.
12. Load 10,000 cpm (Cerenkov) into the "C" and "G" slots, and 20,000 cpm into the "C + T" and "A + G" slots.
13. Electrophorese the samples on a 25% polyacrylamide–7 *M* urea gel as described[51] until the bromophenol blue is approximately half way down the gel (~2 hr).
14. Autoradiograph for 20 hr using Kodak NS-5T film at −20°.

The standard ethanol precipitation steps following the modification reactions inefficiently precipitate the oligonucleotide, resulting in a significant loss of ^{32}P cpm. This has been taken into account in the present procedure by using more ^{32}P-labeled oligonucleotide than usually is required in a standard sequencing reaction.

Procedure III. Preliminary *in vitro* tests for specific priming

A. Primer extension

1. Add to a siliconized 0.5-ml Eppendorf tube: M13 recombinant DNA (0.5 pmol); 5′-end-labeled oligonucleotide (10–30× molar excess over template), specific activity = 2×10^4 cpm/pmol; 1 μl of solution A; H_2O to 10 μl.
2. Heat in a water bath at 55° for 5 min.
3. Remove tube from bath and let stand at room temperature (23°) for 5 min.
4. Add to the annealed DNA: 1 μl of 2.5 m*M* dCTP, 1 μl of 2.5 m*M*

dATP, 1 μl of 2.5 mM dTTP, 1 μl of 2.5 mM dGTP, 0.5 μl of solution A.

5. Mix.
6. Add 1 unit of DNA polymerase (large fragment).
7. Mix and incubate at room temperature for 5 min.
8. Stop reaction by heating at 65° for 10 min.
9. Cool to room temperature.
10. Adjust buffer for appropriate restriction enzyme.
11. Add 1–5 units of enzyme.
12. Incubate at appropriate temperature for 1 hr.
13. Add equal volume of formamide–dye solution.
14. Boil for 3 min; immediately chill on ice.
15. Electrophorese the sample on a 5% polyacrylamide gel under denaturing conditions.[14,40]
16. Autoradiograph.

B. Chain-terminator sequencing. Follow the procedure by Sanger *et al.* for chain-terminator sequencing using M13.[56,57] Substitute the mutagenic oligonucleotide for a universal M13 sequencing primer.

1. Add to a siliconized tube: M13-recombinant DNA (0.1–0.3 pmol) oligonucleotide (10–30× over template DNA), 1 μl of solution A, H_2O to 10 μl.
2. Mix and seal in a glass capillary tube.
3. Heat at 55° for 5 min; cool to room temperature.
4. Follow the procedure of Sanger *et al.*[56]

Use the conditions outlined above for oligonucleotides between 14 and 18 long that bear one or two mismatches. For each experiment, examine a number of different primer : template ratios and priming temperatures.

Procedure IV. Mutagenesis

Steps A–D are carried out in a single siliconized 0.5-ml Eppendorf tube.

A. Annealing

1. Add to a 0.5-ml siliconized Eppendorf tube: M13-recombinant DNA (0.5–1.0 pmol), 5′ phosphorylated oligonucleotide (10–30× molar excess over template DNA), 1 μl of solution A, H_2O to 9 μl total volume.
2. Mix and heat tube in a water bath at 55° for 5 min.
3. Remove tube from bath and let stand at room temperature (23°) for 5 min.

The phosphorylated oligonucleotide is taken directly from the reaction vial in procedure IA. The above conditions generally apply when producing point mutations using oligomers 14–18 long. For shorter oligonucleotides, heat at 55°, then chill to 0–10°. The specific conditions should be determined in the preliminary *in vitro* tests (procedure III).

B. Extension and ligation

1. Prepare solution C in a separate vial. Keep on ice.
2. Add 10 μl of solution C to the annealed DNA.
3. Mix, then add 2.5 units of DNA polymerase (large fragment).
4. Mix and incubate for 5 min at room temperature.
5. Add 1 μl of 10 m*M* dATP.
6. Mix, then incubate at 15° for 12–20 hr.

For shorter oligonucleotides, carry out the initial extension reaction at 0–10°, then continue at 15° as outlined above.

C. PEG–NaCl precipitation. This step is carried out to remove unincorporated [α-^{32}P]dATP from the sample. This allows subsequent determination of the percentage of CC-DNA produced.

1. Add to the reaction tube: 30 μl of H_2O, 50 μl of 1.6 *M* NaCl/13% PEG-6000.
2. Mix and incubate on ice for 15 min.
3. Spin in an Eppendorf centrifuge at room temperature for 5 min.
4. Withdraw all of the solution using a 50-μl glass micropipette.
5. Add 100 μl of cold 0.8 *M* NaCl–6.5% PEG-6000 to rinse the sides of the tube. Do not resuspend pellet at this point.
6. Spin for 30 sec.
7. Withdraw solution as in 4.
8. Resuspend in 180 μl of 10 m*M* Tris-HCl (pH 8)–1 m*M* EDTA.
9. Determine total incorporation of [^{32}P]ATP by scintillation counting (Cerenkov).

D. Alkaline sucrose gradient centrifugation. Prepare one 5 to 20% sucrose gradient for each sample in a 0.5-inch × 2-inch centrifugation tube. This is used with a SW 50.1 rotor. Make a step gradient with 1 ml each of the sucrose stock solutions, starting with the 20% solution. Carefully let the solution run down the side of the tube using either a serological or a Pasteur pipette. Hold the tip close to the top of the preceding layer to avoid mixing. After adding all five solutions, ensure that four schlieren bands are sharply defined. Let the tube sit at 4° for 6–16 hr to linearize the gradient. Alternatively, leave the gradient at room temperature for 2 hr, then at 4° for another 2 hr. After the gradient has formed, prepare the sample procedure from IVC.

1. Add 20 μl of 2 *N* NaOH to the resuspended sample.
2. Mix and incubate at room temperature for 5 min.
3. Chill on ice for 1 min.
4. Place the gradient tube into the SW 50.1 bucket.

5. Apply the sample to the top of the gradient, using a disposable micropipette.

If the top of the gradient solution is too close to the top edge of the centrifugation tube before adding the sample, withdraw 50–100 μl from the top with a Pipetman. The final height of the solution should be within 2–3 mm of the tube edge.

Centrifugation is carried out using a SW 50.1 rotor at 37,000 rpm for 2 hr at 4° with the brake off. After centrifugation, collect aliquots into 1.5-ml Eppendorf tubes by puncturing the bottom of the gradient tube. An apparatus such as the Hoeffer tube holder works well. Collect about 30 fractions (5–7 drops per fraction). Determine the amount of ^{32}P cpm in each fraction by scintillation counting (Cerenkov). The CC-DNA (double stranded) is in the bottom half of the gradient, whereas the single-stranded circular and linear molecules are in the top half. Pool the CC-DNA fraction and neutralize the solution with 1 *M* Tris-citrate (pH 5). Add about 50 μl of Tris-citrate per 300 μl. Check that the pH is between 7 and 8 by spotting a 2-μl aliquot onto pH paper. Measure the amount of ^{32}P-labeled DNA in the pooled CC-DNA fraction and calculate the percentage of CC-DNA formed.

Procedure V. Transformation

A. Prepare $CaCl_2$-treated *E. coli* JM101 cells.[63]

1. Inoculate 5 ml of supplemented M9 medium with a single colony of JM101 from a minimal plate (+ glucose). Incubate with shaking at 37° overnight.
2. Add 0.5 ml of overnight culture to 50 ml of supplemented M9. Incubate with shaking at 37° to an $A_{600} = 0.3$.
3. Harvest cells in two 40-ml centrifuge tubes at 5000 rpm for 8 min.
4. Resuspend cells in 25 ml ice cold 50 m*M* $CaCl_2$. Keep on ice for 20 min.
5. Combine cells into one tube and harvest as in 3.
6. Resuspend in 5 ml cold 50 m*M* $CaCl_2$.

Cells are stored on ice in the cold room and can be used up to 5 days later. However, transformation efficiency increases during the first 24 hr and thereafter gradually decreases.[63]

B. Transformation of $CaCl_2$-treated cells with CC-DNA[26,27,56]

1. Pipette 0.2 ml of $CaCl_2$-treated JM101 cells into culture tubes (13 × 100 mm). Keep on ice.
2. Add 1, 2, 5, and 10 μl neutralized CC-DNA to individual tubes.

[63] M. Dagert and S. D. Ehrlich, *Gene* **6,** 23 (1979).

Prepare three additional tubes for controls. To one tube add no DNA, to the other two add 1 and 0.1 ng of double-stranded M13 vector, respectively.

3. Incubate tubes on ice for at least 40 min.
4. Heat tubes in a water bath at 45° for 2 min.
5. Add 2.5 ml of molten YT top agar to each tube, vortex to mix, and pour onto a YT plate.
6. Allow agar to harden at room temperature for 15 min, then incubate plates at 37° overnight. Plaques appear in 6–9 hr.

The yield of phage plaques range from 10–200 plaques per microliter of CC-DNA solution depending on the age of the $CaCl_2$-treated cells and the percentage of CC-DNA obtained. The controls should result in approximately 1000 and 100 plaques, respectively. In the event that no plaques are obtained from the CC-DNA fraction, dialyze the pooled DNA against two changes of 1 liter of 2 m*M* Tris-HCl (pH 8)–0.01 m*M* EDTA, and transform again with the same volumes of DNA solution suggested in step 2. If this does not result in the appearance of phage plaques, then concentrate the sample 10-fold by extraction with 1-butanol or lyophilization and transform again.

Procedure VI. Isolation of phage.[26,27,56]

Phage-infected cells exhibit a clear "plaque-like" appearance because they grow much slower than the uninfected lawn cells. Prepare phage cultures from 36 individual plaques by the following procedure.

1. Add 0.1 ml of overnight JM101 cells to 5 ml of 2× YT media.
2. Incubate with shaking for 2 hr at 37°.
3. Add this 5-ml culture to 45 ml of 2× YT.
4. Mix and distribute 1 ml to each of 36 culture tubes (13 × 100 mm).
5. Core each phage plaque from the plate with a 50-μl disposable micropipette and blow the agar plug into the culture tube.
6. Incubate tubes with vigorous shaking at 37° for 5 hr. Longer incubation is not recommended.
7. Pour the contents of each tube into 1.5-ml Eppendorf tubes. It is not necessary to remove all the solution.
8. Spin in an Eppendorf centrifuge for 5 min.
9. Pour phage-containing supernatant into fresh 1.5-ml Eppendorf tubes.
10. Add 200 μl of 20% PEG-6000–2.5 *M* NaCl to each tube.
11. Vortex, and let stand at room temperature for 15 min.
12. Spin in an Eppendorf centrifuge for 5 min at room temperature.
13. Carefully pour off the supernatant.

14. Spin each tube again for 1–2 sec and remove all residual traces of PEG solution with a disposable micropipette.
15. Resuspend the phage pellet in 100 μl of 10 m*M* Tris-HCl (pH 8)–0.1 m*M* EDTA.
16. For hybridization screening, stop here and proceed to procedure VII. For sequencing or restriction site screening, continue through step 28.
17. Add 50 μl of Tris-saturated phenol.
18. Vortex for 10 sec, then spin in an Eppendorf centrifuge for 1 min.
19. Remove aqueous (top) layer using a Pipetman or disposable micropipette into a fresh 1.5-ml Eppendorf tube.
20. Extract 2× with 500 μl of diethyl ether.
21. Add 10 μl of 3 *M* sodium acetate (pH 5.5) and 300 μl of ethanol.
22. Precipitate DNA by chilling tubes at −70° for 1 hr or at −20° overnight.
23. Spin tubes in an Eppendorf centrifuge for 5 min.
24. Carefully remove the supernatant.
25. Rinse tube with cold ethanol. Do not disturb the pellet.
26. Spin for 2 min in an Eppendorf centrifuge. Remove the supernatant.
27. Dry under vacuum.
28. Resuspend DNA in 50 μl of 10 m*M* Tris-HCl (pH 8)–0.1 m*M* EDTA.

Procedure VII. Screening procedures

Determine which procedure is suitable for the particular mutation. Use method A if the mutation can be detected by sequencing with an M13 sequencing primer, another synthetic oligonucleotide, or a restriction fragment. Use method B in the case where the mutation creates or destroys a restriction site. The hybridization screen, method C, is the simplest and most versatile procedure.

A. Chain-terminator sequencing.[56,57] Use 5 μl M13-recombinant DNA (from procedure VI,28) as template with only one dideoxyribonucleotide in a chain terminator reaction.[57] The primer can be a universal M13 primer, another synthetic oligonucleotide, or a restriction fragment. Suspected mutants are sequenced again using all four dideoxynucleotide mixtures.

B. Restriction site screening

1. Add to a siliconized Eppendorf tube: 5 μl of M13-recombinant DNA (from procedure VI,28); 1 μl of solution A; M13 universal sequencing primer (1.5 pmol); H_2O to 10 μl.

2. Mix, heat at 55° for 5 min, then cool to room temperature. If a restriction fragment is used as a primer, heat the mixture for 3 min at 100°, hybridize at 67° for 30 min, then cool to room temperature.
3. Add: 1 μl of 0.5 m*M* dCTP, 1 μl of 0.5 m*M* dTTP, 1 μl of 0.5 m*M* dGTP, 0.5 μl of 0.1 m*M* dATP, 5 μCi of [α-^{32}P]dATP, 0.5 μl of solution A.
4. Mix.
5. Add 1 unit of DNA polymerase (large fragment).
6. Mix and incubate for 10 min at room temperature.
7. Add 1 μl of 2.5 m*M* dNTPs.
8. Mix and incubate for 5 min at room temperature.
9. Heat at 65° for 10 min, then cool to room temperature.
10. Adjust the buffer for desired restriction enzyme.
11. Add 1–5 units of enzyme.
12. Mix and incubate for 1 hr at appropriate temperature.
13. Stop by addition of 5 μl of sucrose–dye–EDTA solution.
14. Load the entire sample onto a 5% nondenaturing polyacrylamide gel with 1.5-mm spacers.[51]
15. Electrophorese at 200 V for 16 hr.
16. Autoradiograph.

C. Dot-blot hybridization

1. Spot 2 μl of phage (from procedure VI,15) onto a dry sheet of nitrocellulose.
2. Let air-dry, then bake *in vacuo* at 80° for 2 hr.
3. Prehybridize with 6× SSC + 10× Denhardt's + 0.2% SDS at 67° for 1 hr. This can be done in a sealable bag or in a closed petri dish. Use 10 ml/100 cm^2.
4. Remove the prehybridization solution and rinse the filter in 50 ml of 6× SSC for 1 min at 23°.
5. Place the filter into a sealable bag, add probe, and seal. The probe is prepared according to procedure IB; 10^6 cpm (Cerenkov) are added to 6× SSC + 10× Denhardt's solution (without SDS). Use 4 ml of probe solution/100 cm^2 filter.
6. Incubate at room temperature (23°) for 1 hr.
7. Remove filter and wash in 6× SSC at room temperature. Use 3 × 50 ml for a total of 10 min.
8. Autoradiograph for 1 hr using Kodak NS-5T film at 23°.
9. Wash the filter at a higher temperature with 50 ml of 6× SSC for 5 min. Preheat the wash solution to the desired temperature in a glass dish in a water bath.

10. Autoradiograph as in step 8.
11. Repeat steps 9 and 10, using several wash temperatures.

This procedure can also be carried out using DNA instead of phage. Spot the DNA (1 μl from procedure VI,28) onto a filter and carry out the procedure as outlined above with the exception of steps 3 and 4. Instead, bake the filter, then prehybridize with 5 ml of 6× SSC + 10× Denhardt's solution for 15 min at room temperature. Remove this solution and proceed with step 5. Another variation of the dot-blot method uses an aliquot of the phage supernatant from a 1 ml of culture prior to PEG precipitation of the phage. This requires a plexiglass "dot-blot" apparatus such as the unit marketed by Bethesda Research Laboratories. The device consists of a vacuum manifold with 96 sample wells through which DNA solutions of relatively large volumes are applied to a single nitrocellulose filter. The size of the resulting "dot" is small despite the large sample volume. Fifty microliters of phage containing supernatant from procedure VI,9 are added to individual wells and washed twice with 100 μl of 10 m*M* Tris-HCl (pH 8). The filter is removed from the manifold and treated as described above for phage. This procedure eliminates the PEG–NaCl precipitation step. The remainder of each culture is saved and used for isolation of plaque-purified mutant phage.

The choice of temperatures at which to wash is dependent on the length and particular composition of the oligonucleotide. In addition, the specific mutation constructed will contribute to the ability to discriminate between wild-type and mutant molecules. In the construction of single or double point mutations, increment the wash temperature by 5–10° depending on the length of the oligonucleotides. For other examples, see references cited in footnotes 23, 58, and 59.

Acknowledgments

The authors would like to acknowledge helpful discussions with Dr. Shirley Gillam and Dr. Greg Winter during the development of these procedures. We would like to thank Dr. Alex Markham and Tom Atkinson for synthesizing the two oligonucleotides used in this work. This investigation was supported by grants awarded to M. S. by the Medical Research Council of Canada. M. S. is an M.R.C. Career Investigator.

[33] Mud(Ap, *lac*)-Generated Fusions in Studies of Gene Expression

By JUDY H. KRUEGER and GRAHAM C. WALKER

One of the most powerful tools to be developed in recent years for use in genetic studies of *Escherichia coli* and related bacteria is the operon fusion vector, Mu c *ts* dl(Ap^R, *lac*)[1] hereafter referred to as Mud(Ap, *lac*). This bacteriophage, constructed by Casadaban and Cohen,[1] allows *lac* operon fusions to virtually any genetic locus to be obtained in a single step. Previously, the construction of operon fusions required a somewhat laborious two-step procedure.[2] The introduction of Mud(Ap, *lac*) has not only facilitated the construction of *lac* operon fusions to particular genes of interest,[3–14] but also permitted the conceptually new experimental strategy of screening for genes that are members of a common regulatory network without their functions being known in advance.[15,16]

The basic elements of the Mud(Ap, *lac*) bacteriophage are outlined in Fig. 1. Mu is a bacteriophage that, like λ, can integrate into the chromosome of *E. coli* and make a repressor that prevents the expression of the rest of the phage genes. However, unlike λ, which integrates preferentially at a single site in the *E. coli* chromosome, Mu integrates essentially at random.[17,18] The Mud(Ap, *lac*) bacteriophage is missing some of the

[1] M. J. Casadaban and S. N. Cohen, *Proc. Natl. Acad. Sci. U.S.A.* **76,** 4530 (1979).
[2] M. J. Casadaban, *J. Mol. Biol.* **104,** 541 (1976).
[3] C. J. Kenyon and G. C. Walker, *Nature (London)* **289,** 808 (1981).
[4] A. Bagg, C. J. Kenyon, and G. C. Walker, *Proc. Natl. Acad. Sci. U.S.A.* **78,** 5749 (1981).
[5] J. T. Mulligan, W. Margolin, J. H. Krueger, and G. C. Walker, *J. Bacteriol.* **151,** 609 (1982).
[6] Y. Komeda and T. Iino, *J. Bacteriol.* **139,** 721 (1979).
[7] J. Lee, L. Heffernan, and G. Wilcox, *J. Bacteriol.* **143,** 1325 (1980).
[8] N. Hugouvieux-Cotte-Pattat and J. Robert-Baudog, *Mol. Gen. Genet.* **182,** 279 (1981).
[9] O. Huisman and R. D'Ari, *Nature (London)* **290,** 797 (1981).
[10] M. Fogliano and P. F. Schendel, *Nature (London)* **289,** 196 (1981).
[11] L. N. Csonka, M. M. Howe, J. L. Ingraham, L. S. Pierson III, and C. L. Turnbough, Jr., *J. Bacteriol.* **145,** 299 (1981).
[12] H. I. Miller, M. Kirk, and H. Echols, *Proc. Natl. Acad. Sci. U.S.A.* **78,** 6754 (1981).
[13] P. I. Worsham and J. Konisky, *J. Bacteriol.* **145,** 647 (1981).
[14] S. R. Maloy and W. D. Nunn, *J. Bacteriol.* **149,** 173 (1982).
[15] C. J. Kenyon and G. C. Walker, *Proc. Natl. Acad. Sci. U.S.A.* **77,** 2819 (1980).
[16] B. L. Wanner, S. Wieder, and R. McSharry, *J. Bacteriol.* **146,** 93 (1981).
[17] E. Danielli, R. Roberts, and J. Abelson, *J. Mol. Biol.* **69,** 1 (1972).
[18] A. I. Bukhari and D. Zipser, *Nature (London) New Biol.* **236,** 240 (1972).

METHODS IN ENZYMOLOGY, VOL. 100

ISBN 0-12-182000-9

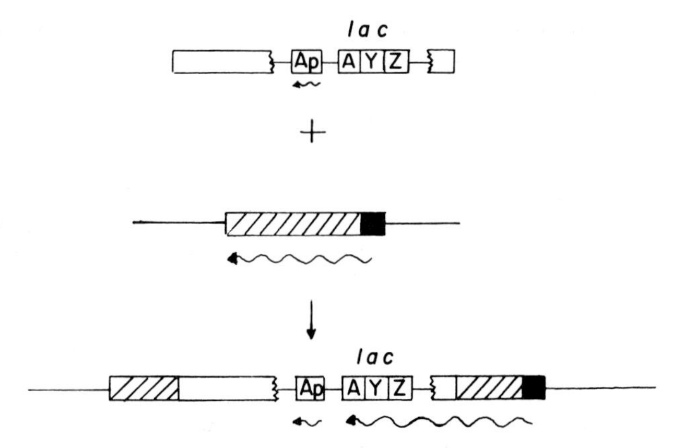

FIG. 1. Insertion of Mud(Ap, *lac*) into a gene to create a *lac* operon fusion.

genes of Mu and in their place carries a β-lactamase gene complete with its own promoter, so that when Mud(Ap, *lac*) integrates into the chromosome that cell becomes resistant to ampicillin. Thus Mud(Ap, *lac*) is similar to transposons such as Tn*5* in that it will insert into DNA with relatively little site specificity, and cells into which the transposable element has inserted acquire a new drug resistance. In addition, the Mud(Ap, *lac*) bacteriophage carries the structural genes for the *lac* operon, but no promoter for these genes; furthermore there is no promoter on the Mu DNA upstream of the *lac* genes. Thus the *lac* genes can be expressed only if Mud(Ap, *lac*) inserts, in the correct orientation, into a bacterial transcriptional unit (Fig. 1). In such a fusion, the expression of β-galactosidase (the product of the *lacZ* gene), permease (*lacY*), and transacetylase (*lacA*) is controlled by the regulatory circuitry of that particular bacterial transcriptional unit. Two points should be stressed. First, the Mud(Ap, *lac*) bacteriophage described here creates operon fusions, not gene fusions; since the *lac* genes have a complete set of translational signals, normal β-galactosidase is synthesized. Second, insertion of Mud(Ap, *lac*) within a gene usually leads to inactivation of that gene and, as a consequence of the polarity exhibited by Mu, also prevents the expression of downstream genes within the same transcriptional unit.

Mud(Ap, *lac*) fusions can be identified by one of two general experimental strategies. The first is initially to regard Mud(Ap, *lac*) as an insertion mutagen and screen or select for particular mutations on the basis of their known phenotypes. For approximately half of such mutations Mud(Ap, *lac*) will have inserted in the correct orientation to create a *lac* operon fusion to the promoter–regulatory region of the gene of interest. For example, in our laboratory we have used Mud(Ap, *lac*) to generate fusions to genes involved in DNA repair,[3] mutagenesis,[4] and methionine biosynthesis.[5] The second general method of identifying fusions of interest is to screen cells carrying random insertions of Mud(Ap, *lac*) for derivatives that express β-galactosidase in response to some specific treatment or condition. Genes identified initially by their regulatory characteristics can then be further characterized on the basis of their map position and other phenotypes. For example, we have used Mud(Ap, *lac*) to enable us to screen for genes whose expression is stimulated by DNA damage.[15]

Media

LB broth contains 10 g of tryptone, 5 g of yeast extract, 5 g of NaCl, and 1 ml of 1 *M* NaOH per liter. LB plates contain in addition 10–15 g of agar per liter. LB/Ca^{2+} broth is the same as LB broth, but contains a final concentration of 2.5 m*M* $CaCl_2$ added after autoclaving. λ-YM contains 10 g of tryptone, 2.5 g of NaCl, and 0.1 g of yeast extract per liter; sterile maltose is added after autoclaving to give a final concentration of 0.2%. M9 minimal medium contains 11 g of $Na_2HPO_4 \cdot 7\ H_2O$, 3 g of KH_2PO_4, 0.5 g of NaCl, and 1 g of NH_4Cl per liter; after autoclaving, the desired sugar is added to 0.2% and $MgSO_4$ to 10^{-3} *M*. When making M9 minimal plates, the agar (15 g/liter final concentration) and salts are made twofold concentrated and mixed after autoclaving. When desired, the β-galactosidase indicator, 5-bromo-4-chloro-3-indolyl-β-D-galactopyranoside (Xgal) dissolved in *N,N'*-dimethylformamide is added after autoclaving to a final concentration of 40 μg/ml. When required, drugs are added after autoclaving at the following final concentrations: ampicillin, 25 μg/ml; tetracycline, 15 to 25 μg/ml; streptomycin, 300 μg/ml; and kanamycin, 25 μg/ml. The optimal concentration for kanamycin may have to be established empirically, as strains seem to vary in their natural resistance. In addition, kanamycin does not always store well; desiccation is important, and refrigeration is recommended.

Preparation of Mud(Ap, *lac*) Phage Lysates

Mud(Ap, *lac*) lysates are prepared by a heat induction of MAL103, a bacterial strain that has both the Mud(Ap, *lac*) bacteriophage and a Mu *c ts* helper phage inserted in its chromosome.[1] A 5-ml LB broth culture of MAL103, inoculated from a fresh overnight, is grown in a flask at 30° with vigorous shaking. At a density of 2 to 3 × 10^8 cells/ml, the culture is shifted to a 43° bath for 25 min. Rapid shaking during phage growth is important to ensure sufficient aeration. The flask is then transferred to a

37° shaking bath and is incubated until the cells have lysed and the culture becomes clear (20–60 min). After the addition of a few drops of chloroform, the culture is left at room temperature for 5–10 min. Further operations are conducted at 4°. A 0.05-ml aliquot of 0.2 *M* lead acetate is added, and the culture is then centrifuged twice to pellet the debris. The supernatant is stored at 4° over a few drops of $CHCl_3$. The titer of Mud(Ap, *lac*) lysates tends to drop fairly rapidly with time. Care in removing the cellular debris will increase the stability of the lysate.

Isolation of Mud(Ap, *lac*) Lysogens

Before attempting a large-scale experiment, the Mud(Ap, *lac*) lysate should first be titered for its ability to transduce cells to ampicillin resistance. Cells grown 1 day for titering can be stored in a refrigerator and used the next day for a large-scale transduction. The strain used for the isolation of Mud(Ap, *lac*) fusions must carry a *lacZ*$^-$ mutation; a *lac* deletion is preferred, since it will not revert and it also allows the greatest degree of flexibility in subsequent manipulations. In addition, it is often useful if the strain is streptomycin resistant, as this makes Hfr "quick mapping" (see below) very convenient. The cells are grown in LB/Ca^{2+} broth to a density of 5×10^8 cells/ml. They are then chilled rapidly, and a portion is stored for use the next day. Serial dilutions (10^0, 10^{-1}, 10^{-2}, and 10^{-3}) of the Mud(Ap, *lac*) lysate are made into LB/Ca^{2+} broth. Aliquots (0.2 ml) of the phage dilutions are mixed with 0.2-ml aliquots of the cells and are incubated at room temperature for 30 min, after which 0.1 ml of each mix is spread on an LB–ampicillin plate. The plates are incubated overnight at 30°. From the results of this experiment the optimal dilution of the Mud(Ap, *lac*) lysate for use in a large-scale transduction can be determined. As discussed above, derivatives having Mud(Ap, *lac*) insertions in genes of interest can be identified by selecting or screening for desired mutant characteristics or by screening for fusions having particular regulatory characteristics. The most sensitive indicator for detecting β-galactosidase activity on plates is 5-bromo-4-chloro-3-indolyl-β-D-galactopyranoside (Xgal).

Mapping Mud(Ap, *lac*) Insertions

The approximate genetic location of a Mud(Ap, *lac*) insertion is most easily established by using the Hfr "quick-mapping" technique. This procedure is adapted from Low[19] and utilizes a series of Hfr strains whose origins of transfer and directions of transfer differ. If the Mud(Ap, *lac*)

[19] K. B. Low, *Bacteriol. Rev.* **36,** 587 (1973).

insertion does not cause a mutation with a phenotype whose loss can easily be detected, the position of such an insertion can be established by taking advantage of the fact that a Mud(Ap, *lac*) lysogen cannot grow at 42° as a consequence of the temperature-sensitive repressor of the phage. Various Hfr strains[19] are inoculated into LB to 10^7 cells/ml and then are incubated at 37° for 2 hr without shaking. A lawn of the Mud(Ap, *lac*) lysogen is prepared by spreading approximately 10^6 cells on LB–streptomycin plates or on minimal–glucose plates containing the appropriate nutritional supplements and streptomycin. A small drop of each Hfr is then spotted onto the lawn of the Mud(Ap, *lac*) recipient. If desired, this procedure can be conveniently carried out by growing the Hfr's in a Microtiter dish and transferring them to the lawn using a multiprong applicator. The plates are then incubated at 34° for 3 hr to allow mating and marker segregation and then are shifted to 42°. Above the background of spontaneous temperature-resistant colonies, a spot of recombinant colonies will grow at the positions of Hfr's that transfer, at relatively high efficiency, DNA that corresponds to the position of the Mud(Ap, *lac*) insertion. The map position of the Mud(Ap, *lac*) can then be mapped more accurately by standard P1 transduction techniques.[20]

Stabilization of *lac* Fusions Generated by Mud(Ap, *lac*)

The Mud(Ap, *lac*) fusions obtained as described above are fairly unstable even at 30° due to Mud(Ap, *lac*) transposing to new sites or catalyzing deletions. Thus, before attempting to isolate regulatory mutations affecting the expression of the fusion of interest, it is necessary to stabilize the fusion. Several methods for accomplishing this purpose are discussed below. An additional benefit of these stabilization procedures is that they allow the fusion strain to be grown at higher temperatures without Mu induction.

Method 1

This method is expanded from the work of Komeda and Iino[6] and results in Mu being replaced by λ in a two-step procedure. The first step is to select for a lysogen of phage λ pl(209),[2] which carries *lac* and Mu DNA. λ pl(209) lacks the phage *att* site and the *int* and *xis* genes and therefore will integrate via homologous recombination into the Mu or *lac* DNA carried by Mud(Ap, *lac*). The cells to be infected are grown to mid-log phase on λ-YM, and a 0.1-ml aliquot is incubated with λ pl(209) at 30° for

[20] J. Miller, *in* "Experiments in Molecular Genetics," p. 201. Cold Spring Harbor Laboratory, Cold Spring Harbor, New York, 1972.

20–30 min in the presence of 10 mM Mg^{2+}. The infected cells are then spread on LB plates containing 10 mM Mg^{2+} and seeded with 10^9 particles of a *c*I mutant of lambda to kill nonimmune cells. The plates are incubated overnight at 30°. The first step can now be carried out more easily using λ pSG1 which is identical to λ pl(209) except that it carries the transposon Tn*9*, which codes for chloramphenicol resistance, inserted in the *lacY* gene.[21] In the second step, the lysogens are purified by streaking to single colonies on LB plates that are incubated at 42°. Some of the colonies in which deletions between the duplicated DNA have occurred will lose Mu DNA and survive at a high temperature. These colonies should be sensitive to ampicillin and resistant to lambda, and the regulatory characteristics of the *lac* fusion in the stabilized strain should be the same as in the original strain. They can also be used to isolate specialized λ transducing phage carrying the fusion.[6]

Method 2

Introduction of the multicopy plasmid, pGW600, carrying a copy of the gene coding for temperature-resistant Mu repressor, into a fusion strain will prevent Mu functions from being expressed and consequently will stabilize the *lac* fusion.[22] Zipser *et al.*[23] cloned the gene coding for Mu repressor into pMB9 (which codes for tetracycline resistance), and pGW600 was obtained by choosing a derivative of their plasmid that coded for a temperature-resistant Mu repressor. Described below are simple, rapid methods for isolation of plasmid DNA, preparation of competent cells and transformation.

Isolation of Plasmid DNA. The procedure for isolation of plasmid DNA is that described by Holmes and Quigley.[24] A 10-ml culture of cells containing pGW600 is grown overnight in LB containing 15 μg of tetracycline per milliliter. The cells are pelleted and resuspended in 0.7 ml of lysis buffer containing 50 mM Tris-HCl (pH 8), 5% Triton, 8% sucrose, and 50 mM EDTA and transferred to a 10 × 75 Pyrex test tube. Lysozyme (0.05 ml; 10 mg/ml) is added, and the tube is immediately placed over a bunsen burner flame until the sample just begins to boil and is then rapidly transferred to a boiling water bath for 30 sec. The cell debris is removed by centrifugation, and the supernatent is extracted with an equal volume of buffered phenol. After centrifugation to separate the layers, the aque-

[21] T. J. Silhavy and J. Beckwith, this series.
[22] J. H. Krueger, S. J. Elledge, and G. C. Walker, *J. Bacteriol.* **153,** in press (1983).
[23] D. Zipser, P. Moses, R. Kahmann, and D. Kamp, *Gene* **2,** 263 (1977).
[24] D. S. Holmes and M. Quigley, *Anal. Biochem.* **114,** 193 (1981).

ous layer is transferred to a second tube and extracted with an equal volume of $CHCl_3$. The layers are separated by centrifugation; the aqueous layer is transferred, and the DNA is precipitated by the addition of an equal volume of isopropanol. The sample is mixed well and placed at −20° for 30–60 min. After a 10-min centrifugation, the supernatant is discarded and the tubes are inverted to dry. The DNA is resuspended in 0.1 ml of 10 m*M* Tris-HCl (pH 8.0) buffer containing 5 m*M* NaCl, and 1 m*M* EDTA, and then 1 μl of a 5 mg/ml solution of RNase A is added. The DNA is stored at −20°.

Preparation of Competent Cells. LB broth (25 ml) is inoculated with 0.25 ml of a fresh overnight culture, and the cells are grown at 30° with vigorous shaking to a density of 2 to 3 × 10^8 cells/ml (a net Klett reading of 55). After being rapidly cooled to 4°, the cells are pelleted and then resuspended in 6 ml of 0.1 *M* $MgCl_2$. The cells are centrifuged again and resuspended in 6 ml of 0.1 *M* $CaCl_2$. After a 20-min incubation on ice, the cells are pelleted a third time. If continuing with the transformation, the cells are resuspended in 1.25 ml of 0.1 *M* $CaCl_2$; otherwise they are resuspended in 1.1 ml of 0.1 *M* $CaCl_2$ and 0.2 ml of glycerol, and are frozen in small aliquots.

Transformation with pGW600. Frozen competent cells are thawed on ice for 30 min; 0.2 ml of competent cells is mixed with 10 μl (approximately 1 μg of DNA) of the plasmid DNA and incubated on ice for 30 min. The cells are then heat-shocked by immersing the tube in a 42° water bath for 2 min. Two milliliters of LB broth are added, and the tube is incubated at 30° for 1 hr with shaking. Aliquots (0.1 ml) of undiluted, 10-fold and 100-fold diluted cells are spread on LB plates containing 25 μg of tetracycline per milliliter.

Method 3

Stabilized strains can also be isolated by selecting directly for a temperature-resistant derivative and screening for ones that are ampicillin resistant and Mu sensitive. These strains should presumably have a deletion that removes both the Mu repressor and *kil* gene but does not extend into the β-lactamase gene of Mud(Ap, *lac*) and therefore does not affect the regulation or expression of the *lac* fusion.[5]

Isolation of Regulatory Mutants

Mutations that affect expression of the genes to which fusions have been obtained can be isolated by looking for colonies with altered β-galactosidase levels. A method is presented here for isolating colonies that constitutively express β-galactosidase in the absence of inducing agents. Simple modifications could be employed to fit the requirements of a particular system. Mutagenesis by Tn*5* is convenient, since the kanamycin resistance marker[25] it carries allows for direct selection of the mutant phenotype, thus facilitating mapping and transfer into new backgrounds. The transposon Tn*5* is particularly useful for isolating mutations because it transposes at a high frequency (10^{-2} to 10^{-3})[26] into fairly random sites[26,27] in DNA. The vector used to introduce the Tn*5*, λ*c*I857 b221 *rex*::Tn*5* *O*am29 *P*am80[26,28] has several desirable properties; it can neither integrate itself into host DNA owing to the deletion that removes the λ *att* site, nor can it replicate on its own in an Su^- host.

The recipient strain (50 ml) is grown in λ-YM to a density of 5×10^8 to 1×10^9 cells/ml (a net Klett reading of 100–150). The cells are pelleted by centrifugation and concentrated 20 fold in λ-YM supplemented with 10 m*M* Mg^{2+}. It often proves practical to perform a trial experiment, saving the majority of the concentrated cells in a refrigerator until the next day. In the initial experiment, a range of dilutions of the infected cells are plated to determine which dilution will result in an appropriate number of colonies per plate (500–1200). Cells (0.1 ml) and a phage dilution (0.1 ml) that will provide a multiplicity of infection of 0.2 are incubated for 45 min at 42°. Dilutions are then spread on M9 minimal glucose plates containing 25 μg of kanamycin per milliliter, 2.5 m*M* pyrophosphate, and 40 μg of Xgal per milliliter. The plates are incubated at 42°. Colonies that are darker blue than their parent strain, indicating increased expression of β-galactosidase, are purified by streaking to single colonies several times. Alternatively one could select for very high levels of β-galactosidase expression by spreading the mix on M9 minimal plates that contain lactose as the carbon source. Linkage of the Tn*5* insertion to the mutations causing the observable phenotype should be checked by P1 transduction.[20] P1 grown on the insertion strain is used to transduce the parental strain to kanamycin resistance. Transductants are screened for the desired phenotype. The linkage should be quite high, although it is rarely 100% owing to transpositions of Tn*5* occurring during the transduction procedure.

[25] D. E. Berg, J. Davis, B. Allet, and J. D. Rochaix, *Proc. Natl. Acad. Sci. U.S.A.* **72,** 3628 (1975).

[26] D. E. Berg, *in* "DNA Insertion Elements, Plasmids and Episomes" (A. I. Bukhari, J. A. Shapiro, and S. L. Adhya, eds.), p. 205. Cold Spring Harbor Laboratory, Cold Spring Harbor, New York, 1977.

[27] K. J. Shaw and C. M. Berg, *Genetics* **92,** 741 (1979).

[28] The double amber-containing phage MJL λ52 was constructed from λ*rex*::Tn*5* by Michael Lichten.

Acknowledgments

This work was carried out in the Biology Department, Massachusetts Institute of Technology, and was supported by grant GM28988 from the National Institute of General Medical Science. G. C. W. was a Rita Allen Scholar.

We express our appreciation to S. Elledge, C. Kenyon, P. LeMotte, W. Margolin, B. Mitchell, J. Mulligan, K.-H. Paek, P. Pang, K. Perry, S. Rabinowitz, and S. Winans for valuable discussions and to L. Withers for help in preparing the manuscript.

Author Index

Numbers in parentheses are reference numbers and indicate that an author's work is referred to although the name is not cited in the text.

A

B

C

D

E

F

G

H

I

J

K

L

M

N

O

P

Q

R

S

T

U

V

W

X

Y

Z

Subject Index

A

B

C

D

E

H

I

K

L

M

R

S

T

X

Y